S0-BLA-820

ALLE·ZEIT·WACH
1842

Robotics and Factories of the Future

Proceedings of an international conference
Charlotte, North Carolina, USA
December 4–7, 1984

Editor: Suren N. Dwivedi

With 263 Figures

Springer-Verlag
Berlin Heidelberg New York Tokyo 1984

Suren N. Dwivedi
Associate Professor in Mechanical Engineering
University of North Carolina at Charlotte 28223
USA

ISBN 3-540-15015-3 Springer-Verlag Berlin Heidelberg New York Tokyo
ISBN 0-387-15015-3 Springer-Verlag New York Heidelberg Berlin Tokyo

Offsetprinting: Color-Druck, G. Baucke, Berlin
Bookbinding: Lüderitz & Bauer, Berlin
2061/3020 5 4 3 2 1 0

Preface

An International Conference on Robotics and Factories of the Future was held on December 4-7, 1984 in Charlotte, North Carolina, USA. The development of technology and the status as it stands today is highly multidisciplinary and complex in nature. It has added new dimensions and challenges in the development of factories of the future. Thus it calls upon an integrated approach of technological, economic and sociological impacts of robots and computers in factories of the future.

Sessions in this conference included: Planning of Automation, Future Trends & Management, Integration of CAD/CAM & Robot Cell, Robots Applications, Manipulation I, Controls I, Manipulation II, Computer Vision and Sensors, Robot Cell and Mobile Robot, Controls II, Artificial Intelligence, Advanced Applications of Robots, Industrial Panel, CAD/CAM and Flexible Manufacturing, Robotics and CAD/CAM Education I, Social and Economic Implications, Robotics and CAD/CAM Education II, and Industry and User Interaction.

142 papers were presented during the conference and participants came from 18 countries.

The plenary session and keynote speakers were:
J. M. Kelly, IBM Quality Institute, Southbury, Connecticut; M. E. Merchant, Metcut Research, Cincinnati, Ohio; Stellio Demark, ASEA Robotics, Milwaukee, Wisconsin; Stanley J. Polcyn, Unimation Inc., Danbury, Connecticut; Carl Ruoff,Jet Propulsion Lab, Pasadena, California; Joseph Tulkoff, Lockheed Georgia Company, Marietta, Georgia; N. M. Swani, Indian Institute of Technology, Delhi, India; James R. Barrett, Computer Vision, Bedford, Maryland; Hadi A. Akeel, GMF Robotics, Troy, Michigan; John M. Vranish, National Bureau of Standards, Washington, DC; Yoshihide Nishida, Mitsubishi Electric Corporation, Janpan; James Clark Solinsky, International Robomation/Intelligence, Carlsbad, California; William R. Tanner, Productivity Systems Inc., Troy, Michigan; M. Jevtic, Institute for Machine Tools & Tooling, Beograd, Jergoslavija; Mr. Lawrence A. Goshorn, International Robomation/Intelligence, Carlsbad, California; Mr. Frank A. Dipietro, CPC Head Quarter, Warren, Michigan; Professor Ezat T. Sanii, North Carolina State University, Raleigh, North Carolina; Mr. Lloyd R. Carrico, Indianapolis, Indiana; Dr. Del Tesar, University of Florida, Gainesville, Florida; and Louis J. Galbiati, Suny College of Technology, Utica, New York. The speakers represented not only universitities but also many government agencies and industries.

The papers in this proceeding provide a comprehensive view of the state of the art in research and development, application and education of robotics and computers in factories of the future.

The outstanding support and the cooperation of the following people are acknkowledged:

Dr. James W. Clay, Director, Urban Institute, UNCC;
Dr. Robert D. Snyder, Dean, College of Engineering UNCC;
Dr. Paul Dehoff, Chairman, Mechanical Engineering, UNCC;
Ms. Mary Pat Young, Conference Coordinator;
Mr. Jim White, IEEE, Chairman;
Mr. Roger Blue, Exhibits Chairman;
Ms. Dori Thurman, Conference Administrative Assistant

David Bourne, Carnegie Melon University, Pittsburg, Pennsylvania; Shyam N. Singh,

Aqua-Chem, Milwaukee, Wisconsin; James B. Canner, Detroit, Michigan; Vinod Nangia, General Electric, Cinncinnati, Ohio; Joseph Elgomayel, West Lafayette, Indiana; Louis J. Galbiati, Suny College of Technology, Utica, New York; William Gerould, McDonald Douglas Automation, Fairfax, Virginia; Ted Hanauer, ASEA Robotics, White Plains, New York; Jim Hansen, Bort Warner Research, Des Plaines, Illinois; H. C. Pande, Ranchi, India; S. P. Puri, Livingstone College, Salisbury, North Carolina; James A. Rehig, Piedmont Technical College, Greenwood, South Carolina; R. Sharan, National Institute of Foundry and Forge Technology, Ranchi, India; J. P. Sharma, UNCC, Charlotte, North Carolina; V. Singh, Institute of Technology, BHU Varanasi, India; Geary V. Soska, Cybotech Corporation, Indianapolis, Indiana; Murali Varanasi, University of South Florida, Tampa, Florida; Ray Asfahl, University of Arkansas, Fayetteville, Arkansas; Lloyd R. Carrico, Indianapolis, Indiana; Lawrence A. Goshorn, International Robomation/Intelligence, Carlsbad, California; Alexander Houtzell, Organization for Industrial Research, Waltham, Massachusetts; R. Janardhanam, University of North Carolina at Charlotte, Charlotte, North Carolina;C. S. Jha, Jawaharlal Neru University, New Delhi, India; K. N. Karna, CCGA Corporation, Matawan, New Jersey; Dilip Kohli, University of Wisconsin, Milwaukee, Wisconsin; Jack D. Lane, Engineering and Management Institute, Flint, Michigan; Richard Rothfuss, McDonald Douglas Automation, St. Louis, Missouri; Ezat T. Sanii, N.C. State University, Raleigh, North Carolina; Nicholas Shields, Jr., Essex Corporation, Huntsville, Alabama; Fred Sitkins, Michigan University, Kalamazoo, Michigan; Mr. Lal Vishin, IBM, Charlotte, North Carolina; and William R. Tanner, Productivity Systems, Inc., Troy, Michigan.

I would like to thank all authors and speakers for their excellent contribution. I am very indebted to the members of the Organizing and Advisory Committee for their support.

Last, but not the least, I thank my wife, Shashi Dwivedi, my daughters, Indira, Iva and Isha for their constant help and encouragement.

Finally, I do appreciate Springer-Verlag's efficiency for the publication of this proceeding in a very short time.

Suren N. Dwivedi
Editor

Committees

SPONSORS:
Robert D. Snyder, Dean,
College of Engineering, UNCC
James W. Clay, Director,
Urban Institute, UNCC

CONFERENCE CHAIRMAN
Suren N. Dwivedi

CONFERENCE COORDINATOR:
Mary Pat Young

CONFERENCE COMMITTEE CHAIRPERSONS:

Publication Committee
J.P. Sharma
Exhibition Committee
Roger Blue
Arrangement Committee
R. Janardhanam
Reception Committee
M.E. Beane

International Organizing Committee (USA):
James S. Albus (DC)
David D. Ardayfio (MO)
Ray Asfahl (AR)
Shyam Bahadur (IA)
A.J.G. Babu (FL)
J. Michael Brady (MA)
James B. Canner (MI)
Lloyd R. Carrico (IN)
John D. DiPonio (MI)
Joseph Elgomayel (IN)
Louis J. Galbiati (NY)
Lawrence A. Goshorn (CA)
K.C. Gupta (IL)
Y. Hari (NC)
Ernest L. Hall (OH)
S.C. Jain (OH)
Kamal Karna (NJ)
Kenneth Knowles (MD)
Dilip Kohli (WI)
Steven Kramer (OH)
A. Kumar (MO)
Jack D. Lane (MI)
Rakesh Mahajan (MI)
Eric Malstrom (IA)
George E. Munson (CT)
E.H. Nicollian (NC)
B. Prasad (MI)
Ramji Raghavan (NY)
W. Edward Red (UT)
Raj Reddy (PA)
James A. Rehg (SC)
Ezat T. Sanii (NC)
R.P. Sharma (MI)
Fred Z. Sitkins (MI)
Ali Seireg (WI)
Nicholas Shields, Jr. (AL)
A.H. Soni (OK)
William Tanner (MI)
K.C. Tripathi (CT)
L. Vishin (NC)
Robert A. Ullrich (TN)
Murli Varanasi (FL)
Alok Verma (VA)
Robert O. Warrington (LA)
Jim White (NC)
William V. Wright (NC)

Bulgaria:
M.S. Konstantinov (Sofia)

Canada:
Alfred Haupt (Toronto)
D.K. Sinha (Manitoba)
D. Necsulescu (Ottawa)
Ahmad Hemani (Montreal)
S. Bala Krishnan (Manitoba)

China:
Yongxing Zhuang (Peking)
Dongying Gan (Chang Chunz)
Hong Cheng (Xuzhou)

Egypt:
A.S. El Sabbagh (Cairo)
M.F. Abdin (Cairo)

Federal Republic of Germany:
H.B. Kuntze (Karlsruhe)
G. Spurr (Berlin)
M. Weokc (Aachen)
U. Dern (Aachen)
D. Zuhlke

France:
Pierre J. Andre (Cedex)
P. Coiffer (Cedex)
Rabischong (Montpe)
Alex Renault (Cedex)
Jean Michel Valade (Cedex)
L. Marce (Cedex)
A. Pruski (Cedex)
B. Mutel (Cedex)

India:
A. Bhattacharya (Jadavpur)
R.K. Dave (Jabalpur)
C.S. Jha (Delhi)
B. Kishor (Varanasi)
R.C. Malhotra (Delhi)
H.C. Pande (Ranchi)
P.C. Pandey (Roorkee)
R. Sharan (Rachi)
N.M. Swani (Delhi)
D. Pershad (Delhi)

Ireland:
Louis Brennan
S. Grewal
Eamonn McQuade

Israel:
Arie Lavie (Jerusalem)
B.Z. Sandler (Beer Sheva)
R. Mozniker (Beer Sheva)

Italy:
Ettore Pennestri (Rome)
Gabriele Vassura (Bologna)

Japan:
Tsutomu Hasegawa (Lbaraki)
Yokio Hasegawa (Tokyo)
Fumihilo Kimura (Tokyo)
N. Sugimato (Tokyo)
Michio Takahashi (Tokyo)
Y. Yamazaki
Yoshihide Nishida (Hyogo)

Mexico:
Jorge Angeles (M. City)
Victor H. Mucino (M. City)

Norway:
H. Christensen (Trondheim)
E. Sodahl (Trondheim)

Romania:
Adrian Davidoviciu (Bucharest)
Gheorghe Dragamolu (Bucharest)

Taiwan:
Tsong Liang Huang (Taipei)

United Kingdom:
Peter H. Jost (London)
Alan Pugh (Hull)
A. Jawaid (Coventry)

U.S.S.R.:
O.I. Semenkov (Moscow)

Yugoslavia:
Vukobratovic (Belgrade)
M. Jeutic
V. Solaja
S. Lazarevic

Advisory Committee (USA):
Ken Allen
Phillip Barone
Tom J. Gregory
Len Gustafson
Robert W. Heald
Larry Heath
Ray Hinson
Atalas Hsie
Daniel M. Hull
Mark Kirkpatrick
Richard Klafter
John Lyman
Neil Madonick
Ronald J. Meyer
Howard Murphy
Salvatore A. Notaro
Carl V. Page
Bob Parent
Sandra L. Pfister
Ronald D. Potter
Les Seager
H. Singh
S.N. Singh
Geary V. Soska
John Tanner
Tom Trozzi
Walter Tucher
Robert A. Ullrich
Muthuraj Vaithianathan
Albert Zachwieja

List of Contributors

M. F. Abdin
Design and Production Engineering Dept.
Faculty of Engineering
Ainshams University
1 EL Sarayat Street
Abdou Basha Square
Abbassia Cairo, Egypt

Om Parkash Agarwal
Dept. of Mechanical Engineering
Temple University
Philadelphia, Pennsylvania 19172
USA

Hadi A. Akeel
GMF Robotics
5600 New King Street
Troy, Michigan 48098-2696
USA

G. S. Alag
NASA Dryden Lancaster
P.O. Box 273
Mail code OFDC
Edwards, California 93523
USA

C. Amarnath
Assistant Professor
Mechanical Engineering Department
Indian Institute of Technology
Bombay 400076, India

Pierre J. Andre
Centre Microsystemes et Robotique
E.R.A. 906
E.N.S.M.M. La Bouloie-Route de Gray
25030 Besancon Cedex - France

Ray Asfahl
University of Arkansas
Industrial Engineering 309
Fayetteville, Arkansas 72701
USA

A. J. G. Babu
University #f Central Florida
Orlando, Florida 32816
USA

S. Balakrishnan
The University of Manitoba
Department of Mechanical Engineering
Winnipeg, Manitoba
Canada R3T 2N2

Jeffrey Balcolm
Prab Robots, Inc.
931 State Route 28
Milford, Ohio 45150
USA

Ramez Ballou
16164 Pine Tree Court
San Josė, California 95131
USA

Han Bao
College of Engineering
North Carolina State University
Raleigh, North Carolina 27695
USA

James R. Barrett
Computer Vision
15 Crosby Drive
Bedford, Massachussetts 01730
USA

James K. Blundell
Dept. of Mechanical Engineering
University of Missouri
Columbia, Missouri 65211
USA

Glen Boston
Manufacturing Engineering
Miami University
Kreger Hall, Room 117
School of Applied Science
Oxford, Ohio 45056
USA

A. Bourjault
Universite De Franche-Comte -
Besancon
Centre De Recherche
Microsystems et Robotique
ERA n 07/0906
ENSMM - 25030 Besancon Cedex
France

David Bourne
Carnegie Melon University
Pittsburg, Pennsylvania 15213
USA

Albert Bowers
Mitre Corp.
1820 Dolley Madison Blvd.
McLean, Virginia 22102
USA

David Brozier
Applicon
32nd Avenue
Burlington, Massachussetts 01803
USA

D. R. Buchanan
Dept. of Textile Materials and
Manufacturing
North Carolina State University
Box 5006
Raleigh, North Carolina 27650
USA

James B. Canner
Sterling Detroit Co.
261 East Goldengate Ave.
Detroit, Michigan 48203
USA

Vikram Cariapa
Mechanical Engineering
Marquette University
1515 West Wisconsin Ave.
Milwaukee, Wisconsin 53233
USA

Lloyd R. Carrico
Director Technical Services
3914 Prospect Street
Indianapolis, Indiana 46203
USA

Pradip Chande
Indian Institute of Technology
Delhi, India

Michael J. Chen
Machine Intelligence Corporation
330 Potrero Avenue
Sunnyvale, California 94086
USA

Hong Cheng
Research Institute of Gengpri HQ
of Coal Mine Construction
Xezkow, China

Charles Chrestman
Program Operations
Itawamba Junior College
653 Eason Boulevard
Tupelo, Mississippi 38801
USA

Margaret Clarke
Clarke Ambrose
11617 N. Monticello Dr.
Concord, Tennessee 37922
USA

Z. J. Czajkiewicz
Wichita State University
College of Engineering
Wichita, Kansas 67208
USA

R. K. Dave
Government Engineering College
Jabalpur, India

Franklin Deaton
Automation Engineering
IBM Corporation
1001 W. T. Harris Blvd.
Charlotte, North Carolina 28257
USA

Stellio Demark
ASEA Robotics
P. O. Box 1560
Milwaukee, Wisconsin
USA

Frank A. Dipietro
Production Engineering Activities
CPC Head Quarter
33001 Vane Dyke Avenue
Warren, Michigan 48090
USA

David Eike
Essex Process Control Group
Alexandria, Virginia 22314
USA

Joseph ElGomayel
School of Industrial Engineering
Purdue University
West Lafayette, Indiana 47907
USA

Louis J. Galbiati, Dean
Engineering Technology
SUNY College of Technology
811 Court St.
Utica, New York 13502
USA

Dongying Gan
Changchun Institute of Optics
& Fine Mechanics
Academia Sinica
Changchun
People's Republic of China

S. Ganesan
Western Michigan University
Kalamazoo, Michigan 49008
USA

William Gerould
Marketing Representative
McDonald Douglas Automation
10530 Ruseheven #500
Fairfax, Virginia 22030
USA

Kalyan Ghosh
University of Quebec at
Trois-Rivieres
P.O. Box 500
Trois-Rivieres (Quebec)
Canada

Lawrence A. Goshorn
International Robomation/Intelligence
2281 Las Palmas Drive
Carlsbad, California 92008
USA

J. S. Gunasekara
Ohio University
Athens, Ohio 45701
USA

Ernest L. Hall
University of Cincinnati
Cincinnati, Ohio 45221
USA

Kenneth Hall
Advance Development and Engineering
Center
101 Chester Road
Swarthmore, Pennsylvania 19081
USA

Geoffrey Hammett
4255 Nora Lane
Duluth, Georgia 30136
USA

Ted Hanauer
ASEA Robotics
4 New King St.
White Plains, New York 10604
USA

Jim Hansen
Borg Warner Research
Wolf Algonquin Road
Des Plaines, Illinois 60618
USA

Yogeshwar Hari
Mechanical Engineering
University of North Carolina at
Charlotte
Charlotte, North Carolina 28223
USA

Fenton Harrison
NASA Langley Research Center
Hampton, Virginia 23665
USA

Ahmad Hemami
Dept. of Mechanical Engineering
Ecole Polytechnique
Campus de l'Universite de Montreal
Case postale 6079, succursale "A"
Montreal, Quebec H3C 3A7

Peter G. Heytler
Industrial Development Division
Institute of Science and TEchnology
The University of Michigan
2200 Bonisteel Boulevard
Ann Arbor, Michigan 48109
USA

Richard Hollowell
Department of Mechanical Engineering
Clemson University
Clemson, South Carolina 29631
USA

Alexander Houtzeel
Organization for Industrial Research
240 Bear Hill Road
Waltham, Massachussetts 02154
USA

A. Hsie
SUNY College of Technology
811 Court Street
Utica, New York 13502
USA

Tsong Liang Huang
National Taiwan University
Taipei, Tawian, R.O.C.

Wilfred Huang
Division of Industrial Engineering
Alfred University
Alfred, New York 14802
USA

Cecil O. Huey, Jr.
Mechanical Engineering
Clemson University
Clemson, South Carolina 29631
USA

Keith E. Hummel
Industrial Systems Division
Center for Manufacturing Engineering
National Bureau of Standards
Washington, DC 20234
USA

Mircea Ivanescu
Automatic and Computer Department
University of Craiova
1loo Craieva
Romania

E. W. Iversen, Manger
Manufacturing Services
FMC Corporation
Highway 31 W. South
Box 9500
Bowling Green, Kentucky 42101
USA

Kalojan B. Jankov
Research and Development Institute
of Robots
Stara Zagora 6000, Bulgaria

A. Jawaid
University of Warwick
Coventry
United Kingdom

A. C. Jha
SRI PRATAP Bhawan
43C Naval Kishore Road
Lucknow 226001
India

C. S. Jha
Educational Advisor
Government of India
New Delhi, India

Michael Joost
College of Engineering
North Carolina State University
Raleigh, North Carolina 27695
USA

K. N. Karna
CCGA Corporation
44 Briarwood Drive
Matawan, New Jersey 07747
USA

J. M. Kelly
IBM Quality Institute
Heritage Road
Southbury, Cincinnati 06488
USA

Vladimir Klimo
Faculty of Mechanical Engineering
Technical University Kosice
041 87 K o s i c e, Czechoslovakia

Dilip Kohli
University of Wisconsin
3201 N. Downer
Milwaukee, Wisconsin 53201
USA

Joseph E. Kopf
Manufacturing Engineering Technology
323 High Street
Newark, New Jersey 07102
USA

Pat Kozlowski
Mega Designs
Roseville, Michigan 48066
USA

Andrew Kunik, Sr.
Cincinnati Milacron
8702 Red Oak Blvd.
Charlotte, North Carolina 28210
USA

Jack D. Lane
GMI Engineering & Management
Institute
Flint, Michigan 48502
USA

John Lynn
Electronics Consultants, Inc.
10260 South West 122nd Street
Miami, Florida 33176
USA

Albert Madwed
A Madwed Company
110 Wedgewood Drive
Easton, Cincinnati 06612
USA

Rakesh Mahajan
6454 East 12 Mile Road
A.P. MES, CIS
GM Technical Center
Warren, Michigan 48090-9000
USA

L. Marce
I.N.S.A. L.A.T.E.A.
20, Avenue Des Buttes De Coesmes
35043 Rennes Cedex - France

Anne Mavor
Mr. H. M. Parsons
Essex Research Group
Alexandria, Virginia 22314
USA

M. E. Merchant
Metcut Research
3980 Rosslynn Drive
Cincinnati, Ohio 45209
USA

Dennis Miller
Control Automation
Princeton, New Jersey 08540
USA

Patrick Minotti
Marc Dahan
Universite De Besancon
Laboratoire de Mecanique Appliquee
Faculte des Sciences
Route De Gray - La Bouloie
25030 Besancon Cedex
France

Mathew Monforte
Forbes Company
2333 White Horse-Mercerville Road
Trenton, New Jersey 08619
USA

James W. Moore
School of Engineering and Applied
Science
University of Virginia
Charlottesville, Virginia 22901
USA

R. Mozniker
Ernst David Bergmann Campus
P.O.B. 1025
Beer-Sheva 84110
Israel

Howard Murphy
Fairchild Camera and Instruments
3440 Hilview Avenue
Palo Alto, California 94304
USA

Timothy Murphy
Design Engineering Dept.
P. O. Box 33189
Charlotte, North Carolina 28242
USA

Akella S. R. Murty
Indian Institute of Technoloy
Kharagpur 721302 India

D. Necsulescu
Engineering Management Program
University of Ottowa
770 King Edward Ave.
Ottawa, Ont. KLN 6N5
Canada

Leon Nguyen
Dept. 220T/251
IBM Corporation
1001 W. T. Harris Blvd.
Charlotte, North Carolina 28257
USA

Bịnh Ninh
Mechanical Engineering Dept.
University of North Carolina at Charlotte
Charlotte, North Carolina 28223
USA

Tuan Ninh
Mechanical Engineering Dept.
University of North Carolina at Charlotte
Charlotte, North Carolina 28223
USA

Yoshihide Nishida
Mfg. Facilities Eng. Dept.
Mitsubishi Electric Corporation
1-1, Tsukaguchi Honmach 1 8 Chrome
Amagasaki, Hyogo, 661 Japan

Sudhakar Paidy
Rochester Institute of Technology
One Lamb Memorial Drive
Rochester, New York 14623
USA

H. C. Pande
B.I.T. Mesra
Ranchi, India

Evan Peele
Hydro-Power Engineering Corporation
P.O. Box 709
Brighton, Michigan 48116
USA

Rafael Peres
Department of Computer Science and
Engineering
University of South Florida
Tampa, Florida 33620
USA

James N. Perry
Schrader Bellows Division
Scovill, Inc.
U.S. Route #1
Wake Forest, North Carolina 27587
USA

D. Pershad
Indian Institute of Technology
New Delhi, India

George Peterson
Engineering Technology Department
Texas A & M University
College Station, Texas 77843
USA

Russell Pfashler
SDRC
Milford, Ohio 45150

Stanley J. Polcyn
Unimation Inc.
Shelter Rock Lane
Danbury, Cincinnati 06810
USA

David Potter
GMF Robotics
5600 New King Street
Troy, Michigan 48098
USA

A. Pruski
University of Metz
Metz, France

S. P. Puri
Division of Business
Livingstone College
Salisbury, North Carolina 28144
USA

Robert F. Puthoff
Digital Design Group #107
834 Tyvola Road
Charlotte, North Carolina 28210
USA

R. Radharamanan
Mechanical and Industrial Engineering
University of Utah
Salt Lake City, Utah 84112
USA

Keith Rathmill
Robotics & Automation
Cranfield Institute of Technology
Bedford MK43 0AL England

H. M. Razavi
Engineering Dept.
University of North Carolina at Charlotte
Charlotte, North Carolina 28223
USA

W. Edward Red
Mechanical Engineering
242 P. Clyde Bldg.
Bringham Young University
Provo, Utah 84602
USA

James A. Rehg
Robotics Center
Piedmont Technical College
P.O. Drawer 1467
Greenwood, South Carolina 29648
USA

Eugene I. Rivin
Mechanical Engineering
Wayne State University
Detroit, Michigan 48202
USA

Richard Rothfuss
Director of Manufacturing Technology
McDonald Douglous Automation
P.O. Box 516
St. Louis, Missouri 63166
USA

Carl Ruoff
Jet Propulsion Lab
Pasadena, California
USA

B. Z. Sandler
Ernst David Bergmann Campus
P.O.B. 1025
Beer-Sheva 84110
Israel

Ezat T. Sanii
North Carolina State University
Raleigh, North Carolina 27695
USA

Ali Seireg
Mechanical Engineering
University of Wisconsin
Madison, Wisconsin 53706
USA

R. Sharan
National Institute of Foundry &
Forge Technology
Hatia, Ranchi - 834 003, India

J. P. Sharma
Mechanical Engineering
University of North Carolina at Charlotte
Charlotte, North Carolina 28223
USA

R. P. Sharma
Mechanical Engineering
Western Michigan University
Kalamazoo, Michigan 49008
USA

C. N. Shen
U.S. Army Armament, Munitions, and
Chemical Command
Armament Research and Development Center
Large Caliber Weapon Systems Laboratory
Benet Weapons Laboratory
Watervliet, New York 12189
USA

Nicholas Shields, Jr.
Essex Corporation
3322 Memorial Parkway South
Suite 8
Huntsville, Alabama 35801
USA

V. Singh
Mechanical Engineering Department
Institute of Technology
BHO 221 005, India

S. N. Singh
Vigyan Research Associates
28 Research Drive
Hampton, Virginia 23666
USA

K. Sinha
Mechanical Engineering
The University of Manitoba
Winnipeg, Manitoba
Canada R3T 2N2

Fred Sitkins
Western Michigan University
Kalamazoo, Michigan 49008
USA

Pepe Siy
Electrical and Computer Engineering
Wayne State University
Detroit, Michigan 48202
USA

Vladimir Solaja
Director of the IAMA Institute
27 Marta 80, P. FHA 802
11001 Beograd, Jugoslavija

James Clark Solinsky
International Robomation/Intelligence
2281 Las Palmas Drive
Carlsbad, California 92008
USA

Geary V. Soska
Cybotech Corporation
P.O. Box 88514
Indianapolis, Indiana 46208
USA

Michael P. Stanislawski
Milwaukee Area Technical College
1015 North Sixth Street
Milwaukee, Wisconsin 53203
USA

Yamazakil H. Suzuki
Production Engineering
Toyohashi University of Technology
Temdaku-Cho
Toyohashi, 440 Japan

M. Swami
Indian Institute of Technology
Delhi, India

Kumar K. Tamma
Mechanical and Aerospace Engineering
West Virginia University
Morgantown, West Virginia 26506
USA

William R. Tanner
Productivity Systems Inc.
1210 E. Maple Road
Troy, Michigan 48083
USA

Howard L. Taylor
Industrial Automation Systems
General Foods Corporation
250 North Street
White Plains, New York 10625
USA

B. Thacker
Universal Computer Applications
Southfield, Michigan 48080
USA

Joseph Tulkoff
Lockheed Georgia Company
Marietta. Georgia 30063
USA

Rainer Uhlig
ESAB
Robotic Welding Division
P.O. Box 2286
Fort Collins, Colorado 80522
USA

Robert A. Ullrich
Owen Graduate School of Management
Vanderbilt University
Nashville, Tennessee 37202
USA

Jean Michel Valade
Equipe Car
L.A.A.S.
7 Avenue Du Colonel Roche
31077 Toulouse Cedex
France

Murali Varanasi
Computer Science and Engineering
University of South Florida
Tampa, Florida 33620
USA

Ing. Gabriele Vassura
40136 Bologna
Viale Risorgimento, 2
Italy

Alok Verma
Mechanical Engineering Technology
Old Dominion University
Norfolk, Virginia 23508
USA

John M. Vranish
National Bureau of Standards
Washington, DC 20007
USA

Wayne W. Walter
Rochester Institute of Technology
Mechanical Engineering
One Lomb Memorial Drive
P.O. Box 9887
Rochester, New York 14623
USA

Robert O. Warrington, Jr.
Mechanical Engineering Dept.
Lousiana Tech.
Ruston, Louisiana 71272
USA

M. Weck
Werkzugmaschienlabor
der EWTH
Aachen, West Germany

Andrew Weilert
University of Kansas
Lawrence, Kansas 66045
USA

Gary Workman
and William Teoh
University of Alabama
Huntsville, Alabama 35899
USA

William V. Wright
Engineering Department
University of North Carolina at Charlotte
Charlotte, North Carolina 28223
USA

K. Yamazaki
Production Engineering
Toyohashi University of Technology
Temdaku-Cho
Toyohashi, 440 Japan

Richard A. Zang
RCA/Technical Excellence Center
13 Roszel Road
P.O. Box 432
Princeton, New Jersey 08540
USA

S. W. Zewari
Mechanical Engineering
Virginia Polytechnic Institute and
State University
Blacksburg, Virginia 24061
USA

Mingfa Zhu
Electrical and Computer Engineering
Wayne State University
Detroit, Michigan 48202
USA

Joseph P. Ziskovsky
GCA/PAR Systems
3460 N. Lexington Ave.
St. Paul, Minnesota 55112
USA

Contents

INTEGRATION OF CAD/CAM & ROBOT CELL
Chairman: William Tanner, Productivity Systems, Inc., Troy, Michigan
Vice Chairman: James A. Rehg, Robotics Center, Greenwood, South Carolina

ROBOTS APPLICATIONS
Chairman: Ray Asfahl, University of Arkansas, Fayetteville, Arkansas
Vice Chairman: Ren-Chyuan Luo, North Carolina State University, Raleigh, North Carolina

MANIPULATION I
Chairman: W. Edward Red, Brigham Young University, Provo, Utah
Vice Chairman: Joseph P. Ziskovsky, GCA Corporation, St. Paul, Minesota

COMPUTER VISION AND SENSORS
Chairman: Lawrence A. Goshorn, International Robomation/Intelligence, Carlsbad, California
Vice Chairman: Murali Varanasi, University of South Florida, Tampa, Florida

ROBOT CELL AND MOBILE ROBOT
Chairman: James B. Canner, Sterling Detroit Company, Detroit, Michigan
Vice Chairman: Jim Hansen, Borg Warner Research Des Plaines, Illinois

CONTROLS II
Chairman: R. P. Sharma, Western Michigan University, Kalamazoo, Michigan
Vice Chairman: R. Sharan, National Institute of Foundry and Forge Technology, Ranchi, India

ARTIFICIAL INTELLIGENCE
Chairman: Kamal N. Karna, CC & GA Corporation, Cliffwood, New Jersey
Vice Chairman: David Bourne, Carnegie Mellon University, Pittsburg, Pennsylvania

ADVANCED APPLICATIONS OF ROBOTS
Chairman: Nicholas Shields, Jr., Essex Corporation, Huntsville, Alabama
Vice Chairman: Geary V. Soska, Cybotech Corporation, Indianapolis, Indiana

Planning of Automation

Chairman: Alexander Houtzeel, Organization for Industrial Research, Waltham, Massachusetts
Vice Chairman: William V. Wright, University of North Carolina at Charlotte, Charlotte, North Carolina

The Design and Planning of Futuristic Factories

Albert Madwed

A. Madwed Company
110 Wedgewood Drive
Easton, CT 06612

Bridgeport Engineering Institute
Robotics and Automation Dept.
P. O. Box 6459
Bridgeport, CT 06606

Summary

The "Third Industrial Revolution" which is the application of computer controlled work devices in manufacturing systems is starting to gain momentum. Although many articles have been written about the Third Industrial Revolution there has not been many articles on the subject of designing and planning a futuristic factory for this new era. The key word in this new era is FLEXIBILITY. Therefore it is only logical that the design and planning of futuristic factories should accommodate this flexibility in automation. It is the purpose of this paper to organize the design principles and requirements for the futuristic factories.

Planning and Building Factories for the 21st Century

According to Webster a factory is defined as "a building or group of buildings with facilities for the manufacture of goods, usually from raw material." This definition is satisfactory for the past but for the future 21st century factory a more explicit definition is required if it is to be a guide for planning a factory. A possible good definition is, "A 21st century factory is a building or space or a collection of buildings and spaces called a system not necessarily located in the same city or country across whose boundaries enter energy, data, capital, people, raw material, components, orders, production intelligence, market and financial intelligence, sales income, and customer feedback and across whose boundaries exits payroll money, profit, taxes, financing money, people, production intelligence, orders to subcontractors, energy both usable and wasted, solid waste, liquid waste, gaseous waste, data and most important, quality products." The factory must function to maximize profit, quality product output and manufacturing intelligence for the society while minimizing payroll, financing money, workers hours, energy consumed and solid, liquid, gaseous and energy waste. The direction of the factory system's activities shall be by a data base central control with the capacity to make modifications of the factories activities by utilizing machine intelligence commonly called artificial intelligence in order to strive for a goal of 100 percent efficiency. A schematic of the factory of the 21st century is present in Figure 1.

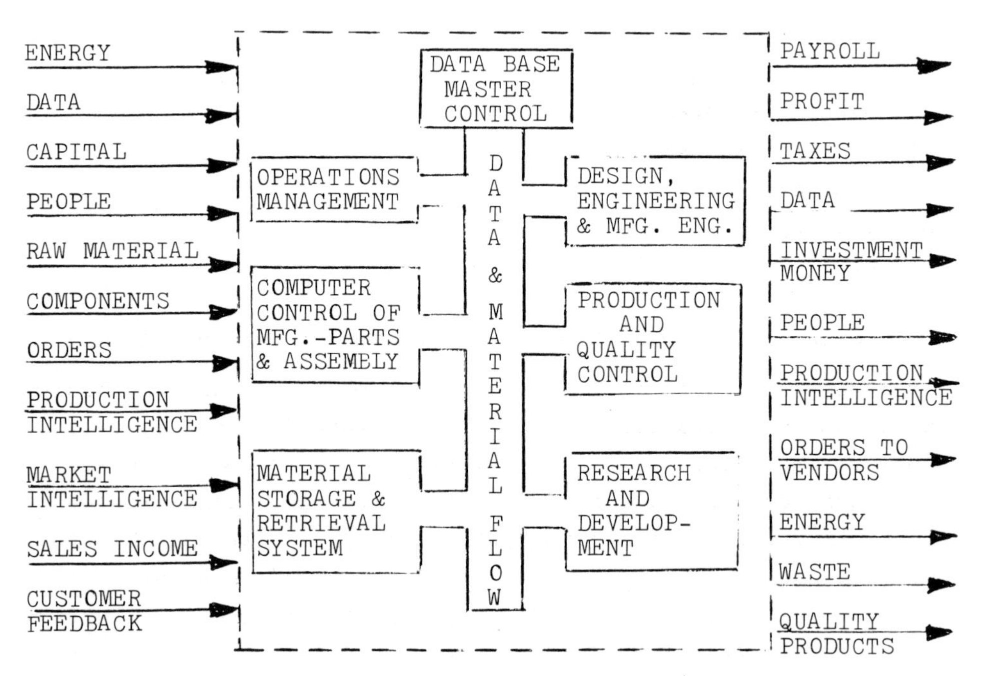

Figure 1 SCHEMATIC OF THE FACTORY OF THE 21st CENTURY

Because of our ability to electrically communicate almost instantaneously, the requirements of having departments of a factory in close proximity is not applicable any more. Therefore a manufacturing organization can have central expert system departments to service many different factories under its management envelope and make corrections and modifications throughout the organization.

Although many articles have been written about this Third Industrial Revolution, there have been very few articles written about planning a new factory building for this new era of manufacturing. The key word in this new High Tech Era of manufacturing is "FLEXIBILITY," therefore the design of futuristic factories should include accommodations so that present as well as future manufacturing systems can be installed with a minimum of effort and physical changes in the factory.

In building a new factory structure for this new age careful consideration must be given to the following:

A) Computer Data Base Department and Communication Network
B) Machining, Processing and Assembly Cells
C) Raw Material and Work-in-Process Storage and Transportation Systems
D) Tooling Department
E) Utilities and Atmosphere Control

A) Computer Data Base Department and Communication Network

From Figure 2 it is seen that the future factory will probably have two communication networks: A hierarchical or supervisory network and a peer to peer network. The hierarchical network will connect and communicate between factory sections without doing the minute details that are usually required in a factory. Many communication problems can be solved by a data bus that connects different pieces of equipment for example a controller and a robot for the purpose of controlling the robot gripper to follow a prescribed path.

Because we are considering a future factory with flexibility capabilities, we have to build the factory with communication bus systems that can service any area of the factory space with only a short easy-to-install connection. The different types of bus systems include serial or parallel systems for the future. Each will have its place but for economy sake probably the most versatile will be chosen for the factory data bus network. Although many companies are designing their own special systems, eventually a standard will be set up and adopted by the industry.

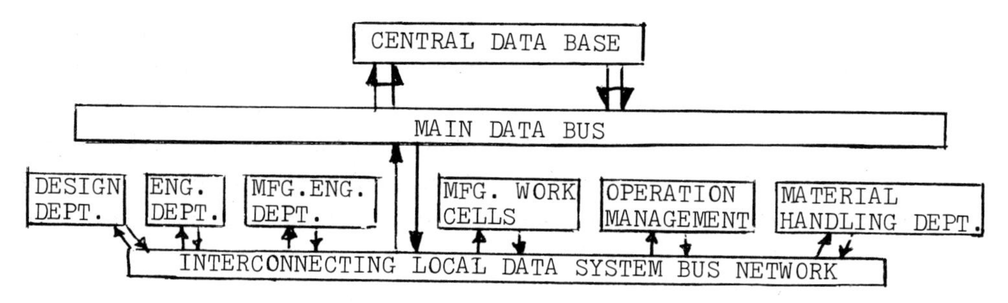

FIGURE 2. DATA BUS SYSTEM FOR THE FACTORY OF THE 21st CENTURY

B) Machining, Processing and Assembly Cells

When designing an automatic factory, the building blocks that are used are intelligent subsystems consisting of flexible manufacturing system cells with mechanical and electrical interfaces. These cells will be directed by a communication network including a central master data base network and a localized interconnecting communication network. Other sections of the automatic factory system which will be connected to the machining, processing and assembly cells will be automatic material flow and storage systems such as robotized mobile work carriers, raw material and work-in-process storage and retrieval systems and tool storage, inspection and retrieval systems and will be discussed briefly in sections C and D. In addition to the usual machines that we are familiar with, the processing cell will include such manufacturing equipment as welding, painting, surface preparation, flame cutting, electroplating and etc. Inspection equipment will be in all sections of the automatic factory to check material and parts into and out of the various systems.

Because there are many design options in flexible manufacturing, design techniques and intelligent data bases will have to be developed to determine the most efficient method of design. For example, it will be necessary to determine whether each cell of the total system should have a random layout, a functional layout, or a modular layout. Also for each option it will be necessary to determine the type and size of buffer storage between components of each cell and between cells to insure the continuity of work flow in spite of the usual mechanical, electrical, tool and processing problems that might statistically occur. Although pages can be written on automatic factory equipment design, a brief summary of the design principles for the cells for the automatic factory of the future is as follows:

1. A data base system should be used as the directing instrument making decisions for all equipment in the factory using computer-aided drafting, designing, engineering, material handling, process control, manufacturing control and artificial intelligence or machine intelligence control.

2. All equipment and process systems should be of modular design with provisions to be updated when new controls are developed or in order to make communication between equipment systems more efficient.

3. Tooling should also be modular and designed so that tool changing and measurement and quality control can be easily maintained and executed.

4. Material, raw parts, and finish parts storage can be integrated into the system without difficulty or complication and can utilize up-to-date AI concepts of the transportation problem in order to minimize work in storage, work in processing and work in buffer storage areas throughout the factory.

5. A comprehensive inspection system should be installed so that any deviation from the quality norm can be detected instantly in order to have as near to 100% acceptable quality output.

6. Careful consideration must be given to the placement of sensory feedback systems in order to keep the total system operating efficiently and to give the AI section of the Data Base Control Computer System time to make necessary corrections in system control in real time. For example it takes time to change

tools, to switch machines or material routing in a factory.

7. The central control should have a problem anticipation capability which should monitor all inputs or potential inputs such as raw material or subcontractor deliveries and make necessary changes in control based on the input intelligence.

8. Every step taken or planned should not be executed until its ability to have complete communication with central control is designed into the installed step.

9. The success of the future automatic factory will depend on the development of automation techniques for assembly. This will require a change in engineering thinking from the concept and design stages of manufacturing in order to make flexible automatic assembly practical.

C) Raw Material and Work in Process Storage and Transportation Systems

Because automatic movement of material is one of the goals of a flexible manufacturing system careful consideration must be given to space planning and the location of storage areas in order to minimize transportation distances. In general the system should be flexible enough to transport raw material, in process parts, finished parts assemblies, tooling devices and tools, fixtures, empty boxes or pallets, wastes, spare parts and materials to be kept in buffer storage in the manufacturing cells.

The storage system can be of two general types. Type 1 is designed so that the transportation device such as an automatic guided vehicle, AGV, can enter the storage area and fetch the material by itself or Type 2 where the storage system itself has a built in system to deliver the materials to a pick up point for the AGV or a delivery point where the AGV delivers the material to be returned to a storage position by the storage systems own transporting device.

In any material handling system for manufacturing the storage of a part after removing from a manufacturing cell is one of the critical value-added-factors in manufacturing. If done properly the handling of the part is greatly simplified in the manufacturing facility. Pallet storage vs. hopper storage can mean a great saving.

In addition to conveyor systems and material handling robots a relatively new device called the Automatic Guided Vehicle (AGV's) is available today. The AGV's that are most popular today are the track guided vehicles primarily because they can travel faster than the tracking guided free moving vehicles, such as those vehicles that optically or magnetically track some line that is attached to the floor. Track guided vehicles make the flexibility of the manufacturing system more difficult because the tracks are usually cut into the floor. What could remedy this difficulty is track networks that can be attached to the floor with bolts.

D) Tooling Department

One of the most critical planning problems of the FMS is tooling. It is not uncommon today to have very sophisticated machinery producing products by complete computer control and having burr problems that can only be handled by people with hand tools. How to tool a job, how to run the tool, how to pick the right speeds and feeds, how to know when burr problems will arise, how to prevent burrs from occurring, how to automatically remove burrs and etc. are all questions that need very serious study and planning in order to get an efficient FMS working. The development of data to prevent or minimize burrs in the first place is essential. Next sensor devices to locate burr and the development of systems to automatically remove burrs is absolutely essential if the FMS concept is to succeed. If this is not possible, then manufacturing processes must be developed and automated to process parts to remove burrs if they are present.

E) Utilities and Atmospheric Control

Because flexibility is the design principle for the factory the installation of the utilities should accommodate possible movement of equipment. The utilities considered here are electric, compressed air, water, drains, and air for ventilation and exhaust. These utilities should be available in all sections of the factory that could possibly require them so that major problems or delays will not occur when a work cell or factory equipment is moved. A bus system similar to the standard electric power systems in the U.S. should be applied to all utilities.

The orientation and floor plan of the factory buildings should be designed to take advantage of solar energy, energy conservation, and long term storage of summer heat and winter cold for future use.

Summary

In summary, we can state that automatic factory automation is a problem requiring great detail, expertise and the application of a tremendous amount of data because the human brain, eye, ear, and hand is not in the line to guide the details of a manufacturing operation.

References

1. Hartley, John: FMS at Work IFS Publications Ltd. VK 1984.

2. Ranky, P.: The Design and Operation of FMS. IFS Publications Ltd.

3. Tanner, W. P.: Industrial Robots Vol. 1 Fundamentals SME. 1981.

4. American Machinist The Computer: The Tool for Today Special Report 746 - June 1982.

The Optimal-Promotion Problem

Wilfred Huang and A. J. G. Babu

Summary

This paper considers the case when the existing location of the facility is no longer at the location of the cheapest transportation cost. A new model, which we termed as the optimal-promotion problem, is introduced to seize the optimality by promoting the existing demand points (markets) such that the profit is minimized.

Introduction

The expense and inconvenience of distributing to distant customers and procuring material from a distance, invite manufacturers to locate near their markets. Therefore the minimization of the total transportation cost is regarded as one of the basic objectives in the industries.
From the transportation aspects of the location problem of the factory, the facility should be set up at a point at (x,y) (assuming it can be located anywhere in a region) where the total transportation cost is defined by

$$\underset{X}{\text{minimize}} \quad \sum_{i=1}^{n} W_i \, d_i(X, P_i) \qquad \text{(P1)}$$

where n is the number of the existing demand points,

X is the location of the facility, with coordinates (x,y),

P_i is i^{th} demand point, with coordinates (a_i,b_i),

W_i is a positive weight indicating the level of interaction between the facility and the ith demand point (e.g. number of trucks per trip),

and $d_i(X,P_i)$ is the distance between X and P_i, which is equal to square root of $[(x-a_i)^2+(y-b_i)^2]$.

Weiszfeld's iterative approach has been used for solving problem P1, see Weiszfeld[8], Kuhn and Kuenne[7] and Cooper[3]. The set of iterative equations below is obtained by setting the derivative of P1 with respect to X equal to zero.

$$X^{k+1} = \frac{\sum_{i=1}^{n} W_i P_i / d(X^k, P_i)}{\sum_{i=1}^{n} W_i / (d(X^k, P_i)} = T(X^k) \quad \text{for } k=1,2,...$$

Usually X^0 is the center of gravity, and the convergence proof of the algorithm is considered in Weiszfeld[8], Kuhn[5,6], Katz[4] and Babu[1].

The problem P1 is being considered as a static facility location problem, because the transportation cost, demands, and locations of demand points are assumed to be fixed. In practice, the demands at the demand points are likely to change over a period of time; new demand points (or markets) may be added. In this regard, Ballou[2], Wesolowsky[10] and Wesolowsky and Truscott[11] consider dynamic facility location models. The authors consider the possibility of relocating the facility in order to maximize profit or to minimize cost over a specified planning horizon. However there are several disadvantages of relocation: first and foremost, relocation costs have to be considered;second, a study may have to be made for labor availability at the relocated site; third, although the new location is the most efficient position, the new labor or the loyal labor may not adjust fast enough to produce efficiently. So instead of investing in relocation, it may be more appropriate to invest in stimulating demands of the markets, for example, advertising. This serves to boost demands of certain markets, existing or new, in such a way that the existing facility will reseize the optimality.

Problem Formulation

Let S_i be the stimulation cost per unit of demand at i^{th} demand point,

R_i be the profit per unit of demand at i^{th} demand point,

and D_i be the net increase in demand at i^{th} demand point.

then the optimal-promotion problem can be mathematically formulated as follows:

$$\underset{D_1,D_2,\ldots D_n}{\text{maximize}} \sum_{i=1}^{n} [R_iD_i - S_iD_i] = \sum_{i=1}^{n} [C_iD_i]$$

$$\text{s.t.} \quad \underset{x,y}{\text{minimum}} \sum_{i=1}^{n} (V_i+D_i)d_i(X,P_i) = \sum_{i=1}^{n} (V_i+D_i)d_i(X^e,P_i) \qquad \text{(P2)}$$

$$D_i \geq 0 \quad \text{for all } i=1,\ldots,n$$

where C_i is the net profit per unit of demand at i^{th} demand point,

V_i is the new demand of the ith demand point, and

X^e is the existing facility location.

The constraint of problem P2 is to pull the optimal solution X^* to the existing facility X^e by means of adding new demands to demand points; in other words, X^e is the solution after solving the "new" problem with new demands, i.e. $X^*=X^e$. Since $X^*=T(X^e)=X^e$, the problem P2 can be reformulated as follows

$$\underset{D_1,\ldots,D_n}{\text{maximize}} \sum_{i=1}^{n} C_iD_i$$

$$\text{s.t.} \quad X^e \sum_{i=1}^{n} [(V_i+D_i)/d(X^e,P_i)] = \sum_{i=1}^{n} [(V_i+D_i)P_i/d(X^e,P_i)]$$

$$D_i \geq 0 \quad \text{for all } i=1,\ldots,n \qquad \text{(P3)}$$

Let $K_i=(P_i-X^e)/d(X^e,P_i)$

Then the constraint can be rewritten as

$$\sum_{i=1}^{n} K_iD_i=-\sum_{i=1}^{n} K_iV_i$$

which shows that the new formulation is a linear programming problem.

Properties of the problem

1. When there are only two demand points (n=2), the existing facility X^e has to be in the convex hull $[P_1,P_2]$ in order to have feasible solution.

The problem to be considered is

$$\underset{D_1,D_2}{\text{maximize}}\ Z=C_1D_1+C_2D_2$$

$$\text{s.t.}\ \min_X\ (V_1+D_1)d_1(X,P_1)+(V_2+D_2)d_2(X,P_2)=(V_1+D_1)d_1(X^e,P_1)+(V_2+D_2)d_2(X^e,P_2)$$

$$D_1\geq 0,\ D_2\geq 0$$

Case 1: Let $V_1=V_2$, so X^* is anywhere from P_1 to P_2

(i) If X^e is somewhere from P_1 to P_2

then $D_1=D_2=0$ and $Z^*=0$

(ii) If X^e is outside of the convex hull $[P_1,P_2]$

then the problem has no feasible solution

Case 2: Let $V_1>V_2$, so $X^*=P_1$

(i) If $X^e=P_1$

then $D_1=D_2=0$ and $Z^*=0$

(ii) If X^e is somewhere between P_1 and P_2 but not at P_1

then $D_2=V_1=V_2$ and $Z^*=C_2D_2$

(iii) If X^e lies outside of the convex hull $[P_1,P_2]$

then the problem has no solution.

Case 3: Let $V_1<V_2$, so $X^*=P_2$

This case is similar to case 2.

2. If the existing facility X^e of problem P2 or P3 lies outside of the convex hull $[P_1,P_2,\dots,P_n]$, there is no feasible solution to problem P2 or P3, since optimal solution always lies within the convex hull, see Wendell and Hurter[9].

3. If the existing facility X^e of problem P2 or P3 coincides with one of the demand points, say P_k, then trivial solution exists:

$$D_k^* = \sum_{i=1}^{n} W_i - W_k$$
and $D_j^*=0$ for $j=1,\dots,n$ but $j\neq k$.

4. A new demand point can be considered in problem P2 or P3 by letting its weight equals to zero.

5. The dual of Problem P3 has only two decision variables, thus it can be solved graphically.

The primal problem P3 is equivalent to

$$\underset{D_1,\dots,D_n}{\text{maximize}} \sum_{i=1,n}^{n} C_i D_i$$

$$\text{s.t.} \quad \sum_{i=1}^{n} K_{ji} D_i = -\sum_{i=1}^{n} K_{ji} V_i \qquad \text{for } j=1,2$$
$$D_i \geq 0 \qquad \text{for } i=1,\dots,n$$

where

$$K_{1i} = (a_i - X^e)/d_i(X^e,P_i) \qquad \text{for } i=1,\dots,n$$
$$K_{2i} = (b_i - X^e)/d_i(X^e,P_i) \qquad \text{for } i=1,\dots,n$$
$$P_i = (a_i,b_i) \qquad \text{for } i=1,\dots,n$$

while its dual is $\underset{Z_1,Z_2}{\text{minimize}} \; -\sum_{j=1}^{2} \sum_{i=1}^{n} (K_{ji}V_i)Z_j$

$$\text{s.t.} \quad \sum_{j=1}^{2} K_{ji} Z_j \leq C_j \qquad \text{for } i=1,\ldots,n$$

$$Z_j \text{ unrestricted}$$

6. At most two D_i^* will have non zero values, and the rest will be equal to zero provided the problem is feasible. It is because there are only two constraints in problem P3. Therefore there can be only two basic variables. However it is possible to have only one basic variable with a non zero value when X^e coincides with one of the existing demand points.

7. Suppose $P_1, P_2, \ldots$ and P_n are not collinear. The optimal solution X^* will not be between any two successive extreme points of the convex hull $[P_1, P_2, \ldots, P_n]$. Hence if the existing facility is between any two successive extreme points of the convex hull $[P_1, P_2, \ldots, P_n]$, then the problem P2 or P3 has no feasible solution.

Extensions

The optimal-promotion model in this paper can be extended to have a convex promotion cost function which is more realistic since the incremental cost of demand is expected to increase. The new problem can be solved using the separable programming technique.

When the optimal-promotion problem is infeasible, we may consider the following choices:

(1) consider promoting a new locality outside the convex hull such that the existing facility lies inside the new convex hull to yield an optimal-promotion problem, or

(2) formulate the problem to make the existing facility as "best" as possible in terms of minimizing the transportation cost, or

(3) consider relocation option.

References

1. Babu, A.J.G., "Some Advances in Simple Location Theory under Certainty and Uncertainty," Ph.D. Dissertation, Southern Methodist University, Dallas, Texas, 1980.

2. Ballou, Ronald H., "Dynamic Warehouse Location Analysis," **Journal of Marketing Research**, V, pp. 271-276, August 1968.

3. Cooper, Leon, "Location-Allocation Problems," **Operations Research**, 11, pp. 331-343, 1963.

4. Katz, I.N., "On the Convergence of a Numerical Scheme for Solving Some Locational Equilibrium Problems," **SIAM J. Appl. Math.**, 17, pp 1224-1231, 1969.

5. Kuhn, H.W., "On a pair of Dual Nonlinear Programs," **Nonlinear Programming**, Abadie Ed. , Chapter 3, John Wiley & Sons, Inc., New York, 1967.

6. Kuhn, H.W., "A Note on Fermat's Problem," **Math Programming**, 4, 1, 98-107, 1973.

7. Kuhn, H.W. and R.E. Kuenne, "an Efficient Algorithm for the Numerical Solution of the Generalized Weber Problem in Spatial Economics," **Journal of Regional Science**, 4, pp. 21-33, 1962.

8. Weiszfeld, E., "Sur le point pour lequel la somme des distances de n points donnes est minimum," **Tohoku Mathematics**, 43, pp. 355-386, 1937.

9. Wendell, R.E. and A.P. Hurter Jr., "Location Theory, Dominance, and Convexity, "**Operations Research**, 21, pp. 314-320, 1973.

10. Wesolowsky, George O., "Dynamic Facility Location," **Management Science**, 19, 11, pp. 1241-1248, July 1973.

11. Wesolowsky, G.O. and William G. Truscott.,"The multiperiod Location-Allocation Problem with Relocation of Facilities," **Management Science**, 22, 1, pp. 57-65, 1975.

Planning of Simple Automation Using Robots

James N. Perry

Schrader Bellows Division, Scovill, Inc.
U.S. Route No. 1
Wake Forest, North Carolina

Summary

The capabilities of todays robots far exceed the needs of most American manufacturing companies. Robotics is not just for large, sophisticated companies but can benefit everyone. A way to systematically examine plant operations is discussed and six generalized applications for a simple pneumatic pick and place robot are described. This is a good way to start automation of factory operations. Experience gained from such simple applications can lead to progressively more complicated and valuable projects. Improved productivity and quality through automation is necessary for any manufacturer to stay competitive.

The Need

The spectre of foreign competition snuffing out one American factory after another is a cliche. Unfortunately, it is true. High costs and poor quality are real problems brought on by obsolete equipment and methods and inadequate application of statistical quality control methods.

This is not a problem of company size or age but of management awareness and capability. Enlightened, effective factories can be any size. A one man shop running one modern injection molding press 24 hours a day with real time monitoring and automatic machine tending is just as impressive as a giant corporation's "factory-of-the-future". Neglect, making do with old equipment, and toleration of substandard quality are similarly not characteristics of size or age.

The problems of uncompetitiveness do not yield to flashy one shot solutions but require continuous effort. Small steps that build on each other. Such progress requires change and openness to new ideas. The concerned person responsible for manufacturing effectiveness may be an owner, superintendent, plant manager, vice president of manufacturing or any title. He (or she) must decide to act and dedicate themselves to this never ending quest for productivity and quality.

When this person decides to take action he may turn to the literature for guidance.

What is found is frequently too advanced for his beginning needs. Vision, tactile sensing, CIM, even FMS are complex and inapplicable to his first order problems. The state-of-the-art as described by trade journals is nearly always far ahead of what the majority of users require. We need a simple X and R or P chart to get a handle on quality - not real-time on-machine gaging. We need work place layout from industrial engineering - not sophisticated operations research. We need accurate bills of materials before the computerized production control system is installed. And we need simple automation and simple robotic applications before we go after the unmanned factory. Walk before we run.

How to Start

Start with where you are. Look at what your factory does. Do a simple flow chart and description of major operations and key pieces of equipment. Identify material flow and inventory points. Describe how material is stored and moved - is it oriented or random. Walk around, talk to people - operators, supervisors, inspectors - what's not working? What causes down time? What are the quality problems? Most factory people know a lot of these things and the act of listening, studying, and putting it on paper will help organize your thinking.

Concurrent with this survey of the factory, determine what objectives are to be met by the factory automation efforts. Objectives fall into two broad interrelated categories - cost and quality:

1. Cost: Is the product labor, material or capital intensive? The kind of automation will depend on which aspect dominates. The tradition is to justify projects on direct labor savings because this is what we know how to calculate. The main benefit, however, may not come from labor savings but from better utilization of material or more up time on a key piece of expensive equipment or easier compliance with safety needs. The benefit must eventually be quantified - a cost improvement assigned. Estimates are much better than ignoring these key intangibles.

2. Quality: Quality means consistency - less variation between parts. Automation almost always improves consistency, and better quality means lower cost. These gains in product cost due to improved quality are difficult to quantify. The unimproved process was not designed to produce rejects so we frequently must compare two theoretically perfect (no defect) processes. Be realistic - estimate. If the present process is producing a known percentage of rejects, then estimate how much the automation may improve this and what it is worth.

It helps to get "new eyes" to help with this survey work. Try to find someone who knows your product or process but hasn't seen your particular operation. Ask them to walk through and comment. Vendors frequently have good experienced manufacturing people and might be worthwhile to walk through with them - show them your ideas and discuss theirs.

At some point you will have enough information to produce a list of possible projects. Write them all down with a brief description. Some will need to be subdivided. Not just "automate product A" but automation of a particular operation. On this list you should now evaluate each potential project against some criteria. For example:

> Cost: How much will the project cost? A range is good enough.
>
> Value: If successful, how will the project help? Lower labor, less scrap, reduced inventory? This evaluation can be a mixture of qualitative and quantitative factors.
>
> Difficulty: How hard is this going to be? Probability of success? Some projects we know can be done - maybe the cost is high and value low - but we know it will work. Others are risky - but if succesful would really help. Only you can decide which kind or project to select.
>
> Help: Who would need to be involved in accomplishing this project? Can we do it ourselves? Will just a robot vendor's applications people be sufficient? Should it be turnkey from a systems house?

Finally, the project should be selected by examining your list. Try ranking all the projects or picking the top ten and bottom ten. Cut the list down to the "best" projects. Best is a value judgment - most savings? Easiest to do? It is recommended that you start with a simple project that can be done primarily by your own people. The cost should be moderate and the probability of success high. Even the simplest project will be more difficult than expected.

After project selection you are ready to do the more detailed analysis that is required for justification and implementation. Automation systems and equipment vendors will need to quote and approval for spending obtained.

Typical First Applications

Below are described six simple robotic applications using the most basic robot - a cylindrical coordinate pick and place pneumatic device. These are real applications. They cost justify and they can be done by in-house people. They are first steps - what is learned can be used on the next project to try something more ambitious. You will learn the importance of raw material consistency - the difficulty of feeding even simple parts - the problems in interfacing with your equipment's electronics:

1. Simple Work Cells: Work is processed first by machine A and then by machine B. A robot unloads machine A and transfers the part to machine B for the next operation. This linking of operations saves labor time and reduces in-process inventories. A complete work cell would also include a method for loading machine A and unloading machine B.

An expanded work cell links machines A, B and C utilizing two robots.

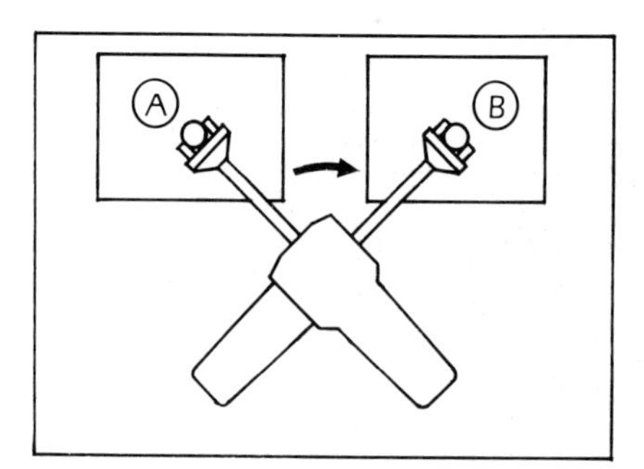

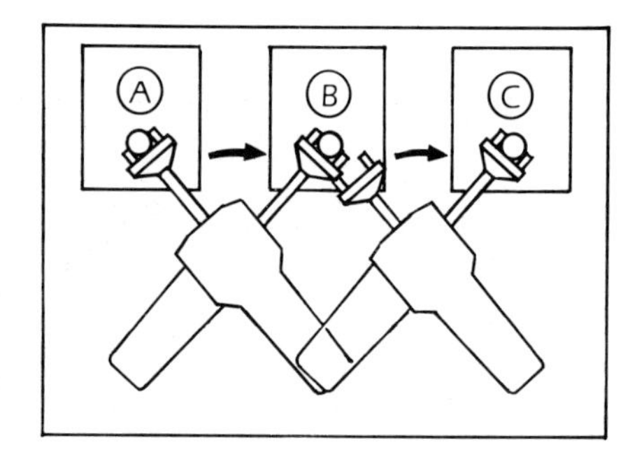

Figure 1: Simple Work Cells

2. Secondary Operations: A plastic part could not be dropped after molding due to cosmetic requirements. A robot removes the part from the opened mold and places it in a trim die for flash removal. The finished part is then placed on a conveyor with no scratches or mars.

Secondary operations on die cast parts are other excellent applications. Heat quenching can be eliminated because, unlike a human hand, a gripper can withstand high heat. The human operator is removed from a difficult environment, a savings in labor and productive man hours.

A mobile cart can be used in this application for flexibility and easy parts changeover.

2. Secondary Operations: (Continued)

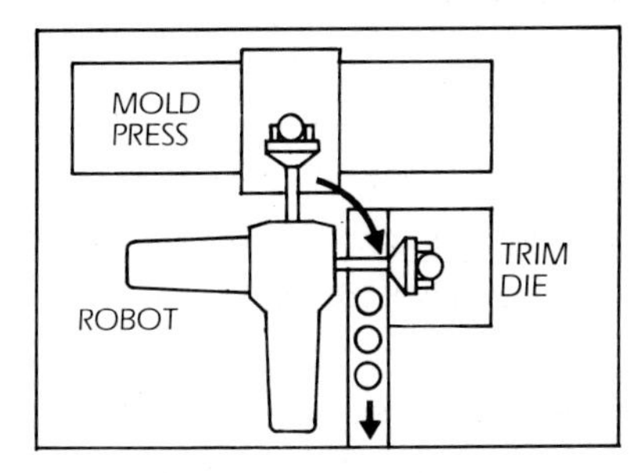

Figure 2: Secondary Operation

3. Loading and Unloading: Two robots are used in this system to load and unload. By coordinating the motions of the two robots, the idle time on machine A was kept to under one second. A variety of sensors can be used to detect reject parts and place them in an alternate location.

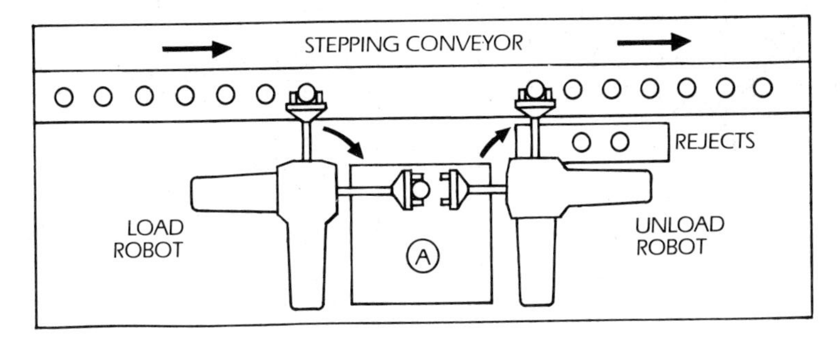

Figure 3: Loading and Unloading

4. Sorting: This system sorts incoming parts into three categories, A, B and C, based on the results of the testing equipment.

4. Sorting: (Continued)

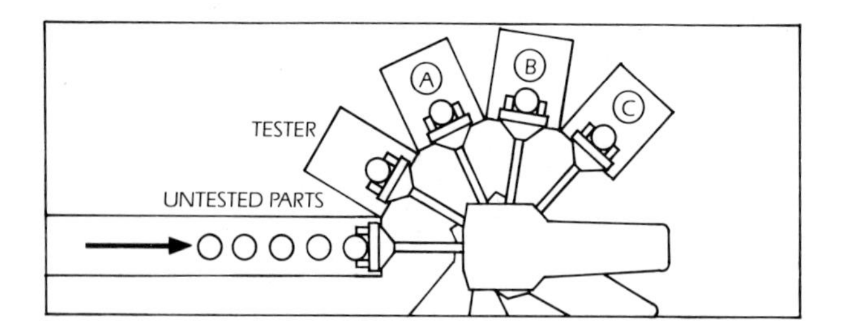

Figure 4: Sorting

5. Rotary Table: Rotary tables are used in many automated processes to bring parts to a variety of different work stations. In this application, a robot loads parts from a delivery track. Only lift and reach were required (no base rotation) for this loading operation. The unloading operation uses information from the operations performed on the parts to place them in either "Good" or "Reject" bins.

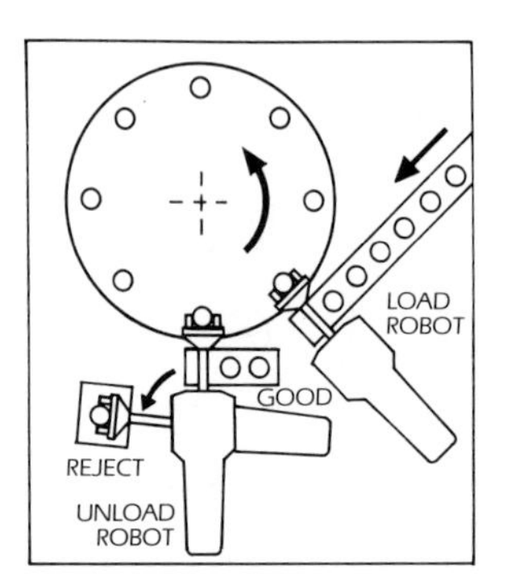

Figure 5: Rotary Table

6. Press Loading: A punch press was used in this application to assemble two cup-shaped parts used to form a wheel for office furniture. Even though the parts supply was fed from a hand-loaded magazine, the robot system tripled output because requirements for human operator shielding and safety harnesses on the press were eliminated. A robot can reach into hostile environments.

The finished part is automatically ejected by the robot arm as the robot moves the unfinished part into position.

6. Press Loading: (Continued)

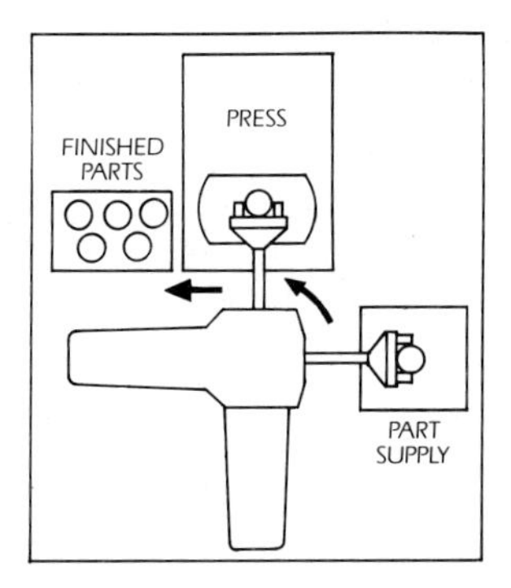

Figure 6: Press Loading

Conclusion

Better quality and productivity is a never ending need. Automation is necessary. Start simple - you will learn a lot and each step will build on the one before.

Group Scheduling for Batch Production

R. Radharamanan

Department of Mechanical and Industrial Engineering
University of Utah, Salt Lake City

Summary

Production scheduling associated with Group Technology is called "Group Scheduling". Under the Group Technology environment, if the families of parts and groups of machines are formed correctly, each job will indicate by its code number which group of machines will be used to process it. From the existing algorithms for Group Scheduling, a heuristic algorithm has been developed and programmed for computer/micro-computer applications. The developed algorithm has been used to determine the optimal group and the optimal job sequence for a batch type production process with functional layout.

Introduction

Group Technology is a rapidly developing productivity improvement tool that can have a significant impact on the development of totally integrated manufacturing facilities and flexible manufacturing systems. For successful implimentation of Group Technology concepts in the major areas of industrial activities such as design, manufacturing, production planning, especially those related to CAD/CAM, suitable Classification and Coding System is an essential prerequiste. Production scheduling is one of the major areas for Group Technology application.

In Group Scheduling, the scope of the problem is reduced from that of a large portion of the shop to a small group of machines. Once the part families and the machine groups have been correctly formed, each job will indicate by its code number which group of machines will be used to process it. Within the group of machines, one needs to simply schedule the given jobs through the machines in the cell. If a machine group/cell is not formed, production scheduling could still be very much simplified by use of part families in assigning jobs to the various machines in the shop. Proper scheduling by Group Technology, combined with reduced set-up time and reduced transportation, will result in a significant cost reduction. Group Scheduling will : reduce set-up time and costs, permit optimal determination of group and job sequence, permit flow line production, optimize group layout and provide overall economic advantage.

The inherent nature of group layout and part family grouping leads to the sequencing problem. The individual components are grouped into part families by Classification and Coding Systems. A plant layout is designed so that each part family is assigned to a specific machine group. In the group layout, subsequent required work-stations are directly adjacent to preceeding operations and hence the transportation time is minimum and it can be considered negligible in the sequencing problem. The waiting time for a part to be processed through a particular work-station, becomes predictable and control of the manufacturing process is maintained. Finding an optimum operation sequence for a given part family and a machine group is not an easy task. For "n" jobs and "m" machines, all possible combinations of operation sequences are $(n!)^m$. For example, if three different jobs are to be processed through four different work-stations, then $(3!)^4=1296$ different sequences are possible, assuming that the operations can be performed in an arbitrary order.

When jobs are grouped into part families and processed as groups through either machine group or a conventional layout, the problem of optimum operations sequencing and machine loading become much simpler compared to the general job shop condition. When part family is assigned to a given machine group, the problem size is reduced to n!. However, there still remains the extremely complex problem of resolving just how to schedule these jobs, once they are assigned to a particular machine group for processing.

Group Scheduling

There are many heuristic algorithms developed for general job shop application. Many of them are not readily applicable for practical shop conditions, due to various limitations based on unrealistic hypotheses, complicated computations involved etc. Also, most of these algorithms are not addressed to "Group Scheduling", but to a general job shop situation which is more complicated and difficult to solve. However, several efforts have been made by Petrov [2], Hitomi and Ham [3], Taylor and Ham [5] to deal with "Group Scheduling problems. There are two basic models of optimum sequencing for Group Scheduling : scheduling for a single part family and scheduling for a set of part families. Regardless of a single group or multi-group, Group Scheduling problems can be represented as unidirectional method and non-unidirectional method. The most important requirement for Group Scheduling is to have "part families", and to process them as a part family or a set of part families. However, the machines/operations for these jobs do not necessarily have to be grouped as a machine group/cell. The two different cases for Group Scheduling are : conventional layout (functional layout) and group layout (cellular layout). In Group Scheduling, there are two different methods of transferring the parts from one machine to the next. They are by successive transfer (lot by lot) and overlap transfer (piece by piece).

Algorithm for Group Scheduling

Group Scheduling can be analysed in a multi-stage and multi-product manufacturing system. An essential requirement is that jobs are grouped into part families, thus both optimal group and optimal job sequence can be determined such that the total throughput time is minimized. There are various scheduling methods such as branch and bound, heuristic and others. In this paper, a heuristic method will be presented and discussed for determining optimal group and optimal job sequence. The heuristic algorithm was initially developed by Petrov [2] and then modified by Ham [4] for Group Scheduling. The algorithm developed in FORTRAN language which slightly differs from that of Ham has been used to find the optimal group and the optimal job sequences for a set of part families in a batch type production process [1]. First, a simple example is given to illustrate the steps involved in finding the optimal job sequence. The data needed for Group Scheduling a single part family is given in Table 1.

Table 1 Data for Group Scheduling a single part family

Job	Machine time in hours M_1	M_2	M_3	M_4
J_1	30	20	34	26
J_2	42	14	16	12
J_3	30	8	32	28

The processing time matrix is given by

$$A = \begin{bmatrix} 30 & 20 & 34 & 26 \\ 42 & 14 & 16 & 12 \\ 30 & 8 & 32 & 28 \end{bmatrix}$$

Matrix A is sectioned into two equal time matrices T and T'.

$$T = \begin{bmatrix} 30 & 20 \\ 42 & 14 \\ 30 & 8 \end{bmatrix} \quad ; \quad T' = \begin{bmatrix} 34 & 26 \\ 16 & 12 \\ 32 & 28 \end{bmatrix}$$

When there are odd number of columns in the time matrix, for example five, the first three columns form the T matrix and the last three form the T' matrix, having the third column common to both the matrices. The sums and the difference of the sums of the sectioned matrices are given by,

$$\Sigma T = \begin{bmatrix} 50 \\ 56 \\ 38 \end{bmatrix} \quad ; \quad \Sigma T' = \begin{bmatrix} 60 \\ 28 \\ 60 \end{bmatrix} \quad ; \quad \Sigma T' - \Sigma T = \begin{bmatrix} 10 \\ -28 \\ 22 \end{bmatrix}$$

To determine the optimal job sequence, the following heuristic rules must be observed:

Rule I: For those jobs in which $(\Sigma T'-\Sigma T)$ is positive or zero, that is $(\Sigma T'-\Sigma T)\geq 0$, the job sequence is in the ascending order of ΣT values. For those jobs in which $(\Sigma T'-\Sigma T)$ is negative, that is $(\Sigma T'-\Sigma T)<0$, the job sequence is in the descending order of $\Sigma T'$ values.

Rule II: The job sequence is assigned in the descending order of the numerical value of $(\Sigma T'-\Sigma T)$.

Rules III and IV: Rules I and II are applied to those problems where each job is processed by a set of machines/operations in the same order. However, when there are jobs with omission or absence of some operations, rules III and IV must also be applied. Essentially, rules III and IV are the same as rules I and II respectively, except that in the application of rules III and IV, for each line of the sectioned time matrices T and T' its mean is determined. These mean values of the time $\bar{T}$ and $\bar{T}'$ are found by dividing the sectioned time matrices ΣT and $\Sigma T'$ respectively by number of operations performed in each line of the sectioned matrices, T and T'.

Special Cases: If the different values of $(\Sigma T'-\Sigma T)$ and $(\bar{T}'-\bar{T})$ are exclusively positive or negative, rules I and III respectively will alone supply a unique solution for optimal job sequence, and rules II and IV are not used. When a number of identical values are present in ΣT and $\Sigma T'$ or $\bar{T}$ and $\bar{T}'$ rules II and IV respectively are used to determine the adequate sequence for the identical values.

Applying the appropriate rules, the optimal job sequences are:

Rule I: $(\Sigma T'-\Sigma T)\geq 0$, $(\Sigma T)\uparrow$; $(\Sigma T'-\Sigma T)<0$, $(\Sigma T')\downarrow$: $3\to 1\to 2$

Rule II: $(\Sigma T'-\Sigma T)\downarrow$: $3\to 1\to 2$

The optimal job sequences by rules I and II are found to be the same. After obtaining the job sequence, the total processing time of the jobs in the sequence must be determined.

Processing time calculation: The processing time calculation for the job sequence is given in table 2. The group of machines used for processing the jobs in the part family are given in column 1. The processing time values for job 3 (first job in the job sequence) in different machines are found in column 2. The cumulative time values for job 3 corresponding to each machine are given in column 3. The processing time values for the second job in the job sequence (job 1) are given in column 4. The cumulative time values for job 1 are found in column 5. The first value in column 5 is the sum of the first values in column 3 and 4. This value added with the second value in column 4, (60+20=80) is compared with the sum of the time value accumulated by the previous job

Table 2 Processing time calculation

Machine	Job sequence					
	3		1		2	
1	2	3	4	5	6	7
M_1	30	30	30	60	42	102
M_2	8	38	20	80	14	116
M_3	32	70	34	114	16	132
M_4	28	98	26	140	12	152

upto machine 2 and the processing time needed for job 1 in that machine, (38+20=58) and the greater of this two values is taken as the cumulative time value for job 1 upto machine 2. In this case 80>58 and hence 80 is chosen. This procedure is repeated in determining the other cumulative time values for job 1 with the rest of the machines in which it is processed. The complete procedure used for job 1 is repeated for the other jobs in the job sequence. The cumulative time of the last job with the last machine (job 2 with machine 4) is the optimal processing time for the job sequence using the group of machines in column 1. In this case it is 152 hours. In general, the complete procedure of the processing time calculation should be applied to each job sequence determined by each heuristic rule and the minimum value of the total cumulative time values found by the rules will give the optimal processing time and the job sequence corresponding to this optimal time is the optimal job sequence.

Table 3 shows the total processing time values for all possible sequences of the example under consideration. It is seen from this table that the job sequence obtained using the heuristic rules is optimum with minimum total processing time of 152 hours. Comparing this value with that of the other sequences, it is found that the difference between the minimum and the maximum time is 34 hours. This simple example shows the efficiency of Group Scheduling using the heuristic method for minimizing the total time required for manufacturing the jobs in a part family. Figure 1 shows the optimal job sequence by Gantt Chart with the total processing time of 152 hours.

Table 3 Job sequence with total processing time

Job sequence	Total processing time in hours
J_1-J_2-J_3	170
J_1-J_3-J_2	156
J_2-J_1-J_3	186
J_2-J_3-J_1	182
J_3-J_1-J_2	152
J_3-J_2-J_1	182

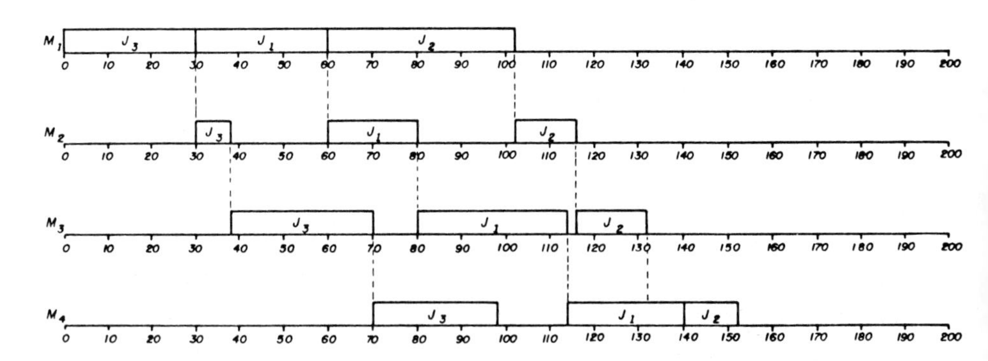

Figure 1 Processing time for the optimal job sequence by Gantt Chart

Application to batch production

The case study has been carried out in a medium size batch production industry where axial hydraulic pumps are manufactured in batches to meet out the market demand. The industry maintains a functional layout and the jobs are processed in lots between one machine to other. The flow of material is non-unidirectional. The axial pump consists of 67 different components out of which 43 components are being manufactured and the other 23 are received from the suppliers. The components that are being manufactured are alone considered for group scheduling application. First, the components are arranged in the chronological order of their manufacturing codes and then analysing their design specifications, shape, material, machining processes, accuracy etc., they have been classified and coded according to Opitz Classification and Coding System. The 43 components are then grouped into four different part families, (A, B, C, D) considering their similarities in shape, machining characteristics, dimensions etc. Based on the machining characteristics of each part family the machines in the functional layout are grouped into four different groups G_1, G_2, G_3 and G_4 and due to the limited number of certain machines which are needed by more than one part family, those machines are grouped in more than in one group, thus creating the necessity to find out the group sequence for processing. For more informations regarding how the part families and the machine groups are formed, the reader may refer to [1].

After forming the part families and the machine groups to process them, the set-up time for each machine to process each job and the processing time for each job in each machine are measured using time study procedures and the expected time element for each activity is estimated. These values are used for finding the optimal group and job sequences using the heuristic algorithm developed and programmed in FORTRAN for computer/microcomputer applications. To present some of the results obtained using the algorithm, part family B and machine group G_2 in which it is processed are chosen. The time elements for part family B is given in table 4. Making use of the program, the optimal job

Table 4 Time elements for part family B and machine group G_2 (in minutes)

Job	Machine group, G_2						
	M_1	M_2	M_3	M_4	M_5	M_6	M_7
1	0	0	0	720	0	0	0
2	0	630	1260	4200	0	210	0
3	0	75	0	0	0	0	0
4	0	1080	0	4320	270	6480	1080
5	0	0	0	0	210	3360	630
6	0	0	1800	4500	0	0	0
7	1785	0	0	37230	0	0	0
8	0	0	0	0	600	0	0

←--------- T ----------→ (M_1–M_4)

←----------- T' ---------→ (M_4–M_7)

sequence by each basic rule and the processing time for each job sequence are calculated and they are given in table 5. The minimum processing time is obtained by rule 1, (51540 minutes) and this determines the optimal processing time and the job sequence corresponding to this minimum time is the optimal job sequence. Similar results have been obtained for the other part families [1]. For determining the machine group sequence, the set-up time of the machines are also included in forming the time matrix [1].

Table 5 Job sequence and processing time by different rules

Rule	Job sequence	Total processing time in minutes
I	5→8→1→4→7→6→2→3	51,540
II	4→5→8→1→3→2→7→6	52,050
III	5→8→1→2→4→6→7→3	52,140
IV	7→5→6→8→4→2→1→3	55,665

Conclusion

The algorithm for this heuristic method is far simpler and easier to compute, compared to other similar heuristic algorithms and certainly in comparison to other optimization methods such as branch and bound method. The program developed for computer/ micro-computer applications proved its versatility and applicability for practical production scheduling problems in a Group Technology environment. In future, if batch production industries go for group layout and group scheduling concepts under cellular or flexible manufacturing systems with computer applications, robots can successfully be integrated for material handling and machine loading and unloading operations with higher productivity and overall economic advantages.

References

1. Grützmacher, E.F.: Tecnologia de Grupo para aumento da produtividade. Tese de Mestrado, UFSM, Brasil, 1984.

2. Petrov, V.A.: "Flowline Group Production Planning" (English translation from Russian). Business Publications, Ltd., London, 1968.

3. Hitomi, K.; Ham, I: Product-mix and machine loading analysis of multistage production systems for group technology. ASME, New York, 1982, 6 p.

4. Kyu-Kab Cho; Emory Enscore, E. Jr.; Ham, I.: A heuristic algorithm for multi-stage group scheduling problem to minimize total tardiness. Society of Manufacturing Engineers, May 1982, p:514-517.

5. Taylor, J.F.; Ham, I.: The use of micro computer for group scheduling. Society of Manufacturing Engineers, May 1981, p: 483-491.

Future Trend[illegible]

Future Trends and Management

Chairman: Robert A. Ullrich, Vanderbilt University, Nashville, Tennessee
Vice Chairman: J. P. Sharma, University of North Carolina at Charlotte, Charlotte, North Carclina

World Trends in the Automation of Manufacturing

M. Eugene Merchant

Director, Advanced Manufacturing Research
Metcut Research Associates Inc.
Cincinnati, Ohio 45209, U.S.A.

Summary

The overriding and all-encompassing world trend in the automation of manufacturing today is that toward computer-integrated manufacturing (CIM). It is this technology which holds the capability to accomplish overall automation of the total system of manufacturing. Thus, understanding of the nature, benefits, societal implications, current trends, and potential of CIM is the key to understanding the nature of world trends in the automation of manufacturing. These subjects are therefore the focus of this paper. Although pursuit of research, development, and implementation of CIM technology has not yet resulted in realization of full computer-integrated manufacturing anywhere in the world, nevertheless the industrialized nations and much of the world manufacturing industry are today pursuing a variety of programs having that as their goal. Of these, the integrated automation of the production activities which take place on the factory floor -- integrated automation in the form of flexible manufacturing systems (FMS) -- has proceeded further than that in any other sector of the overall system of manufacturing. The resulting performance benefits being demonstrated by such systems are most impressive and foreshadow the power and capability which full CIM can eventually bring to the future factory.

Computer-Integrated Manufacturing

The overall and all-encompassing world trend in the automation of manufacturing today is that toward computer-integrated manufacturing (CIM). Already, even in its current early stage of development, it has demonstrated greater potential to improve manufacturing capability than the combined potential of all other known types of advanced manufacturing technology.

What is the technological basis for this overriding importance of computer-integrated manufacturing? It is primarily the unique potential which the computer has to provide manufacturing with two powerful capabilities, never before available, namely:

1. online variable program (flexible) automation
2. online moment-by-moment optimization

Further, and very importantly, it has the capability potential to do that not only for the "hard" components of manufacturing (the manufacturing machinery and equipment) but also for the "soft" components of manufacturing (the information flow, the databases, the control, etc.). In addition, and of utmost importance, it has the capability potential to do that not only for the various bits and pieces of manufacturing activity but also (since the computer is a systems tool) for the entire system of manufacturing. Thus, it has tremendous potential to integrate that entire system, producing what may be called the computer-integrated manufacturing system. This generic concept of the computer-integrated-manufacturing system is shown in Figure 1. It is the key to accomplishing overall automation of manufacturing, right from the design of the product through all aspects of its production.

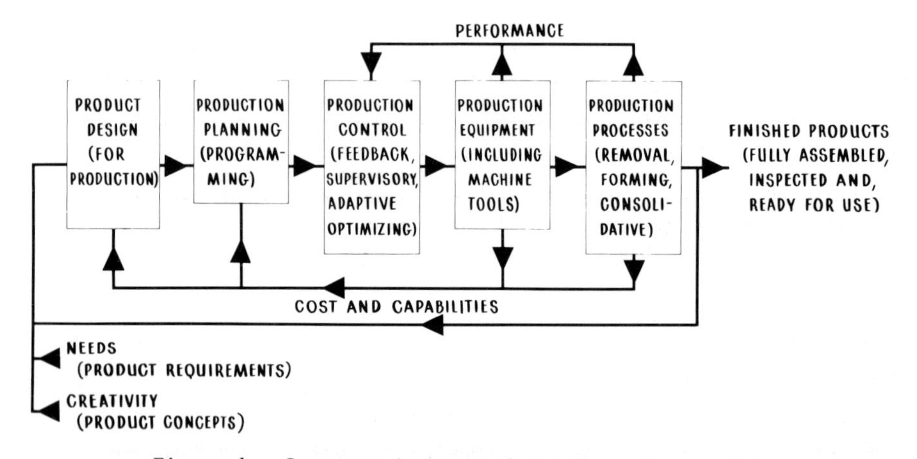

Figure 1. Computer-integrated-manufacturing system.

The basic importance of this generic integration concept for the overall automation of manufacturing is being increasingly recognized by industry today as the key to meeting, successfully, the challenges posed by the economic and social pressures impacting manufacturing today. In fact, as long as five years ago, a group of top corporate U.S. executives, interviewed by Drozda [1], stated, "systems integration is, for many companies, 'the key to survival in the next decade.'" It is this concept which provides the background guiding today's strong technological trend toward development and implementation of computer-automated manufacturing.

The primary reasons for this powerful, overriding trend are the long-term socioeconomic benefits which this technology promises to bring to nations which develop and implement it in their manufacturing industry. Thus, it is important to understand the nature of these benefits if the technological trend is to be properly guided and the resulting technology properly used.

Overall Socioeconomic Benefits

The major, long-term, overall socioeconomic benefits realizable from the development and implementation of CIM in the manufacturing industries of industrialized countries are those which stem from its potential to reduce the cost of creating the primary, real, tangible wealth of such countries. The reason for this is that manufacturing is the principal real-wealth-creating activity of such nations today. Although manufacturing accounts for about one-third of the gross national product (GNP) of the typical industrialized country today, services account for about one-half of that GNP. While services are essential to the support of a high standard of living and quality of life, they do not create primary, real, tangible wealth. Yet such wealth is, historically, the basis and source of all other wealth in a nation. Therefore, subtracting out the half of the GNP coming from services, it can be seen that manufacturing then accounts for two-thirds of the remainder -- two-thirds of the real-wealth-creating sector of the economy. Thus, it follows that manufacturing is responsible for about two-thirds of the primary wealth-creating activity of a typical industrialized country, the remainder of the real wealth creation coming from the extractive (agriculture, fishing, and mining) industries and the construction industry.

Taking a specific example, in the case of the United States today, manufacturing accounts for about 24 percent of the gross national product and the extractive and construction industries account for about 13 percent. On the other hand, the service industries account for about 63 percent of the GNP. However, only the 24 percent of GNP represented by manufacturing and the 13 percent represented by the extractive and construction industries result in creation of primary, real, tangible wealth. Thus, in the case of the United States, manufacturing accounts for about 65 percent of the direct real-wealth-creating activity of the country, or again about two-thirds.

Now let us relate these facts to the question of the potential of computer-integrated manufacturing to provide overall benefit to a country and society in general -- its effects on employment, quality of life, standard of living, etc. Since, as indicated above, creation of primary, real, tangible wealth is the basis and source of all other wealth in a nation, increases in the standard of living, quality of life, employment, and the general economic well-being of a country stem directly from decreases in the cost of creating such wealth. With manufacturing accounting for two-thirds of that wealth creation in industrialized countries, it follows that reduction of the cost of manufacturing must be a top priority item in such countries. CIM-based automation, even in its present early stage of development, has already demonstrated far greater potential to reduce manufacturing costs than anything that has appeared on the scene since the onset of the Industrial Revolution.

CIM and Factory-Floor Automation

As yet, development and implementation of full computer-integrated manufacturing has not been realized anywhere in the world. Thus, no significant performance data yet exist which can document the benefits made possible by the automation resulting from such full integration. However, on the factory floor, such integration has proceeded further than in any other sector of the manufacturing system. This has taken the form of flexible manufacturing systems (FMS). Their importance stems from the fact that such systems are not stand-alone pieces of equipment, but are fairly sophisticated integrated systems of a number of software and hardware modules. They include integrated automation of not only such "hard" functions as workpiece and tool transport, loading, unloading, and changing, but also of such "soft" functions as planning, scheduling, and control of workpiece and tool flow and workstation loading. Thus, in a sense, they are "microcosms" of a portion, at least, of the future computer-automated, -optimized, and -integrated factory. Therefore, a look at the performance benefits already being demonstrated by such systems can provide some insight into the nature and magnitude of the still greater potential benefits to be reaped eventually from CIM-based automation of the overall system of manufacturing. In addition, such a look reveals how amazingly large some of the benefits already being obtained with these microcosms of a part of the total automated manufacturing system really are. We will illustrate the nature and magnitude of these benefits by means of a few examples.

The first of these systems, described by Dronsek [2], is the FMS now operating at Messerschmitt-Bolkow-Blohm (MBB) in Augsburg, West Germany -- one of the more sophisticated FMS operating in industry in the world today.

At Augsburg, MBB builds the center section ("wing box" and related structure) of the Tornado swing-wing fighter. In 1975, when the plans were made for going into production on the Tornado at Augsburg, they decided to use CIM as fully as possible and to start with as fully automated machining of the titanium parts as possible. They then developed their long-range master plan for computer-integrated manufacturing called Computer Integrated and Automated Manufacturing (CIAM). The long-term, overall goal is reduction of manufacturing costs by integrating design, production planning and control, and production operations into a single computer-automated system.

The initial major development in the CIAM program was their flexible manufacturing system, fully operational for the past several years, for the machining of titanium parts for the central section of the Tornado. The basic elements of this system are: (1) 28 numerically controlled (NC) machining centers and multispindle gantry and traveling-column machines, (2) fully automated systems for tool transport and tool changing, (3) an automated workpiece-transport system, and (4) coordinated computer control of all these elements. The components of this system are illustrated in Figure 2.

The system has demonstrated remarkable benefits. They find that the machines in the system are cutting metal, on average, about 75 percent or more of the time, that is, machine utilization is 75 percent or better. Lead time for production of a Tornado is only 18 months, compared to about

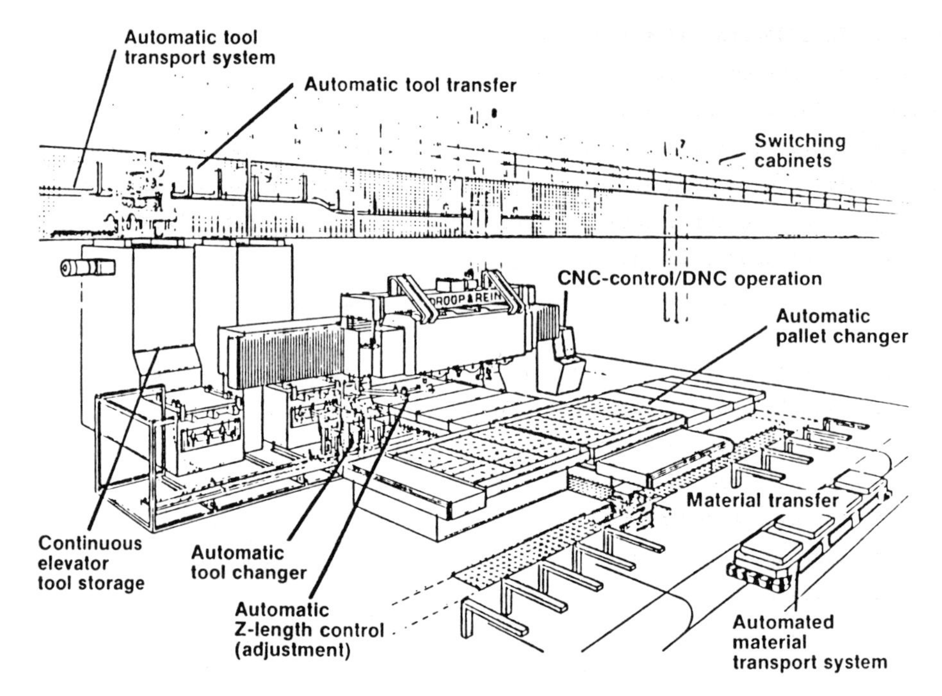

Figure 2. Components of MBB machining facility: 5-axis CNC multiple-spindle, bridge-type milling machine with automated peripheral equipment.

30 months for planes produced by more conventional means. The system reduced the number of NC machines required (compared to doing the same job with unintegrated stand-alone NC machines) by 52.6 percent, required personnel by 52.6 percent, required floor space by 42 percent, part through-put time by 25 percent, total production time by 52.6 percent, tooling cost by 30 percent, total annual costs by 24 percent, and capital investment costs by 10 percent. This last fact alone illustrates that using CIM can free up the idle capital associated with machines that are normally underutilized. This capital is more than sufficient to provide the additional sophisticated capital equipment needed to implement flexible manufacturing systems. Further, the increased productivity represented by all these figures is most impressive.

In addition, nonquantifiable benefits were experienced with this system. Quality has been improved, manifesting itself in the form of higher accuracy and reproducibility, lower rework costs, and lower scrap rates. This quality improvement in turn has resulted in lower quality-assurance costs. In addition, adherence to production schedules has improved greatly and the usual flood of paper has been decreased considerably (no order chasing). Furthermore, working conditions have improved due to the decreased risk of accidents, the relief from heavy physical labor, and the more challenging nature of the work. Increased flexibility has made the manufacturing operation more independent of batch size, of the types of parts, and of production quantities.

The next examples are drawn from Japan. With the implementation of flexible manufacturing systems already advancing quite rapidly there, efforts are now being directed toward developing capability for these systems to run unmanned at night. The first such system to attain that capability was one for machining cylinder heads for marine diesel engines at the Internal Combustion Plant of the Niigata Engineering Company in Niigata, Japan. This FMS was put into service in 1981 and is probably the world's first full system capable of running unattended for long periods of time [3]. It is machining 30 different types of cylinder heads in lot sizes ranging from 6 to 30 parts and is obtaining impressive results. The system runs 21 hours a day, including unmanned operation at night. The unmanned operation even extends to automatic mounting and demounting of many of the simpler types of workpieces on pallets by a robot. The number of machines required to produce these parts has been reduced from 31 (including 6 NC machines) to 6 (an 81-percent reduction). The number of operators has been reduced from 31 to 4 (an 87-percent reduction). The time during which parts are being produced has been increased from 9 hours to 21 hours per day, and the lead time for these parts has been reduced from 16 days to 4 days. This again illustrates the tremendous capability of computer-integrated manufacturing to increase the utilization of machines, and thus the capital investment which they represent, as well as to increase manufacturing productivity in general.

A number of other Japanese companies are in various stages of utilizing computer-controlled FMS running unattended on night-shift operations. With that capability, many of these companies are now beginning to take advantage of not only 24-hour operation, but also the great versatility of these computer-controlled machines and systems, to accomplish just-in-time (kanban) production. To accomplish this, they let the requirements of the assembly floor of the manufacturing facility "pull" the needed parts through the factory, rather than depending on scheduling to "push" them through. This is accomplished by loading these highly flexible cells and systems in one 24-hour period with just the mix of parts that will be needed on the assembly floor during the next 24-hour period. Consequently, work in progress is reduced virtually to zero, thus freeing tremendous amounts of idle capital.

In light of the foregoing examples, it is important to emphasize one tremendously significant fact about the economics of CIM-based factory automation. This fact, simply stated, is that such automation is essentially free if properly utilized! The flexible manufacturing system, as a microcosm of the future computer-integrated and automated shop, demonstrates this. First, the reduced capital investment in FMS workstations (due to the much smaller number required because of greatly increased utilization when compared to stand-alone, unintegrated workstations) more than pays for the required additional supporting, integrating facilities. These facilities include not only such entities as the control computer and the work-and-tool-transport equipment but also the software system. Secondly, the drastic reduction of work-in-process inventory and stock waiting to be assembled (virtually to zero due to the capability of these flexible systems to produce just whatever mix of parts is required for immediate assembly) frees up sufficient capital (previously lying idle in inventory) to more than pay for the total FMS. Another way of looking at this latter fact is that the just-in-time production, which is made possible by truly flexible automation, allows a plant to turn over its total inventory many more times per year than is possible in a conventional plant.

References

1. Drozda, T. A.: Manufacturing's sweeping computer revolution. Production 84/6 (1979), 74-82.

2. Dronsek, M.: Technische und wirtschaftliche probleme der fertigung im flugzeugbau. Proceedings, Produktionstechnisches Kolloquium Berlin 1979. Munich: Carl Hanser Verlag, 1979.

3. Editorial Report: Paying a visit to FMS plant. Metalworking Engineering and Marketing 4/1 (1982), 72-76.

Robotics – Its Influence on Productivity

Stanley J. Polcyn

Unimation
Danbury, Connecticut

Robotics -- Its Influence on Productivity

During the 1970's declining capital investments, higher labor costs and decreasing R&D spending had an adverse effect upon the productivity performance of American industry. Unfortunately, these factors coupled with rising inflation and high energy costs, have significantly affected our nation's balance of trade and ability to compete effectively in the world marketplace.

In the 1980's, a significant shift in manufacturing is occurring on a global basis, due to the new realities of intense global competition to overseas manufacturing and sourcing. The result is that American manaufacturers are being faced with four alternatives:

1) To go out of the manufacturing business;

2 To risk a continuing loss of market share to domestic and foreign competitors;

3) To produce overseas with foreign labor, but this requires complex political and business risks.

4) To automate and imnprove their operations at home.

If America's vital manufacturing sector is to remain healthy into the next decade and beyond U.S. producers must accellerate their use of advanced forms of automation in manufacturing. We are part of a global economy and we compete in global markets -- like it or not.

A problem involved in the decreasing American productivity is the reluctance of U.S. companies to take advantage of available technology -- too often where we are beaten today, we are being beaten by American technology which has been adopted successfully by foreign competition. This is certainly true of robotic technology. If our foreign competition can recognize the advantages of these technologies and has been willing to accept the risks in adopting them, why haven't American companies done the same thing? There are several significant reasons I believe:

Adoption of technology is impeded in the U.S. by the financial analysis and control systems used by management to evaluate investment. The traditional investment evaluation methods have been developed to deal with simple extension of existing technology. However, any technology which has the potential to produce a significant change in total productivity, profitability or ability to compete must represent a substantial change from current technology. Therefore, I believe the use of traditional evaluations inhibit and impede new technology acceptance.

Secondly -- it is characteristic of too many American companies that senior managers are only interested and willing to use only the technologies with which they are already familiar. Unfortunately, this results in sponsorship of new technology being driven by young, enthusiastic, impatient, middle-level technicians and managers -- and more unfortunately they must struggle to push acceptance up through an unsympathetic, unbending, conservative bureaucracy with little or no encouragement from the top.

A third factor of significance is the lack of direction and commitment -- a lack of an "industrial plan" by our government. While many foreign governments are seeding, nurturing, growing, motivating and protecting their new developing technologies and consolidating, re-aligning, and reorganizing the older, more sedate ones, our government representatives are arguing where to provide government support and direction and help. Be it new high-tech industries or save the old mature -- and the answer really is both. We have a need as a nation to organize our strenghts and efforts for success if we expect to retain an industrial manufacturing dominance.

The increased utilization of computer controlled machines and robots in the factories of the future is essential if manufacturers are to meet the requirements of an increasingly competitive era; the ability to make products cost-effectively; and the flexibility to respond rapidly to changes in the marketplace. Current manufacturing processes are inadequate to meet the challenges.

Advanced automation will not be a panacea for U.S. manufacturers and will not be easy. Enhancing American manufacturing productivity will be just as dependent on improving the management of human resources, if not more so. It should also be recognized at the outset that the use of advanced automation does not present an immediate crisis for the American workforce -- despite the prevalent fears, computers and robots have not yet taken over the factories. The widespread introduction of the advanced technologies is likely to be gradual, perhaps taking decades or more. If effective, the infusion of the technologies will create efficiencies that increase overall workplace competitiveness and job opportunities will be increased overall.

Industrial robots are one of the principal tools for improving manufacturing productivity and product quality in the years ahead and one of the key ingredients in the revitalization of manufacturing operations throughout the United States. The technology impacts all phases of engineering and manufacturing and the way we manage our factories and businesses today. With increased computer control and advances in sensory perception, tactile and visual, robots are "getting smarter" and their application to new and more challenging tasks is increasing.

Originally in the early unimation days we recognized that one could automate large production runs by specific design. However, short and medium run production was not cost effective to automate and often did not lend itself to "hard automation" design. What was needed was a flexible approach whereby the automation or system could be quickly taught and easily changed to accommodate short production runs, varied products, varied line speeds or variable cycle times -- and we called this flexible automation a robot.

It had and has definite advantages. By being pre-engineered:

1) The reaction time from decision to use is shorter than hard automation.

2) The debugging costs are substantially less.

3) It can be re-used with change and does not become obsolescent.

Let's take another slant at it and consider what robots are doing today. This list is not exhaustive but it does cover the bulk of established, proven applications and thereby represents no technical risks for the users.

DIE CASTING --

This application is at an independent die casting facility that uses 75 robots to unload their machines. They also quench, trim, lubricate dies, and set inserts into molds.

SHEET METAL TRANSFER --

20 robots are used transferring parts press to press. These make 6 meter transfers in about 12 seconds, but today technology can do it in approximately 3½ seconds.

ASSEMBLY --

This is an assembly job at a major robot user in the telecommunications field. It places between eight and twelve pins into a plastic base depending upon the model connector being produced in that run and then places it into a press that molds the assembly together.

ELECTRONIC TESTING --

This is an example of electronic testing. A diode is picked up from the bowl feeder escapement and places the diode into a fixture on the testing machine which tests the diode for 67 different parameters. The robot then presents the diode to a tap machine which puts a different colored dot on it according to the tests it failed.

ARC WELDING --

This is a machine arc welding parts for a large motor assembly. Cosmetic welds were required that were multi-pass. This was a six month payback.

MACHINE CELLS --

The robot loads and unloads the machine tool, inspects the parts and uses a buffer to balance the cycle time.

INVESTMENT CASTING --

Ten robots are on line dipping wax molds to make ceramic molds that will be used to cast aircraft turbine blades. This is a natural for robot application due to the fragility of the parts and the high scrap rate and rework needed when done manually. Today robotic technology is also being applied to assemble and trim the wax molds.

SPOT WELDING --

Over 4000 unimate robots are used on automotive respot lines around the world. Shown here is the General Motors line with 60 units installed.

PALLETIZING --

The sophistication of computer controlled robots makes palletizing easy. To palletize a 50' by 50' matrix as few as 23 program steps and two taught locations are required. A non-computerized palletizer would require 15,000 programmed steps.

Catalytic converters are palletized into shipping containers. The benefit to the user was the elimination of serious workman's compensation problems.

MATERIAL HANDLING --

A material handling application is this flourescent tube palletizing in which 30 tubes either 4' or 8' long are handled in each cycle.

FOUNDRY --

15 robots are used to carry a torch to "bake out" large deep molds and 15 to spray them.

As you have just witnessed, most robots today are used in metalworking. For the most part these robots are "blind" and are adequate in highly organized factory environments where parts positions are changed relatively infrequently or where extensive software packages have already essentially recorded positions of all parts. However, the majority of American industry is just not so orderly.

To open the market to rapid growth throughout the widest ranges of industries -- robots must be automatically adaptive -- capable of recognizing re-orienting and have adaptive abilities. To accomplish this a need for technology breakthroughs exist -- particularly in sensory-feed-back capabilities which is moving very slowly -- certainly not as fast as developments in robotic control technology. Today, the new sophisticated controls that have been developed are moving robot technology into new markets with applications in pharmaceuticals, electronics, textiles, food processing and some day into service industries such as hospitals, gasoline stations, fast food restuarants, refuse pickup and the household.

The new generation of robotic controls use the latest microprocessor technology and have a number of attributes:

-- High speed computing power

-- A wide range of control functions

-- Enhanced editing for faster program changes and updates

-- More amounts of available memory

-- Real time communications with computers and other production equipment

-- External communication ports that provide high speed sensory interfaces direct to the controller

-- And the ability to run multiple programs concurrently.

This increased control power is expanding the work capabilities of our robot arms and is permitting them to be successfully applied to tasks today that could not be done a year ago.

A considerable amount of research is currently being devoted to the areas of visual, lazer and tactile sensing -- processing speed and accuracy must be increased and costs decreased -- mobility for the robot is also still in the laboratory and is needed to apply robotics to the non-industrial world. Voice recognition would make it easier to use the robotic and voice synthesis has its virtues in application opportunities. It is our opinion that these robot peripheral areas are now the most challenging needs and opportunities for our technologists to conquer.

According to Peter Drucker, innovation is flourishing in the United States. We are supposedly way ahead of anyone who may be second -- and that's the good news. The bad news is that it's really our last bastion. We'd better stay ahead in innovation or we will lose the entire game. The United States is leading the rest of the world into a new era. This new era, like the industrial one which preceded, is driven by a combination of historical social and technological forces. However, of these, the impact of the technological change is the most important. Those who understand and adapt to this change will survive and prosper -- those who will not will fail.

Technological change can, to a great extent, be controlled and manipulated. The survivors in the new era will be those who are best able to do so. The development of new technologies, the vigor of a healthy economy, the drive that produces growth -- these come not from bureaucratic corporations or the government, but from committed, passionate and sometimes slightly crazy individuals who will overcome absurd obstacles in order to bring their dreams to fruition. If you are one of these -- stand up and be counted because American industry needs you. It is up to those of us in technology -- you and me -- to make technology development and its application successful in our country -- in our industry -- and in our business.

If I am successful today -- I will have persuaded you to accept this as your personal challenge.

Magnetoresistive Skin for Robots

John M. Vranish

JOHN M. VRANISH, NATIONAL BUREAU OF STANDARDS, U.S.A.

A tactile imaging skin for robot grippers based on magnetoresistive technology is proposed. In the design considered here, the skin would consist of a thin film magnetoresistive array with sensor elements 2.5 mm apart (density can be increased an order of magnitude) covered by a sheet of rubber and a row of flat wires etched on a mylar film. Linear pressure and compression relationships are expected over a 20 dB range. By varying rubber stiffness, the pressure range could be set anywhere between 30 N/m^2 to 3000 N/m^2 for applications requiring sensitivity and 2.0 X 10^4 N/m^2 to 2.0 X 10^6 N/m^2 for more rugged industrial uses. By varying rubber thickness the skin could be constructed to detect different compression ranges. For example, a thin skin (2.5 mm) could sense compression from .0025 mm to .25 mm whereas a thick skin (7.5 mm), which is more compliant and conformal, could sense compression from .025 mm to 2.5 mm. This paper describes design, operation and expected performance of the skin.

I. INTRODUCTION:

The design of a magnetoresistive skin for robot grippers which yields high performance (great sensitivity, wide dynamic range, and linear response) from a simple, high density sensor array appears feasible. A linear response is expected over a 20 dB range. A design with sensor elements 2.5 mm apart and 2.5 mm thick is considered herein. The principles of operation of such a skin and its theoretically expected performance are described in this paper.

II. DESIGN GOALS

The overall design goal is to explore the practical performance limits of magnetoresistive skin for robots. Figure 1 shows the proposed construction of the skin. As an interim measure it is useful to create a functional specification for the first prototype, as follows:

Dynamic Range

A dynamic range of 20 dB i.e., maximum allowable pressure divided by minimum detectable pressure = 100.

Sensitivity

a. Force sensitivity

A threshold pressure of 40 N/m^2 is detectable using .2 mm thick open celled sponge rubber between the magnetoresistive elements and the flat wires generating the field (dynamic range of 40 N/m^2 to 4000 N/m^2). By making rubber of stiffer material the threshold pressure can be increased to 2.0×10^4 N/m^2 with a corresponding dynamic range of 2.0×10^4 N/m^2 to 2.0×10^6 N/m^2.

b. Spacial sensitivity

A threshold detection of .025 mm is desired. The skin should be able to detect an encounter with an object when it has been compressed .025 mm. This assumes a skin thickness of .45 mm from the surface of the magnetoresistive elements to the external surface of the skin. Making the skin thicker will cause a proportionate decrease in spacial sensitivity.

Element Spacing And Skin Thickness

Magnetoresistive sensing element are 2.5 mm apart with 64 elements on a 25 mm x 25 mm area and the skin can be made as thin as 2.5 mm including the rubber, sensing elements and alumina (Al_2O_3) substrate. As explained in the section on spacial sensitivity above, the rubber layers can be made as thick as possible but with the tradeoff of decreased spacial sensitivity.

Power Dissipation

One hundred (100) mA and 5 watts are the maximum allowed current and power in the 25 mm x 25 mm chip.

Cross Talk

Less than 5% magnetic field intensity cross talk between adjacent elements is desired.

Durability

The system must be able to withstand 10,000 cycles of 33% compression of the rubber with 10% or less plastic deformation.

III. CONSTRUCTION

The proposed construction of "Magnetoresistive Robot Skin" is shown in Fig. 1. 50 ohm active elements of the magnetoresistive material "permalloy" 81-19 Ni-Fe are fabricated by etching on a substrate of Al_2O_3 along with gold shorts and thick film edge conductors to form an active element network array. Above this array is a thin film of rubber typically on the order of .22 mm thick. Over the top of this rubber sheet is thin film mylar which in turn has a pattern of flat wire copper conducting wires etched on it. These wires are typically 6 μm wide and provide the magnetic fields which will selectively affect the permalloy elements on the substrate located underneath. The covering sheet of rubber which includes the tread pattern (and if necessary magnetic shielding material) can be thicker, perhaps on the order of 2.5 to 5.0 mm thick. Fig. 1b shows a top view of the raster scan geometry. Fig. 2 shows a photograph of an array fabricated with the elements spaced 2.5 mm from each other for an array of 64 elements on a 25 mm X 25 mm square. Fig. 3 shows a photographic enlargement of this array. The small chevrons are the magnetoresistive sensor elements and the larger rectangles are shorts connecting the elements.

IV. DATA ACQUISITION

Fig. 4a shows a possible strategy for achieving a raster scan and extracting maximum sensitivity from the permalloy elements. In this electronic scanning scheme, the permalloy rows are all excited continuously by a square wave signal. Since the permalloy elements require a warm up to prevent noise spikes, a continuous wave signal is required. The 6 μm wide flat conducting wires can be pulsed, however, and these are on a row by row basis. That is, the first wire may be pulsed 8 times, then the second and on through all 8 of the wires (or columns). Each time a flat wire is pulsed, the first element of each permalloy row is affected. Thus, if the permalloy network is sampled row by row during the time a flat wire is being pulsed, the array can be read on an element by element basis. Each individual permalloy row is amplified before being multiplexed out thus making the system as sensitive and noise free as possible. Fig. 4b shows the electro mechanical operation of the system. Fig. 5 shows details of the electronic circuit.

Each permalloy element encounters an $\vec{H}$ magnetization field from the flat wire above it equal to:

$$H = \frac{I}{2\pi S} \left\{ \tan^{-1} \frac{[b+S/2]}{R} - \tan^{-1} \frac{[b-S/2]}{R} \right\} \qquad (1)$$

where: H is in amps/meter
S = width of conductive strip in meters.
R = distance of the conductive strip above the permalloy in inches.

b = distance of center of the conductive strip from the center of the permalloy element at the permalloy. surface in meters.
I = current in amperes for the selected skin geometry.

We also know that the permalloy element shows a change in resistivity that is linear with respect to the H field impinging on it up to $\pm \frac{1.5 \times 10^3}{4\pi}$ A/m (R/R_o = 1%).[2] A 10% variance from linearity is expected at fields up to $\pm \frac{1.5 \times 10^3}{4\pi}$ A.m. Since $\frac{.4 \times 10^3}{4\pi}$ A/m is the maximum magnetic field intensity expected at the magnetoresistive elements, linearity variance should be in the 2% range.

Experience with the magnetoresistive material to be used shows that $\pm \frac{1}{4\pi}$ A/m should be easily detectable.

For the skin to yield a linear response with respect to pressure, it is essential that the magnetic field at each permalloy element be inversely proportional to its distance from the flat wire above it or $\frac{H(R)}{H(R_o)} = \frac{R_o}{R}$ (2)

R_o = distance of the conductive strip above the permalloy element in mm with no pressure applied.

H_o = magnetic field at the permalloy element with no pressure applied.

The skin is constructed with each permalloy element having a flat wire directly above it thus b=o and equation (1) reduces to

$$\frac{H(R)}{H(R_o)} = \frac{\tan^{-1}\left(\frac{S}{2R}\right)}{\tan^{-1}\left(\frac{S}{2R_o}\right)} \qquad (3)$$

---

[1]The concept and initial studies of a crosstie random access memory (Cram) by L.J. Schwee, R.E. Hunter, K.A. Restorff and M.T. Shepard J. Appl Phys 53(3) P. 2T762 March 1982

[2]Private communication from Len J. Schwee.
To determine the limits of permalloy linearity set $H = H_k \sin \phi$ where ϕ is the angle between the easy axis and the direction of magnetization. Easy axis is the direction M (magnetization will take if there is no field applied).
H_k is the anisotropic field of the film. $R = R_o + \frac{\Delta R}{2} \cos 2\theta$; θ = angle between current and magnetization. Set $\theta = 45°$ to easy axis. Solve for R and if $\theta = 45 - \phi$. We can now solve the 2 equations simultaneously. Thus $R \approx R_o + R\, H/H_k$ for $H < .25\, H_k$ [since $H = H_k\phi$ for small angle approximation.]

In the design proposed here

S is typically on the order of 6μm

R_o is approximately .225 mm

R can vary from .150 mm to .225 mm (33% allowable compression of the rubber).

Thus the small angle approximation can be used and

$$\frac{H(R)}{H(R_o)} \approx \frac{\frac{S}{2R}}{S/2R_o} = \frac{R_o}{R} \quad (4)$$

The maximum error introduced by using this small angle approximation for our design is that of a 1.2° angle or 1.5/100 of 1%.

V. RUBBER FORCE ANALYSIS

The rubber structures (Fig. 6) play a key role in linking the force encountered by the robot gripper and resistivity effects in the permalloy elements. Rubber acts essentially like an incompressible fluid in that its volume must be conserved. However, unlike an incompressible fluid which passes pressure to the walls of its container, rubber experiences internal shear stresses. For the model shown in Fig. 6b, the rubber is free to change its shape (no confining side walls) so that all force applied to the top of the rubber block goes into increasing the internal stresses in the rubber. Accordingly, the following equation may be applied.

$$\frac{G}{\lambda^2 - \frac{1}{\lambda}} = K_{oust} \quad (5)$$

G = shear stress (psi)
λ = h/ho where
h = rubber thickness under compression.
h_o = rubber thickness before compression.

In the tactile sensor array the rubber can be compressed as much as 1/3 without degredation. The discussion that follows examines the relationship between the deformation of the rubber and the force per unit area causing the deformation. A linear relationship is desired.

Since the rubber of the skin has no confining side walls, G (shear stress) is directly proportional to downward tactile force per unit area.

G = KF/A; where F/A = Force per unit area.

Thus K_{oust} (λ^2 - 1/λ) = KF/A;

And it is clear that the force and hence the pressure F/A F is a function of λ .

The linearity of this relationship will now be explored.

Expanding in a Taylor's series about λ_o we have,

$$\frac{F(\lambda)}{A} = \frac{Koust}{K}\left(\lambda_o^2 - \frac{1}{\lambda_o}\right) + \frac{Koust}{K}\frac{\partial\left(\lambda^2 - \frac{1}{\lambda}\right)}{\partial\lambda}\Bigg|_{\lambda_o = \lambda}(\lambda - \lambda_o) + \frac{Koust}{K} R_n \quad (6)$$

Where R_n = remainder term.

But $$\frac{F(\lambda)}{A} = \frac{Koust}{K}\left(\lambda^2 - \frac{1}{\lambda}\right) \quad (7)$$

So $$\left(\lambda^2 - \frac{1}{\lambda}\right) = \left(\lambda^2 - \frac{1}{\lambda}\right) + \frac{\partial\left(\lambda^2 - \frac{1}{\lambda}\right)}{\partial\lambda}\Bigg|_{\lambda = \lambda_o}(\lambda - \lambda_o) + R_n \quad (8)$$

and the first two terms represent the region of linearity and R_n is the deviation from linear.

We know $\lambda_o = \frac{h}{h_o} = 1$,

and $$\lambda = \frac{h}{h_o} = 1 - \delta \text{ where } \delta = \frac{\Delta h}{h_o}. \quad (9)$$

Substituting the conditions of equation 9 into equation 8 and solving for Rn we get $1 + \delta + \delta^2 - \frac{1}{1-\delta} = Rn$ (10)

Recalling δ max = 1/3, Rn max or % deviation from linear = 5.6%

Since the sensor elements measure λ, the signal processing can assume a linear relationship between $\frac{F(\lambda)}{A}$ and λ and be at most 5.6% in error through 1/3 deformation of the rubber.

VI. PHOTO LITHOGRAPHY PROCESS

Some of the essential details of the photo lithography process (Fig. 7) can now be discussed. The objective of the photolithography process in this instance is to provide smooth, quiet, low cost junctions between the permalloy elements, the gold shorts and the electronics. This is accomplished by first putting thick film pads on the edges of the Al_2O_3 substrate (to carry the signals on and off the permalloy/gold columns). Ultimately the electronics shown in Fig. 5 will be in the reverse side of the Al_2O_3 substrate). Following this, the permalloy elements will be deposited on the substrate, etched to the proper shape and magnetically oriented with a hard and easy axis. The final step will be to vacuum deposit and etch gold film shorts which form low noise junctions with both the permalloy elements and the thick film edge pads. This permits consistent low noise current flow.

[3]Louis J. Zapas NBS, private communication

VII. EXPECTED PERFORMANCE

In this section the theoretical capabilities of the skin are compared with the design goals to give expected performance. As shown in Figs. 1, 2 and 3, the permalloy magnetoresistive sensor elements are 2.5 mm apart and an array of 64 elements is on a 2.5 cm x 2.5 cm Al_2O_3 board. Experience with the 81-19 Ni-Fe "permalloy" has shown that magnetic field changes of $\pm \frac{1}{4\pi}$ A/m are detectable. If 30 mA is pulsed through a flat wire .22 mm above a permalloy sensor element, that element will receive a magnetic field of 20.885 A/m. If the wire is moved 33% closer through tactile compression of the skin, the element will now receive a field of 31.324 A/m for a maximum $\Delta\vec{H}$ of 10.439 A/m. Thus the skin has a theoretical dynamic range of 10 log $10.439/1/4\pi$ = 21.2dB. Tests have indicated that open celled sponge rubber can be ground to a thickness of .22 mm and can be repeatedly compressed 1/3 of its unstressed thickness without physical degredation. This 1/3 compression is a linear force-compression relationship to within a 10% variance. The same tests on the sponge rubber indicate that in its linear region it compresses .25 mm for each pressure increment of 1370 N/m^2. Thus the sensor element can detect pressures ranging from 29.3 N/m^2 to 3843.6 N/m^2 with pressure sensitivity of 29.3 N/m^2.

As was mentioned in Section III above, a stiffer rubber can be used. If rubber of 65 durometer hardness is used and grooves cut in it to allow it to compress without sidewall constraint (Fig. 6), the operating range of the skin is approximately 2.0×10^4 to 200×10^6 N/m^2 with 2.0×10^4 N/m^2 being the threshold value. These values are calculated assuming .22 mm rubber thickness and the grooves as shown in Fig. 6 are cut the length of the 2.5 cm strip. These grooves leave blocks of rubber 2.5 cm x 2 mm x .22 mm.

$$\alpha = \alpha_{55} \frac{E_{55}}{E} \frac{h\beta}{\sqrt{A}})^{2/3} \text{ is an} \qquad (10)$$

Equation relating rubber compression to stress and the geometry of the rubber assuming no sidewall constraints.[5]

α = percentage of deflection of rubber used in skin

α_{55} = percentage of deflection of 1 inch cube of 55 durometer rubber used as a standard.

E_{55} = compression modulus of elasticity in psi of 1 inch cube of 55 durometer rubber used as a standard.

E = compression modulus of elasticity in psi of rubber used in skin.

---

[5]Pages 223,4 design of machine elements, MF Spotts Fifth Edition, 1978. Prentice-Hall, Inc.

h = skin thickness (inches).

β = ratio of length to width of rubber blocks used in skin.

A = area of rubber block normal to force (inches).

For the skin, α_2= 33.3% maximum, h = .22 mm (.009 inch), β = 1/.08 and A = .5 cm^2 (.08 in^2).

$$\alpha_{33.3} = \alpha_{55} \frac{(105)}{145} \frac{[(.009)1/108]^{2/3}}{\sqrt{.08}}$$

α_{55} = 55.8% corresponding to a load of 2.0 X 10^6 N/m^2 (300 psi) (shown in a graph in Spotts page 224). Thus the maximum load the skin can oppose and still be in the linear region (33.3% compression) is 2.0 X 10^4 N/m^2 (300 psi). With its 20 dB dynamic range, the minimum load will be 2.0 X 10^4 N/m^2 (3 psi).

Spacial sensitivity is a function of the minimum magnetic field change that the mangetoresistive elements can detect. As such it is independent of rubber stiffness; but dependent on rubber thickness, the location of the flexible wires with respect to the magnetoresistive elements and the amount of current flowing through the flexible wires. The minimum detectable change in magnetic field is $\pm \frac{1}{4\pi}$ A/m. The magnetic field at a sensor element when the skin is uncompressed H = 262.45/4 A/m, and thus the field threshold of spacial sensitivity is H = $\frac{263.45}{4\pi}$ A/m

Using the equation (1) and recalling b = o for a sensor element directly below flexible wire we have

$$H = \frac{I}{\pi S} \tan^{-1}\left(\frac{S}{2R}\right) \qquad (11)$$

I = 30 mA
S = 6 μm
H = $\frac{263.45}{4\pi}$ A/m

R calculates to .224250 mm. Thus the maximum spacial sensitivity possible is .00075 mm assuming that the entire skin rubber is only .22500 mm thick. As rubber tread is added on top of the mylar the rubber compresses as shown in Fig. 6b. Using the example of Fig. 6d.

$$\frac{d_2}{d_1} = \frac{h_2}{h_1}; \qquad \frac{h_2}{h_1} = \frac{22425}{.2250}$$

And assuming we wish to begin detecting at a spacial deformation of .025 mm we have: $h_1 - h_2 = .025$ mm. h_1 calculates to be 7.5 mm. Thus the skin can be approximately 7.5 mm thick, with the flat wires embedded in the rubber .225 mm from the sensor elements, and still detect a deflection as small as .025 mm at its surface. It can compress as much as 7.5 mm (33.3%) and give readings in which tactile force is linearly related to skin compression to a 10% or less variance. Using the tread pattern shown in Fig. 6a, pressure at the skin is transmitted directly down toward the sensor element below and does not dilute in a conical manner. This serves to preserve skin spacial sensitivity despite increasing rubber thickness.

VIII. CROSS TALK

Magnetic field intensity cross talk between adjacent sensor elements is calculated below. The worst case cross talk occurs when the flat wire generating the magnetic field is .225 mm above the sensor element directly below it. The nearest sensor element in a neighboring column is 2.5 mm to one side and so the relationship between the field it encounters and that which the sensor directly below the flat wire encounters is:

$$H_1 = \frac{I}{2\pi S}\left\{\tan^{-1}\frac{(b_1+S/2)}{R_1} - \tan^{-1}\frac{(b_1-S/2)}{R_2}\right\}$$

$$H_2 \quad \frac{I}{2\pi S}\left\{\tan^{-1}\frac{(b_2+S/2)}{R_2} - \tan^{-1}\frac{(b_2-S/2)}{R_2}\right\}$$

where H_1 = Magnetic field at sensor element directly below wire

H_2 = Magnetic field at nearest sensor element in adjacent column

$b_1 = 0$

$b_2 = 2.5$ mm

$S = 6\mu m$

$R_1 = R_2 = .225$ mm

$$\frac{H_1}{H_2} = 124.4$$

Thus the nearest element in an adjacent column encounters cross talk of .8%. Cross talk in the two elements adjacent to the element being measured is even less. This is because the magnetic field encounters these adjacent elements at an angle not parallel to their surface specifically $\theta = \tan^{-1}\frac{(.009)}{.1} = .09$ radians. Thus the cross talk is reduced to $2(1.6\%) \sin\theta = .14\%$.

IX. CURRENT AND POWER

In this section the maximum current and power in the skin are calculated. In a 2.5 cm x 2.5 cm section, there are eight (8) columns of eight (8) permalloy elements, each of which is a 50 ohm resistor. Each column of 50 ohm resistors is balanced by a 400 ohm resistor to complete the bridge circuit into a differential operational amplifier (Fig. 5). Since 5 volts are used to drive the electronics, there are $5V/800\Omega = 6.25$ mA in each column and 50 mA in the total 8 columns. Since the 400 ohm balancing resistors are located off the skin, the total power dissipation in the permalloy is 1.25 watts, with 2mW dissipated in each sensor element. The flat wires on the mylar film are 6 μm wide, 2.5 cm long and 2 μm thick. Made of copper they are 48 ohms each and consume $(30 \times 10^{-3})^2$ 48 = 43.2 mW. Each line has a duty cycle of 1/8 so the effective heating is much less than the 43.2 mW might imply. The lines will end in a common resistor which is variable, set at approximately 120 Ω (since 30 mA current and 5V are used). This resistor is off the skin, however, so the total power dissipation in the skin is 1.25 watts in the permalloy and 43.2 mW in the flexible wires. The total current is 50 mA in the permalloy and 30 mA in the flat wires.

X. SHIELDING

Magnetic shielding does not appear to be necessary. It seems simplest to use a signal processing technique in which the resistance of a column is measured just before the flat wire is pulsed, then during the pulse. The difference between the two readings is a measure of the force on the skin above the element regardless of stray magnetic fields. This is provided that the stray filds plus pulsed signal stay within the $\pm \frac{1.5 \times 10^3}{4\pi}$ A/m linear region of the permalloy the anticipated stray fields should not greatly exceed $\frac{10^3}{12\pi}$ A/m, earths field, and the pulsed signal $\frac{10^2}{\pi}$ A/m. Thus the total will be well within the permalloy linear region. The speed of pulsing, 1 kHz is much faster than robot movement so stray fields can be considered constant during a pulse.

XI. SUMMARY

It has been shown how magnetoresistive technology might be used to develop a skin for robots. This skin is theoretically able to perform tactile imaging with a linear dynamic range of 20dB and a threshold of .025 mm compression. The pressure threshold relating to the .025 mm compression can be made as low as 30 N/m^2 with a dynamic range of 30 N/m^2 to 300 N/m^2. Using a stiffer rubber, this threshold and dynamic range could be increased to 2.0×10^4 N/m^2 to $2.0 \times 10^6 N/m^2$. The proposed thin film magnetoresistive array has 2.5 mm spacing between sensor elements; but easily can be made an order of magnitude more dense. Construction appears simple and economic and a skin which could survive repeated use (10,000 cylces to 33% compression) with 10% or less permanent deformation should be achievable. Further development and prototype construction are being undertaken.

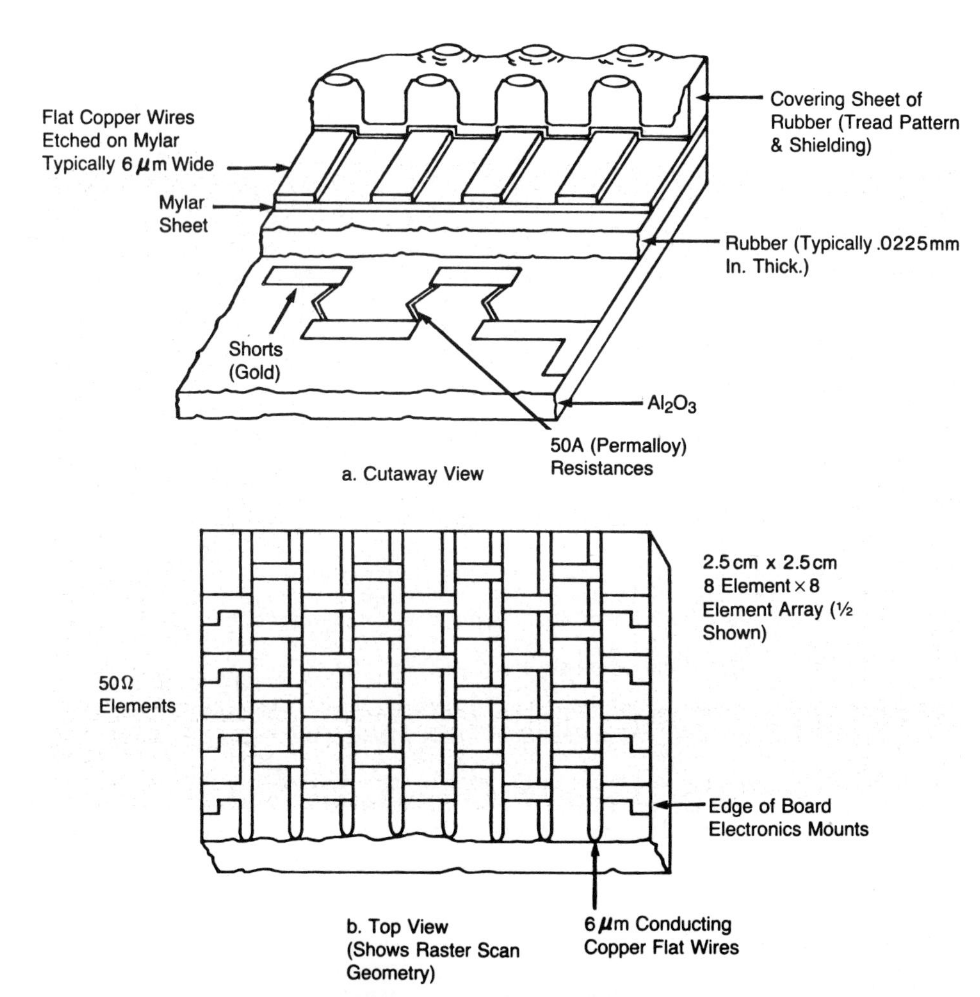

Fig 1 Construction of Magnetoresistive Skin

Fig. 2 Magnetoresistive Array

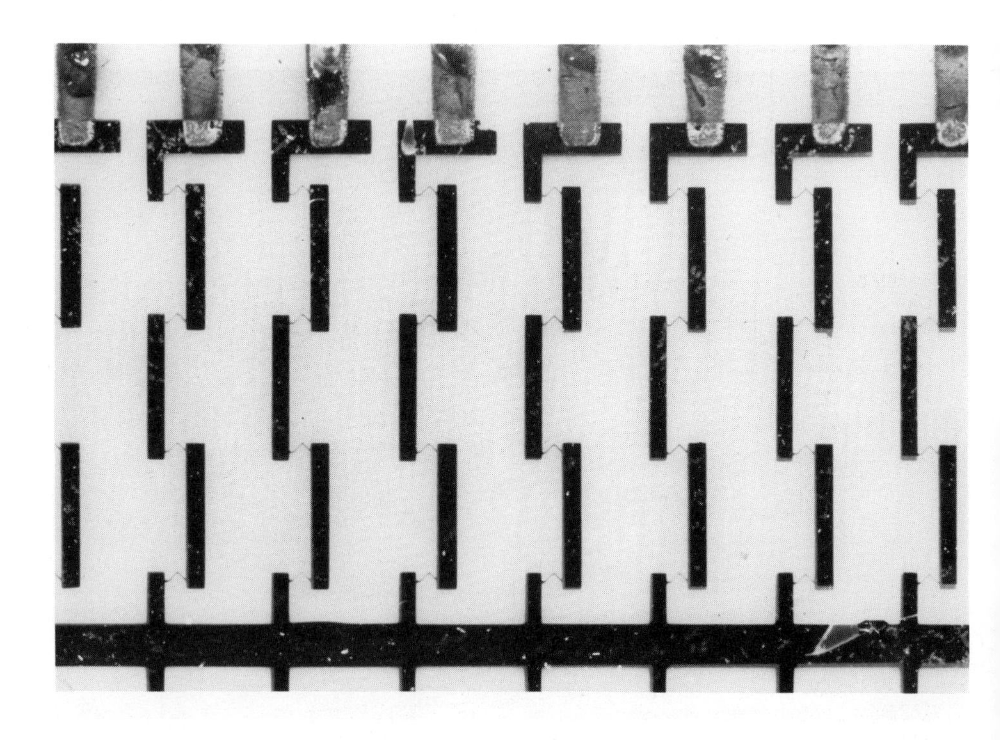

Fig. 3 Magnification of Array

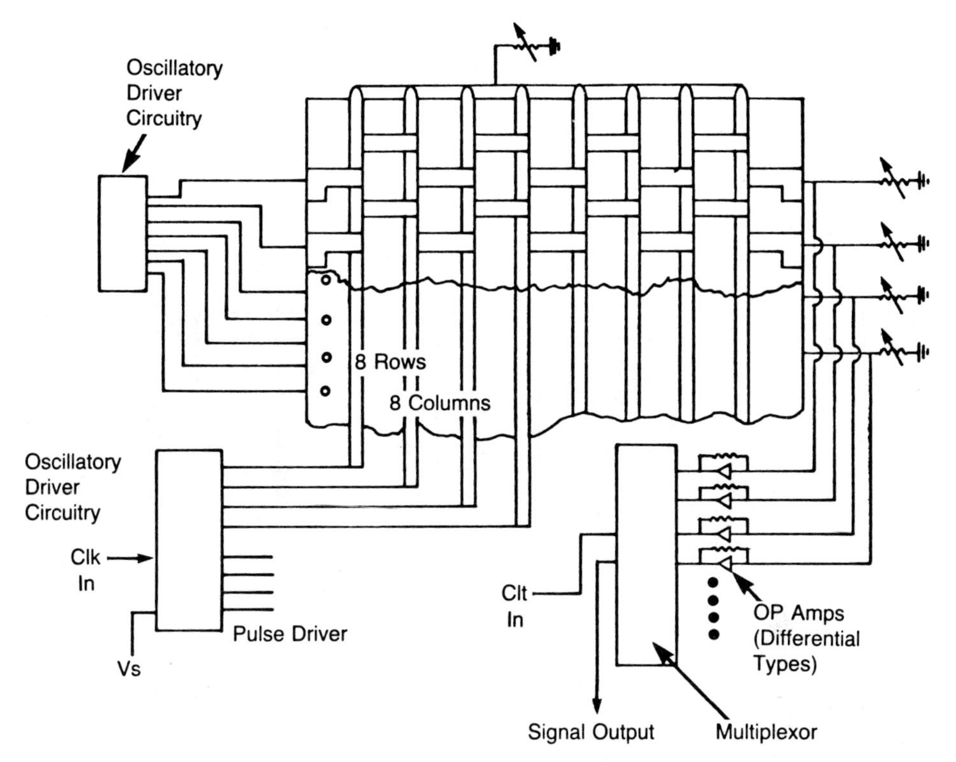

a) Raster Scanning Scheme

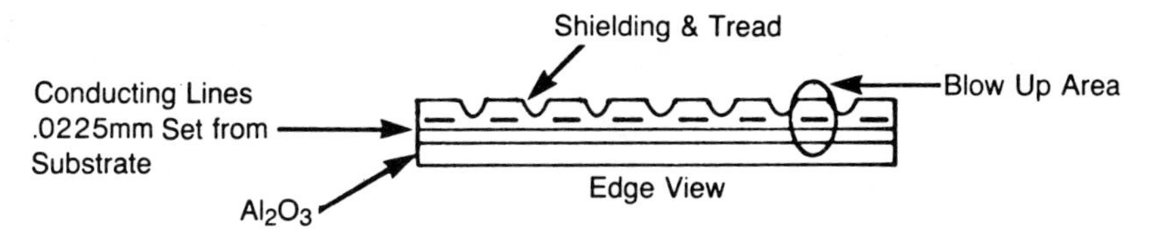

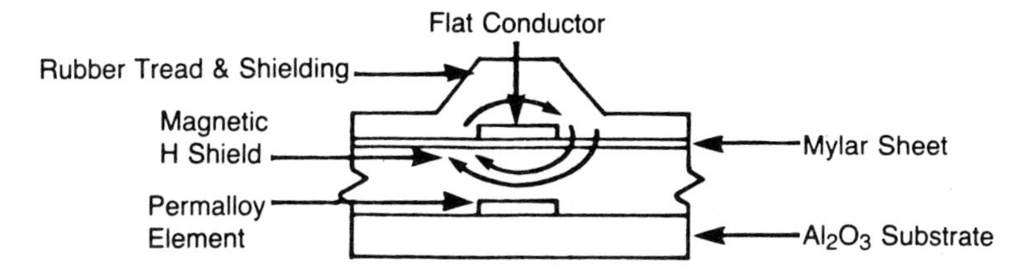

Fig 4 How the Magnetoresistive Skin Works

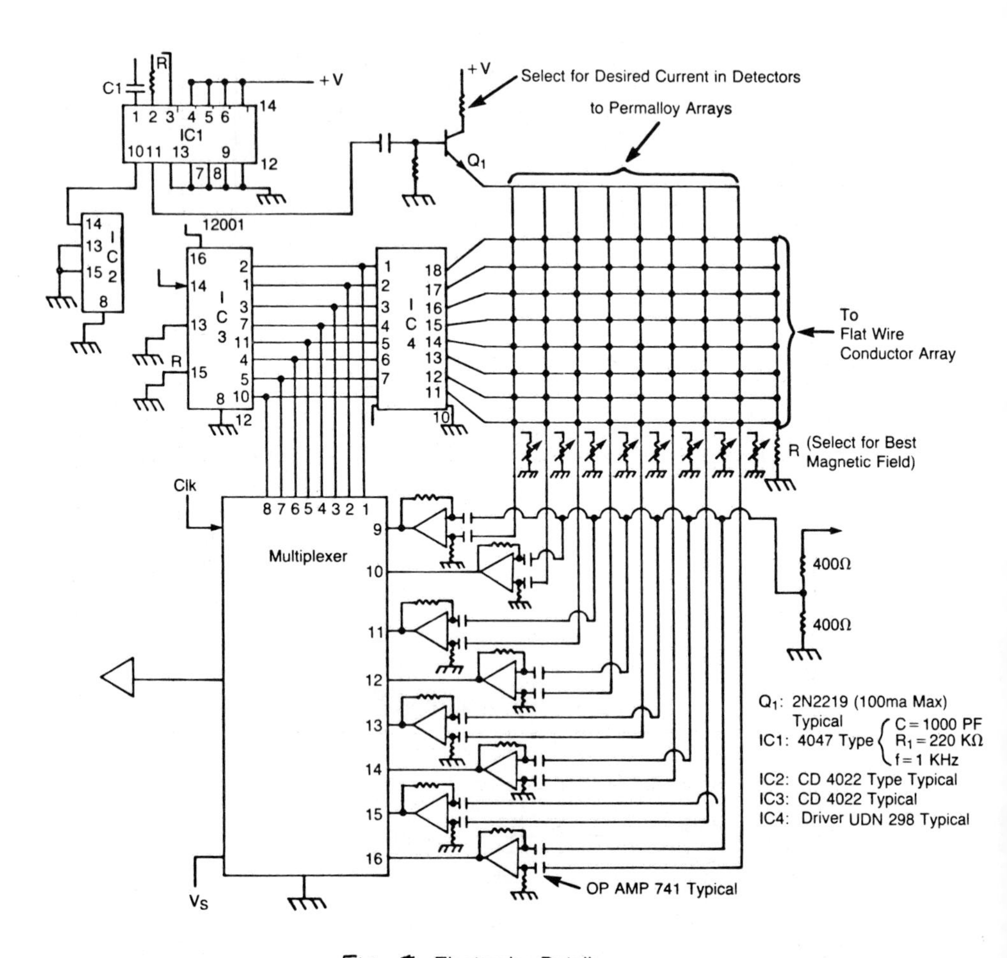

Fig 5 Electronics Details

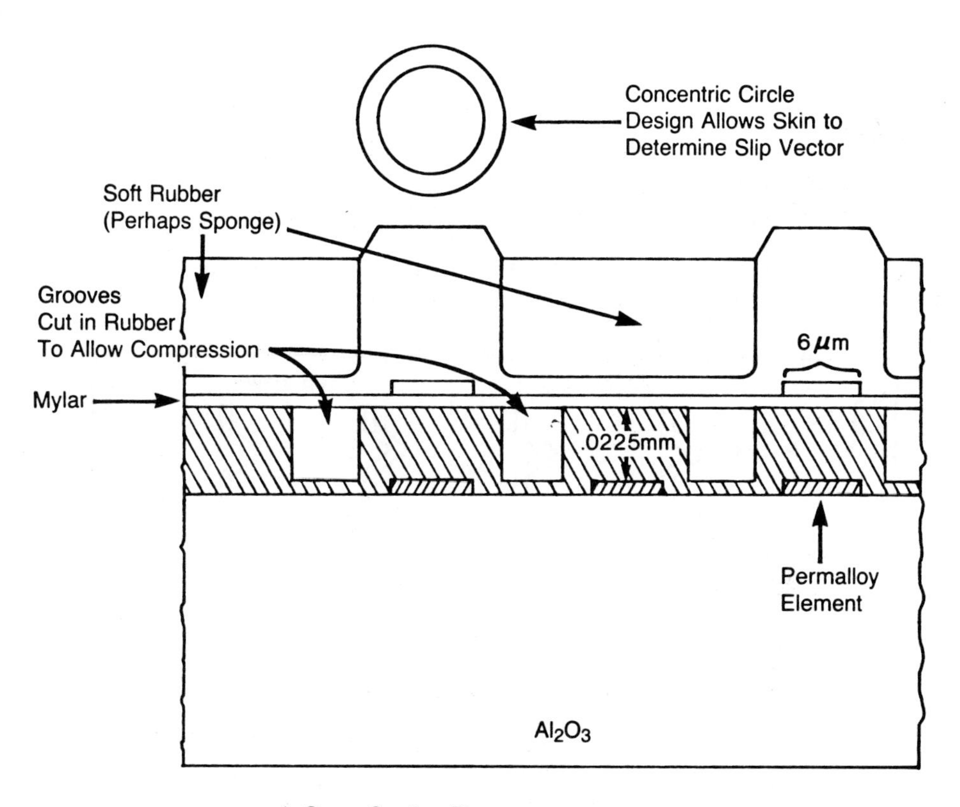

a) Cross Section Blowup

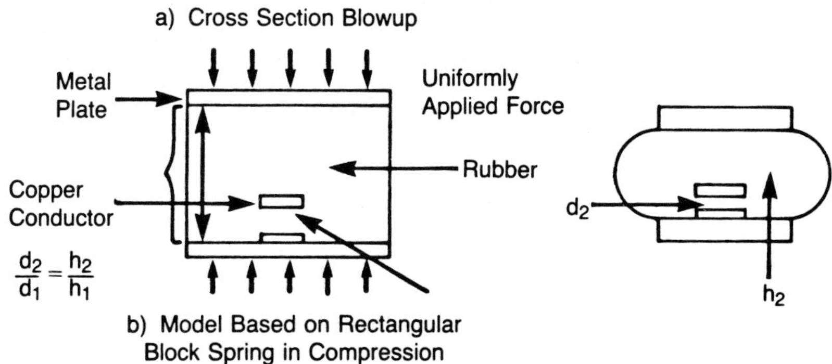

b) Model Based on Rectangular Block Spring in Compression

Fig 6 Tread and Rubber Details

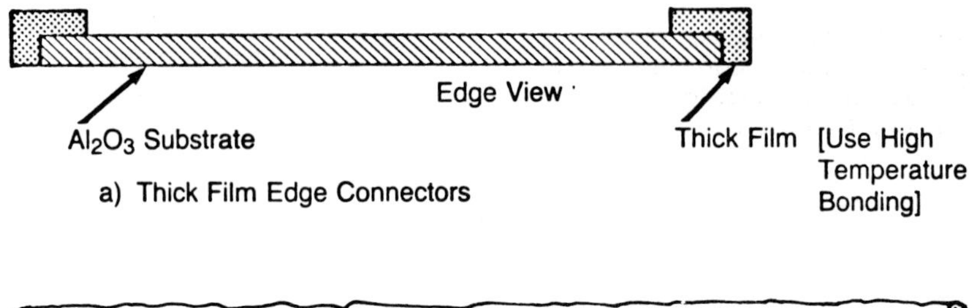

a) Thick Film Edge Connectors

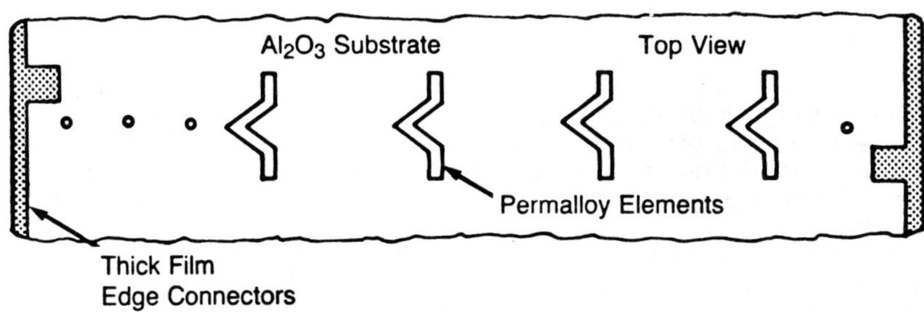

b) Permalloy Elements Added to Substrate

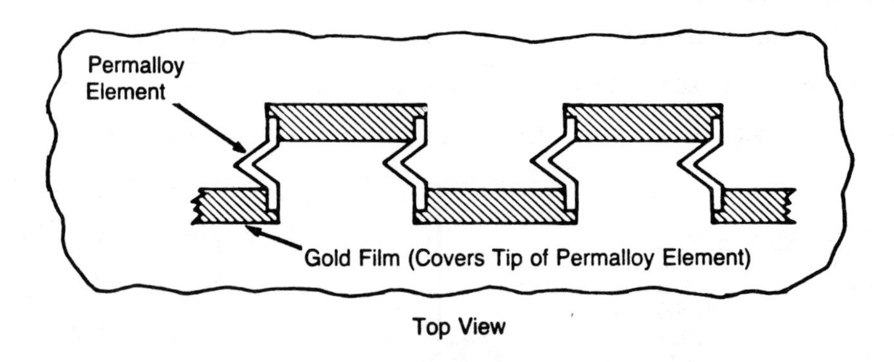

Top View

c) Add Gold Film Connecting Shorts

FiG 7 Photo Lithography Process and Current Flow

The Industrial Robot: 1985–1995

Peter G. Heytler

Industrial Development Division
Institute of Science and Technology
The University of Michigan
Ann Arbor, Michigan

Summary

During 1984, an extensive survey utilizing the Delphi technique was conducted among robot manufacturers and users. Divided into three sections, the survey extensively covered major topics associated with the use of robotics, including technology, social, and marketing issues. The purpose of the study was to determine future trends in robotics through 1995. The Delphi process is an iterative response format; the final iteration was under way at the time of publication. However, important conclusions are contained in the preliminary data. Below is an overview of some of the more important early results.

Marketing Trends

By RIA estimates, U.S. manufacturers purchased a total of some 2800 robots in 1983. This is an increase of 22% from the year before, and over 40% above 1981. Robots are clearly a burgeoning aspect of automation; thus, a logical starting point in our survey was to determine what form this trend will take over the next decade, to 1995. The consensus, shown in Table 1, reflects a widespread opinion held at the time of the first round of the survey (late 1983 and early 1984) that there would be a backslide into recession by mid-year 1984, leading to a reduction in capital investment from 1983 levels. However, this was seen to be a brief downturn, followed by another vigorous recovery. Thus, while the 1984 sales figure of 3175 units represents only a modest 13.4% increase over

Table 1
Robot Unit Sales

Year	Units
1983 (act.)	2,800
1984	3,175
1985	4,000
1986	4,951
1987	6,440
1988	7,250
1989	8,500
1990	10,000
1995	16,000
2000	20,000

1983, the next three years average out to nearly 27% per year. After 1987, a combination of a new recession and the beginnings of market saturation are seen as leading to a permanent reduction in demand growth. This will be the time of shakeout in the robot industry.

The automotive industry in the U.S. has been the predominant purchaser of robots; GM and Chrysler were the first purchasers of commercially-produced units in 1961. As of last year, the "Big 4," plus Nissan USA, accounted for just over half of all robots purchased domestically (Table 2). By 1995 this percentage will be halved; this, despite a continued sharp increase in the actual number of units bought, from 1400 to 3900. A major upsurge in robot usage is foreseen to occur in the electronics industry, from a current market share of about 7% (or roughly 200 units) to 17% (1700 units) by 1990, and to 18% (2900 units) by 1995. Significant growth is also forecast in the consumer nondurables and aerospace industries.

Table 2

Robot Unit Sales, by Industry

	1983	1990	1995
Agriculture	1%	1%	2%
Mining and extractive	1	2	2
Construction	0	1	1
Electricity generation	1	1	1
Consumer non-durables	2	6	6
Non-metal primary commodities	3	3	4
Primary metals	5	3	6
Fabricated commodities	6	6	6
Machinery	11	12	12
Electronics	7	17	18
Automotive	51	35	24
Aerospace	6	7	12
Other transportation equipment	6	6	6
	100%	100%	100%

Throughout the time-frame of this study, material transfer robots are expected to be the largest application sector, at a roughly 21-23% market share (Table 3). As the auto industry's presence in the market dwindles, so too will the percentage of spot welding units. Instead, sophisticated assembly robots will find increasing use across the board in industry, especially as more intricate software and better vision and tactile sensors become available.

Table 3

Robot Unit Sales, by Type

	1983	1990	1995
Material Transfer	23%	23%	21%
Spot Welding	29	17	11
Arc Welding	12	12	11
Spray Painting	12	12	11
Processing (routing, grinding, drilling)	6	6	5
Electronics Assembly	6	12	16
Other Assembly	6	12	16
Inspection	6	6	8
Other	0	0	1
	100%	100%	100%

Trends in Robot Installations

Outside of the auto and appliance manufacturing industries, most investment in robot equipment has been on a rather small scale. Frequently, purchases of only one or two robots at a time have been made, with these machines fairly simple and dedicated to a very limited range of operations. Existing production facilities have been modified to accommodate these "islands of automation," without a unified program of integration into a system.

As robotic technology becomes more widespread and the experience base associated with it grows, more sophisticated and unified approaches will be taken (Table 4). New production facilities will replace obsolete plants, with robots installed as components of a computer-integrated manufacturing operation, often by a single supplier. By 1995, the units purchased in an "island of automation" approach will be in the minority.

As the current round of the survey continues, it is expected that this trend will become more pronounced.

Table 4
Future Facilities Utilizing Robots

New robot purchases:	1983	1990	1995
Incorporated into existing facility	80%	60%	50%
Greenfield facility	20	40	50
Total "turn-key" package	30%	50%	50%
Robot and delivery only	70	50	50
"Island" workstation	89%	59%	47%
Part of CAD, CAM, or CIM	11	41	53

Robot Features

In general, future robots can be expected to have many of the following characteristics:

- Increased strength, as actuators become more compact and powerful.
- Faster and more accurate, primarily due to the increased use of electric units.
- Computer controlled.
- Increased use of sensors with feedback.
- More user-friendly, as CAD/CAM databases become more wide- and sophisticated.
- Modularized, to allow more custom applications.
- Cheaper, due to improved economies of scale and improvements in technology.

Currently, most robots are programmed using some form of a point-to-point teaching method, which requires an operator to lead a robot manually through a series of steps, with some form of controller interpolating the intermediate points. The result is not always the most efficient path. As the emphasis on robots as part of a total computer-operated manufacturing process grows, it will be the computer which tells the robot how to perform its tasks. Point-to-point teaching will be relegated to older, less efficient "island" stations. This trend will also be reflected in the reduced use of programmable controllers (PCs) in favor of small computers (Table 5).

Table 5
Types of Controllers

	1983	1990	1995
Micro-computer	49%	56%	59%
Mini-computer	21	23	23
Mainframe computer	1	6	6
Programmable controller	28	15	12
Other non-computer	1	0	0

The robots themselves will become more sophisticated. As shown in Table 6, relatively simple four- and five-axis models, each of which currently represents about a third of robot sales, will give way to more complicated six-axis units by 1995. One-fifth of these future robots will have two or more arms (Table 7). Speeds are foreseen as improving by approximately 50% over the next decade, while repeatability margins will be lowered to around one-third of current levels. As might be expected, electronics assembly units will lead the way, with speeds of 60 inches per second, and with repeatabilities in the neighborhood of ±.001".

Sensor technology will most likely represent the single most important near-term advancement in robots. A majority of the survey respondents, when asked to list the most active and well-funded areas of robotics-related research and development over the next decade, cited one aspect or another of sensors. While much sensor R&D work is now being performed both commercially and in academia, there seems to be a rather distinct division of labor. Machine vision, perhaps because of its vast potential in areas other than robotics, is receiving much attention from the business sector, while work on tactile and other sensing devices, with their more limited range of applications, is being left more to the universities and research institutions. This arrangement probably will not change dramatically in the near future.

Table 6

Robot Axes of Motion

	1983	1990	1995
Four or fewer	35%	27%	24%
Five	31	30	28
Six	34	41	45
Seven	0	2	3
Eight or more	0	0	0

Table 7

Robots With Multiple Arms

1983	1990	1995
2%	10%	20%

Today, roughly one in ten robots in the U.S. is equipped with some type of sensor (Tables 8-11). Most often, if a sensor is incorporated into the work cell at all, it is usually located on ancillary equipment, such as a feeder. As robot cells become more complex, with the robot performing more delicate tasks, sensory feedback will become a much more important feature to the robot. Thus, the one-in-ten figure will rise to at least eight-in-ten, and possibly higher. Machine vision and contact sensors each will be found in about 30% of all robots, up from a current figure of 3%. Other types of sensors, including temperature detectors, sonar or radio wave devices, etc., will go from 7% to 36% over the same period. As might be expected, inspection and assembly robots will be the primary beneficiaries of machine vision. However, with on-going research into seam tracking, 30% of arc welding robots will "see" by 1995, while another 30% will utilize another tracking method. While many of the vision units will be supplied by the robot manufacturers themselves, half will come from companies specializing in machine vision units.

Table 8

Robots Equipped with Vision

	1983	1990	1995
Machine Tending	1%	10%	10%
Material Transfer	5	14	20
Spot Welding	0	5	10
Arc Welding	5	25	30
Spray Painting	0	10	10
Other Processing	1	10	15
Electronics Assembly	5	30	50
Other Assembly	5	25	40
Inspection	10	40	60

Table 9

Robots Equipped with Touch/Force Sensors

	1983	1990	1995
Machine Tending	2%	15%	20%
Material Transfer	5	20	30
Spot Welding	0	10	15
Arc Welding	2	10	15
Spray Painting	0	0	0
Other Processing	5	20	45
Electronics Assembly	10	50	60
Other Assembly	5	30	40
Inspection	5	15	20

Table 10

Robots Equipped With Other Non-Vision, Non-Contact Sensors

	1983	1990	1995
Machine Tending	10%	30%	40%
Material Transfer	10	30	50
Spot Welding	5	10	20
Arc Welding	5	30	50
Spray Painting	1	7	15
Other Processing	5	15	20
Electronics Assembly	10	30	40
Other Assembly	10	25	30
Inspection	10	25	40

Table 11

Sensor Location*

	1983	1990	1995
Internal Sensors	41%	67%	77%
External Sensors	65	51	41

*Columns total more than 100% due to the use of both locations on some robots.

By 1995, there will be a number of firms specializing in software time-sharing for robot programming (Table 12). One of the major limitations to this concept is the current lack of standardization in robotic software. Even by 1995, our panelists only rate chances for a standard language only at abut 50-50 (Tables 13-14). Obviously, much more work needs to be done in this area.

Table 12

Robot Programming Utilizing Commercial Time-Sharing

1983	1990	1995
1%	10%	15%

Table 13

Probability of Having A Standardized Off-Line Programming Language

1983	1990	1995
0%	20%	50%

Table 14

Off-Line Robot Programming Utilizing CAD for Debugging

1983	1990	1995
3%	30%	50%

Software Engineering Management: An Engineering Manager's Nightmare?

Howard L. Taylor

General Foods Corporation
250 North Street
White Plains, New York 10625

Abstract

Automation technology requires integration of computers with equipment. These computers rely on increasingly sophisticated software. Until recently, engineers have been able to use outside consultants, business programmers, and software packages for the required software. As engineers learn software skills, and understand the limitations of these alternatives, they are beginning to write their own computer software.

The engineering manager is presented with a difficult challenge: how to manage an engineer that is now writing software 15% or more of the average work day? This paper describes the engineering manager's challenge and suggests a method to meet it.

I. Background

Computers are becoming an essential tool for engineers. The processing power and their applicability to engineering activities are increasing while the costs are decreasing. One of the main features of computers is its flexibility; its ability to change its function quickly and inexpensively.

This flexibility is possible because the computers can be programmed, rather than be hard-wired. The general term for programming is software.

Software engineering is a relatively new discipline for engineers. As recent graduate engineers begin to write computer software, their managers are faced with a dilemma: How to support this new tool and also manage it?

Main Idea of Paper

Engineering managers should be trained to manage software development and maintenance.

II. Engineers Should Develop Software

To suggest that engineers should stick to engineering and leave software to programmers avoids current reality. Software is a necessary part of engineering.

A. <u>The Needs for Software Are Great</u>

In industrial automation, there are at least three levels of control software.

- Control of a unit of operation, such as a Filler, Boiler or Oven.
- Control of a line or process consisting of several unit operations. Examples are a Packaging Line, an Assembly Line, or Power Generation.
- Supervisory monitoring and control of a building or a group of lines or a group of processes.

The support of control software also requires that other software be developed.

- Engineering design
- Engineering analysis of collected process data
- Administrative engineering processes, such as estimating, scheduling, purchasing and project management.

B. <u>Engineers Are Best Able To Develop This Software</u>

The single most inhibiting factor to increased application of computers is neither computer costs nor the availability of programmers; rather it is the availability of professionals able to bridge the business need (in this case engineering) with the computer technology. The engineer is best able to provide that bridge.

There are three types of resources to supplement an engineer's software development.

- Using Information Services professionals requires that they understand a reasonable amount of engineering disciplines and that detailed functional specifications be prepared. These activities are both time consuming and expensive.
- Software Packages are useful, but incomplete in meeting needs. They usually require modifications, and also must be maintained.
- Outside consultants are also useful, but they are difficult to manage and their work often results in low quality software.

These resources are useful for supplementing the engineer's software development, but are inadequate if used as substitutes. The engineer should be in control of software in the same manner as all other aspects of engineering design.

III. <u>Current Engineering Software Is Low Quality</u>

Software quality is defined by these characteristics.

- Reliable - It performs its designed functions when needed.
- Testable - Its logic can be easily verified.
- Portable - It can be moved from one computer to another.
- Maintainable - It can be quickly and inexpensively modified by someone other than the original developer.

These are quality characteristics accepted by the software industry. In applying these characteristics to a review of current engineering software, the result is that the software is overwhelmingly low quality.

- Contracts with outside consultants often result in later, overestimated cost software that does not meet any of these four quality characteristics.
- Company developed software usually meets its designed functionality, but is often late, unsecured, undocumented and not maintainable.
- There is a lack of objective testing, back-up, security, change control procedures as well as insufficient maintenance resources.

At the same time there are well developed software industry methods for developing and managing high quality software. These methods are directly applicable to engineering software development.

IV. Engineering Managers Should Be Trained to Manage Software Development

The situation described above will continue until the engineering managers begin to provide good software management. The choice is not whether engineers should develop software; neither is it whether engineering managers should be trained to manage software development. The choices are when and how. Most of the current engineering managers graduated before computers were an integral part of the university curriculum. Today the universities teach how to write software, but are seriously lacking in management techniques. Software management has been primarily learned through experience and on-the-job training for Information Services professionals as they move up the organization. There are many books and publications on the subject, but few focused training courses are available.

The next paragraphs recommend the appropriate skills to be learned and the content of an engineering managers training course.

A. Skills To Be Learned

There are three primary skills that an engineering manager should have.

- Ability to control software development in an engineering environment.
- Ability to organize for software development and maintenance.
- Ability to manage outside consultants in developing software.

B. Training Course Content

A training course focused on these three skills should be based around an industrial automation case study so that behavior changes would be likely to result from the course. The course content should include these subjects.

1. Software Concepts
   . Differences between software and hardware
   . Definition of Software Quality
   . Software cost profile
   . Overwhelming costs of maintenance
   . Types of maintenance
   . Software life cycle
2. Project Life Cycle
   . Business Need
   . Requirements Definition
   . Design
   . Construction
   . Start-up
   . Evolution
3. Key Software Techniques
   . Software Functional Specification
   . Structured Design
   . Testing Levels
   . Testing Methods and Guidelines
   . Coding Guidelines
   . Structured Programming
4. Management Tools
   . Documentation Standards
   . Structured Walk-throughs
   . PERT
   . Phase Check Point Reviews
   . Audits
   . Status Reports
5. Management Responsibilities
   . Training of Software Professionals
   . Job Descriptions
   . Development Systems
   . Planning
   . Estimating
   . Change Control
   . Make vs. Buy
   . Separation of Development from Operations
6. Organization
   . Different Types of Teams
   . Team Members Responsibilities

7. Prototyping
   - User Driven Computing
   - Definitions
   - Techniques
   - Life Cycle for Prototyping
8. Contracting
   - Time and Materials Contracts
   - Fixed Price Contracts
   - Vendor's Project Organization
   - User Organization

V. Training Course Development Steps

There are a few outside resources for providing software management training for engineering managers. The most effective training programs are those that have been integrated into the company's systems and engineering procedures. This paper recommends that a customized training course should be delivered initially to the current engineering managers, and also repeated frequently as new engineering managers are identified. These steps are recommended for developing a training course.

- Establish a team with representation from Engineering, Information Services, and the Training departments.
- Develop a formalized software life cycle that combines systems activities with the engineering construction activities.
- Detail the content outlined above.
- Develop a Case Study using one of your recent or current projects.
- Develop a training course focused on your Engineering Managers.

Computers are now vital and necessary tools for engineering. Their effective use depends on quality software. Engineers can, if properly managed, develop high quality software. This will begin to happen as the engineering managers acquire the software management skills and provide the proper support to the design and project engineers.

The best way to acquire the software management skills is to develop a training program for engineering managers, and deliver the course as needed.

Operations Management Models for Industrial Robots

D. Necsulescu and A. S. Krausz

Department of Mechanical Engineering, University of Ottawa,
Ottawa, Ontario, Canada.

Summary

The management of the operation of a robotic manufacturing system requires important modifications of the currently available techniques for productive systems management. Long term planning and scheduling techniques should be updated for representing the effect of computerization of the production process. The paper presents structural definitions of some robotic manufacturing systems, a model for robotic manufacturing cell reliability applied for the evaluation of the overtime requirements and a model for controlling the smooth transition of the personnel of a company from a traditional to a computerized manufacturing system.

Configurations of Robotic Manufacturing Systems

Present day robots consist of an arm, a wrist and an end effector. The arm configurations depend on the type of joints of the arm component used for facilitating the movements of the end effectors. A rectangular configuration leads to high repeatability and accuracy and is attractive for overhead mounting in workshops. A revolute configuration is widely employed when the robot is replacing human operators. Multi-arm robots are in development but less used now in industrial applications due to control complexity and the resulting slow movements [1].

The common end effector, the gripper, is poorly imitating a human hand; other types (vacuum cups, magnets, hooks) or simple tools can be efficient in applications.

The introduction of robots in manufacturing systems has the purpose to eliminate hazardous working conditions for human operators or to generate higher profits. The perspective of using more robots for increasing productivity already generated labor concern due to the prospect of job displacements. Present day evaluations of the magnitude of the loss of jobs is based on the currently available robots and current skills; robot technology will evoluate but also further training of human operators will enhance human competitivity in the process of job displacement. This situation requires adequate personnel planning models for the transition from the present to the future, computerized, manufacturing systems. The models should reflect characteristics of these new flexible systems. In a robotic manufacturing cell a robot can tend several machine tools. Flexibility in such a cell is enhanced by using machining centers rather than dedicated machine tools.

In a robotic manufacturing cell (RMC) a fixed base robot can serve several machines located on a circle with the robot in the center. One type of robotic manufacturing system contains overlapping machine cells in which a robot tends a particular cell and also transfers workpieces from and to the adjacent cells. Another type of system has a conveyor for transferring workpieces from one cell to another. In a more advanced configuration, a flexible machining system, a random access is possible to any machine of the system under the control of the computer of the system. The characteristics of the modern configuration of the robotic manufacturing system built a high redundancy in the system in the form of a "r out of n system" or of a parallel system from reliability viewpoint.

Managing Computerized Manufacturing Systems

The operation of an unmanned manufacturing system requires complete programming of all movements of workpieces and tools in accordance with the required machining for the production batches. Process planning and scheduling attracted less attention and only some very complex automated systems are currently available [2]. Production planning over a horizon of 6 to 12 months can be in some situations profitable by smoothing production rate subject to seasonal demand variation. In a computerized manufacturing system the production cost is no more linearly dependent on the amount of regular time and overtime of employees; most employees are programming (numerically controlled machine-tools material handling devices, robots, etc.), supervising, maintaining and, after failures, repairing the equipment. Optimal producting plan, in such conditions, is the result of balancing the increasing setup cost of processing a large variety of products, the opportunity cost resulting from machines in idle time and inventory holding cost [3]. A result could be the decision to build up batches over a medium term in the form of a "dispatch queue" in order to reduce mean processing time of jobs, in particular when operating close to the production capacity [4]

Any management technique required for the operation of robotic manufacturing cell should take into account the effects of random factors, as for example machine failure, on the productive activity.

Stochastic Modelling of Robotic Manufacturing Cell Operation

In this paper a queuing model of the operation of a job shop is embedded in a Markovian model of the machines reliability. [5]

The Markovian model of the RMC state probabilities is based on the following assumptions:

- identical machining centers
- exponential probability density function of failure time
- exponential probability density function of repair time

We denote:

$I =$ system state (i.e.RMC state)

$i =$ number of machines in operating state

$S =$ total number of machines installed in RMC

$\ell =$ failure rate of a machine

$m =$ repair rate of a machine

$D =$ transition rate matrix

$P(t) = [P_o(t) \ P_1(t) \ldots P_S(t)]_t$ = system state probability vector

$v = [v_o \ v_1 \ \ldots v_S]_t$ = limiting system state probability vector (t = subscript for matrix transposition)

$c_A^a =$ Combination of A objects taken a at a time

The dynamics of the probabilities of the system states is described by the following differencial equation in matrix form:

$$\frac{d}{dt} P(t) = DP(t) \tag{1}$$

For $S = 3$, the transition rate matrix is

$$D = \begin{bmatrix} -3m & \ell & 0 & 0 \\ 3m & -\ell-2m & 2\ell & 0 \\ 0 & 2m & -2\ell-m & 3\ell \\ 0 & 0 & m & --3\ell \end{bmatrix} \tag{2}$$

For S machines, the limiting state probabilities are:

$$v_i = C_S^{S-i} \ v_S \ (\frac{\ell}{m})^{S-i} \quad \text{for } i = 0,1,2 \ldots . S-i \tag{3}$$

$$v_S = (1-\frac{\ell}{m})^{-S} \tag{4}$$

The operation of the RMC can be described by a single phase multiple channel M/M/i queueing model.

We denote:

$\lambda =$ arrival rate per day per system

μ = processing rate per shift per machine

β = number of shifts per day $(1 < \beta < 3)$

α = $\lambda/(\mu\beta)$

i = number of machines currently in operation

P_n = probability of n workpieces in the system

L_q = number of workpieces in the queue

Overtime, for example $\beta = 2.5$, means two shifts and four hours overtime per day.

A steady state solution can be obtained when

$$\lambda < i\,\beta\,\mu \tag{5}$$

In this model $i\beta$ is assumed constant, i.e. three machines one shift will process as much as one machine three shifts.

This model can be used for the stochastic evaluation of the requirements for overtime in a RMC.

We will analyse a RMC having installed $S = 3$ machines tools. We will consider two situations:

situation A = overtime and/or multishift operation is not permitted

situation B = overtime and/or multishift operation is permitted.

In the situation A, $\beta = 1$ and, in accordance with the condition (5) a steady state operation is possible if:

$$\frac{\lambda}{\mu} < i \tag{6}$$

The probability of successful steady-state operation of such a system are:

$$v_1 + v_2 + v_3 \quad \text{for} \quad \frac{\lambda}{\mu} < 1 \tag{7}$$

$$v_2 + v_3 \quad \text{for} \quad 1 \leq \frac{\lambda}{\mu} < 2 \tag{8}$$

$$v_3 \quad \text{for} \quad 2 \leq \frac{\lambda}{\mu} < 3 \tag{9}$$

where v_1, v_2, v_3 are given by equations (3) - (4).

It is obvious that a system with $S = 3$ that is normally operated with $\lambda < \mu$ will be most of the time idle. If the system is normally operated at $\lambda > 2\,\mu$ and no overtime is permitted $(\beta = 1)$, the system will have steady state operation only when all three machines are in operation, the event having the probability v_3.

For the situation B we will assume that

$$\lambda < 3\,\mu \tag{10}$$

i.e. the daily arrival rate in the system is less than three times the processing rate of each machine in a shift. This system can have a steady-state operation in the event of having currently available only one or two machines, by extending their utilization to more than one shift. This is a reasonalbe assumption for a system that is normally operated only one shift. The condition (5) gives the minimum requirements for production time per 8 hours shift.

$$\beta = \frac{\lambda}{i\mu} \quad \text{for} \quad \lambda < 3\mu \quad \text{and} \quad \beta \leq 3 \tag{11}$$

For the extreme case $\lambda = 3\mu$ the condition (11) gives

$\beta i = 3$

The probability distribution of the overtime requirements $(\beta-1>0)$ is:

$$v_3 \qquad \beta-1=0 \tag{12}$$

$$v_2 \qquad \beta-1= \frac{\lambda}{2\mu} \tag{13}$$

$$v_1 \qquad \beta-1= \frac{\lambda}{\mu} \tag{14}$$

Optimization Model for the Retraining of the Personnel.

During the transition of a traditional manufacturing system to a computerized one we assume the following controllable factors in personnel planning: hiring, promotion, retraining and firing.

For this problem, the requirement of workforce in each trade (dependent upon the rate of automation and demand for the final products of the industry) is assumed given and the rate of employee turnover is considered an uncontrollable random factor.

For the optimization problem we denote:

x_{ijt} = number of workers retrained from trade i to trade j in time 't'

x_{irt} = number of workers who resigned from trade i in time period 't'

x_{ift} = number of workers from trade i fired in time period 't'

x_{hit} = number of workers hired for trade i in time period 't'

For the same activities, in the next time period, in the subscript t + 1 is substituted for 't'

C_{ijt} = the cost of retraining one worker from trade i into trade j in time period 't'

C_{jrt} = the cost of the resignation of a worker from trade j in time period 't'

C_{jft} = the cost of firing a worker from trade j in time period 't'

a_{it} = the number of persons required in trade i at time 't'

We assume the total number of trades in time periods 't' and 't+1' be m and n respectively ($n > m$).
Out of the n trades, we assume that trades k, ℓ are not required in time period 't+1', so that

$$\sum_{i=1}^{m} x_{ikt} \quad \text{and} \quad \sum_{i=1}^{m} x_{i\ell t} \quad \text{do not exist}$$

Similarly, x_{hkt} and $x_{h\ell t}$ do not exist (for these variables, a large penalty is attached in the objective function).
The problem can be formulated in linear programming format (transportation) for two time periods as

$$\text{Min} \sum_{t=1}^{x} \sum_{j=1}^{n} \sum_{i=1}^{m} C_{ijt}\, x_{ijt} + \sum_{j=1}^{m} (C_{jr1}\, n_{jr1} + C_{jf1} n_{jf1} + C_{hj1} x_{hj1})$$

$$+ \sum_{j=1}^{n} (C_{jr2}\, x_{jr2} + C_{jf2} x_{jf2} + C_{hj2} x_{hj2} \tag{15}$$

Subject to:

$$\sum_{j=1}^{n} x_{ij1} + x_{if1} + x_{ir1} - x_{hi1} = a_{i1} \quad \text{for } i=1 \text{ to } n \tag{16}$$

$$\sum_{i=1}^{m} x_{ij1} + x_{hj2} - x_{jr2} - x_{jf2} = a_{j2} \quad \text{for } j = 1 \text{ to } n \tag{17}$$

and

$$x_{ij1} \geq 0 \quad \text{for } i=1 \text{ to } m \text{ and } j=1 \text{ to } n \tag{18}$$
$$j=1 \text{ to } n$$

$$x_{jr1} = z_{j1} \quad \text{for } j= 1 \text{ to } m \tag{19}$$

$$x_{jf1} \geq 0 \text{ for } j=1 \text{ to} \tag{20}$$

$$x_{hj1} \geq 0 \text{ for } j=1 \text{ to } m \tag{21}$$

$$x_{jr2} \geq Z_{j2} \quad \text{for } j=1 \text{ to } n \tag{22}$$

$$n_{jf2} \geq 0 \quad \text{for } j=1 \text{ to } n \tag{23}$$

$$x_{hj2} \geq 0 \quad \text{for } j=1 \text{ to } n \tag{24}$$

where

Z_{j1} and Z_{j2} are random variables

The model can reflect the strategy of retraining rather than firing and hiring for meeting the personnel requirement of a computerized system by setting a very high penalty for firing.

Conclusion

The state of art in robotic manufacturing systems management shows various levels of development. The control of robots and NC machining centers is well developed and readily available from manufacturers. Process planning and scheduling is still in development and, if performed with the traditional non automated techniques, can reduce the production capacity of the system. Long term planning for the computerized manufacturing system cannot be performed with the techniques used for traditional systems and require modifications for taking into account effects of the machine failure in an unmanned job shop. Personnel planning, in an environment where job displacement is a permanent factor, requires a systematic approach for a smooth transition from a traditional to a computerized manufacturing system.

References

1. Necsulescu, D.S.; Krausz, A.S.: Philosophic Background of the Attitude Toward Computerized Manufacturing Systems. 1984 SME World Congress on the Human Aspects of Automation, Montreal, Technical Paper MM 84-627, 1984.

2. Bourne, D.A.; Fox, M.S.: Autonomous Manufacturing: Automating the Job-Shop, Computer, 9 (1984) 76-86.

3. Necsulescu, D.S.; Krausz, A.S.: Updating Production Management Techniques for Industrial Robots, Proc. of the 7th Symposium on EAM, CAE and Robotics, Toronto, 1984 39-44.

4. Buzacott, J.A.; Shanthikumar, Y.G.: Models for Understanding FMS, AIEE Transactions, December (1980) 339-349.

5. Necsulescu, D.S.; Krausz, A.S.; Damlaj, A.: Reliability Modelling of a Robotic Manufacturing Cell, IASTED Journal on Robotics and Automation (submitted for publication).

Exponential Manufacturing Systems: Do Robots Reproduce Themselves?

Ray Asfahl

Industrial Engineering
University of Arkansas
Fayetteville, Arkansas 72701

Abstract

Robots do not reproduce in the sense that living organisms do, but a sort of human-assisted mode of robot reproduction is already happening. As robot machines become more flexible and intelligent, strategies for diverting their capabilities to the reproduction of their own kind become feasible. This paper examines the potential enhancement of long range manufacturing capacity by diverting a portion of robot capacity to reproduction of that capacity. The concept applies not only to robots but to other flexible automatic manufacturing processes as well. One of the principal benefits of exponentiating manufacturing systems is their capability of recovering from damage without irreplaceable loss of capacity. The first arena for exploitation of the concept may be for manufacturing processes in outer space, where life support systems are costly and bulky to lift into orbit. But the worldwide pressure to reduce manufacturing costs here on earth may drive some conventional manufacturing facilities to employ exponentiation as a way to perpetuate and extend plant capacity without adding labor cost.

Perhaps the greatest mystery and miracle of life is the phenomenon of reproduction. In higher forms of life sexual reproduction is an even more special miracle, adding an important dimension to the meaning of life. It is science fiction to imagine robots as ever being capable of sexual reproduction, but startling as it may seem, robots are today engaged in the manufacture of other robots like themselves. The most outstanding example is in a GM-FANUC plant in Japan. Human beings are in complete control, of course, and the process is only partial; that is, the robots perform only a portion of the tasks required to produce a new robot. Also, the diversion of the capability is likewise partial. Some of the robots so produced are used in other unrelated tasks, and only a portion are retained for the manufacture of more robots.

Exponentiation

The strategy of diverting a portion of manufacturing capacity into the reproduction of that capacity is called manufacturing exponentiation. Butow, et al (Ref. 2) credits Von Neumann with the concept that automata mimic biological systems. NASA has investigated the subject thoroughly as a part of summer study at NASA Ames Research Center in Mountain View, California.

Conventional manufacturing is by linear systems which essentially convert raw materials into useful finished products while utilizing energy and generating some by-products, waste, or scrap. The system is linear in that the greater the manufacturing capability that is installed, the greater will be output. Five identical factories will produce five times the output of a single factory, when the system is linear. But suppose a sixth factory is built in which some of the manufacturing capability is used to produce additional manufacturing capacity. This sixth factory is an exponentiating system, and the manufacturing capacity it spawns will eventually surpass the combined output of the other five. The time required will depend upon the

$$100 = .25 \left[\frac{1.75^n - 1}{.75}\right]$$

$$300 = 1.75^n - 1$$

$$301 = 1.75^n$$

$$\ln 301 = \ln 1.75^n$$

$$\ln 301 = n \ln 1.75$$

$$n = \frac{\ln 301}{\ln 1.75} = \underline{10.20 \text{ years}}$$

Conclusions and Applications

The robot case study may seem farfetched, but since robots are today engaged in the manufacture of other robots it seems appropriate to consider the implications of such activity in this International Conference on Robotics and Factories of the Future. Today's exponentiating systems employ very low r and K_R coefficients and the offspring robots are SOLD, not retained in the factory for either productive or reproductive efforts. At least the intent of robot manufacturers has been to sell their products. Reluctance on the part of management of conventional manufacturing plants to purchase the growing supply of industrial robots may lead to further exponentiation at the source.

The existence of exponentiation among robots on a small scale today and the presence of forces which may accelerate the development raises the question of ethics. Manufacturing workers are already sensitive to their potential replacement by robots on a one to one basis, or even a three or four to one basis. But the concept of exponentiation seems a great deal more threatening to the worker. The limits of a conventional decision to install an automatic process are usually known and analyzed prior to the investment. If personnel are to be replaced, the number and identities of the personnel to be replaced are known. If the objective is to increase market share or total sales of product instead of replacing personnel, the numbers and types of additional workers who will not be employed (due to the installation of the automated process) are generally known, provided the automated process is not an exponentiating one. Exponentiating manufacturing systems have the potential of growing to eventually consume the jobs of an undetermined number of human workers engaged in the activity being automated at first on a partial scale.

One of the more exciting characteristics of an exponentiating system is recovery. Damage or injury to a conventional linear system spells a decrease in irreplaceable manufacturing capability, but an exponential system is able to recover to the level of full capacity. The recovery characteristic is analogous to healing in a living organism.

One interesting potential application of exponentiating machines, especially exponentiating robots, is in space stations. Outer space is a very hostile environment for humans, and the utilization of robots in outer space has even more promise than does the use of robots on earth. The difficulty comes in the task of getting the station started and manned with productive robots. Exponentiation is certainly not the whole answer because raw materials in outer space are an obvious problem. But exponentiating systems provide an interesting research model for manufacturing systems both here on earth and in outer space.

References

1. Asfahl, C. Ray. Robots and Manufacturing Automation. New York: John Wiley and Sons, Inc. 1985.

2. Butow, Steve, Stan Kent, Janet Major, and Anthony Matthews, "The Get Away Special," Robotics Age, Vol. 5, No. 5 September-October 1983, p. 40.

Integration of CAD/CAM & Robot Cell

Chairman: William Tanner, Productivity Systems, Inc., Troy, Michigan
Vice Chairman: James A. Rehg, Robotics Center, Greenwood, South Carolina

Computer Aided Plastic Part Design

Russell R. Pfahler

Structural Dynamics Research Corporation
2000 Eastman Drive
Milford, Ohio 45150

ABSTRACT

Design of plastic parts or complete plastic products can include all the elements of today's computer-aided-engineering (CAE). These CAE tools range from the most up-to-date and powerful 3-dimensional geometric modeling program, through finite element stress analysis programs and specialty programs for both mold filling and mold cooling analysis, to color computer graphics for display of results, and finally to state-of-the-art CAD/CAM equipment to produce final part drawings and NC tapes for tool making. Special expertise is required to synthesize all these tools into a new plastic part design; but this expertise is readily available and simple to use for those firms wishing to commit to improved product quality, faster entry into the market, and improved product cost and manufacturability.

THE DESIGN PROCESS

The computer aided design of plastic parts follows a general series of product development steps; beginning with market evaluation and progressing through system definition, concept development, concept evaluation, detailed design, design development and production engineering.

In the market evaluation a potential product is reviewed for similarity to other products, improvements or changes needed to those products, manufacturing or distribution constraints, and other concerns pertinent to the firm's business plan. Market planners review all these, gather additional market information, evaluate it, and search for strength and weaknesses in the overall market.

Once the market planners identify a market position, the system is defined in detail to categorize the new product's physical and functional specifications, production costs, manufacturing, delivery, support and

CAE Development Process for Plastics Part and Mold Design.

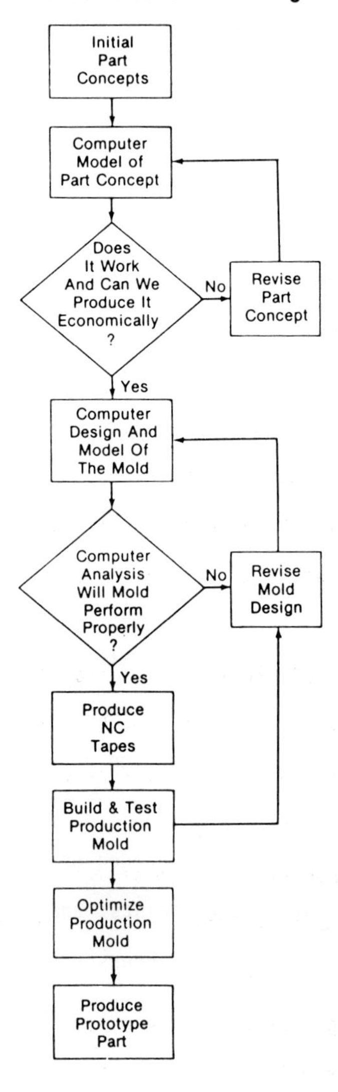

service. These become the basis for the design of the part or product.

Addressing all system specifications, the engineers generate a broad array of alternative concepts, narrowing the possibilities to the most likely three or four concepts. Then they compare each for functional and structural performance. The two or three most viable concepts are usually chosen to undergo extensive evaluation.

Using computer modeling techniques, such as finite element analysis and solid modeling, the engineers analyze the merits of each of the two or three concepts. The system is broken into components which are individually analyzed. Then the engineers select the concept which is most likely to satisfy the program goals and objectives.

Concept sketches are transformed into detailed layouts defining components and subsystems. Using the computer, the engineers develop component models to predict performance. Iterations and changes to the model are much simpler than physically changing a prototype. The components form a system which undergoes close scrutiny for interactions as the process of design development begins.

Using computer model predictions, details are developed to provide data necessary to construct the initial prototype. Thorough testing of the prototype usually confirms the accuracy of computer-predicted system and component performance. The design can be quickly brought into spec when necessary using the understanding gained from the analysis and quantitative results available. The prototype is then "fine-tuned" for manufacturing and production concerns.

In the final step, design engineers take a closing look at prototype details for manufacturability and product functionality as they develop the production process plan.

Lets now look at some of the details comprising the steps involving the engineering staff in the computer aided design of a plastic part. This process usually begins after the completion of the market evaluation and the system definition, although it is most desirable to have the engineers participate in these phases as well.

ENGINEERING STEPS

Computer aided design of plastic parts often begins with the designer's thoughts of the part's geometry and the various functions it is to perform. The use of a tested solid modeling program can speed-up this process many times and eliminates potential problems, such as interferences of separate parts. Many of the CAD/CAM systems do not offer a modeler sufficient to perform all the tasks required. The engineer should be able to create in free form, not simply use an automated "old fashioned" drafting approach many systems offer. He should further be able to utilize any shape or contour, add or subtract openings or appendages as needed, and not be limited to creating a wire frame drawing of his part. In more advanced solid modeling programs, like SDRC GEOMOD, the engineer can perform all these functions and address numerous and varied design concerns, such as ergonomics and visual aesthetics; and utilize major design visualization enhancements like transparency, multiple light sources, and three dimensional rotational views.

Once the part is taken to the concept evaluation stage, additional geometric modeling is performed to begin to create a working computer model. The process begins with the creation of 3-D geometry of an initial plastic part design in the computer using solid modeling module. Styling, interference analysis and tolerance build-up are reviewed in the assembled 3-D model. This initial model, which may also be kinematically exercised and evaluated, establishes a data base for all future analyses.

FINITE ELEMENT ANALYSIS

Once the part geometry is defined, the data can be transferred to the finite element software module. This module provides the basic information for adding load cases and viewing results of stress analyses. After a short iterative process, the plastics specialists arrive at a part design which meets the required performance and manufacturing characteristics. The team also analyzes the economic feasibility of manufacturing the part.

Even after the analysis portion of the project is completed the computer-aided-engineering (CAE) system of plastic part design provides valuable services to the engineer. The finite element modeling system should be able to effectively present results in the most understandable graphics format. Here the use of computer generated color contour plots is accepted at the best format for display. These plots can show high stress areas or areas of unreasonable part deformation. Combining this with capabilities from the solid modeling program, cutting planes through the structure at any angle can be generated to show internal or hidden stresses and deflection. If there are multiple parts, a mechanism program can be activated to show the interaction of the parts. The parts can even be animated so the complete motion process can be viewed to pinpoint any potential interferences or undesirable movements in the process.

TESTING

Where similar parts/components are available the computer again becomes a valuable tool in the part development. Using modal analysis software, the part can be excited in a manner similar to operating conditions and the vibratory deformed shape and resultant stresses can be calculated and displayed. Here, as in finite element analysis, the value of the results display cannot be over-emphasized. Instead of reviewing stacks of print-out the animated display quickly shows the engineer where his design is functioning properly and where improvements are needed.

SYSTEM SIMULATION

Perhaps the greatest contribution of the computer in plastic product design is in its ability to synthesize the finite element analysis results and testing results into a realistic structural model of the product. In this process the results of both steps are combined into a single model, which exhibits color contours and animation to close the gap between the two techniques. This provides the designer with a clear picture of his product - again without sorting through reams of computer paper.

Once the designer sees these results he can simply change his input data (from the test or finite element analysis) to determine their effect on the final product. After the desired results are attained, the individual inputs that went into the model can be "worked backwards" through the process to see what changes to the part or product are necessary to obtain the desired condition - thicker walls, larger radii, different materials, or other engineering parameters.

PRODUCTION ENGINEERING

These steps have taken us through the concept evaluation, the detailed design, and the design development up to the actual construction of a prototype. Parallel to this last step are the newer computer aided manufacturability tools for mold filling and mold cooling circuit analysis. These comprise much of the production engineering step of the process.

MOLD FILLING ANALYSIS

An interactive computer program can be used to design injection molded plastic parts and molds. This mold filling analysis program simulates the molding of a part by predicting how the plastic will flow in a mold.

With this software you can choose the proper flow path and balance the runner system and cavity gates to control the flow of plastic. Expensive mold debug time is reduced and product development cycles compressed.

Molds designed using mold filling analysis fill with a smooth, uniform flow front. The desired polymer orientation pattern is obtained without areas of overpack. Residual stress, part warping and sticking in the mold are reduced. Molds fill at a pressure suited to the machine capacity. The program considers frictional heating to ensure a good melt temperature at the end of the flow, without overheating. The runner freezes off at the required time, improving cycle efficiency and reducing overpack.

If a mold in production is not performing well, this method is a cost-effective way to rapidly troubleshoot mold filling problems. Applying the analysis to a new design before problems are built into a mold, can result in even greater savings in reduced mold debug time and fewer prototype molds.

Once the mold filling analysis has been processed an interactive review of the analysis results takes place with the analyst reviewing color contour, criterion and X-Y plots of the results. Specific displays include views of filling patterns, contour plots of temperature, pressure, shear rate and shear stress at the instant of fill. You can efficiently review this information to determine acceptability of knit line placement and evaluate other molding problems such as overpack, freeze off, material burning and degradation and residual stress. Based on this thorough review, you can modify input parameters including part geometry to improve the filling characteristics of the part/mold combination.

MOLD COOLING CIRCUIT ANALYSIS

Using PolyCool, an interactive software program for optimizing cooling circuitry in thermoplastic injection molding, you can predict the efficiency of mold cooling system designs and reduce part processing cycle times. Typically, locations of cooling passages are dictated by the availability of space in the mold after all other working parts are accommodated. Sizing of coolant channels is often determined by the experience of the mold designer and flow path hookups are determined by the mold set-up technician through trial-and-error.

By using this software you can analyze mold cooling efficiency and optimize the cooling system before molds are built. For existing molds, proposed adjustments for improving cooling can be rapidly evaluated, and costly shop floor trial-and-error avoided.

The program eliminates guesswork in the design of cooling systems. It predicts cooling time for specific mold and part configurations. Adjustments to cooling line locations, circuit hookups and flow rates can be made in the computer to reduce cooling times, to balance core-cavity cooling and control cooling induced warping. With this software cooling cycle reductions of up to 35% have been realized.

Mold cooling analysis is simple in the method. You can define heat removal problems in terms of six cooling characteristics: mold material properties, plastic part thickness and material properties, cooling circuit pressure drop, the geometric form of the coolant passages, and the location of each passage relative to the surface of the molded part. PolyCool then analyzes the heat removal process by calculating the coolant flow rate, the overall heat transfer rate, and the cooling time as cooling fluid temperature and pressure are varied through predetermined ranges.

GOING TO PRODUCTION

After the previous steps have been completed and the design is ready, the information necessary to produce the part must be able to be used in subsequent operations, such as drafting and the production of NC tapes to machine the tools necessary to manufacture the product.

Returning to the solid modeling program, the data recorded to define all the 3-dimensional surfaces of the part can be converted to splines and an IGES file or specific description to become an NC tape for any machine. It is important that NC machine operating parameters be known before taking this last step, since all machines have specific needs in preparing the tool paths.

COST AND PAYBACK

The most cost-effective applications for this technology are part and mold designs representing a degree of "challenge" due to part complexity, end-user structural requirements, close tolerance, a desire for reduced molding cycle times, etc. In such cases, the return on investment comes quickly:

- The first prototype is a computer model, less tooling costs and quicker turnaround.
- Reduction of product development cycle by 30% savings in time and engineering costs through the integration of design and manufacturing analyses.

Engineering decisions and investments can be made earlier in the product development cycle as design and manufacturing alternatives are evaluated upfront in a computer model. This model is easier to change than engineering drawings and tool steel molds. More importantly, you're not making alternative adjustments on the production floor. A production mold is delivered eliminating most of the expensive and time-consuming shop troubleshooting.

CONCLUSION

Computer aided design of plastic parts is not a black art - it is a true engineering tool. All the elements are commercially available and are less expensive than once thought. Most firms have much of the required hardware in place and need only purchase the proper software. Software can be made available on a trial basis and any viable software vendor will offer fully documented training for engineers slated to use the tools. Start-up times vary with the complexity of the products manufactured, but an engineering staff with normal education and training can become proficient in 4-6 months. The improved output, accuracy, and speed of developing new plastic products often make the payback less than a year for all the software and related items.

Perhaps the best way to enter this technology is to take a trial project to a consultant who possesses extensive experience in the process and all the software tools required. The results of such a project in terms of cost and time savings can be the guide to potential savings from the process to your organization.

There is no reason for any firm to hesitate to obtain these tools. Those who wait for next months development will always wait. Those who want to improve their product quality, increase their engineering output, send their new products to the market faster, and improve product cost and manufacturability will act on this technology and profit from it. Others will watch them prosper from it. The choice belongs to the firms who manufacture plastic products.

Integrated Vision – Based Work Cell

Michael J. Chen, Dennis McGhie, and Chris Eisenbarth

Machine Intelligence Corporation

330 Potrero Avenue

Sunnyvale, California, U.S.A.

ABSTRACT

This paper describes the application of state-of-the-art computer vision technology for a manufacturing process. The concepts of FMS - Flexible Manufacturing Systems, work cells, and work stations are illustrated in this paper. Several computer-controlled work cells used in the production of thin-film magnetic heads are described. These cells use vision for in-process control of head-fixture alignment and real-time inspection of production parameters. The vision sensor and other sensors, coupled with transport mechanisms such as steppers, x-y-z tables, and robots, have created complete sensorimotor systems to perform various functions. These systems greatly increase the manufacturing throughput as well as the quality of the final product.

INTRODUCTION

There are two key features in advanced manufacturing technology - programmable automation [24] and intelligent sensors [26]. Robotics, which couples artificial intelligence to sensing and manipulation, offers the potential for a much broader range of applications than traditionally defined [25,32,33,35]. The predominant noncontact, nonintrusive sensing currently used in the industry is machine vision [1, 10,11,12,31,34]. It has evolved from academic research to a versatile and practical industrial tool. A previous survey [28] listed seventeen application areas for machine vision in sensor-controlled manipulation and inspection. The current scope [29] of computer vision covers binary, grey scale, three-dimensional, stereo, grey-scale color, and binary color [8,23]. Other sensing modalities, such as tactile [26], force [27], and torque [26], have been widely investigated in the laboratory and will gradually be incorporated in manufacturing processes. This paper illustrates several examples of utilizing robotics and computer-vision technology for use in the microelectronics industry.

Thin-film magnetic recording heads contain thin layers of materials. The materials are deposited and shaped on a ceramic substrate using a variety of techniques, including photolithography. These techniques have made it possible to create thin-film read/write heads with precise geometries. The mass production of thin-film heads has been accomplished by combining novel technologies such as etching, sputtering, and plating. The handling of thin-film heads is facilitated by slicing the wafer into strips. These strips, which are known as head-bars, contain multiple thin-film heads. These bars are mounted on fixtures, precisely aligned, ground and lapped to their specifications. The work stations described below provide strict control of the electrical and dimensional parameters for pre- and post- grinding processes. These procedures are among the most labor intensive in the thin-film-head manufacturing process. A series of MI proprietary automated workstations have been developed for in-process measurement and inspection of thin-film magnetic heads. These units coordinate the production process, allow tighter control of process parameters, document defects and enhance yield. The work cells are robotic systems with advanced sensing as the core technology - in particular, computer vision. Other forms of optoelectronic sensing have also been applied in the construction of these units. These workstations have been delivered and installed by Machine Intelligence as functional turnkey systems as well as networked autonomous production units.

HUMAN AND MACHINE INSPECTORS

Most natural visual scenes contain an overabundance of hierarchical classes [30,36] of information - spatial (position, orientation, shape, depth/range etc.), temporal (time-invariant/time-varying), and spectral (intensity, frequency/color [8,23]). Human beings utilize their visual sensing in pursuing multiple goals in an unstructured, constantly varying world with multiple objects; they work with a general vision system. Manipulators in manufacturing perform a small number of repeated tasks with few objects in constrained environments engineered to simplify those tasks . This difference, in preconditioned scenes and predefined tasks, constitutes an important concept in the engineering of vision-based systems. It simplifies robotic world modeling which is normally a computationally expensive process.

There are several advantages associated with utilizing computer vision systems in the production process -
(1) Consistency - permits process parameterization and characterization (2) Objectivity - avoids human physical, physiopsychological variations (3) Throughput - high speed operation (4) Accuracy - high degree of precision (5) In-Process Inspection - identifies process trends (6) Early warning - detects early symptoms, avoids value-added waste
(7) Indefatigability - tirelessness, suitability for monotonous scenes (8) Tolerance - for unfavorable environments (9) Automatic data base generation (10) Reliability (11) Cost Effectiveness (12) Serviceability - diagnostic/debugging tools aid field service (13) Flexibility - allows easy changeover to new process parameters

WORK STATIONS, WORK CELLS, FLEXIBLE MANUFACTURING SYSTEMS

There are previous reports of the utilization of machine vision for industrial applications in workstation, workcell, and FMS environments. One recent example is the British Scamp FMS project [18] which used two vision stations for automatic part identification and orientation. Machine Intelligence Corporation has designed and implemented automated work stations for three of the most critical stages in thin-film head fabrication associated with grinding and lapping. The stations are built respectively for pre-grinding fixture alignment, post-grinding dimensional inspection (for integrity of electronic components), and post-grinding head rail (aerodynamic surface) inspection. From here on, the **work stations** described will be referred to as the fixture alignment station, the dimensional inspection station and the rail inspection station. These stations contain electrical, mechanical, optical, processing and software components.

The work stations are individually controlled by their respective **work cells.** Each work cell has a cell control computer for coordinating the multiple event sequences (Figure 1). For example, the dimensional inspection work cell described in this paper controls the sequencing of a grinding cell. This cell is composed of a cell control computer, a robot manipulator with a transverse axis, a pair of dimensional inspection work stations and four grinding stations. The programmable, automated manufacturing work cell is the basic unit of a flexible factory. The work cells are controlled and monitored by the plant host computer (refer to Figure 1 for the hierarchical control [2,17] structure). This configuration allows future expansion to other work cells to perform additional functions. Simply stated, **FMS (Flexible Manufacturing System)** provides the flexibility to change the parameters in inspection, assembly or material handling. It also provides the ability to produce various quantities of different parts using the same automated equipment.

To integrate robots with sensors or automatic test (or measurement) equipment, the system engineer must consider not only the testing equipments and the robot, but also buffer storage, material handling systems, shared data-base resources and most importantly, the information exchange among the system components. Regarding material exchange, the work stations and cells cannot be considered as fully automated until the workpieces can automatically move into the cell, around the cell, and out of the cell. Of the many types of movers, either a belt conveyor, a cart system or robots can be used for transportation. We have custom-designed a transverse axis for positioning the Motoman L3 robot to move the head-bar fixtures between the grinders, the buffer storage positions, in and out of the station for the dimensional inspection cell.

Major Systems Components

The control strategy for the robotic workstations, workcells, and factory host computer is **hierarchical** in structure. The components of these work stations and cells are **modular** in architecture. The microscope and optical subassemblies are almost identical among the workstations. The micropositioning stages and stepper controls have common interface specifications. The human interface panels are similar and

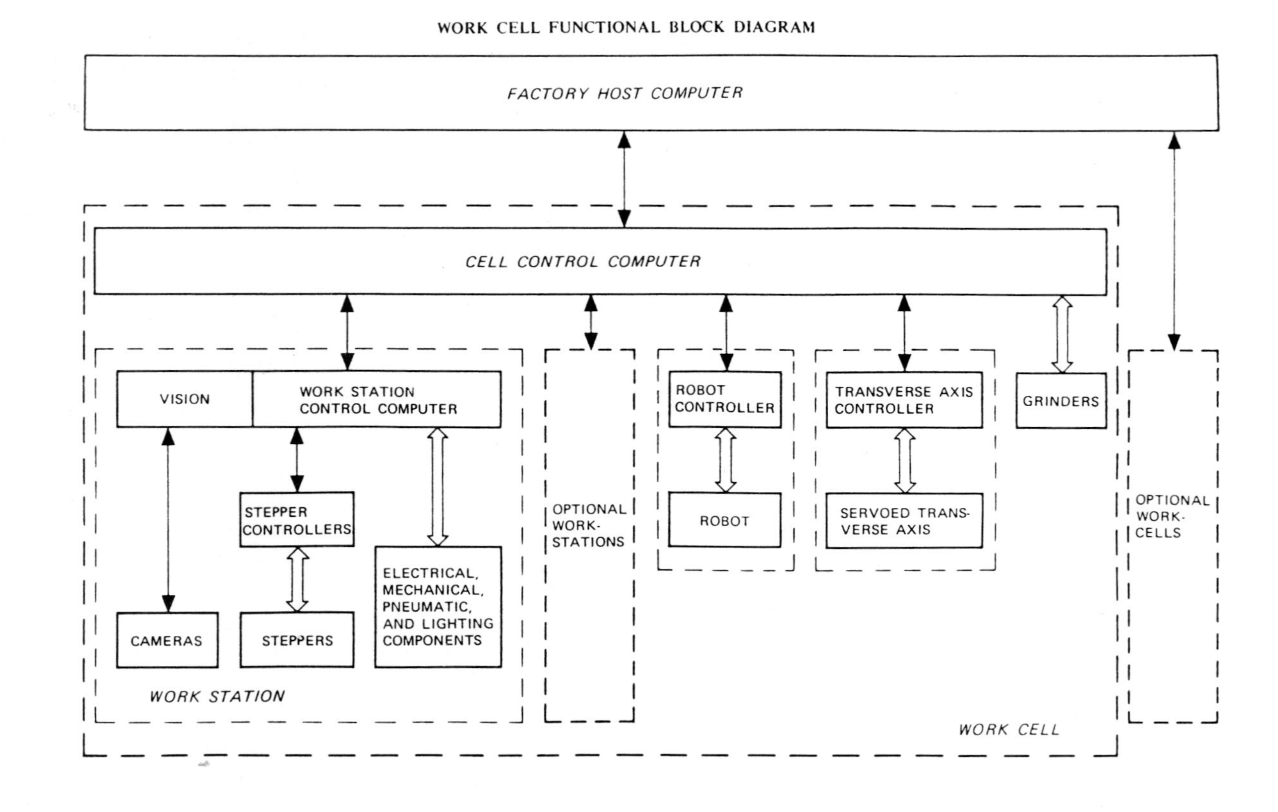

Figure 1. Hierarchical control architecture for thin-film-head automation.

share the same designations. The software modules for alignment, measurements, calibration and process sequencing are implemented from a single comprehensive library. The common characteristics of the stations are fully utilized for clean and uniform design. Nevertheless, substantial customization has been done to optimize the performance of each application.

The system for real-time image-processing and scene-analysis is the MI VS-100 vision system. This system contains greatly enhanced software based on the SRI vision module [16]. The application-specific software for system-control and event-sequencing is developed using the MI DS-100 Machine Vision Development System [7]. This system is used to develop and test software for the target VS-100. The DS-100 enables users to use the disjoint 'blobs' as the conceptual framework in image and pattern processing, instead of constantly coping with more complicated pixel-level structures. It provides a library of vision subroutines and special utilities for serial I/O and parallel I/O (lighting, electro-optical sensor controls etc.), steppers, x-y table and robot control. In performing visual alignment and measurement, the predominant scene-analysis technique involves extensive use of windowing for checking and verifying local features. This concept falls into the structural approach of syntactic pattern recognition [15] as opposed to a pure statistical [13] one. Nevertheless, it does not resort to the establishment of complete syntactic grammar. The advantage gained is increased speed and simplication in the programming concept.

Modes of Operation

The stations operate in three modes: Automatic, Manual and Maintenance. The Manual mode allows manual loading and unloading of the head fixture for inspection and calibration. This manual feature allows the utilization of the system as a stand-alone human-assisted work station. In Automatic mode, the station receives commands from and returns status to the cell controller. The cell controller is in turn controlled by the plant/factory host computer. The total configuration constitutes a network for full automation. In Maintenance mode, the control-buttons on the workstation front-panel can be used to activate x-y-z stage, motor, lighting, or pneumatic devices. This facilitates functional verification and maintenance.

Fixture Alignment Cell

There are numerous alignment and inspection procedures in semiconductor or integrated circuit manufacturing [3,4,19,20,21]. The former includes photomask, wafer or chip alignment used in pattern exposure, probing, or wire bonding. Computer-controlled, automatic alignment of discrete or overlapped parts by x-y-θ table for image-masking and differencing was also reported and made possible by utilizing wide/narrow-angle cameras [9] viewing overlapped fields.

The fixture-alignment cell operates in the following context.

Subsequent to the separation of individual head bars from the wafer, they are mounted on fixtures for further processing and handling. The grinding process, which consists of different types of operations, is an important link in head fabrication. It relies on an alignment step to assure correct positioning of the head bar relative to the fixture so that the grinders remove a precise amount of deposited material. The fixture heights must be adjusted in an alignment cell. This height determines the depth of the subsequent grinding process. Each fixture alignment cell consists of a cell controller, a robot, and one or two fixture alignment stations. During the aligning process, linear landmarks created in the photolithography process on the head-bar are used as the visual reference. The fixtures are adjusted by two robotic screw-drivers which are driven by high-resolution stepper motors. The alignment is a measure/control iterative process until the alignment is brought under the preset tolerance. Figure 2 shows the fixture-alignment station (right) and its cell controller (left). Shown in Figure 3 is the robot part-transfer operation in the fixture aligning process. Also shown is the custom-designed gripper with built-in compliance. The fixture alignment workstation performs this critical, yet human-error prone alignment process several times faster than human beings. Statistical results indicate that the alignment process is repeatable to a standard deviation of 10 microinches (0.25 micron). The data show that this automated station surpasses human operations by over ten times in speed and five times in accuracy.

Dimensional Inspection Cell

A grinding cell is the basic operating unit for performing thin-film head bar grinding and inspection operations prior to rail inspection and lapping. A grinding cell consists of a cell controller, several precision grinders, dimensional inspection stations, and a mobile robot on a transverse axis. Following the slicing of the head-bars from the thin-film wafer and the grinding process, the dimensional inspection station (Figure 4) inspects their deposition edge. Inspection consists of quantitative gaging and measurement of critical dimensions and qualitative determination of integrity and completeness based on the quantitative results. The qualitative results are displayed on the monitor and passed to the cell controller that commands the robot to take suitable measures. The cell controller also sends these results to the host computer for plant process-monitoring and statistical purposes. Measurements are made to extract designated aspects of the internal head geometry and are compared to stored nominal ranges. Typical techniques include line and edge detections and centroid measurements, from which distances are derived. Also measurements are made for detection and quantization of aberrations in selected zones. Simultaneous scanning and processing are performed to maximize throughput.

From an operational standpoint, the workstation acquires a part from the robot load position and transports it to the optical subsystem under a combined clamping (for stability) and translation (for part transport) mechanism. The 3-D clamping method guarantees steady motion and also maintains fixed depth-of-field under the high-magnification microscopic optics. The features of a selected subset of read-write heads are scanned under the programmed translational motion control. The measured values are compared with the master values and preset tolerances. The inspection is performed under a sampling scheme. A subset of the bar is viewded in each cycle. This subset varies in a periodical fashion to permit all head positions to be sampled at intervals. Once the inspection for a given head-bar is completed, the results are displayed on the monitor. The pass/fail signal light is lit correspondingly. The inspection results are also communicated to the cell controller and from there to the host computer (if connected). The part is then moved to the unload position.

Rail Inspection Cell

Following the grinding cell operation and prior to the lapping operation, a rail inspection cell inspects head bar samples for defects on the rail (air-bearing) surface. The rail surface condition is crucial for the proper settling and riding of the head over the disk surface during head read/write cycles. The results from this station determine whether the lengthy lapping process would be productive. Rejects discovered at this stage can be discarded, thereby releasing the lapper for good parts. Each rail inspection cell contains a cell controller, a robot, and multiple rail inspection stations (Figure 5). Under software control, the rail inspection station looks at the top surfaces of the rails and performs measurements of the amount of chipping. It then compares these values with the specified master values and tolerances. An accept/reject decision is then made and the robot is directed to the next action.

DISCUSSIONS

Each cell is automatically controlled by the cell controller, based on information from the host computer. If the cell controller detects an error such as a bad interlock status or a machine time-out, it suspends operation and reports that status to the host computer. In addition to the normal automatic operation, each cell is capable of manual mode operation permitting operator access to individual stations, or maintenance mode operation permitting access to all actuators. To maintain accuracy, the work stations periodically calibrate themselves with a master calibration fixture and perform a self-diagnostic procedure. Efforts have been made in instructing the system to determine many classes of defects, including rare defects and those not encountered in training, and to distinguish cosmetic from rare defects.

There is no trivial way to create a black-box with grey-scale, binary, color, or structured-light processing and bring it into the factory and apply it immediately. Lighting must be controlled for optimal operating speed and efficiency, especially in a microscopic environment. In the dimensional inspection station, discerning different features with varying reflectivities and from different angles requires three kinds of lighting. Three threshold settings are utilized to maximize the visibility of the features under back, top high, and top low lightings. The scenes are basically optimized under different grey scales for different levels of illumination. Automatic thresholding is performed at intervals to cope with variations in ambient light intensity.

One of the most critical bottle-necks in developing near-gigabyte disk drives lies in thin-film magnetic read/write head technology. Thin-film-head manufacturing traditionally suffers from low-yield and highly-stringent requirements. Although thin-film-heads are unapproached by ferrite-heads in performance, their manufacturing is more labor-, capital-, and technology-intensive. To improve the manufacturing throughput and to ease the constraints, we have designed and implemented several workstations to help enhance the yield, reduce the cost, and provide dramatically better control over the critical electrical and dimensional parameters. These work stations are systems built to solve real production problems. They are not laboratory prototypes or experimental tools. They represent the fruition of the combined, multidisciplinary efforts from systems, mechanical, electrical, optical, and computer engineers. We have listed below several points which we consider significant from the conception to completion of these workcells.

(1) For practical application of sensors and robots, the manufacturing process has to be analyzed and ordered logically for robotic implementation. Although sensor develpoment increases the ability of robot systems to be integrated into an overall automation scheme, sensor implementation should not needlessly complicate the implementation of robotic systems.
(2) The architectures of the robotic systems are designed to work in an integrated manufacturing environment and support programmable automation. The programming language allows the user to combine manipulation, sensing, computational, and data processing functions provided by the system.
(3) Although the stations described in this paper automate specific portions of the head-bar manufacturing process, the concepts discussed in this paper can be extended to other systems or other manufacturing processes. The concept of utilizing machine vision for in-process adjustment, quality inspection and control have been illustrated previously.
(4) The described systems represent the coordinated effort between MI, as a vendor of the robotic systems, and a microelectronics company. To bring these systems to fruition, MI has put in substantial efforts in learning the thin-film production process. This is the only way to design automatic machinery with optimal performance.

REFERENCES

[1] **Agin, G.**:" Computer Vision System for Industrial Inspection and Assembly", IEEE Computer, May, 1980.
[2] **Albus, J.; McLean, C.; Barbera, A. J. and Fitzgerald, M. L.**: " An Architecture for Real-Time Sensory-Interactive Control of Robots in a Manufacturing Facility"; Proceedings of the 4th IFAC/IFIP Symposium, 81-90, 1983.

[3] **Baird, M.**: " SIGHT-I: A Computer Vision System for Automated IC Chip Manufacture"; IEEE Transactions on Systems, Man, and Cybernetics, Vol. SMC-8, No. 2, Feb. 1978.
[4] **Berry, D. et. al.**: " Appl. of Automatic Alignment to MOS Processing in Projection Printing"; Solid State Tech., 1983.
[5] **Binford, T.**: " Survey of Model-Based Image Analysis Systems"; Int. Journal of Robotics Res. Vol.1, No.1, 1982.
[6] **Bourne, D. A. and Fussell, P.**: "Designing Languages for Programming Manufacturing Cells"; Electro/82 Conf. Rec. 23-3/1-9, 1982.
[7] **Chen, M. J. and Milgram, D.** : " A Development System for Machine Vision", Proc. of PRIP, IEEE Computer Society, 1982.
[8] **Chen, M. J. and Milgram, D.** : " Binary Color Vision"; Proc. 2nd Int. Conf. Robot Vision and Sensory Control, 1982.
[9] **Chen, M. J.**: " A Vision Guided X-Y Table for Automatic Inspection"; Handbook of Industrial Robotics, John Wiley and Son Inc., 1984.
[10] **Chin, R. T.**: " Machine Vision for Discrete Part Handling in Industry, A Survey"; IEEE Computer Soc. Conf. Record, Workshop on Industrial Applications of Machine Vision, 1982.
[11] **Chin, R. T. and Harlow, C.**: "Automated Visual Inspection: A Survey"; IEEE Trans. on PAMI, Vol. PAMI-4, No. 6, Nov.,1982.
[12] **Davis, L.**: " Industrial Vision Systems, Future Trends"; in Proc. Industrial Applications of Image Analysis, organised by IRSIA/IWONL and CETEA, Antwerp, Oct., 1983.
[13] **Devijver, P. and Kittler, J.**: Pattern Recognition - A Statistical Approach. Prentice Hall Inc., London, 1982.
[14] **Duda R. and Hart, P.**: " Pattern Classification and Scene Analysis"; A Wiley Interscience Publication, John Wiley and Sons, Inc., 1973.
[15] **Fu, K. S.**: Syntactic Pattern Recognition and Applications. Prentice Hall, Inc., N. J., 1982.
[16] **Gleason, G. and Agin, G.**: " A Modular Vision System for Sensor-Controlled Manipulation and Inspection"; 9th Int. Symp. on Industrial Robots, Washington, D.C., 1979.
[17] **Graupe, D. and Saridis, G.**: " Principles of Intelligent Controls for Robotics, Prosthetics, Orthotics"; Workshop Res. Needed to Adv. the State of Knowledge in Robotics, 1980.
[18] **Heywood, P. W.**: " Vision Sensing for FMS - The Scamp Project"; in Proc. Industrial Applications of Image Analysis, organized by IRSIA/IWONL and CETEA, Antwerp, Oct., 1983.
[19] **Hsieh Y. and Fu, K.**: " A Method for Automatic IC Chip Alignment and Wire Bonding"; IEEE Proc. Comp. Soc. Conf., August, 1979.
[20] **Horn, B.**: " Orienting Silicon Integrated Circuit Chips for Lead Bonding"; CGIP, vol. 4, pp. 294-303, Sept., 1975.
[21] **Huang, G.**: " A Robotic Alignment and Inspection System for Semiconductor Processing"; Intelligent Robots: 3rd Int. Conf. Robot Vision and Sensory Controls, SPIE Vol. 449, 1983.
[22] **Kanade, T.**: " Geometrical Aspects on Interpreting Images as a Three-Dimensional Scene"; Proc. IEEE., July, 1983.
[23] **Milgram, D. and Chen, M. J.**: " Binary Color Vision for Industrial Automation"; Final Report, AFOSR, 1983.
[24] **Nitzan**, D. **and Rosen, C.**: " Programmable Industrial Automation"; IEEE Trans. on Comp., Vol. C-25, No. 12, 1976.
[25] **Pugh, Alan, ed.**, : "Robot Vision"; U.F.S. Publications, Ltd., U.K., and Springer-Verlag, Berlin, 1983.
[26] **Rosen, C. A. and Nitzan, D.**: " Use of Sensors in

Programmable Automation"; IEEE Computer, Dec., 1977.
[27] **Raibert M. and Craig, J.**: "Hybrid Position/Force Control of Manipulators"; Robot Motion,MIT Press,Cambridge,Mass.,1982.
[28] **Rosen, C. A.**: " Machine Vision and Robotics, Industrial Requirements". in G. C. Dodd and R. Rossal(eds.), Computer Vision and Sensor-Based Robots, Plenum, N.Y., 1979.
[29] **Rosenfeld A. and Kak, A.**: Digital Picture Processing. Vols. 1 & 2, Second Edition, Academic Press, 1982.
[30] **Sanderson, A.**: " Robot Vision and Industrial Automation"; in Proc. Industrial Applications of Image Analysis, organized by IRSIA/IWONL and CETEA, Antwerp, Oct., 1983.
[31] **Skaggs, F.**: " Vision-Based Parts Measurement System"; in Intelligent Robots - Third International Conference on Robot Vision and Sensory Controls, SPIE Proc 449, 1983.
[32] **Suetens, P. and Chen, M. J.** : " Artificiele Intelligente, Visiesystemen en Beeldverwerking" Technisch Management, Nov., 1983.
[33] **Suetens, P. and Chen, M. J.**: " Le lien entre l'intelligence artificielle et les systemes de vision"; Technique & Management, Nov., 1983.
[34] **Winkler, G.**: " Automatic Visual Inspection"; in Industrial Applications of Image Analysis, Proc. of a workshop organized by IRSIA/IWONL and CETEA, Antwerp, Oct., 1983.
[35] **Zimmerman, N. J.**: " Robot Vision in Holland"; D.E.B. Publishers, The Netherlands, 1982.
[36] **Zimmerman, N. J., Van Boven G. and Oosterlinck, A.**: " Overview of Industrial Vision Systems"; in Industrial Applications of Image Analysis, Proc. of a workshop organized by IRSIA/IWONL and CETEA, Antwerp, Oct., 1983.

Figure 2. Thin-film -head fixture alignment station/cell controller.

Figure 3. Fixture alignment station with the dual microscope setup and the robot with compliant gripper operating in background.

Figure 4. Thin-film-head dimensional inspection station with robot transverse axis.

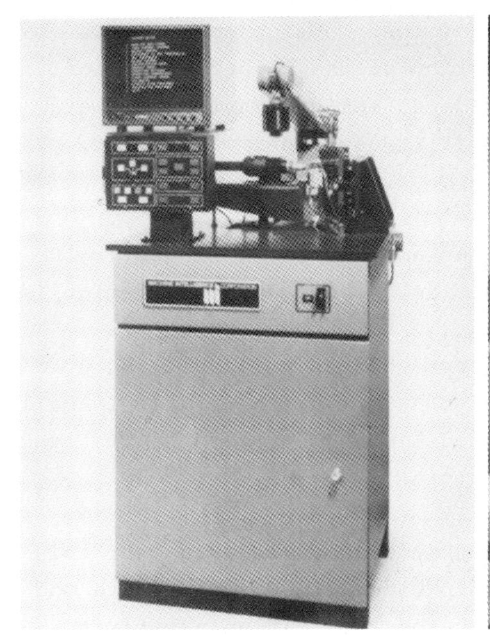

Figure 5. Thin-film-head rail inspection station.

Figure 6. Closeup view of the microscope and the X-Y-Z stage.

Design of an Application Database on a Computer-Aided Design System

Timothy Murphy

Duke Power Company
P.O. Box 33189
Charlotte, N.C. 28242

Abstract

This paper presents the design of an application database for an electrical circuit design program implemented on a Computer-Aided Design system. A step-by-step design process is given beginning with initial data identification and ending with the final logical schema. Problems encountered during the design along with the solutions used to overcome these problems are discussed. The vendor supplied interface between the graphics software and the data base management system is also discussed since its limitations imposed restrictions on the design of the application.

Introduction

The design of a computer application is divided into two design tasks, the program design and the database design. A well-designed implementation of both of these design tasks is essential to the efficiency and operation of the computer application. However, the database design is dependent on the type of Data Base Management System (DBMS) that is used to implement it.

Conceptual Design

The first step in the design of an application database is the development of the conceptual design. The conceptual design is the abstract view of the application data from the user's perspective. A data model is first choosen which will be used to represent the conceptual design. The design is then formally described in terms of the data model without regard to the type of DBMS that is used to implement the database.

The data model used to describe the conceptual design for this project is the Entity-Relationship Diagram. For this data model the types of entities required by the application are first determined. Then the attributes describing these entities are listed. Finally, any relationships between data entities are indicated.

The Entity-Relationship Diagram for the circuit database is illustrated in Figure 1. Four types of entities are included in the design:

1) The PANEL entity describes all electrical panels or enclosures which house devices with electrical connections.
2) The DEVICE entity includes all electrical devices that are located in or mounted on the panels listed in the PANEL entity.
3) The WIRE entity lists each wire used in a circuit. A wire is the continuous physical wiring between active devices and may include many individual conductors.
4) The TERMINAL entity identifies all electrical terminals which are connected by the wires in the WIRE entity. These terminals are part of the devices listed in the DEVICE entity.

Four relationships between the above entities are included in the conceptual design as shown in Figure 1. The MOUNTED_ON relationship identifies the devices in the DEVICE entity that are mounted on the electrical panels in the PANEL entity. It is a one-to-many relationship. The PART_OF relationship identifies the terminals in the TERMINAL entity which are part of the devices listed in the DEVICE entity. It is a one-to-many relationship. The TERMINAL entity also has a relationship with the WIRE entity. The terminals in the TERMINAL entity that are connected to the wires listed in the WIRE entity are identified by the WIRED_TO relationship. It is also a one-to-many relationship. The last relationship is the CONNECTED_TO relationship and identifies all connections between terminals in the TERMINAL entity. These terminals may be on the same electrical device, on different devices within the same panel, or on devices in different panels. It is the only many-to-many relationship in the conceptual design.

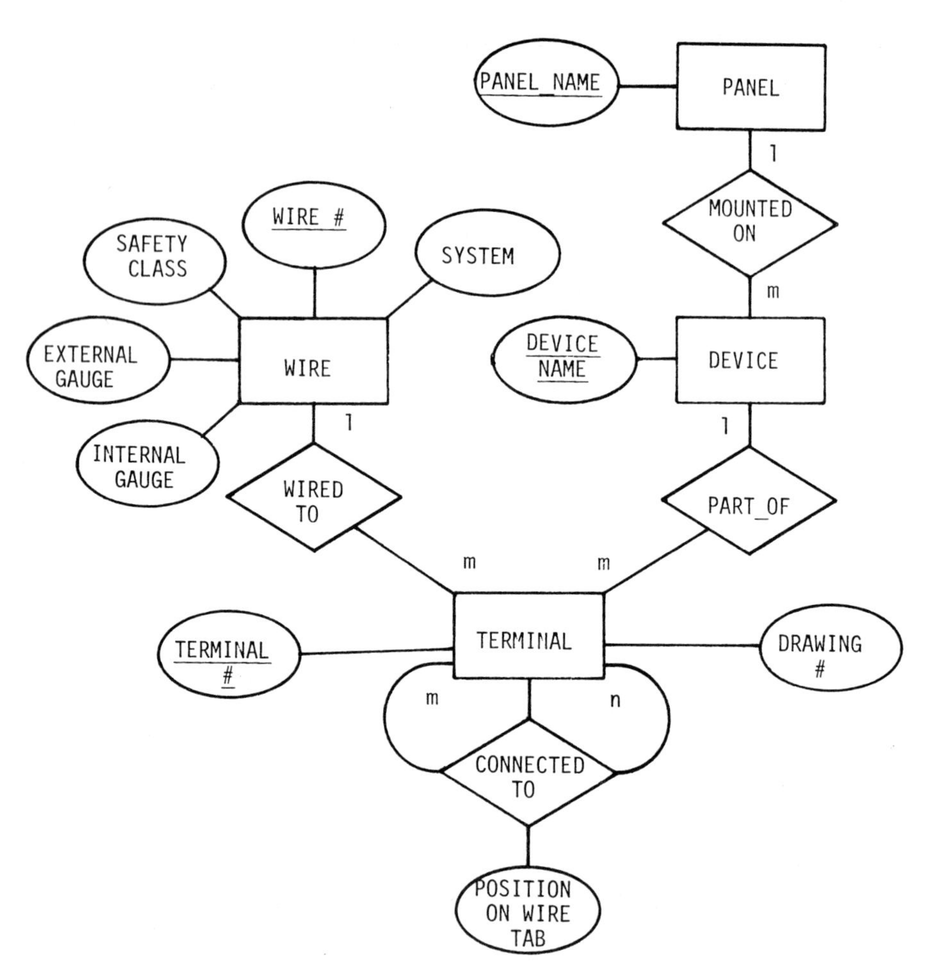

Figure 1 Entity-Relationship Diagram of the Circuit Database

Logical Design

The second step in the design of an application database is the development of the logical design. It defines in detail the logical record types and data attributes of the database. The logical design is first written as a modified form of the conceptual design taking into consideration the type of DBMS that is to be used to implement the database. This is referred to as the logical schema. The type of DBMS used in this project is a hierarchical DBMS.

The logical schema for the circuit database is illustrated in

Figure 2. Each entity of the conceptual schema is represented by a record type in the logical schema. Attributes are shown as fields in the logical record format with key fields identified with an underline. Relationships are identified by a line connecting the related record types with double arrows pointing to the child record type and a single arrow pointing to the parent record type.

Two additional record types are shown in the logical schema. One is shown as a child record type of WIRE. As the Entity-Relationship Diagram illustrates, the TERMINAL entity is the child of two different parent entities. However, this cannot be implemented directly since the DBMS requires a hierarchical structure.

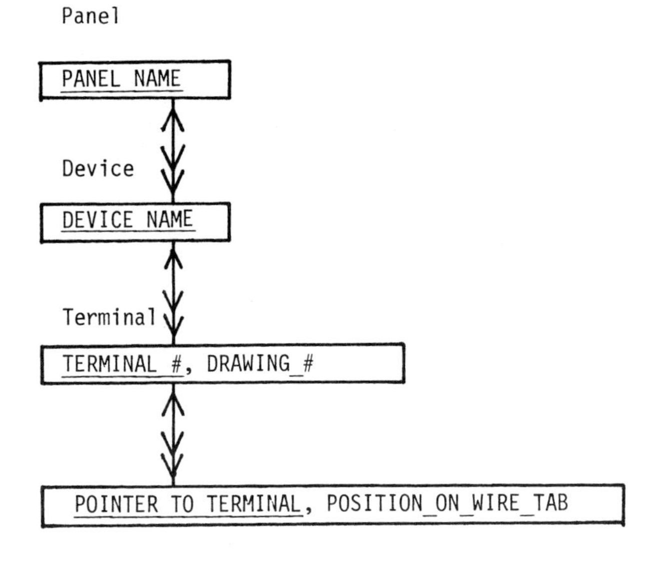

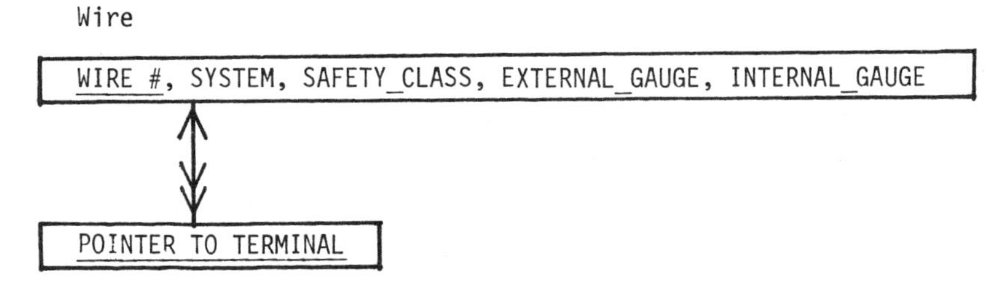

Figure 2 Logical Schema of the Circuit Database

Therefore, the TERMINAL entity is shown in the logical schema as a child of the DEVICE entity and a new record type inserted as the child of the WIRE record type. This new record type contains pointers to the specific occurrences of the TERMINAL record type that are related to the wires in the WIRE record type. This method, although indirect, allows a network data structure to be implemented on a hierarchical DBMS (1).

The second additional record type is used to implement the CONNECTED_TO relationship. Since, in a hierarchical DBMS, a record type cannot be a parent or child of itself, a new record type is used to contain pointers to the related occurrences of the original entity. In this case the occurrences of the new record type point to the terminals that are connected to the parent terminal.

Modified Logical Design

There are several limitations of the DBMS and graphics language used in this project which impose restrictions on the design of the application database. Two of these limitations affect the logical schema shown in Figure 2. Therefore, it is necessary to modify the logical schema in order to provide a database structure that is consistant with the capabilities of the DBMS and permits efficient interactive use by the application software.

The first restriction on the database design is imposed by the graphics language. The capability to manipulate attributes within the database from the graphics language is provided only for record occurrences that are linked to a graphic symbol by the system software. Therefore, any record type in the logical schema which does not have a graphic counterpart in the design file cannot be updated directly by the graphics language.

The second restriction on the database design involves record types which are added to the logical schema to serve as pointers to other record types in the database. Although this is a valid technique for designing a logical schema, the hierarchical DBMS does not have the capability to use this pointer to relate occurrences. All relationships must be defined as a hierarchy in order for the DBMS to recognize the relationship.

Three steps are required to modify the logical schema for the circuit database shown in Figure 2.

1) The TERMINAL and WIRE record types are both associated with graphic symbols on the circuit. The remaining four record types cannot be updated directly by the graphics language as described previously. There are two alternatives for solving this problem:
   a) A FORTRAN task can be initiated by the graphics language which would use an interface to the DBMS to update these attributes. However, a batch program is not a very efficient method for an interactive application.
   b) The PANEL and DEVICE record types can be combined with TERMINAL to form one record type. This method will produce redundant data for the Panel-Name and Device-Name attributes which should be avoided in order to eliminate data discrepancies. However, since there is only one attribute from each of the two eliminated record types, this small amount of data redundancy is a small price to pay to gain efficient interactive operation of the application software.

   Using the second alternative, the logical schema is modified as shown in Figure 3.
2) The record which was added as a child of WIRE to point to the TERMINAL record type violates the second restriction described previously. However, since TERMINAL no longer has a parent, this record type can be eliminated and WIRE can be added as a parent of TERMINAL. The modified logical schema is shown in Figure 4.
3) The last problem with the logical database design is the record type added as a child of TERMINAL. This record type violates both restrictions described earlier. It is not linked to a graphical symbol in the design file and therefore cannot be updated directly by the graphics language. Also, the DBMS cannot relate occurrences in this record type to occurrences in

Terminal

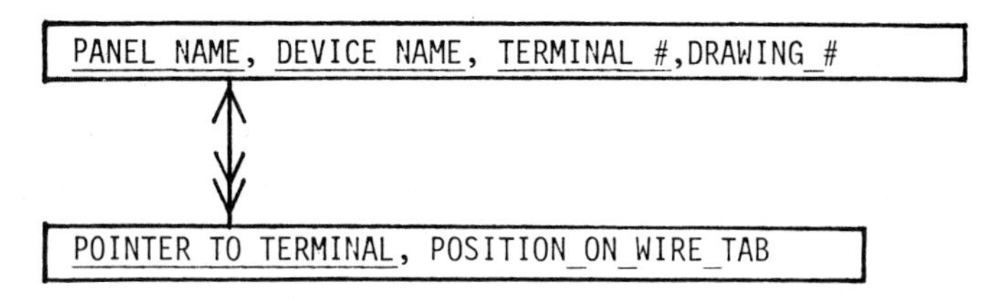

Wire

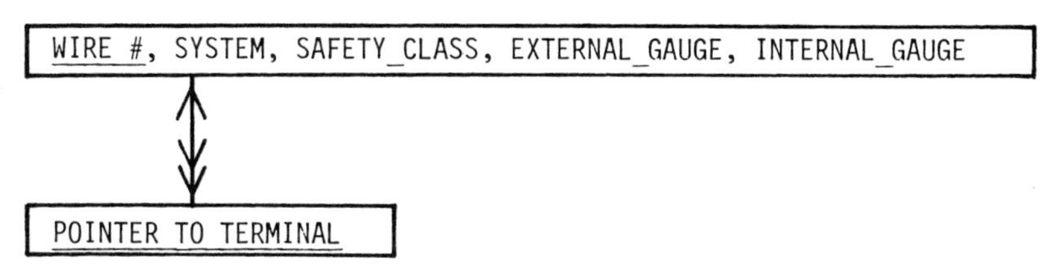

Figure 3 Modified Logical Schema of the Circuit Database-Step 1

Wire

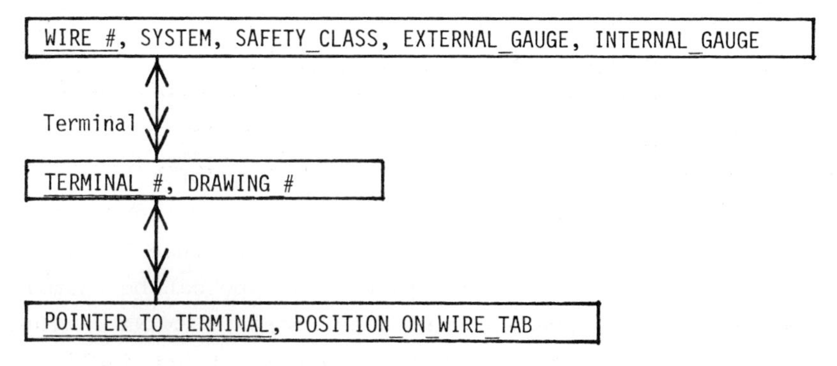

Figure 4 Modified Logical Schema of the Circuit Database-Step 2

TERMINAL other than the parent occurrences.

The second restriction in Step 3 cannot be solved using this DBMS. Special software would be required to relate occurrences in this record type to occurrences in TERMINAL. However, for the circuit design problem, this capability is not required. Therefore, only the first restriction must be resolved. To accomplish this, the attributes of this record type can be added to TERMINAL to form one record type. The Pointer attribute is actually the key to the TERMINAL record type and is listed in this manner. Since a maximum of three connections can be made to one terminal, these fiedls only need to be duplicated in the record type three times. However, since external and internal connections are shown in separate locations on the Wire Tab, it is easier for the report generator to print this format if a separate set of fields is established for the two types of connections. In this case, the Internal Panel attribute is not required since it is a duplication of the Panel attribute in the key. The final logical schema for the circuit database is shown in Figure 5.

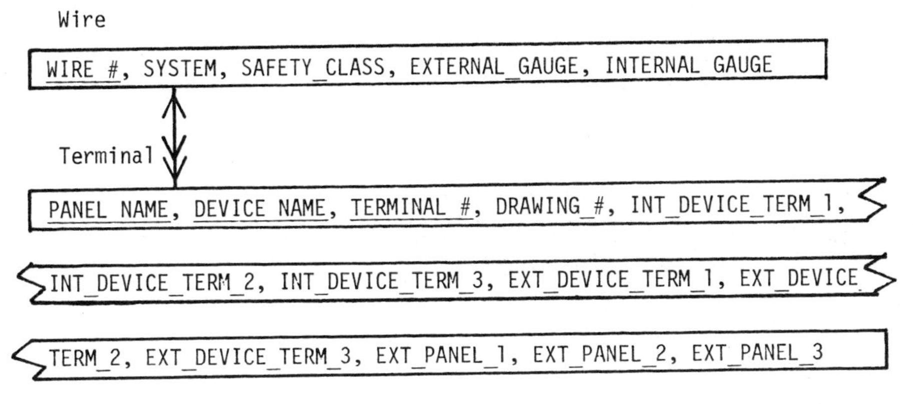

Figure 5 Modified Logical Schema of the Circuit Database-Step 3

In terms of relation schemes, the WIRE record type is in third normal form (1) since each attribute is fully functionally dependent on the key. However, the TERMINAL record type is not even in first normal form since the Int_Device_Term, Ext_Device_Term, and Ext_Panel attributes are repeating groups. This leaves open the potential for several problems.

If one of the attributes of a repeating group is to be updated or deleted, a search must be made of all attributes of the group in order to find the value that is to be changed or deleted. If the attribute is being entered for the first time then a search must be made for the first available blank attribute.

Another problem exists with the number of connection attributes that may be updated. Since there are three internal device-terminal attributes and three external device-terminal attributes, it is possible to enter six connections for the terminal occurrence. However, a maximum of only three connections are permitted. Therefore, each time a connection is entered a search must also be made to ensure that no more than three connection attributes have non-blank values.

One last problem exists because of the definition of a connection. Each device-terminal which is identified in a connection attribute must also exist as the key to another occurrence. The device-terminal which formed the key to the original occurrence must be identified in a connection attribute in this second occurrence. This requirement must be assured each time a connection is updated in the database.

To ensure that these problems do not occur in the circuit database, algorithms have been included in the circuit design program which updates the database. The program will always verify that a connection being updated has not already been entered. The program has also been coded so that no more than three connections are ever made to one terminal. As it updates each connection it also creates a second occurrence with the contents of the connection device-terminal attribute as a key. The key of the original occurrence is then used to update the connection device-terminal attribute of the second occurrence. With these algorithms included in the program, the integrity of the database is preserved.

Conclusion

Database technology associated with Computer-Aided Design systems is generally lagging commercially available state-of-the-art products. Computer-Aided Design systems have provided exceptional capabilities in graphical techniques that in some cases can produce images that resemble photographs. However, the links between these graphics products and their associated Data Base Management Systems are extremely weak. This limits the effectiveness of these systems to be used in Computer-Aided Engineering applications and adds complexity to those applications which are pioneering this new technology in spite of these limitations.

References

1. Jeffrey D. Ullman, Principles of Database Systems, Computer Science: Potomac, MD, 1980, pp. 91-95.

Interface Robot to a Micro-Computer

A. Hsie, H. Parsaei, and A. Rabie

State University of New York College of Technology
Utica, New York

SUMMARY

As the popularity of computer controlled robots keeps increasing, we can definitely see the need of interfacing a robot with micro-computers in industry or in school laboratories increases as well.

This paper uses a Minimover-5 and a Vic-20 as an example to demonstrate the indepth of the basic techniques of interfacing a robot to a micro-computer.

INTRODUCTION

The purpose of this paper was to interface the MiniMover-5, a Microbot product, robot arm to an 8-bit parallel port. The user 8-bit parallel port chosen was the user I/O port of the Vic-20 computer (a Commodore product).

Before the MiniMover could be connected to the computer an interface connector had to be made and the decoding circuit modified on the MiniMover. A 24 pin connector was fitted to the computer and a ribbon cable then was attached to a proto-board. Wire wrap posts were than connected to the board to provide test points for the 308 Logic Analyzer. Another set of wire wrap posts were connected to the 40 pin edge on the proto-board. The 40 pin edge of the proto-board could then be connected to the MiniMover. Data lines 0 - 7 and a ground were the lines needed to move the MiniMover. The schematic for this is in Attachment 1. To modify the decoding circuit on the MiniMover several jumpers had to be installed as shown in the schematic.

PROCEDURE

The procedure for outputting data to a motor of the MiniMover is as follows:

1. The driver software determines the particular 4 bit pattern which is to be output to a specific motor.
2 The driver software places this pattern in bits 1, 1, 2 and 3 of a register which will eventually be output to the parallel port.
3. The driver clears the most significant bit of the register (latching pulse).
4. The driver software places the number of the particular motor (1-6) to be driven in bits 4, 5 and 6 of the register.
5. The register is output to the partial output port. This sets up the data bus and the address select lines.
6. The most significant bit of the register should then be set and the register output again. This provides a positive going signal on the out line of the MiniMover. This will strobe the four bits of the motor pattern into the latch addressed by the motor number.

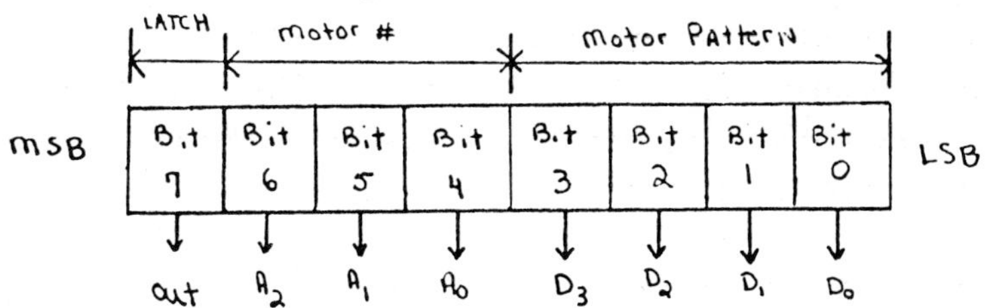

Bit connections to MiniMover-5 interface card.

The motors would be selected using the following patterns for bits 4, 5, and 6:

000 = base motor
001 = shoulder motor
010 = elbow motor
011 = wrist right
100 = wrist left
101 = gripper

DATA PATTERN

The motor pattern could be either clockwise or counter-clockwise depending on which order they were latched in.

Bit 3	Bit 2	Bit 1	Bit 0
0	1	0	1
0	1	0	0
0	1	1	0
0	0	1	0
1	0	1	0
1	0	0	0
1	0	0	1
0	0	0	1

Clockwise ↓   Counter C.W. ↑

This represents one pattern. It takes 96 patterns to move the motor from one limit to the other. Bit 7, the MSB (most significant bit) had to be continuously changing from 0 to 1 throughout the process to ensure the decoding circuit is being latched correctly.

APPLICATION

The user port of the Vic-20 gives you control over port B on the VIA chip #1. Eight lines for input or output are available as well as two lines for handshaking with an outside device. The port can be made an output by Poking all 1's or 255_{10} to the Data Direction Register (DDR). This is located at address 37138_{10}. The address of the port itself is 37136_{10} and this is where our program outputs the selected patterns to the Mini-Mover-5. A copy of the program is available by contacting the author.

It is preferred to keep the operation of the MiniMover as simple as the Apple II Interface. Each of the motors can be moved either clockwise or counter-clockwise. The motors are visibly marked on the Mini-Mover:

1 = Base motor
2 = Shoulder
3 = Elbow
4 = Wrist Right
5 = Wrist Left
6 = Gripper

These numbers relate to the keys on the Vic-10 keyboard. The keys directly below each of these keys moves the motor counter-clockwise.

Example: 3 moves the elbow motor clockwise
E moves the elbow counter-clockwise

These directions are repeated when the program is run.

An asterisk * halts all motor movement in either direction.

Since there is no feedback from the MiniMover-5 care must be taken not to overdrive the motors and stretch the cables on the arm.

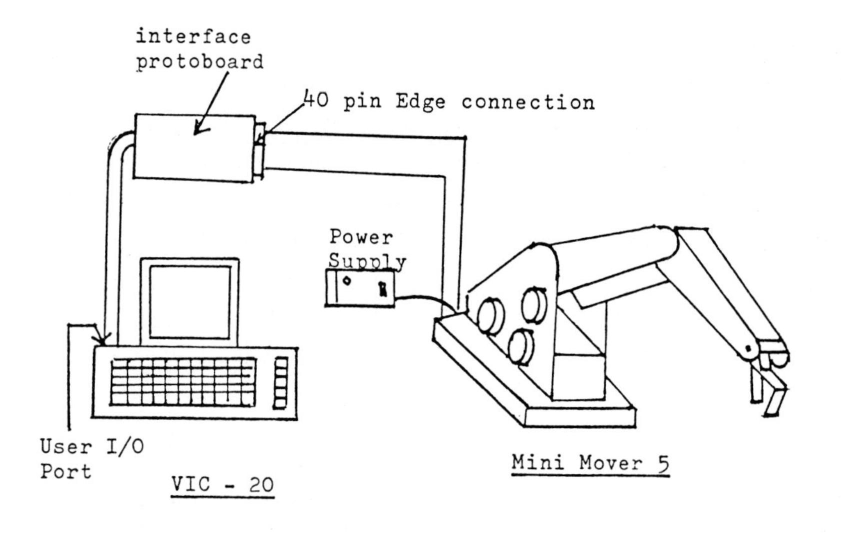

Electrical Connections between VIC- 20 and Mini Mover 5

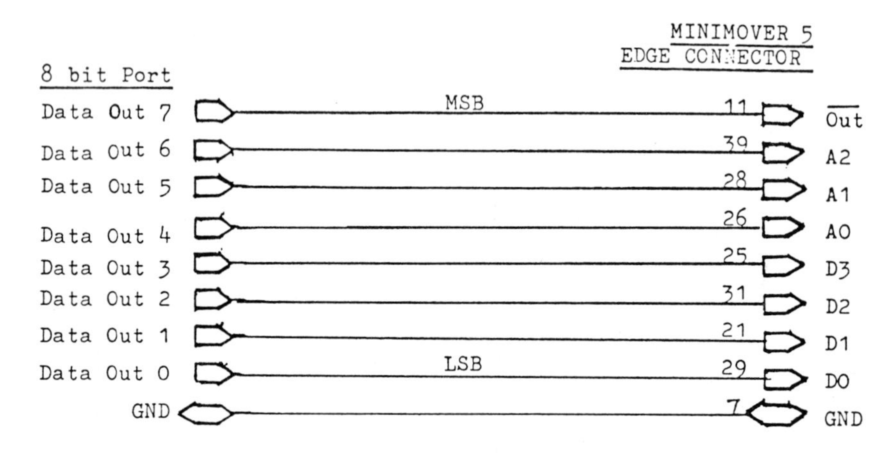

Connections for 8-bit Parallel I/O Port

Note: Jumping the 74LS00 outputs in this way as shown bellow will always enable the 74LS138 3 of 8 decoder. The address lines A0--A2 are used to designate the stepper motor.

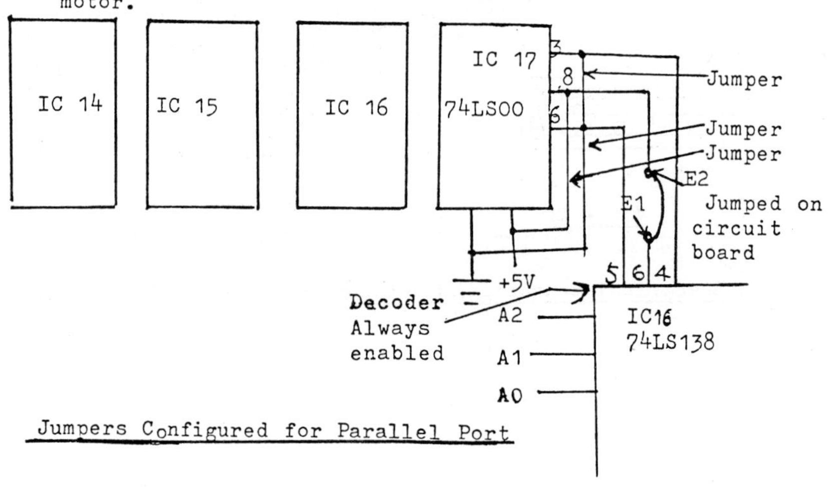

Jumpers Configured for Parallel Port

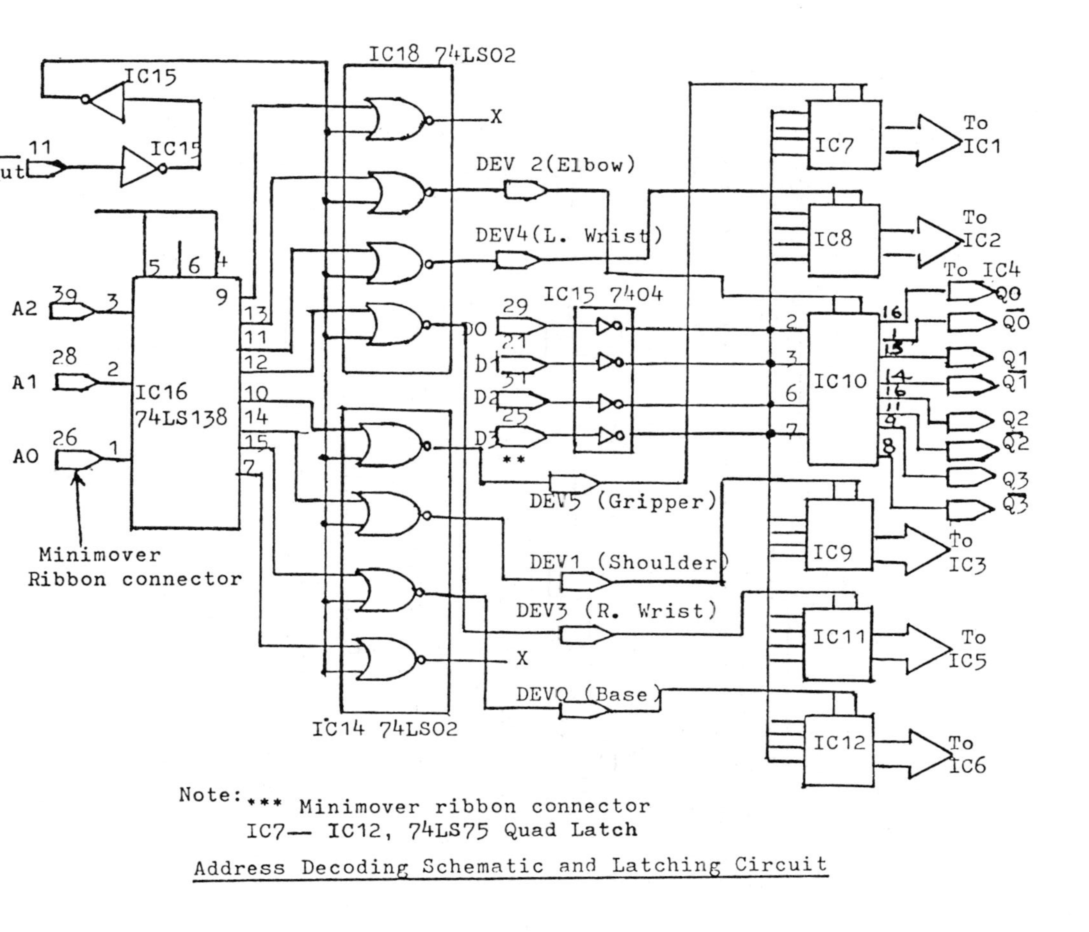

Note: *** Minimover ribbon connector
IC7— IC12, 74LS75 Quad Latch

Address Decoding Schematic and Latching Circuit

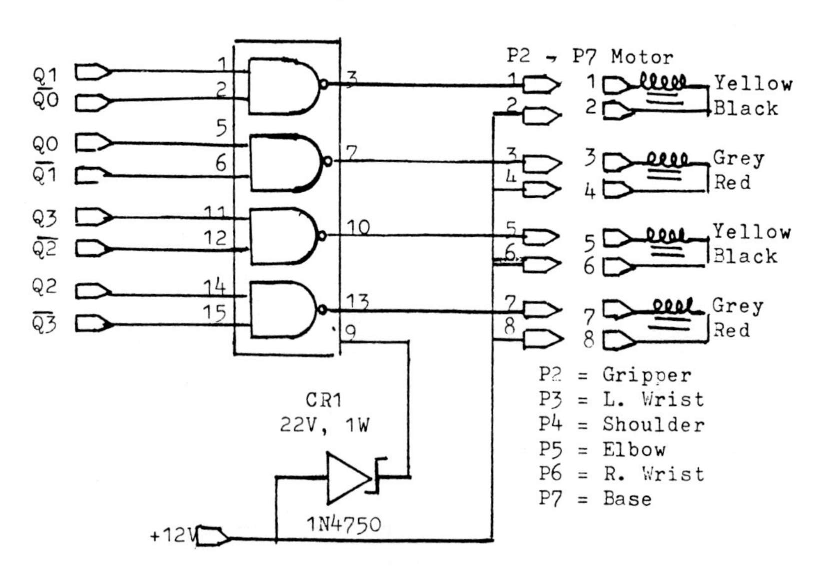

Stepper Motor Driver Schematic

Explanation:

IC1-,IC6 : Sprauge UDN 5707A Quad 2 input NAND drivers prevent more than one motor from being selected regardless of what valve is output from computer. This prevents possible harm to the motors or power supply from over heating.

MOTOR KEY

1 - Base motor C.W.
Q - Base motor C.C.W.
2 - Shoulder C.W.
W - Shoulder C.C.W.
3 - Elbow C.W.
E - Elbow C.C.W.
4 - Wrist right C.W.
R - Wrist right C.C.W.
5 - Wrist left C.W.
T - Wrist left C.C.W.
6 - Gripper C.W.
Y - Gripper C.C.W.

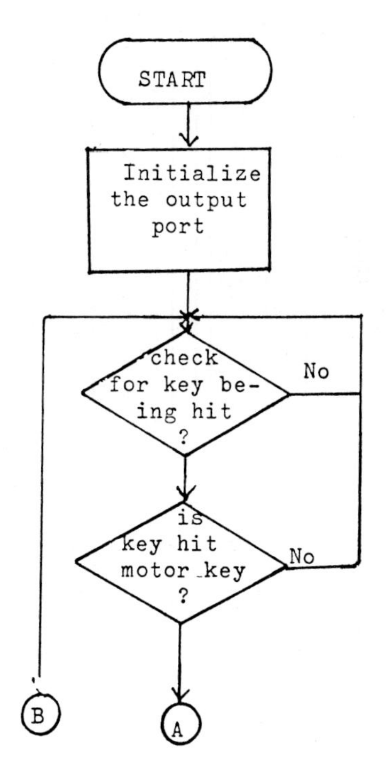

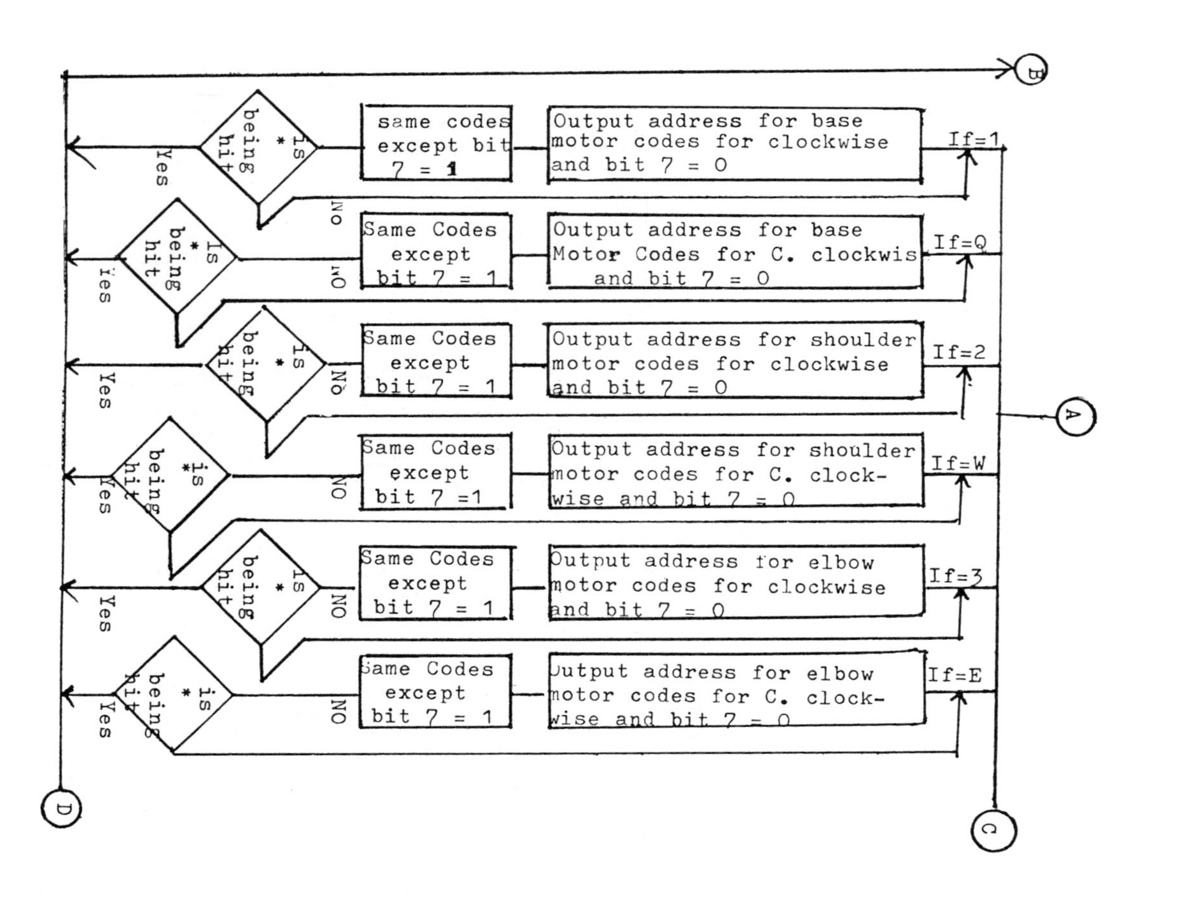

A
B
C
D
If=1
Output address for base motor codes for clockwise and bit 7 = 0
same codes except bit 7 = 1
is * being hit
No
Yes
If=Q
Output address for base Motor Codes for C. clockwis and bit 7 = 0
Same Codes except bit 7 = 1
Is * being hit
NO
Yes
If=2
Output address for shoulder motor codes for clockwise and bit 7 = 0
Same Codes except bit 7 = 1
is * being hit
No
Yes
If=W
Output address for shoulder motor codes for C. clock-wise and bit 7 = 0
Same Codes except bit 7 =1
is * being hit
NO
Yes
If=3
Output address for elbow motor codes for clockwise and bit 7 = 0
Same Codes except bit 7 = 1
is * being hit
NO
Yes
If=E
Output address for elbow motor codes for C. clock-wise and bit 7 = 0
Same Codes except bit 7 = 1
is * being hit
NO
Yes

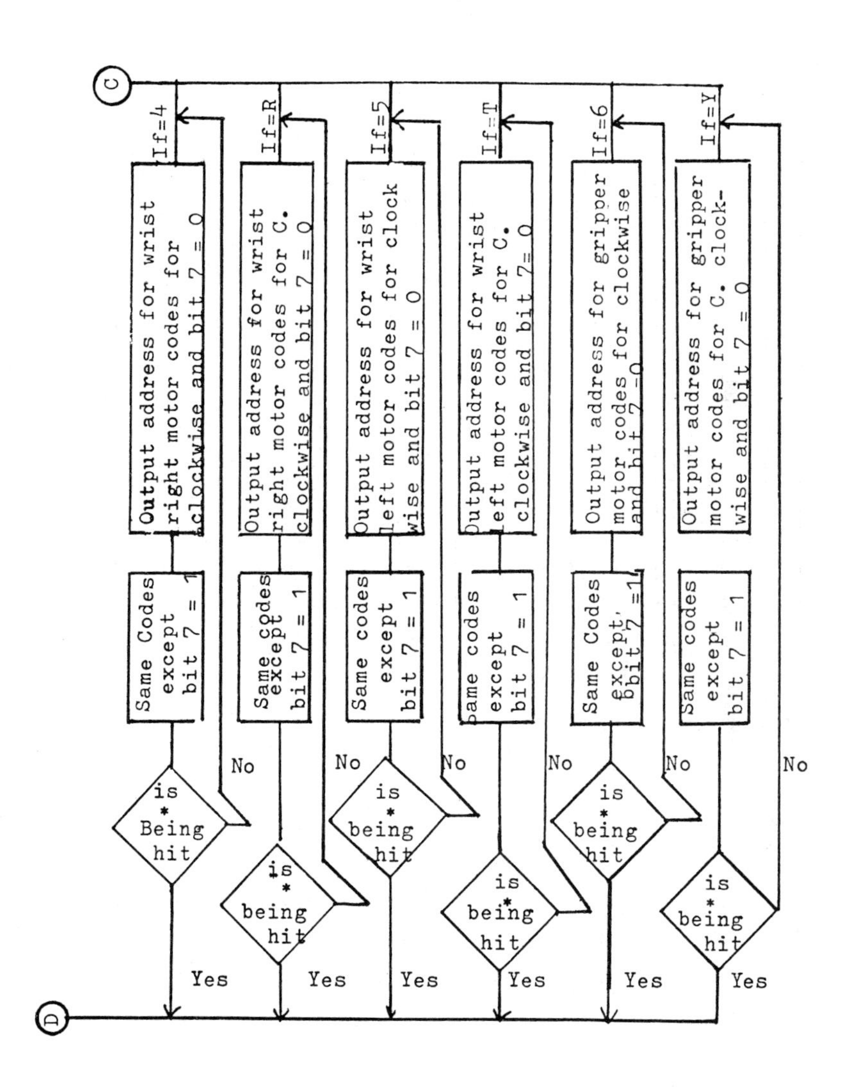
C
If=4
Output address for wrist right motor codes for clockwise and bit 7 = 0
Same Codes except bit 7 = 1
is * Being hit
No
Yes
If=R
Output address for wrist right motor codes for C. clockwise and bit 7 = 0
Same codes except bit 7 = 1
is * being hit
No
Yes
If=5
Output address for wrist left motor codes for clock wise and bit 7 = 0
Same codes except bit 7 = 1
is * being hit
No
Yes
If=T
Output address for wrist left motor codes for C. clockwise and bit 7= 0
Same codes except bit 7 = 1
is * being hit
No
Yes
If=6
Output address for gripper motor codes for clockwise and bit 7 =0
Same Codes except bit 7 =1
is * being hit
No
Yes
If=Y
Output address for gripper motor codes for C. clock-wise and bit 7 = 0
Same codes except bit 7 = 1
is * being hit
No
Yes
D

System Studies of CAD/CAM and Robotics Integration for Productivity

R. K. Dave, J. P. Sharma, and Suren N. Dwivedi

Mechanical Engineering
Govt.Engineering College
JABALPUR, M.P., INDIA.

Mechanical Engineering & Engineering Science Deptt.
University of North Carolina at Charlotte
CHARLOTTE, NC 28223, USA.

Summary

'Productivity' is very important and extremely sensitive to any industrial set up. So far the interphasing of design, process utilization, and manufacturing has been done by individuals associated with each system separately. Then introduction of computers and automation changed the trend and has made a tremendous impact on productivity. Though it added new dimensions in solving automation problems of manufacturing, it has to go a long way before the CAD/CAM integration with robots is applied in the factory of the future for productivity. In this paper a system analysis is presented linking productivity through CAD/CAM integration with robots in small scale, medium scale, and large scale industries.

Introduction

With the development of chip-technology, the computers are becoming smaller and smaller in size. Several hundreds and thousands of bytes memory are being now available with desk top computers weighing only five pounds. The interface accessories like disk-drive, modem, mouse trap, flexible, multicolored monitors, and word processor all included weighs within 30 - 40 lbs. The desk top computers can now communicate fast and print information within fraction of a second. These computers are flexible and have potential to interact with the mainframe sitting hundreds and thousands of miles away in one central location. The miniaturization of computers, reduction in weight, multiple choice of programming and activities, and cost reduction have put these computers within the reach of individual or groups to perform their work and business more effectively. Small and large business houses, manufacturing, and production organizations are progressively finding these computers very useful in managing the fast development of their business and productivity.

Development of high technology made it possible to interface computers with machines for high productivity. Robotics, CNC, DNC, and FMS are becoming more and more popular with industries for automation and increased productivity. Introduction of computer graphics have made the designer's tasks easier. Any changes to be made in design can now be easily incorporated by writing on the screen itself. All necessary components of computer aided design can be discussed at the computer site to review the design to meet all standards for maintenance, manufacturing and assembly. As such, it has now become imperative to go into system studies of computer applications for integration of design, process, manufacturing and productivity.

A system approach shows that integration of computers for design, manufacturing, robotization and productivity depends upon the following factors:

(a) sensitivity of human behavior in a given socio-economic environment.

(b) nature of industries

(c) work environment, and

(d) investment priority

Sensitivity of human behavior in a given socio-economic environment: The developed countries with high per capita income do realize the importance of increasing productivity through application of computers and robotization. This is because the high tech development will promote their standard of living. Computers are now becoming a way of life in their day-to-day work. Their socio-economic set up is totally integrated with technology. They can not think of their lives without technology. On the other hand in developing countries, the sensitivity to socio-economic structure is much different. The per capita income in these countries is very low. With the spread of education, these countries have a surplus of scientific and technical manpower which is increasingly growing every day. They are looking for new jobs and employment opportunities. They fear that increasing use of computers will take away their jobs, so the society as a whole community resent computerization and robotization of existing and future upcoming industries. They are fully conscious and aware that the new technology will improve their standard of living and create more job opportunities for the future generation, But at present, economic considerations and the race for earning more money are their immediate necessity of life. It makes them feel that the introduction of computers and automation will stop their participation in work places to earn high wages. Thus, they insist on incentives and immediate jobs and believe in increasing productivity through mass participation. Industrialists, however, are keen for automation as they see immediate growth of productivity.

This clash of ideas between two classes of societies is bound to delay the automation program, but appreciation which prevails at both ends will certainly allow computerization and robotization of industries at a slow rate in the developing countries. In their opinion, robots are unfriendly substitutes for human skill.

The countries which are just beginning to organize their industry and technical education are not passing through this socio-economic crises. They are merely looking at these computers and robots which may bring happiness to their life style in the near future. They are not totally aware of their socio-economic implications and crisis which these developing countries are passing through. These countries have little to offer towards sensitivity of human behavior unless a dynamic socio-economic environment is set forth.

The quantum of sensitivity of human behavior in a particular socio-economic environment is a very important and critical factor which should be considered before integration of computers with robots, manufacturing and productivity is planned in these countries. The interaction of socio-economic factors with human sensitivity for automation through CAD/CAM and robots is shown in figure 1.

Nature of Industries

There are varieties of industries in which computer aided design (CAD)/ computer aided manufacturing (CAM) and robotization integration is very vital for productivity. The increase in productivity through use of modern high technology will mostly depend upon the nature of industries and their response which is shown in figure 2. These are as follows:

(i) small scale industry
(ii) medium scale industry and
(iii) large scale industry

Small scale industry

This type of industry is generally owned by a single person as the investment outlay is in the order of Rs 30,00,000 ($300,000) or less. This is further classified depending upon the number of persons employed and the nature of the job, e.g.

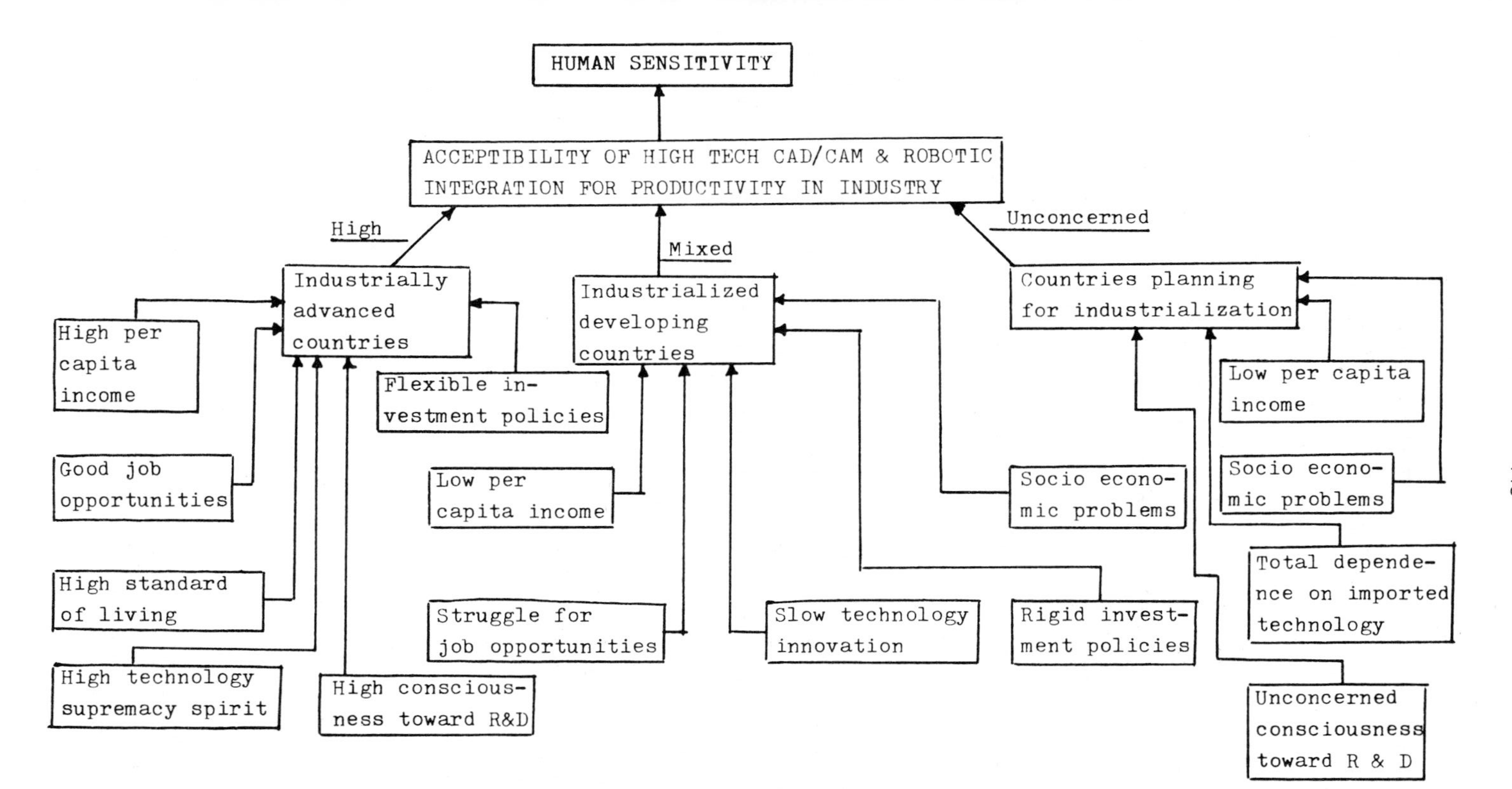

FIG 1 : CONTROLLING FACTORS ON HUMAN SENSITIVITY RESPONSE TOWARDS ACCEPTANCE OF CAD/CAM & ROBOTICS IN INDUSTRY

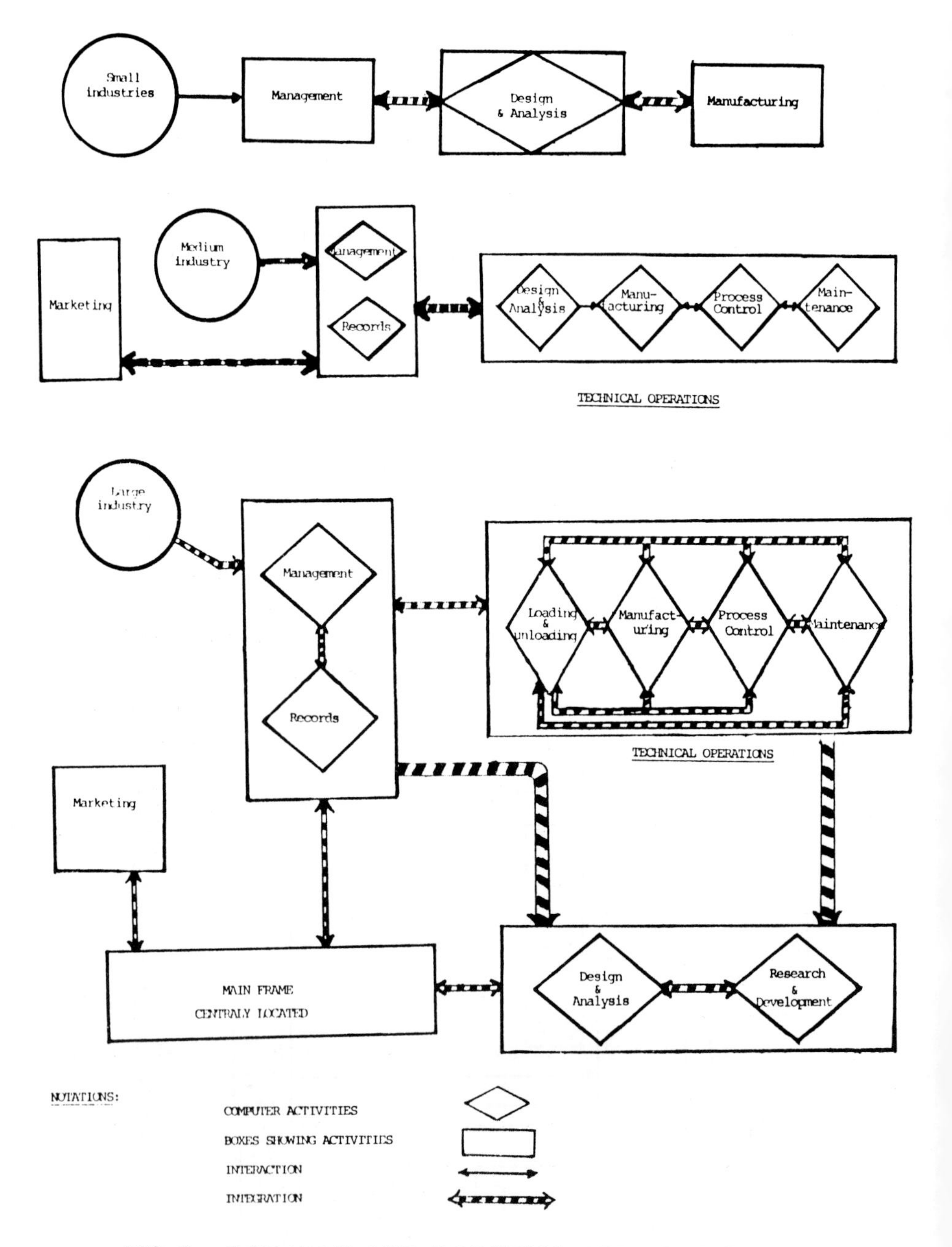

FIG 2 : CAD/CAM AND ROBOTICS APPLICATION AND INTEGRATION SHOWN IN VARIOUS INDUSTRIES

(a) Individual business industry and

(b) Small scale industry

The individual business industry usually will not employ any permanent person on their roll. In this all industry functions like management, design, manufacturing sales and purchase will be done by one or two persons of the family. The small scale industry will have employment potentials to employ five to ten persons. Mostly, at large, management funcitons will be looked after by the employer and under his direct instructions, and technical operations will be carried out by the employee.

The mode of communications in this type of industry is excellent with less chances of misunderstanding. All business functions are under one roof with close interaction and understanding of management and technical functionaries. As such the management hardly see any CAD/CAM integration and deployment of robots in small scale industries. However, use of desk top computers will help them keep the business records, inventory, and designs for expansion of their industries and businesses. Today, desk top computers in developed countries are increasingly becoming important among small businesses, because of large memories and various scientific and engineering functions that can be performed by these computers. These computers are becoming more popular because of expanding capabilities to perform several functions like design, drafting, word processing, data storage, data retrieval, record keeping, and communication with mainframe through modem. Because of its light weight, compactness, and low capital cost, these computers will provide several hands to small scall industries with easy access to all information connected with the nation's network.

Medium scale industry

The investment outlay in this type of industry is much higher than that of small scale industry. The employment potential is generally in the range of 20 to 100 persons and sometimes it may go up to 500 persons. This type of industry is organized and managed by leaders of their respective functionary groups. The leaders of these administrations and accounts group, the design and analysis group, and the various divisions of technical operation groups reports the top management of their productivity achievements, future development and planning activities. In addition these different groups also effectively communicate with one another for synchronizing their tasks and work plan.

In this category of industry, the technical operations like manufacturing and production are always a continuous process. Several investigators have reported that only 5% of time in the machine shop is spent on machining operation and the balance of 95% of time is spent on moving and waiting of materials. This shows that there are enormous potentials on the part of the computers and robots to cut down the time for moving the materials for process and help increasing the productivity with considerable saving. The machine utilization factors can be increased substantially by way of CAD/CAM and robot integration.

The other problem of medium industry is the effective communication among the management and production personnel. This can be synchronized by development of simple software programs for use in computers. The various functions and activities of medium scale industry can be linked together by use of microcomputers for effective communications and integration of CAD/CAM functions for increased productivity. Robots can also play an important role in medium industries where critical and sensitive products are to be manufactured. Deployment of these robots will help in automation and increasing productivity if cost, investment and social factors justify their use.

Large scale industry

In large scale industry sector, the various departments are headed and entirely controlled by the specialized management structure. The representatives of these managements meet at corporate level to discuss their strategy for productivity and goals for development. This type of industry employ more than 1,000 persons, and the investment outlay is very high of the order of millions of dollars and more. The various functions in technical division is so large that it require special attention, communication, synchronization and integration at all levels. For this, CAD/CAM and robot integration can effectively help in increasing their productivity.

This type of organization generates enormous technical and business management data at all levels which is difficult to handle effectively without the help of computers. The fast data retrieval is required to perform various functions without loss of time. Thus, it is necessary to plan and install several work stations operated by microcomputers in different divisions in relation to technical operations, R&D, marketing, administration, accounts, planning and management. These workstations should be connected to each other through mainframe having large memories and flexibilities for data entry and retrieval at any time of the day.

These large companies also have several units and divisions situated in different parts of the country. The multi-nationals also have their branches all over the world. For the effective interaction and communication of these units and branches, it may be desirable to have a central powerful mainframe connected with telecommunication links or through satellites for data entry, data processing and data retrieval. These large industries do require considerable support for effective control and reduction of waiting time for movement of materials for effective utilization of machine time through computer controlled automation. These companies also produce large quantities of products which should pass through a quality control program. Presently, industries do not offer 100% inspection of the products in process. Use of CAD/CAM integration with robots can effectively help in 100% inspection of products after each process to increase the reliability of their products. The CAD/CAM and robotic applications and integration of robots for small, medium and large industries for increasing their application productivity is shown in figure 3.

While planning the future industries or modernization activities in existing industries, it is desirable to group the various industries based on the nature of output products, and technology involved in its operations. These are classified into the following groups:

1. Manufacturing
   - (i) Engineering
   - (ii) High tech
   - (iii) Spacing
2. Processing
   - (iv) Process
   - (v) Metullurgical
   - (vi) Mining
   - (vii) Petrochemicals and
   - (viii) Utilities

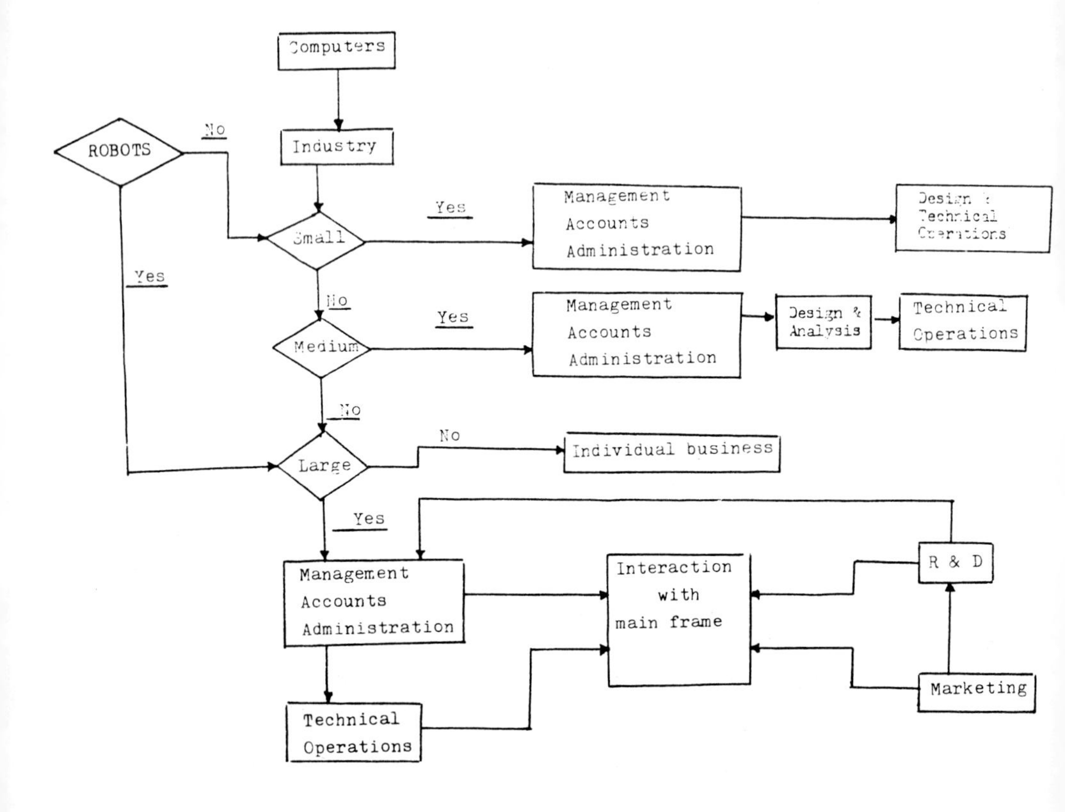

FIG 3 : SYSTEM INTEGRATION OF COMPUTERS & ROBOTS FOR PRODUCTIVITY

The first three basically categorized as manufacturing group and the rest from (iv) to (vii) as processing group. Utilities comes in a separate catagory because of large investment and energy feeder to both manufacturing and processing industries. However, from the analysis point of view, it is being put into the processing group.

(i) Manufacturing:

In this type of industry, manufacturing is carried out with a view to provide specified shape and size of a product from materials. Most of the operations in manufacturing includes cutting, forging, drawing, deep drawing shaping, machining, milling, drilling, welding and assembly to a close tolerance. When material goes through these operations, considerable time is also lost in handling materials, setting time for tools to perform a particular operation, flow of materials and inspection. In order to cut down the time lost in manufacturing and maintenance down time, the machine manufacturers incorporated automated concepts in sequential tool operations. With the advances made in computer technology and controls, the industry started paying attention and efforts in the following manufacturing areas for increasing the productivity.

(a) Automation through computer control.
(b) Group technology and cellular manufacturing.
(c) Computers in workstations for inspection.
(d) Multistation manufacturing systems
(e) Integrated manufacturing software for communication.
(f) Computer-integrated maintenance and automation.
(g) Robotization for handling and movement of job materials including operations and inspection.
(h) Design and design alteration for manufacturing requirements.

Computer controlled machines and their integration with design and manufacturing operations have helped to increase the machine utilization time, reduction in work-in-process inventory, reduction in number of drawings and design, reduction in set-up time and several other time factors sensitive to business and productivity.

(2) Process industries: In this group the industries are mostly related with chemical processing, textiles, food and beverages, extraction of metals from ore, mining, oil refining, and fertilizers. The main important operating factors of process industries are time, process temperature, and operating pressures. The movement of raw materials and mixing of various elements in proper size, weight and volume do require assistance from computers. The quality of the final product will depend upon the control of process temperature and pressure. This requires careful inspection, feedback control, and monitoring. This can be achieved only by computer con-

trol instrumentations and controls which can monitor these parameters to yield product of specified composition. Computers in conjunction with controls have provided good support in on-line process controls and helped quality production. These computers are also to be interlinked with mainframe and CAD system to provide all necessary functions like process control and monitoring, maintenance, inspection and communication. The communication system is very vital and should be simple to be followed by both the machine and man.

Robots can interact with both machine and man to perform operational activities in difficult environment of the process industries. Apart from this, these robots can receive instructions to carry out mechanical jobs like material handling, work performance, repairs and servicing. The potential applications of computers and robots in an integrated way in various types of manufacturing and processing industries are shown in tables 1 and 2.

Conclusions

The system analysis approach should be adopted first to identify industries and its work requirement. Based on this information, it is possible to access the requirement of computers, robots and their integration for productivity. Microcomputers with expandable memory and flexibility to perform both technical and business function will be useful for small scale industry. Microcomputers at several work stations and its network with technical and management personnel will be useful for management communication and productivity in medium scale industry.

In large scale industry, the microcomputers can effectively be used for manufacturing process control, quality control, inspectrum, calibration and sensitivity check. The network of these computers with mainframe will help in communication for both productivity and business operation. The integration of CAD/CAM and robots will help in advancing the productivity technology for factories of the future.

Robots can effectively be deployed to do mechanical jobs and at workspace in difficult environments. The robots if used as compliments to both machine and man will help in increasing productivity and plan factories of the future for creation of new job opportunities.

ENGINEERING	HI - TECH	PROCESS	METALLURGICAL
Structural Machine tools Mechanical components Electrical-Mechanical-Components Automobiles etc	Electronics Instrument Hi-tech materials Computers etc	Chemicals Textiles Pharmaceutic Food & Beverages etc	Iron & Steel Ore extraction etc
Computer for	**Computer for**	**Computer for**	**Computer for**
Design & Analysis Manufacturing Assembly Automation Inspection Maintenance Quality-Control Communication	Design & Analysis Manufacturing Assembly Automation Inspection Maintenance Quality Control Communication Calibration	Design & Analysis Process Control Automation Maintenance Quality Check Communication	Process Control Automation Maintenance Communication Quality Inspection
Robots for	**Robots for**	**Robots for**	**Robots for**
Loading Unloading Positioning Job Movement Welding Inspection Repetitious work	Positioning Assembly Calibration Sensitivity Check	Loading Unloading Job Movement Safety	Loading Unloading Job Movement Safety Repetitious work Inspection High Temperature

TABLE 1 :

TYPES OF INDUSTRIES SHOWING POTENTIAL APPLICATIONS & INTEGRATION OF COMPUTER & ROBOT FOR PRODUCTIVITY

MINING	PETROCHEMICALS	SPACE	UTILITIES
Coal Mineral processing Oil drilling etc	Fertilizers Refineries etc	Communication Satellite etc	Power Nuclear power etc
Computer for	**Computer for**	**Computer for**	**Computer for**
Process Control Automation Maintenance Inspection Communication	Process Control Automation Maintenance Inspection Communication	Communication Movement Positioning Control Maintenance	Communication Maintenance Distribution control
Robots for	**Robots for**	**Robots for**	**Robots for**
Loading Unloading Work performance Safety Environments Servicing & Repair	Material handling Work performance Safety Environment Inspection	Operation Work performance Safety Servicing & Repair	Distribution Repair

TABLE 2 :

TYPES OF INDUSTRIES SHOWING POTENTIAL APPLICATIONS & INTEGRATION OF COMPUTER & ROBOT FOR PRODUCTIVITY

References

1. Ayres, R.U. and S.M. Miller: Robotics Application & Social Implications, Ballinger Publishing Co., Cambridge, Mass. 1977.

2. Rembold, U.; Seth, M.K. and Weinstein, J.S.: Computers in Manufacturing. Marcel Dekker Inc., New York 1977.

3. Coffet, P. (Editor): Robot Technology-Modelling & Control Vol. 1, Prentice Hall, New Jersey 1983.

4. Houtzeel, A.: American Machinist- The Many Faces of Group Technology, January 1079.

5. Carter, C.F.: Trends in Machine Tool Development and Applications, Proc 2nd Int. Conf. Product Dev. & Manu. Technology, MacDonald, London 1972.

6. Merchant, M.E.: Economic, Social and Technological Consideration in the Development of Computer Aided Manufacturing Engineering November 1980.

7. Dallas, D.B.: The Advent of the Automatic Factory, Manufacturing Engineering, November 1980.

8. Sharma, J.P. and Dwivedi, S.N. Application of Computers in the Factories of the Future, Proc SME, Inter. conference Computers in Engineering, Las Vegas 1984.

Robotics – the First Step to CIM and The Factory of the Futurie

Joseph P. Ziskovsky

GCA Corporation
3460 No. Lexington Ave.
St. Paul, Minnesota 55112

Summary

The modern industrial robot is just one component of the Factory of the Future and the first step toward implementation of most CIM (Computer Integrated Manufacturing) systems. There are various types of robots which are defined through a combination of configuration/design, classification, power systems and type of control.

Guidelines versus standards are used in application and safety planning. An application can be broken down into three phases for both the application itself and in the area of safety planning. Robots do not end here, they are just the first step. Research continues to enhance the capabilities of flexible automation.

ROBOTS THE FIRST STEP

There are six (6) basic elements that go into making up the FMS, CIM, or FOF systems:[1]

Computers
CAD/CAM
Controls
Smart Processes
Information
and, of course, Robots.

Because robotics can be done today, they become the first step toward implementation of an overall CIMPLAN™ (GCA Corp) leading toward the realization of FMS, FOF, or CIM.

Each of these entities FMS, FOF, and CIM can be characterized by the 'E^3 CHARACTERISTICS' of Flexible Automation, as defined by Dennis Wisnosky in his 1982 article "Planning for the Factory of the Future".

1. **EFFICIENCY**- These systems are efficiently integrated and continuous.

2. **ECONOMY**- They are flexible and economical in the face of change.

3. **EFFECTIVENESS**- They are effectively organized to do the right thing with maximum productivity.

In essence this is part of classical marketing: building the right product, with the right materials, at the right time, at the right price, and with the right resources.

WHAT IS A ROBOT?

RIA has defined a robot as a:

"... reprogrammable multi-functional manipulator designed to move material, parts, tools, or specialized devices, through variable programmed motions, for performance of a variety of tasks."

This makes the robot look just like any other material handling device, which it is; but for our purposes, a robot could be better defined as:

"...a reprogrammable manipulator that possesses some degree of intelligence that will allow it to interact with its environment."

The key to Computer Integrated Manufacturing is the word **INTEGRATED**. Robots to some extent can be integrated and can interact with other components of the work cell. In fact, they must be capable of this if they are to be part of the

Factory of the Future. This is not to say that today's robots are intelligent and rational. In actuality, they are only computer peripherals with the level of intelligence, i.e. interaction capability, dependent upon the sophistication of the controlling computers' hardware and associated software. The robot is NOT the ultimate answer to productivity problems. It is only part. All the elements of the system must be considered.

ROBOT CATEGORIZATION

Robots can be categorized by various elements: type of operation, design/configuration, power system, or type of control system. Each element can be considered independently or in conjunction with others.

There are four (4) classifications for robot operation:

1) Fixed Sequence
2) Variable Sequence
3) Playback
4) Intelligent

There are four (4) core design/configuration categories:

1)Jointed Arm: articulated robots resembling the human arm.
2)Cartesian - XYZ - Rectangular: gantry and pedestal
3)Cylindrical Coordinate
4)Spherical

There are three (3) classifications for power systems:

1)Hydraulic 2)Pneumatic (air) 3) Electric

As was stated above, the robot is nothing more than a computer peripheral; it is really the control system that makes the robot what it is. The more sophisticated the control system, the more sophisticated the robot. There are four (4) basic types of control systems:

1)Bang-Bang: reacting to limit switches with little

or no interaction.

2) Programmable Controllers: more interaction capability but limited in sophistication.

3) Micro-Processors: ease of reprogrammability, and capable of a high level of interaction.

4) Micro-Computers: even more interaction capability. Capable of supporting a high level language. Can be tied to other parts of the total system in a real time environment.

ROBOTIC IMPLEMENTATION

There are three (3) phases of implementing any robotic application. They are:

- Application Planning (Pre-work)
- Installation and Start-up
- Continued operations [2]

Actual application planning involves a four-step process: [2]

1) Selection of the application.
2) Examining of what is to be done and how it is done now.
3) Selection of the robot system.
4) Making a working system.

Selecting the application is not a simple as it may seem. You can't just walk through the factory and say, "This looks like a good application." Involve the people that may doing the work now. They know the process best and can be a great asset in identifying and implementing the robot.

Get the total view once a candidate application has been selected. Look upstream and downstream and evaluate how the robot will affect the line.

Evaluate the product to be handled. Is its design conducive to being handled by a robot? It may have to be redesigned, to be handled in a blind, deaf, and for the most part senseless environment. Involve the design and industrial engineers in the application planning. If the robot is to interact with other equipment, determine if they will need to be modified or replaced.

Selecting the robot should be a concurrent process with evaluating the application. As was seen above, robots come in various types and capabilities. A robot vs. application requirement comparison should be done comparing the payload, speed, reach, repeatability, control capabilities, etc. that are required for the application.

It is the micro-processor based control system that allows the robot to interact and communicate with the other elements of the system. Attention must be given to the overall plan and its requirements. The robot controller should be upgradable to allow for future needs. Even with all this, DO NOT overestimate the capabilities of the robot.

Each and every application will require end-of-arm tooling, a gripper, a hand, or an end effector. End effectors are application dependent. It is the end effector that really makes the robot capable of doing its job. In fact, an ineffective end effector can make the best application useless. Attention must be given to how the part or device is to be grasped, manipulated, or acted upon.

<u>SAFETY PLANNING</u>

One of the most neglected areas in application planning is that of safety. To minimize danger, a risk assessment is performed in each of the three operating modes:[3]

- -Automatic or normal
- -Programming or teaching
- -Maintenance

Each mode is examined for potential hazards. Hazards fall into three areas:[2]

- * IMPACT- being struck by the robot or an object the robot is handling.
- * TRAPPING POINTS- between the robot and another object.

* OTHER– which can result from the application itself.

Hazards are created from various sources. These include **Control Errors**, both hardware and software; **Human Error**, a result of complacency or lack of respect for the robot during the time of human interaction; and **Unauthorized Access**.

Safety is common sense, and safety planning can be reduced to a basic formula:

ROBOT SAFETY = R^3 (Robots Require Respect)[2]

Respect for what the robot is , what it can do, and what it cannot do.

After the system is installed, a continual awareness training program should be established. A goal for a safety awareness program should be to instill the following maxims:

* If the robot is not moving, DO NOT assume it is not going to move.
* If the robot is repeating a pattern, DO NOT assume it will continue with the same pattern.
* Maintain respect for what the robot is and what it can do.

The robot is bringing us to the beginning of the next productivity revolution. It is only the beginning. Technologies are changing more rapidly than most companies can keep up with. Remember, the robot is just one part of the whole system. The principles discussed here can be equally applied to the whole.

REFERENCES

1) Wisnosky, Dennis; "PLANNING FOR THE FACTORY TO THE FUTURE", ENTERPRISE, August, 1982.

2) Ziskovsky, Joseph P., "The R^3 Factor of Industrial Robot Safety", Proceedings Robot 7, Vol. 1,, Robotics International of SME, pp 9-1 to 9-12, 1983.

3) Ziskovsky, Joseph P., "The Factory of the Future, FMS and Robot Application Planning", Proceedings Automach Australia '84 Conference, SME, 1984.

Robots Applications

Chairman: Ray Asfahl, University of Arkansas, Fayetteville, Arkansas
Vice Chairman: Ren-Chyuan Luo, North Carolina State University, Raleigh, North Carolina

Premachining Processing of Bearing Carriers

George Peterson

Engineering Technology Department

Texas A&M University

College Station, Texas

Summary

It was the intent of this project to develop the jigs, fixtures and robot programming necessary to demonstrate the feasibility of automating the bearing carrier preconditioning process currently being used by Lufkin Industries. This process is presently being done by hand and is used to prepare bearing carriers for short-term storage and finish machining.

The project started with a review of the operation as it was being done and proceeded through the design and implementation of a totally automated system. The automated process begins with the parts being delivered from a shot blaster on a pallet and progresses through the preconditioning operation to the loading of the part on an NC lathe.

Introduction

On Tuesday, May 10, 1983, several members of the Engineering Technology Department visited the Lufkin Industries Foundry facilities located in Lufkin, Texas. The purpose of this trip was to examine the possibility of assisting in the automation of a bearing carrier preconditioning operation.

Automation of the process appeared to be feasible and it was anticipated that a more efficient, higher quality, more consistent process would result. The process developed consists of the same sequence of steps, but is fully automated. During the design and construction of the necessary hardware and software, consideration was given to the expansion of the process to include the full range of bearing carrier sizes. However, initial construction was done of the Hyatt A5218TS.

Project Development

The initial activity undertaken was to explore the current

research and implementations of robotic finishing operations, specifically with respect to fettling of castings. The Lufkin bearing carrier preconditioning project (LBCPP) was influenced to a large extent by information gained through the literature.

The LBCPP system implements a Cincinnati Milacron HT^3-586 industrial robot with a customized machining cell to perform the bearing carrier preconditioning. The system is flexible enough to accomodate a variety of bearing carrier sizes and also to pick and place bearing carrier's from a number of different sources (conveyors, pallets, etc.). A diagram of the system is shown in figure 1.

Justification

The first part of the project included determination of the justification for the implementation of a robot. Articles on robotics justifications and recommended standards for these justifications explain the motivations most common for implementing robots. Examples are: reduced labor costs, elimination of dangerous jobs, increased output rate, improved product quality, increased product flexibility, reduced materials waste, reduced labor turnover (5). The domino effect of justification is explained (13), and a brief overview of some standards being developed in the robotics area has been presented (10).

Grinding: Tool wear is a problem

The process of grinding in the LBCPP was the most important consideration. Various types of grinding and other metal removal processes were necessary to finish all the surface of the bearing carrier. For instance, a belt grinder removes metal from the large surface area of the outer diameter, a grinding wheel takes off large protrusions like gates and

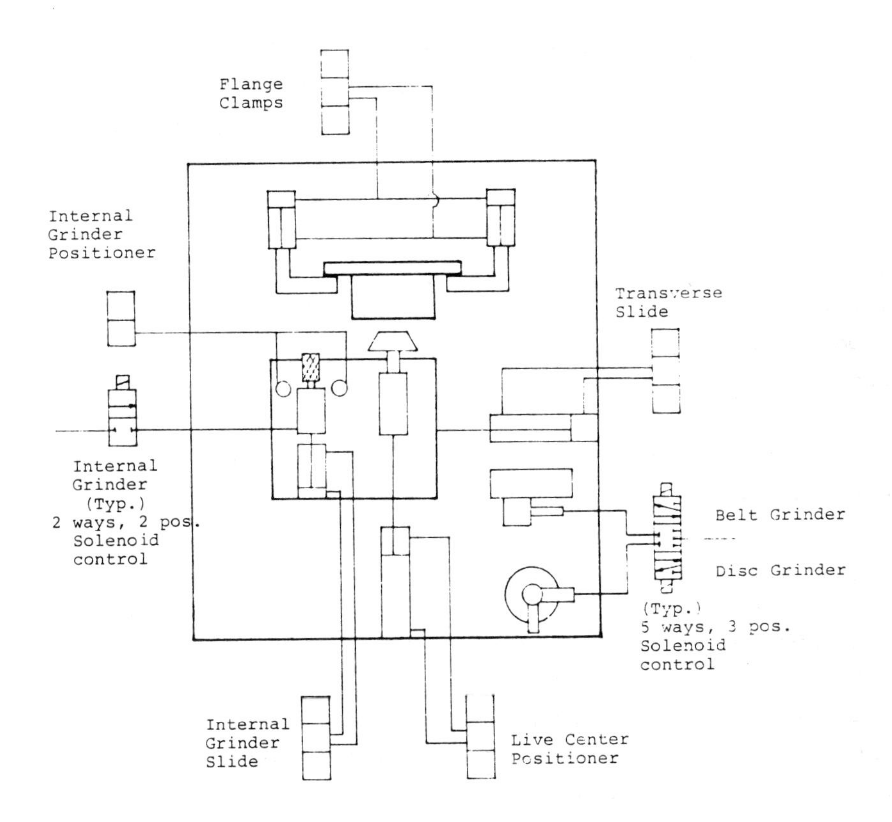

Figure 1 Work Station Layout

flashing, and an internal grinder removes sand inclusions and particulate material on the inner diameter. Examples of methods of implementing these types of operations have been found in three articles (1,3,4). These articles proved to be instrumental in defining the specifics of the grinding operation. Specific examples of automated grinding operations were investigated (1,7) and these helped solidify some of the initial ideas. Many sources were used in the construction of the project, however, a problem arose when considering how to give the robot the dexterity it needed to perform the necessary operation.

The Question of Grippers

The design of the end-effector required for the project was a major consideration. The gripper had to be flexible enough to grasp the bearing carrier using the outside surface of the gripper, and also be able to manipulate the belt grinder and the disk grinder using the inside surface. The design had to be functional, yet, not too elaborate. Using the basic gripper configurations found in a review of the literature (9,11), a configurations inside diameter gripping and outside diameter gripping a dual function gripper was designed to accomodate the bearing carriers and the tooling. This design allowed enough flexibility so that the bearing carrier could be taken to almost every point in the working envelope of the robot, in any orientation.

Loading/Unloading

With flexible grippers the robot could load peripheral equipment such as internal grinders, CNC lathes or other machining centers. In the case of the LBCPP, it was desirable to load and unload the preconditioned bearing carriers onto an NC lathe. This could be accomplished by "rear loading", and example of which is given in the literature (2). using this operation and the preconditioning operation performed by the LBCPP, bearing carriers could be taken from the raw casting stage to the finished stage implementing a single robot program. Other machines such as internal grinders could also be loaded/unloaded in this manner but those possibilities, were not explored (6).

Adaptive Control: Sensors are needed

A weakness in the LBCPP is the amount of adaptive control used. In a grinding application it is inevitable that one will have to deal with tool wear. Adaptive control is the answer to this problem but can be difficult to implement. With no sensor feedback, the robot can be inefficient in an operation such as this. The metal removal rate is not controllable, nor is compensation for tool wear. Without some sort of sensory device, the robot cannot adjust itself to accomodate different conditions on a particular workpiece, for example an extra large inclusion.

Adaptive control methods, particularly for the T^3-586 robot, are presented (4) in the literature along with specific examples for the Cincinnati Milacron (8,12). It is highly recommended that these types of adaptive control be implemented in any other advanced type of grinding/fettling operations.

It should be mentioned that the Cincinnati Milacron is completely capable of providing these types of controls and is a suitable robot to perform projects like LBCPP given the necessary software.

System Description

The automated preconditioning system designed for Lufkin Industries consists of two major components. A Cincinnati Milacron HT^3-586 industrial robot which provides motion, control, and part orientation for the preconditioning, and a positioning fixture which provides clamping and peripheral equipment. These two components working together, complete the preparation of a fully preconditioned casting from a raw casting, straight from the foundry.

The Cincinnati Milacron robot provides complete control of the operation along with the parts handling and grinding functions. The robot used the internal processing unit to provide output signals which control various pneumatic cylinders and motors in sequence with the various operations being performed by the robot. The signals sent to the positioning fixture provide sequencing which allows the transfer of the bearing carrier from the robot to a secure position in the fixture. All part movement is done by the robot.

In addition to parts handling the robot executes two of the three grinding operations. The fact that a major portion of the grinding/preconditioning is being done by the robot and that the robot is readily reprogrammable adds to the flexibility inherent to this design. Adding to the flexibility the grippers on the robot are capable of providing multifunctional capabilities, that of picking up the bearing carrier in two positions and grasping both grinders.

The Process

There are basically two ways to feed the bearing carrier into the work space of the robot. One is through a pallet arrangement that must be resupplied from time to time. The other is by means of a conveyor that has an attachment which allows the bearing carrier to be confined to a certain area on the conveyor allowing the robot to pick it up at the same place everytime. For the sake of demonstration this application was done using a pallet and the "Index" function in the robot software.

The beginning of the operation occurs with the picking up of a bearing carrier off of the supply pallet. The robot uses the outer most grippers along the axis of the hand to pick up the bearing carrier by expanding the jaws to clamp the internal diameter.

After having grasped the bearing carrier the robot transports it to the orientation station. At the orientation station the robot uses the roll axis of the carrier to turn the casting against a limit switch which is fastened to the table. The limit switch rides on the outside surface of the casting and completes a circuit which in conjunction with the "perform on interrupt function" sets the part down whenever the larger of the two slots running along the axis of the bearing carrier is encountered. The wheel on the limit switch arm is of sufficient size to allow it to fall into the large slot, but will not allow the wheel to fall into the smaller slot. This allows the robot to distinquish one side from the other and determines part orientation. If the robot completes the roll axis movement prior to tripping the limit switch it will put the casting down, index the roll axis, and pick the bearing carrier up. It then

8. Mortensen, A., "Automatic Grinding", ISIR/Robots, Vol. 1, Chicago, Illinois, 1983, pp. (8-1) - (8-11).

9. Multer, R.F., "Effective Interfacing Through End Effectors", ISIR/Robots 7, Vol. 1, Chicago, Illinois, 1983, pp. (4-1) - (4-11).

10. Oitinger, L.V., Stauffer, R.N., "Update on Robotic Standard Development", Robotics Today, October, 1983, pp. 25-29.

11. Reed, C.K., "Two Hands Are Better Than One", ISIR/Robots 7, Vol. 1, Chicago, Illinois, 1983, pp. (4-12) - (4-23).

12. Schraft, R.D., Schweizer, M., Abele, E., Struz, W., "Application of Sensor Controlled Robots for Fettling of Castings", ISIR/Robots 7, Vol. 2, Chicago, Illinois, 1983, pp. (13-44) - (13-57).

13. Van Blois, J.P., Andrews, P.P., "Robotic Justification: The Domino Effect", Production Engineering, April, 1983, pp. 52-54.

Robotic Assembly of an Electric Switch

Kenneth Hall

Gulf+Western Advanced Development and Engineering Center

Swarthmore, Pennsylvania, U.S.A.

SUMMARY

The paper describes how a robotic system can assemble a typical electromechanical product which functions as a switch.

Several system features are incorporated, including adaptive assembly to accommodate minor random product variants, a unique multi-function gripper which progressively assembles diverse components, and integration of the robot with other equipment.

The gripper concept efficiently performs the assembly while eliminating the need for external fixturing and significantly reducing robot travel.

INTRODUCTION

This paper describes part of a system designed and built to demonstrate robotic assembly of a typical generic electromechanical product.

Although not a production system, several typical features are incorporated, such as integration of the robot with other equipment, adaptive programming for product mix, conveyor transport with two assembly stations, component feeding stations, a fastener installation station, and a functional test station.

In its implementation, part of the assembly posed an interesting problem because of the interlocking nature of the components and led to a unique concept of progressive assembly within a multipurpose robot-mounted gripper.

PRODUCT AND TASK

The "product" is comprised of several components of a domestic circuit breaker which, in its simplified and modified from, functions as a switch. It is shown with and without the cover in Figure 1, and with components separated in Figure 2.

The task was to assemble the coil (1), lever (2) and handle (3) into the box (4), then install the cover (5) and secure it with a fastener (6). The terminal (7) and contact (8) were assumed to have been previously installed.

Assembly of components (1), (2) and (3) into the box was the interesting problem because of the interlocking nature of the three items as installed. It would be easy to place components (2) and (3) but installation of the coil is not then possible (see Figure 3).

ROBOT

The robot available was a fairly simple model having programmable motion about three vertical axes and a two position (up or down) motion at the tool axis.

GRIPPER DESIGN

For Internal Components

Although it would have been possible, and even conventional, to make a subassembly of components (1), (2), and (3) in a pre-fixture, a more satisfying solution evolved in the concept of progressive assembly within a multipurpose gripper.

The resulting design is depicted in Figure 4. The body has a fixed rib and incorporates separate pneumatically-operated fingers on each side. One finger (1) grips the coil and the other (2) grips the lever. At the end of the body is a pneumatically-operated plunger (3) which captures the pivot hole in the handle.

The principle of operation is shown pictorially in Figure 5. First the robot moves to position the coil gripper above the coil then lowers and picks it up (a). Next the lever gripper is positioned over the lever then lowered. With the lever gripper still open, the next robot movement inserts the coil into the lever aperture (b), after which the lever gripper closes (c). The coil/lever subassembly is then lifted and taken to a point adjacent to the handle. After lowering, the next robot movement inserts the coil into the handle aperture. After this the plunger is lowered into the pivot hole in the handle, thus capturing it (d). The coil/lever/handle subassembly is then transferred to the assembly station where a simple lowering into the box and simultaneous release of all three components accomplishes the installation.

Photographs of the gripper and component subassembly following each insertion are shown in Figure 6.

For Cover

The gripper design for cover installation is shown in Figure 7. The cover is picked up by a vacuum orifice with a resilient rubber pad (1). After installation of the cover at the second assembly station as shown in Figure 8, the outer pair of pneumatically-operated fingers (2) are activated to grip the body. The robot then transfers the assembly to the fastening station, where a screw is installed to retain the cover, and then to a discharge chute.

Complete Gripper

The grippers for internal components and cover installation are incorporated into an inverted-U assembly as shown in Figure 9 and are utilized as required by 180° rotation of the robot tool axis.

ADAPTIVE PROGRAMMING

To demonstrate robotic flexibility, boxes are identified by a label denoting a 10 amp or 20 amp rating and can be randomly introduced into the system. The labels are detected by fiber-optic photo-sensors and, by program branching, the robot is instructed to collect appropriate (10 amp for 20 amp) components from alternate feed stations, and to discharge the finished assembly at an appropriate exit chute.

FUNCTIONAL TEST

Prior to installing the cover, the handle is mechanically activated and a continuity check made across the two terminals. In the event of noncontinuity, another program branch instructs the robot to omit the cover installation and remove the incomplete assembly from the conveyor for rework.

FEEDERS

The component feeders are gravity-feed magazine type units with pneumatically-operated singulation escapements which are operated on demand by the robot controller.

SYSTEM EVALUATION

The system is shown in Figure 10. It operates reliably and demonstrates what can be achieved using a simple robot without recourse to complex electronics and mechanics.

The concept of assembly within a robot gripper has proved eminently practical and reliable. It efficiently accomplishes an assembly task that would otherwise be very difficult.

Although gripper design is necessarily tailored to the components involved, the concept offers the advantage of eliminating external fixtures and mechanisms.

Figure 1 Product to be Assembled

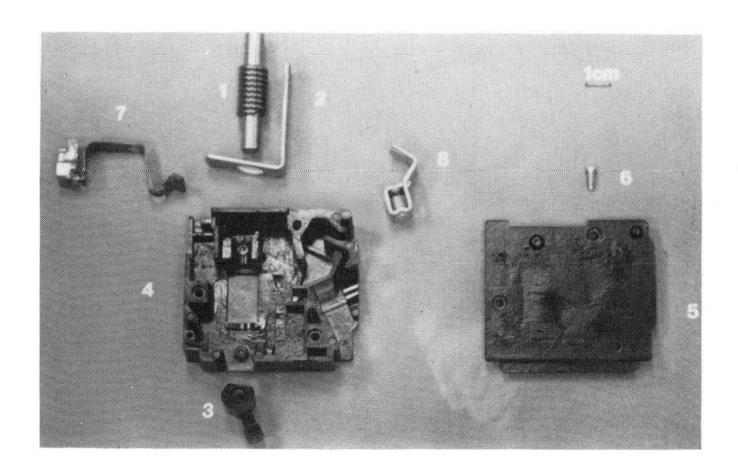

Figure 2 Product Components

Figure 3 Assembly Problem

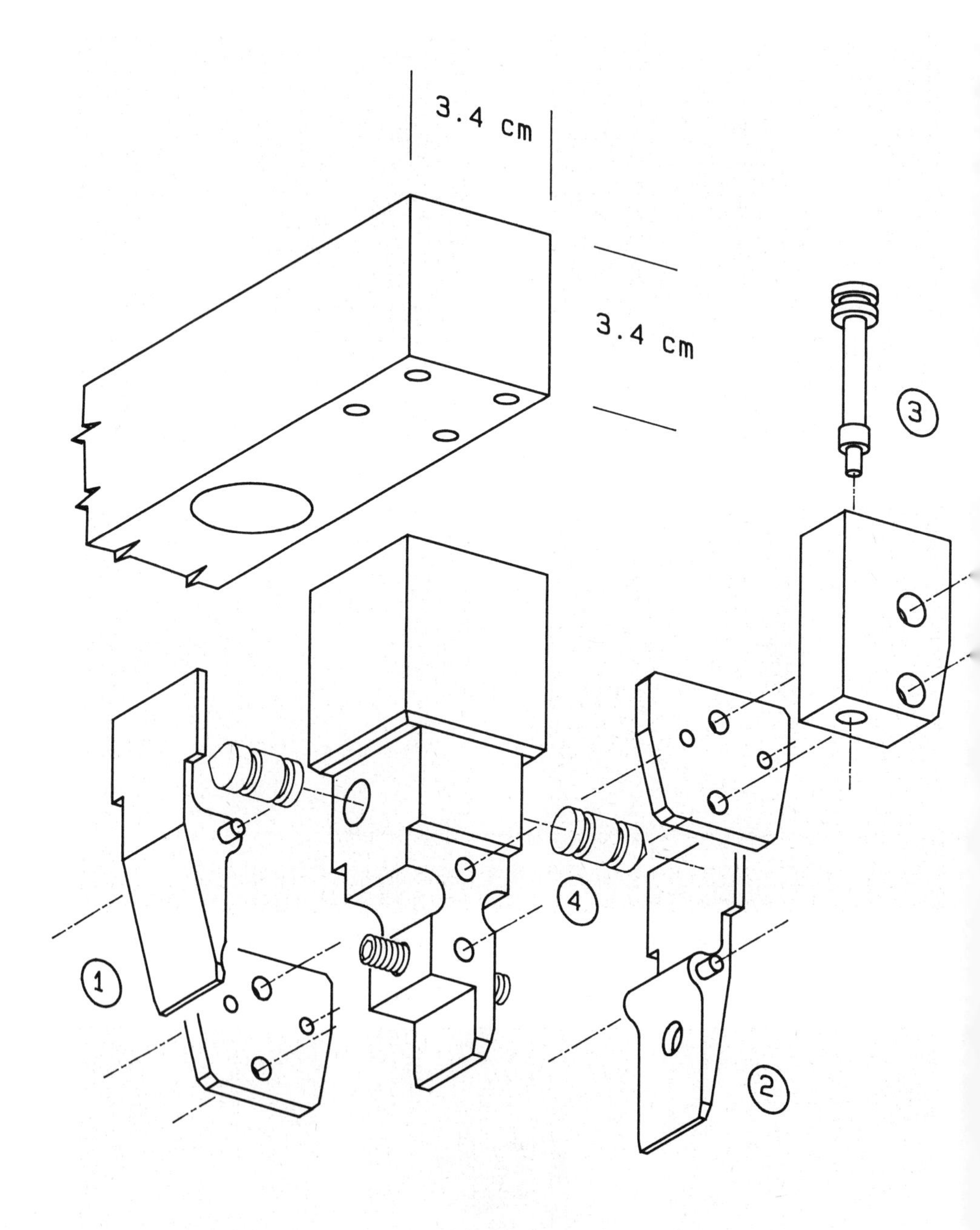

Figure 4 Progressive Assembly Gripper

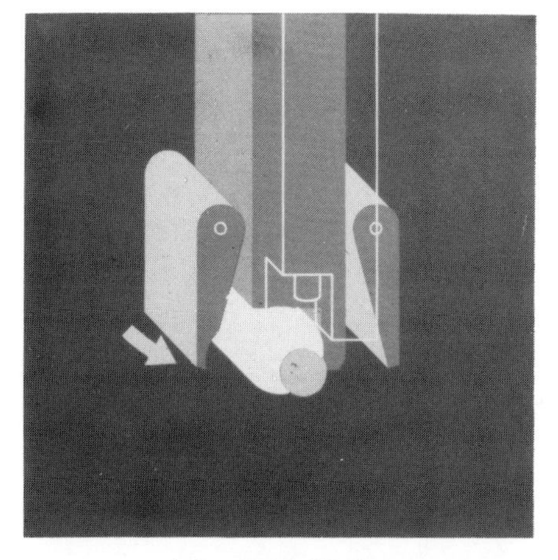

(a) Coil Pickup

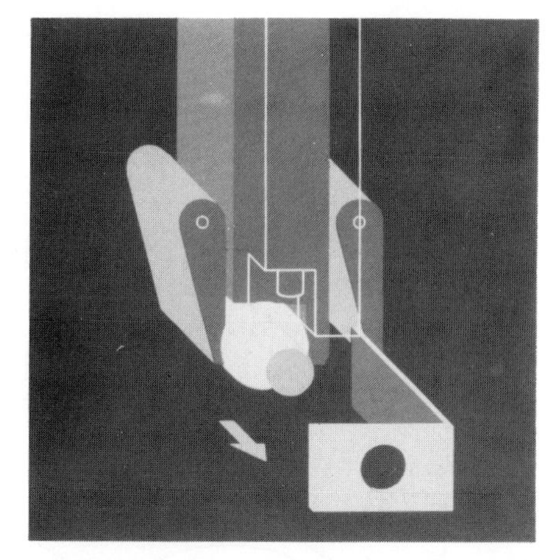

(b) Coil/Lever Insertion

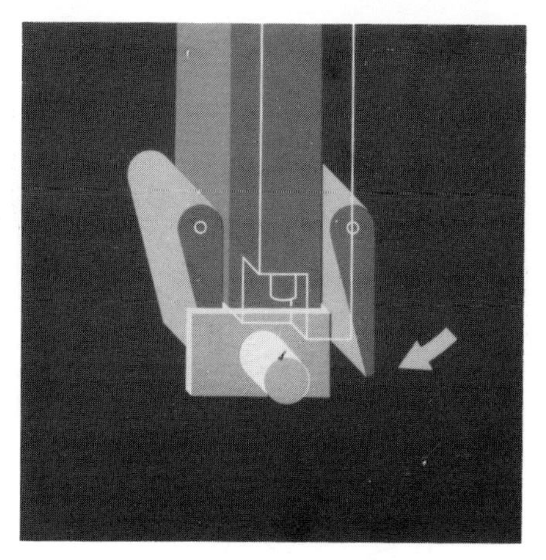

(c) Lever Pickup

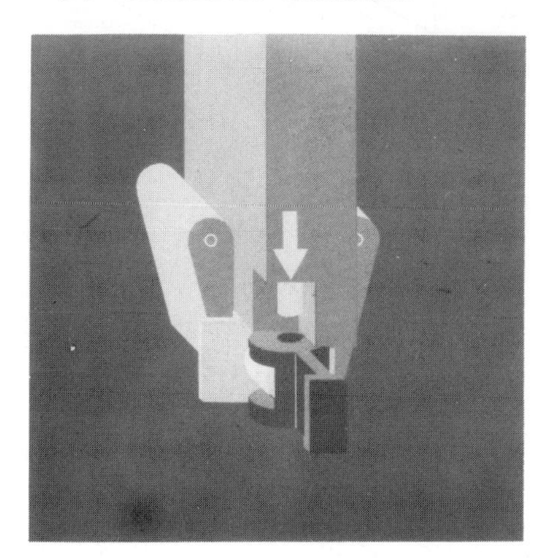

(d) Coil/Handle Insertion & Pickup

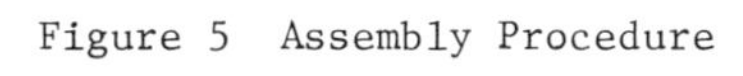

Figure 5 Assembly Procedure

(a) Coil

(b) Coil/Lever

(c) Coil/Lever/Handle

Figure 6 Progressive Subassemblies

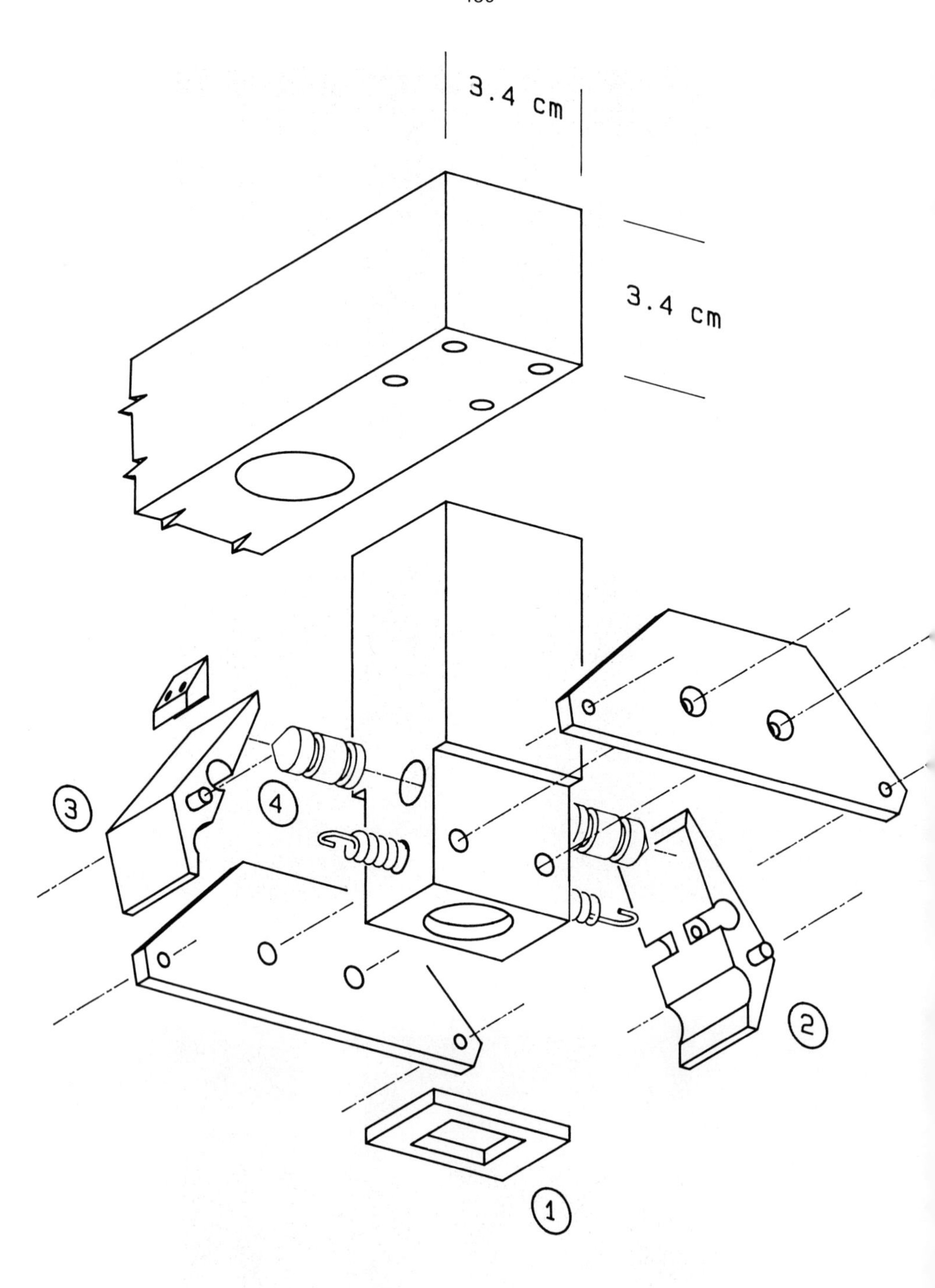

Figure 7 Cover Installation Gripper

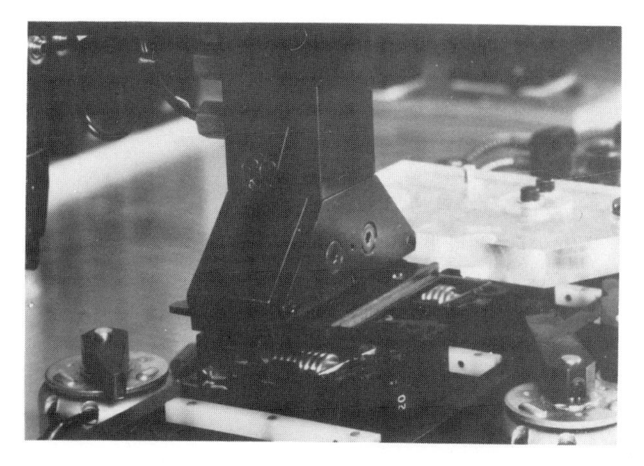

Figure 8 Cover Installation

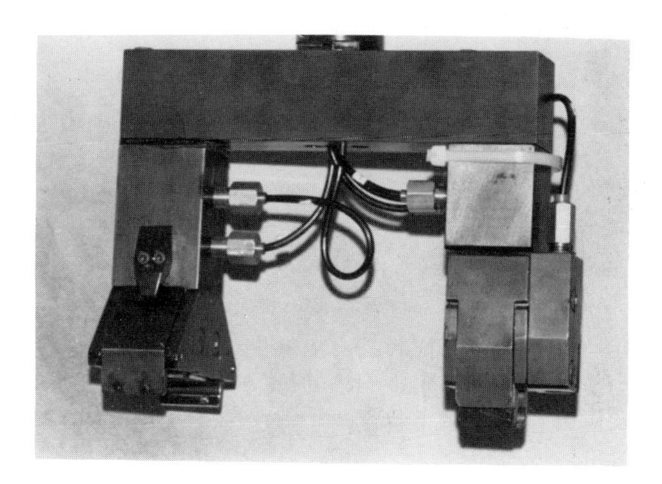

Figure 9 Complete Gripper Assembly

Figure 10 Assembly System

Robotics and Scheduling in Flexible Manufacturing System

V. Singh

Mechanical Engineering Department
Institute of Technology
Banaras Hindu University
Varanasi - 221 005
India

SUMMARY

Intelligent robots with senses may execute different sequences of operations. Robotized manufacturing systems (RMS) permit one to automate typical assembly operations in FMS to enhance the flexibility and efficiency of the system. Loading and sequencing, as a composite of interdependent tasks, constitute important aspect of scheduling problem in FMS. Mathematical formulations for scheduling problem with the objective of workload balance and job lateness are presented and solution methodologies are discussed. Applications of robotics in the present problem environment are indicated.

INTRODUCTION

Intelligent robot systems have the ability to adapt to their environment. This implies powerful real time computer systems and advanced sensors. The sophisticated robots should have visual and tactile sensors, various manipulators, multiped walking mechanisms and flexible hands with fingers. Such robots will be used where high-level decision-making capabilities are required to initiate appropriate swift, skillful and complex working operations. It may have many actuators to be controlled at once with the help of optimal-position control system. On the factory floor the economic justification, reliability, ease of programming and robustness are prerequisities. Industrial robots (IR) and flexible manufacturing systems (FMS) are the foundations of robotized manufacturing systems (RMS). It includes production preparation, planning, scheduling and control by computer. Tool changing in machine tools working in a FMS is on the whole automated but their delivery and insertion in the machine tool magazine is in the majority of cases done manually [3].

Fig. 1 depicts a typical FMS. Jobs are selected for a planning period and their routing is decided based on the objectives of loading problem. Movement of jobs on the main conveyor and to and from various machines can be controlled by robots. Similarly picking of tools from the tool bank, loading and unloading of tools can be carried out by robots at each machining centre. In this paper is presented the loading problem in random FMS with bi-criterion objective of balancing the workload amongst the machining centres and meeting the due dates of the jobs. Following Stecke [5], Shanker and Tzen [4], a mathematical model is developed and results are compared for exact formulation and heuristic procedures developed by them. With the help of a simulation model the effects of loading on system performance under different dispatching rules are examined.

LOADING PROBLEM IN FMS

Assuming that part type selection, machine grouping, production ratio and resource allocation problems have been solved, the loading problem is specified as selecting a subset of jobs from the job pool, and assigning their operations to the appropriate machines in the ensuing planning period so as to achieve certain specified objectives as much as possible while meeting the system constraints.

Formulation :Consider a FMS with n machines. Each machine has a known tool slot capacity of its tool magazine. The jobs arrive in a random order. For a given scheduling period, assume that there are m jobs to be loaded. For each job, the processing time, tool slot requirement, due date are known with certainty.

Nomenclature

Subscripts : i = job, k = operation, j = machine.

H = length of scheduling period;

p_{ikj}, s_{ikj} = processing time and number of slots required respectively for processing operation k of job i on machine j ;

y_i = number of operations required by job i ;

$B(i, k)$ = set of machines;

t_j = tool slot capacity of machine j;

w_{di}, w_{oj}, w_{uj} = relative weight of job i for meeting the due date, overload and under load respectively on machine j;

O_j, U_j = overload and underload on machine j;

D = parameter;

Z = number of slot duplication;

R_i = remaining time of job i;

$x_i = \begin{cases} 1 & \text{if job i is selected,} \\ 0 & \text{otherwise;} \end{cases}$

$x_{ikj} = \begin{cases} 1 & \text{if operation k of job i is assigned on machine j,} \\ 0 & \text{Otherwise.} \end{cases}$

Constraints

The constraints necessary for loading are developed individually.

(a) Tool slots $\sum_{i=1}^{m} \sum_{K=1}^{Y_i} s_{iKj} x_{iKj} \leq t_j$

The possible duplication and/or overlap for the tool slots can be incorporated into the tool slot constraint as :

$$\sum_{i=1}^{m} \sum_{K=1}^{Y_i} s_{iKj} x_{iKj} - \sum_{i_1=1}^{m-1} \sum_{i_2=i_1+1}^{m} \sum_{K_{i1}=1}^{Y_{i1}} \sum_{K_{i2}=1}^{Y_{i2}} (Z_{i_1K_{i1}; i_2K_{i2}})(x_{i_1K_{i1}j})(x_{i_2K_{i2}j})$$

$$+ \cdots + (-1)^{p+1} \sum_{i_1=1}^{m-p+1} \sum_{i_2=i_1+1}^{m-p} \cdots \sum_{i_p=1}^{m} \sum_{K_{i1}=1}^{Y_{i1}} \cdots \sum_{K_{ip}=1}^{Y_{ip}} (Z_{i_1K_{i1}; i_2, K_{i2}; \cdots i_p, K_{ip}})$$

$$(x_{i_1K_{i1}j}) \cdot \cdots \cdot (x_{i_pK_{ip}j}) \leq t_j \quad ; \qquad (1)$$

(b) Unique Job Routing $\sum_{G \in B(i,K)} x_{iKG} \leq 1$ (2)

(c) Non-Splitting of the Job $\sum_{K=1}^{Y_i} \sum_{j=1}^{n} x_{iKj} = x_i \cdot Y_i$ (3)

(d) Machine Capacity $\sum_{i=1}^{m} \sum_{K=1}^{Y_i} P_{iKj} x_{iKJ} + U_j - O_j = H$ (4)

$U_j > 0, \; O_j > 0$

(e) Integrality of Decision Variables $x_i = 0$ or 1 (5)

$x_{ikj} = 0$ or 1

$i = 1, 2 - - -, m$

$j = 1, 2 - - -, n$

$k = 1, 2, - - -, y_i$

Objective Functions

(a) Balancing the Workload

$$\text{Minimize } Z_1 = \sum_{j=1}^{n} W_{oj} O_j + \sum_{j=1}^{n} W_{uj} U_j \qquad (6)$$

(b) Balancing the Workload and Minimizing the number of Late Jobs

Minimize $$Z_2 = \sum_{j=1}^{n} W_{oj} O_j + \sum_{j=1}^{n} W_{uj} U_j - \sum_{i=1}^{m} \frac{W_{di} X_i}{\max(D, R_i - 2H)} \qquad (7)$$

where $x_i = \begin{cases} 1 & \text{if job } i \text{ is already late,} \\ 0 & \text{otherwise} \end{cases}$

SOLUTION METHODOLOGY

The above formulations contain nonlinear terms which can be linearized. Heuristic algorithms proposed for the two formulations are

A. Loading Algorithm 1 (LA1), for Balancing the Workload

(1) For each job, consider all possible routes and for each route, sum up the processing time on each machine.

(2) For each machine, rearrange the route-processing-time (RPT) in descending order. If tie, the job with higher overall total processing time will be placed first.

(3) From the machine with the most remaining capacity, select the first job from the list in step 2.

(4) If both feasible and acceptable, select the job on that route, and delete all the RPT that relate to the job selected. Otherwise, delete the RPT from the list of the machine being examined. The job, however, will be retained on the RPT lists of other machines.

(5) If RPT list is exhausted, go to step 6, otherwise update the remaining capacity of all the related machines by subtracting the processing time of the job selected from the previous values of the remaining capacities and go to step 3.

(6) Do pairwise interchange between the sets of selected and unselected job by removing one selected job out and moving one unselected job in.

B. Loading Algorithm 2 (LA2), For Balancing Workload and Minimizing the Number of Late Jobs.

(1) Classify jobs into four classes according to due dates as overdue (date over), very urgent (Within 2H), urgent

(between 2H and 3H), and normal (beyond 3H).

(2) Overdue jobs are selected on the first priority. If a job has more than one route, the route with the maximum total remaining capacity on the machines it will pass through, should be selected.

(3) Treat very urgent jobs according to the LA1.

(4) Urgent jobs will be arranged and selected together with the normal jobs by the same heuristic for balancing the workloads, except that to their RPT will be added weights depending upon the relative importance of meeting the job due dates.

SYSTEM DESIGN

The scheduling problem in FMS comprises of allocation of machines to the jobs and sequencing of jobs on allocated machines. Based on informations [2] available from various sources, 4 machines are included in simulation model. An independent and identically distributed exponential inter-arrival time is selected with an arbitrary chosen mean of 70 minutes. The batch size is assumed to be uniformaly distributed between 5 and 15. The due dates of jobs are assigned according to TWK method [1] using the definition, $D_i = A_i + KP_i$, where for job i, D_i is due date A_i is arrival time, P_i is total process time and K (≥ 1) is a parameter. The number of operations for each job is assumed to be uniformaly distributed over 1 to 3. Each operation may have upto 3 alternative machines to process it. Processing time for each operation is assumed to be uniformly distributed over 6 to 30 minutes. 5 slots are assumed in each magazine. The percentage distribution of tool slots needed by each tool in terms of number of slots 1, 2, 3 is 80, 15, 5 respectively. Each operation needs only one tool and a 5% of tool duplication or tool slots overlap is assumed. Scheduling period considered is 8 hours, equal to one normal shift of a working day. Machine utilization is taken as the criterion of system performance.

The five loading policies examined are :

(1) Loading in the sequence of job arrivals.

Loading for workload balancing using solution obtained, (2) analytically from IBM MIP/370 package, and (3) from heuristic algorithm LA1 respectively.

Loading for workload balancing and minimization of the number of late jobs using solutions obtained, (4) analytically from IBM MIP/370 package, and (5) from heuristic algorithm LA2 respectively.

Loading policies (1) and (3) are further examined with dispatching rules (a) FIFO (first-in-first out), (b) SPT (shortest processing time first), (c) LPT (Longest processing time first), (d) MOPR (most operations remaining first).

CONCLUSIONS

It is observed that a random FMS requires balanced workload to achieve better system performance. The heuristic method proposed seems appropriate to this point. The rule SPT, which is a standard dispatching rule in many situations, performs the best on an average. An experimental robot system can be assembled to simulate the operations discussed here. Interactive, master and slave modes of operations are possible. To perform unpredictable tasks, semiautonomous robots manipulated by human operators may also be planned.

ACKNOWLEDGEMENT- The author thanks Prof. K. Shanker [4] for his assistance.

REFERENCES

1. Conway, R. W., Priority dispatching and job lateness in a job shop, The journal of industrial engineering, July 1965, pp. 228-237.

2. Ito, Y., Japanese FMS - Present and a future view. McGraw Hill, 1981.

3. Panov, A. A., Production improvement through the introduction of flexible manufacturing systems, Soviet engineering research, Vol. 3 No 7 1983, pp. 50-52.

4. Shanker, K., and Tzen, Y. J., Loading and dispatching problems in random flexible manufacturing system, Technical report No.. IE & M/55, March 1984, AIT, Bangkok.

5. Stecke, K.E., Formulation and solution of nonlinear integer production planning for flexible manufacturing systems, Management Science, 29, 1983, pp. 273-288.

Table 1 : Job Descriptions for Simulation (first production cycle)
Descriptors :

JN=Job Number, AT=Arrival Time, DD=Due Date, BS=Batch Size, ON=Operation Number, UPT=Unit Processing Time, SN=Slot Needed, RN=Route Number, MN=Machine Number, TDO=Tool Slot Overlapping or Tool Duplication.

JN	AT	DD	BS	ON	UPT	SN	RN	MN	TDO
1	122	698	8	1	18	1	1	3	
2	266	2822	9	1	25	1	1	1	
							2	4	
				2	24	1	1	4	
				3	22	1	1	2	J502R1
3	278	2202	13	1	26	2	1	4	
							2	1	
				2	11	3	1	3	
4	361	1153	6	1	14	1	1	3	
				2	19	1	1	4	
5	391	2083	9	1	22	2	1	2	
							2	3	
				2	25	1	1	2	J203R1
6	396	2156	10	1	16	1	1	4	
				2	7	1	1	4	
							2	2	
							3	3	
				3	21	1	1	2	
							2	1	
7	447	3087	12	1	19	1	1	3	
							2	2	
							3	4	
				2	13	1	1	2	
							2	3	
							3	1	
				3	23	3	1	4	
8	476	3388	13	1	25	1	1	1	
							2	2	
							3	3	
				2	7	1	1	2	
							2	1	
				3	24	3	1	1	

Table 3 : CPU time* and Workload Unbalance for Policies 1 and 3
Sequential Loading: Policy 1 Heuristic Loading : Policy 3

Run	Unbalance (min.)	CPU (sec.)	Unbalance (min.)	CPU (sec.)
Average	376.7	0.097	225.2	0.6140

Table 4 : Machine Utilization for Policies 1 and 3

Policy 1 (Policy 3)

Run	FIFO	SPT	LPT	MOPR
Average	0.722 (0.802)	0.707 (0.844)	0.673 (0.798)	0.699 (0.840)

Table 2 : Results of the 5 Loading Policies in First Production Cycle

	Loading Policy				
Item	1	2	3	4	5
CPU (Sec.)*	0.049	41	0.2	11.2	0.3
Tool Slots Used:					
Machine 1	2	4	1	2	5
2	3	4	4	2	3
3	4	2	3	4	2
4	4	4	5	3	3
Workload Unbalance:					
Machine 1	45	12	270	45	248
2	141	34	13	13	141
3	123	11	12	54	252
4	234	24	70	10	75
Total System Unbalance	543	81	365	122	761
Jobs Selected	1,2,3,5,6	1,5,7,8	4,5,6.7	1,2,4,5,6	1,2,4,5,8
Late jobs	4	4	1	-	-

* CPU at IBM 3031, AIT Bangkok

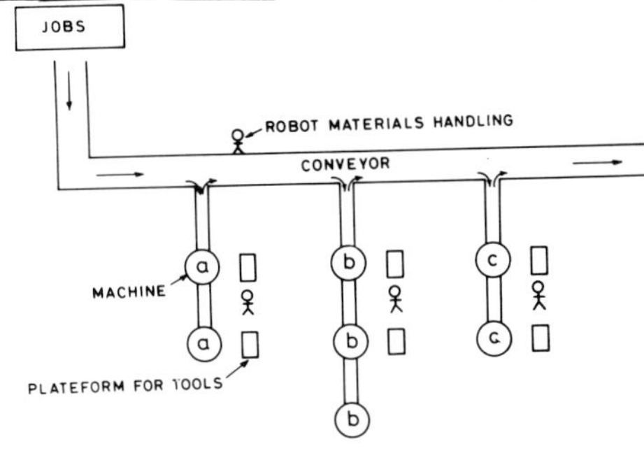

FIG.1 - A TYPICAL F M S

Abstract-Assembly Type Industrial Robots Equipped with an Intelligent, Self-Adjusting Tool-Changing End Effector Perform a Variety of Assembly Tasks in One Robot Cell

Mathew Monforte

Monforte Robotics, Inc.
2333 Whitehorse-Mercerville Road
Trenton, New Jersey 08619

Summary

A series of flexible intelligent robotic tool-changing end effectors able to perform an endless variety of assembly tasks in one robot work cell is described. In order to achieve multiple assembly tasks effectively, the end effector system must be able to automatically reconfigure itself to duplicate tasks normally performed by humans, multiple robot cells, or multiple automatic assembly equipment. In addition, the system must be able to sense a variety of size parts, exert a predictable amount of force for each part or tool handled, and have the intelligence to detect errors in cell functions and respond in a variety of predetermined and desired manners. The systems described are a series of pneumatically operated end effectors with quick retooling capability and a range of sensors and features capable of duplicating a variety of assembly tasks commonly found in industry.

INTRODUCTION

Typical industrial assembly processes involve multiple tasks performed on one product in various states of production. The introduction of robotics commonly demands one task per robot since industrial robotic end effectors are usually custom designed to perform a specific task in the assembly process. Often, a change in product design or model will warrant removal or replacement (or both) of the end effector to be compatible with current running configurations. Sensor capability designed into each end effector should be considered an aid in monitoring cell functions. The principal advantage of using robotics in assembly over other traditional methods is versatility. Although the capabilities of a robot in the work cell are limited to functions performable by the current end effector in use, this paper deals with The FOREMAN Hand, a tool-changing end effector system developed by Monforte Robotics, Inc., in Trenton, New Jersey, which is capable of performing multiple, diversified tasks intelligently, in one robot work cell, using only one robot. Over 70% of products manufactured in the United States are produced in batches. Although the robot can easily run a variety of motion programs, the end effector is almost always dedicated and must be retooled to meet current assembly line requirements. The FOREMAN Hand eliminates this consideration by not only utilizing a large set of uniquely styled adapter plates designed to fit onto a universal hand, but changes the different adapter plates by itself according to the changing needs of the work environment.

DESIGN CONSIDERATIONS AND GUIDELINES

In order for a multifunctional, intelligent end effector to be practical and effective, the following guidelines were taken into consideration when designing the FOREMAN Hand.

The end effector must

A. be durable in construction, but light in weight to be compatible with the largest selection of robots.
B. retain as much accuracy as possible from the host robot by having the fingertip to mounting clearance as small as possible.
C. be able to exert a maximum gripping force attractable to higher payload robots.
D. be able to grasp a variety of parts or tools using desired gripping forces where needed.
E. be able to acquire work pieces or tools in a minimum of six distinct modes of pick-up.
F. be able to acquire work pieces or tools larger than maximum finger opening.
G. be able to retool itself automatically under robot control.
H. be able to work with parts or tools in a work cell in a manner similar to custom-designed end effectors.
I. be able to engage or disengage custom toolings or work tools in less than ½ a second.
J. have a cost-effective retooling method.
K. have an interface method compatible with the largest selection of robots.
L. be as user friendly as possible.
M. have a selection of sensors to monitor gripper duties.
N. have a selection of features to perform complex assembly.

In order to appeal to a wide range of assembly operations, The FOREMAN Hand's sensors and features were designed in a modular fashion. Taking all of these factors into consideration, Monforte Robotics designed and developed The FOREMAN Hand. The final design outcome was a series of 12 end effectors, all meeting the design guidelines (this paper only encompasses the end effectors incorporating the highest level of sensors and features), with common internal mechanisms, sensors, and other features; only added were predicted assembly tasks. The series also offers an upgrading feature. This, then, is The FOREMAN Hand.

GENERAL DESCRIPTION

The FOREMAN Hand's main frames were designed of magnesium because of its strength to weight ratio. Total weight of end effector ranges from 2.45 lbs. to 3.35 lbs. Using one pneumatic cylinder, the fingers operate in a true parallel fashion via a rack and pinion. Two body sizes were designed to accommodate the full range of sensors and features (Figs. 1a, 1b).

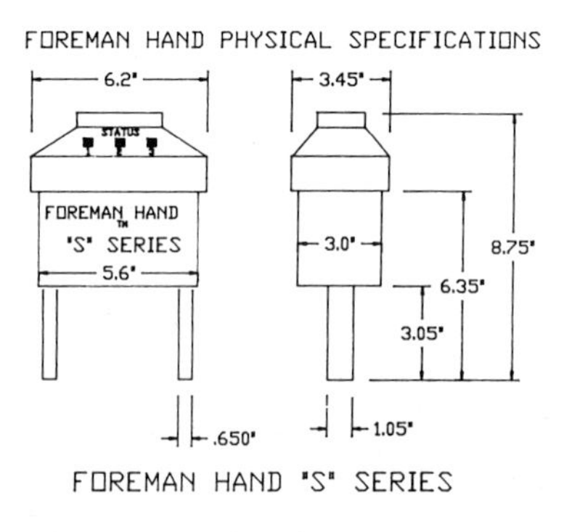

FOREMAN HAND "S" SERIES

Fig. 1a

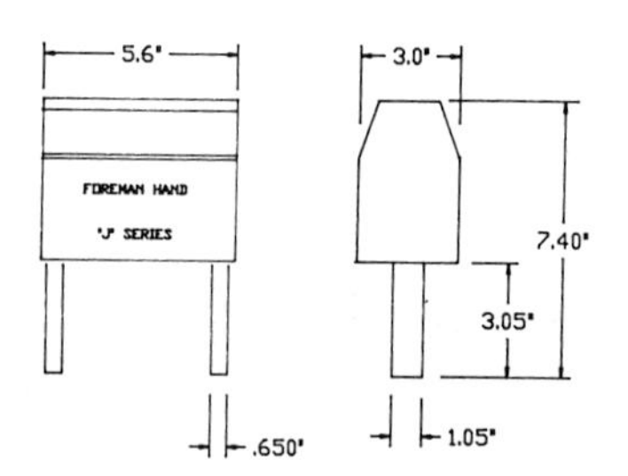

FOREMAN HAND "J" SERIES

Fig. 1b

SIX MODES OF PICK-UP, TOOL-CHANGE SENSING

Expanding the capabilities from a two-finger I.D./O.D. end effector to a multifunctional type was achieved by the introduction of adapter plates.

FOREMAN HAND ADAPTER PLATES

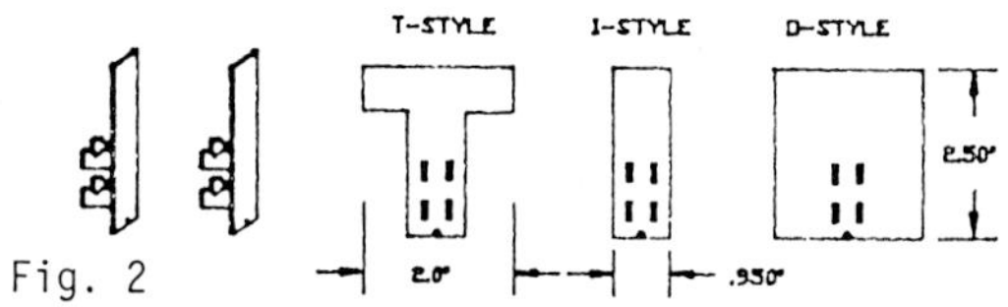

Fig. 2

The MRI adapter plates are the key to the uniqueness of the FOREMAN Hand quick-change tooling system. Offered in three styles, T, I, and O, adapter plates can easily be configured to meet custom application needs.

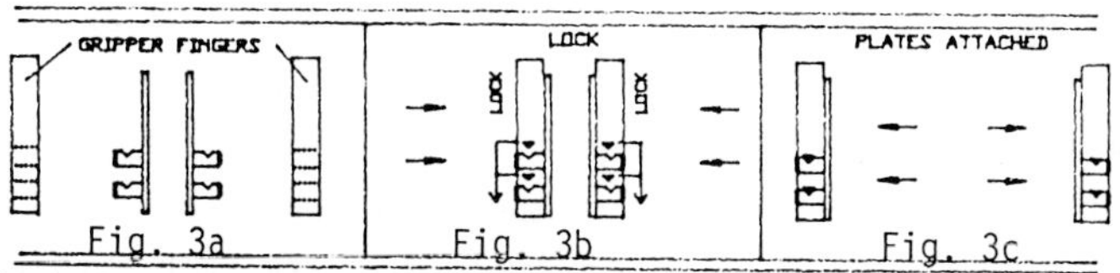

Fig. 3a Fig. 3b Fig. 3c

IN LESS THAN ½ OF A SECOND, FOREMAN Hands can attach or release a set of MRI adapter plates to either the inside or outside of the gripper's fingers. A signal from the host robot I/O activates the pneumatic locking mechanism (requires 65-85 psi lubricated or non-lubricated air). Once locked, the adapter plates are mechanically seated to a fixed position. Optional optic sensors located inside the fingers verify adapter plate acquisition and proper position.

FOREMAN SIX MODES OF PICK-UP

Fig. 4

UNLIMITED TOOLING ARRANGEMENTS

CONFIGURE ADAPTER PLATES TO MEET YOUR APPLICATION NEEDS

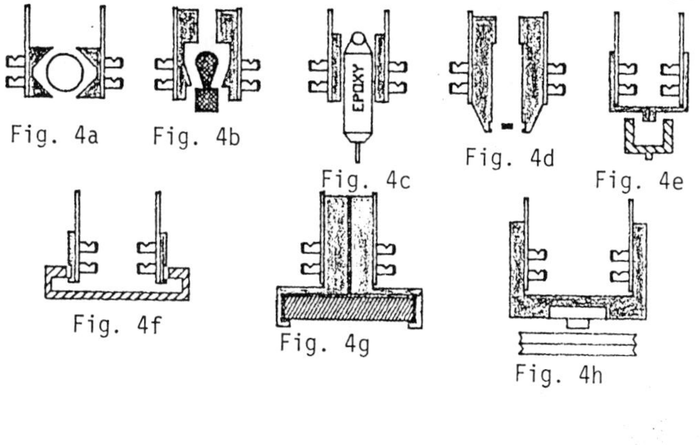

EPOXY

Fig. 4a Fig. 4b Fig. 4c Fig. 4d Fig. 4e

Fig. 4f Fig. 4g Fig. 4h

(See Fig. 2. Adapter plates with prongs or keys were designed to be compatible with the thru-holes pattern in each finger. Once inserted, either into the inner or outer surfaces of the finger (see Fig. 3), the four prongs, or keys, are locked and held in place (see Fig. 3b) via a pneumatically driven mating device in each finger. Because of The FOREMAN Hand's ability to acquire the adapter plates, the use of the end effector in the work cell is greatly expanded. Once attached, the adapter plates can either sweep with the motion of the fingers (Figs. 4a, 4b, 4d, 4e, 4f, 4g) for O.D., I.D. type pick-ups, or the plates can be mounted to equipment for a clamp type pick-up (Figs. 4c, 4h). The adapter plate technique offers the flexibility of six distinct modes of pick-up. Optical sensors in each finger verify acquisition and proper positioning of each adapter plate. The signal generated from the sensors can either halt a program in the case of an error, or let a program proceed for a successful pick-up. A typical system cycle to complete the process with sensing verification is as follows:

1. Ready for pick-up. (Fingers open for I.D. acquisition, closed for O.D. acquisition; see Fig. 3a.)
2. Position finger over adapter plate prongs (or keys).
3. Activate pneumatic locking device (Fig. 3b).
4. Monitor sensors status (acquisition position).
5. Move fingers to ready position--tooling attached (Fig. 3c).

Time tests for the entire cycle described are typically 425 milliseconds for acquisition and 450 milliseconds for releasing. End users of the FOREMAN system need only to custom design and attach to adapter plate face configurations compatible to assembly tasks. O.D., I.D., large, small, power tool, multiple parts, and a variety of others were built and successfully tested. A positive position dead block arrangement designed to upper faces have also been proved to be quite useful in handling delicate objects (Figs. 4b, 4D, 4g). Since each part handled has a custom type pick-up, finger positions can easily be controlled mechanically.

A variety of sensors and features were added to the end effectors to aid in monitoring cell functions.

An infrared emitter detector arrangement located .100" from the fingertips acts as a non-contact sensor (Fig. 5). Opaque objects interrupting the beam produce a signal verifying part in grasping range (Fig. 5b). Standard thru holes in the adapter plates permit the use of this sensor with plates attached to fingers. Rhomboid prisms, fiber optics, etc. mounted to adapter

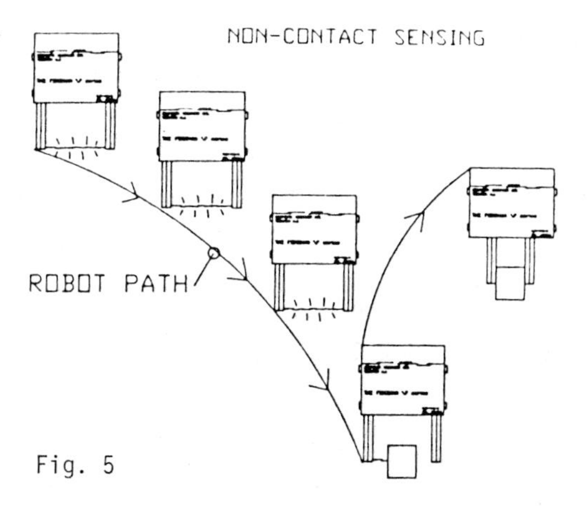

Fig. 5

NON-CONTACT SENSING WITH TOOL KEYS ATTACHED

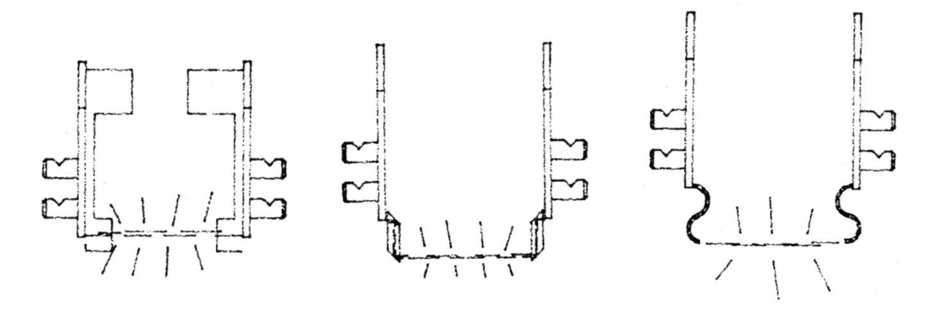

Fig. 6a. Non-contact sensor used with toolings attached.

Fig. 6b. Re-routing beam using Rhomboid Prism.

Fig. 6c. Re-routing beam using fiber optics.

FINGERTIP SENSING

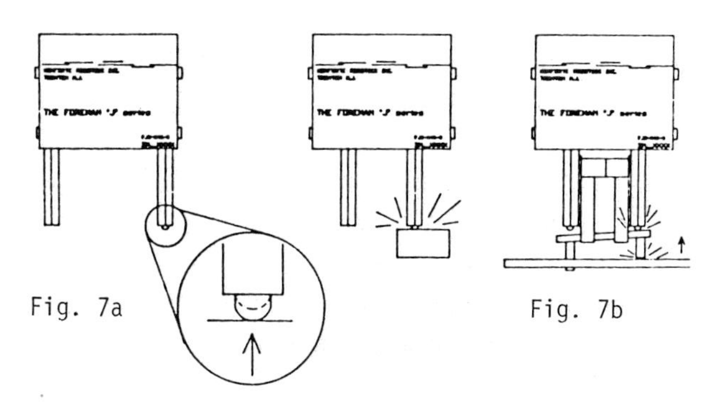

Fig. 7a

Fig. 7b

plates can re-route the beam to a variety of desired positions as assembly tasks warrant (Figs. 6a, 6b, 6c).

FINGERTIP SENSING

A button-type arrangement located at the fingertip ends provides fingertip tactile sensing (Fig. 7a, touch verification). Expansion of this sensor can be accomplished with the use of the adapter plates. Compliance built into plate toolings can in many instances monitor insertion of parts by acting upon forces during insertion (Fig. 7b).

GRIPPING FORCES

The input psi to the actuator is run thru a bank of valves and regulators directly interfaced to the Host Robot I/O. Users pre-set regulators to desired needs and can operate the valves' opening and closing via the robot program thru the I/O.

COMPLIANCE-COLLISION

The lower 2/3 of The FOREMAN end effector are built on a suspension, held in its base position by four expansion springs (Fig. 8). Eight V slots permit offset of lower portion at any angle (Fig. 8b). Optical sensors provide a signal to verify offsets in lower housings. Removal of forces causing the offset will permit returning of locator pins to their base position.

PROGRAMMABLE FINGER POSITIONS

A position sensor located in The FOREMAN Hand's upper housing reads finger positions. Under pneumatic control the fingers can be positioned and taught 256 positions throughout their 3-3/4" stroke. A teach pendant is used to jog fingers to a desired position (Fig. 9a). Once positioned, pressing the "teach" button on the pendant places the current position into The FOREMAN Hand Controller's memory. Up to 2048 positions can be held in memory per program with recall position repeatability of $\pm$.004". In a gauging fashion this feature is used as a contact tactile sensor. As The FOREMAN Hand is taught part sizes, it can be taught part sizes out of spec and respond to them as the Host Robot program dictates.

Status L.E.D.s on end effectors and controls were added to aid in program writing and program monitoring.

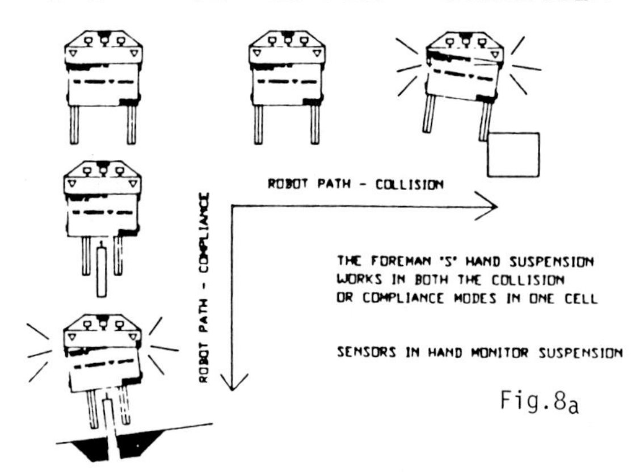

Fig.8a

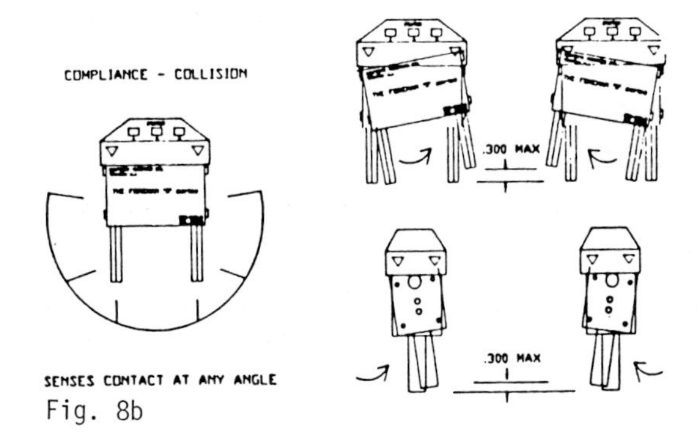

Fig. 8b

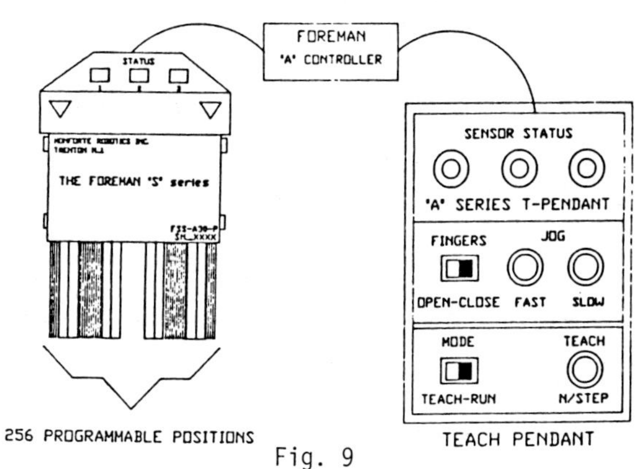

Fig. 9

To aid in minimizing down time, an audible alarm in controllers was added to alert operators via Host Robot program of unwanted sensor status, while stopping program running.

Most all industrial robots have available I/O's to access instructions to and from the program written. Although other methods do exist, this is the most common available. All sensors and most features in this system directly interface thru I/O's in a parallel fashion.

SUMMARY

When it comes down to which end effector to use, the main considerations are cost effectiveness and how user friendly the device is; in short, just what the end effector can do, how well it can do it, and how well it does this at what price. The FOREMAN Hand, developed by Monforte Robotics in Trenton, New Jersey, champions all these considerations: it can perform almost any task, provided with the proper adapter-plate fingertips; since it can adapt itself to a new task by changing fingertips, its work efficiency is extremely high. Going hand in hand with this efficiency is its cost effectiveness; since one robot can do the work of several, expense levels automatically plummet.

An ideal, practical application demonstrating The FOREMAN Hand's attributes is on a hand-held calculator assembly line. The conventional approach is to have a series of robots along the line, each dedicated to a specific task; glueing, screwing, etc. If such a study line required eight robots, with three system engineers to monitor the line, the total price for the whole system might add up to roughly $500,000. Add to this the estimated average downtime should one robot break down or a part feeder get stuck; the entire system would have to stop until a worker either fixed the glitch or stood in for the malfunctioning machine.

With The FOREMAN Hand integrated into this system, only three tool-changing robot cells would be required. The same gripper can pick up a power tool, manipulate it, change the fingertips to enable it to adapt to a glueing tool, manipulate the glueing tool, and so on. With one engineer and one technician, the total price for the system works out to approximately $250,000, half the price of the conventional set-up. The estimated output for The FOREMAN Hand gives a 28% lead over the conventional assembly line in terms of number of units manufactured per hour. Thus, in cost effectiveness, level of user friendliness, and in total practical effectiveness, The FOREMAN Hand stands head and shoulders above conventional assembly systems.

Practical Research and Development applications for The FOREMAN Hand demonstrate the advantage of The FOREMAN'S universal hand. Before The FOREMAN, when an alternate gripper was required, a new hand would have to be completely designed and constructed for each new application. The design would have to include new fingers, a mechanism for actuating them, designs for electronic sensors for the fingers; designing a way of mounting the finalized gripper onto the robot, and then providing for electric, pneumatic, and communication lines for the gripper to communicate to the controller

With The FOREMAN Hand, a single gripper is all that is necessary. Instead of having to redesign an entirely new gripper each time, all you have to do is manufacture an appropriate set of fingertips for the application. Because the nature of The FOREMAN Hand requires just one universal hand, all the extra considerations of redesigning the connections and monitoring systems are dispensed with; all that needs be redesigned are the fingertip adapter plates. As described before, time and cost are cut dramatically.

Among other practical applications, The FOREMAN Hand expecially improves the level and quality of research when working on robotic manipulation of materials. Cost effectiveness and time considerations are also especially important in educational applications. Students on a set time schedule must be able to utilize the quickest method of end effector changing. With The FOREMAN Hand, students can make the most of their time on the class robot, easily and effectively. They can merely go to the machine shop, design the particular sets of fingertips necessary for their projects, and let The FOREMAN Hand do the rest of the work.

Thus The FOREMAN is the ideal end effector for any application. It is the only end effector on the market capable of adjusting itself to changing demands and applications. When added in are its flexibility, lightweight design, quickness of operation, plus time efficiency, user friendliness, and cost effectiveness, The FOREMAN Hand is the logical choice for any system. It is truly "the only hand your robot will ever need."

to ensure that the end effector remains firmly locked in the quick change in the event of an interruption of either electrical or pneumatic power.

e. Physical Dimensions

The physical dimensions of the "quick change" are to be minimal. 7 1/4 inch diameter, 7 inch axial length and 20 lbs. weight are the maximum dimensions permitted.

f. Repeatability of Mating

Mating is to be repeatable sufficient to meet the misalignments discussed in A. Maximum Payload above.

g. End Effector Holster Design

The holster design must position the end effector with sufficient accuracy that the robot can successfully change end effectors (see Fig. 2 for photograph of holster). It must also provide compliance sufficient to compensate robot servo repeatability and programming inaccuracies. This compliance must have at least 1/8 inch X, Y, Z axial movement with less than 400 lbs. force buildup in any direction (Fig. 1a) and 2° tilt or more about X, Y, Z axis with less than 1,000 in-lbs of torque.

III. OVERALL MECHANICAL DESIGN AND LOCKING MECHANISM

The mechanical design consists of an "A" plate mounted on the robot wrist and "B" plate attached to the end effector (Fig. 3 shows detail on the "A" and "B" plates). The "A" plate contains the locking mechanism and acts as a manifold for the hydraulic and pneumatic fluids which enter radially. The "B" plate holds the taper for the locking mechanism. The electronics and fluid channels are arranged in a circle about the locking mechanism. Many combinations of electronics and fluid channels are possible. One arrangement is, for example, seven (7) 1/4 inch diameter hydraulic or pneumatic channels, three (3) 1/4 inch diameter locating pin and pneumatic channels and 38 electronics/fiber optics channels (using two (2) plugs of 19 channels each). The "quick change" is 7.25 inch diameter, 6.5 inch long and weighs 20 lbs.

The locking/unlocking mechanism must meet demanding requirements. It must be robust to withstand hydraulic back pressure. It must lock and unlock with exceptional reliability. It must be small and light and must operate on shop air (90 psi). (Fig. 4 shows the commercial device on which it is based "Rapidapt" by Younger Tool Corp.) Hydraulic fluid flowing through a channel in the "quick change" causes a reactionary force acting to separate the "A" and "B" plates (306 lbs per channel for 1,000 psi fluid pressure). If several channels are activated simultaneously, for example four (4), the total separation force will exceed 1,200 lbs. The locking mechanism shown in Fig. 4 has been tested safe to 6,000 lbs. (tested by John Vranish, Lew Ives NBS 11 Mar 1983) and thus has sufficient strength. The physical dimensions (Fig. 4) of 24 oz weight, 1.83 inch height and 3 inch diameter are satisfactory. What remains to be shown is how this commercially based, hand operated system can be augmented to lock and unlock reliably by a shop air powered actuator.

Kinetic energy principles are used to magnify the capabilities of a small air vane actuator to ensure reliable unlocking and firm locking. The discussion below gives a more detailed description.

Enroute to understanding the locking/unlocking operation it is useful to review the manual version (Fig. 4). In performing the unlocking process, the operator rotates the knurled ring counter-clockwise. The cam lock-pin assembly moves with it. The pin (from the cam lock-pin assembly) moves along the groove of the mating taper until it comes to the straight vertical section at which point the taper can be pulled straight out. In performing the locking process, the operator merely releases the knurled ring. Being spring loaded (22 lbs. at open), the ring accelerates clockwise, driving the pin of the cam lock-pin assembly up the grooved ramp in the taper. This forces the taper up into a close mate. This also forces the taper to rotate a few thousandths of an inch until the straight vertical section of its groove locates in rotation against the positive locking pin. Thus the taper (and tool attached to it) is firmly locked in. This system tends to stick in the locked position occasionally requiring use of a strap wrench. This is no problem for a human operator. For robot operation however, sticking must be eliminated.

Unlocking will be discussed first (Fig 5). A step function of air impacts the rotary actuator driving the steel ring in a counter clockwise direction. The steel ring, feeling no resistance except its own inertia and the friction between it and the graphite impregnated delrin ring bearing, accelerates. After 10° of motion it collides with the steel anvil shoe which is bolted to the cam locking-pin assembly. But this assembly is fixed to the locking pin which in turn is wedged in the ramp section of the taper groove. Thus, at collision the anvil shoe absorbes the kinetic energy of the rotating sleeve. This impact jars the locking pin loose and the system opens. The calculations below provide a first order estimate of the maximum impact force available for unlocking. The rotational inertia (I) of the rotary actuator and sleeve is calculated as 6.48×10^{-2} poundal in^2. Assuming a point collision, which occurs where R = 1.564 in., the rotational system can be set equal to a thin walled rotating cylinder ($I = mR^2$) where I = 6.48×10^{-2} poundal in^2 and R = 1.564 in. Such a thin walled rotating cylinder has an effective mass (m_{eff}) of 2.648×10^{-2} poundals.

Assuming no friction
$KE = 1/2\ Iw^2$.

The force generated by the collision between the rotating sleeve and the anvil shoe can be estimated to be a linear force in a direction tangential to the circumference of the rotating sleeve and normal to the surface of the anvil shoe.

Thus $KE = 1/2\ Iw^2 = 1/2\ m_{eff} \cancel{R^2} \frac{V^2}{\cancel{R^2}}$

R = 1.564 in

V = Tangent velocity vector of the thin walled rotating sleeve at impact.

a = Tangent acceleration vector of the thin walled rotating sleeve.

T = Torque of rotary actuator from product specification.

RO_i = Arc distance sleeve circumference moves to impact (Fig. 5).

$T = 16.1$ in-lbs for 70 psi air

$\frac{T}{R} = m_{eff}\, a \qquad a = 4{,}663 \text{ in/sec}^2$

$V = at; \quad R\theta_i = 1/2\, at^2 = 1/4$ in.

where $\theta_i = 10°$ (.17 radians)

$t = 10.35 \times 10^{-3}$ sec; $V = 48.3$ in/sec

The kinetic energy is absorbed by the sleeve and anvil shoe. Examination of these surfaces indicates that the collision is elastic.

Thus $KE = 1/2\, F_{dmax}\, \Delta\ell_1 + 1/2\, F_{dmax}\, \Delta\ell_2$

where

F_{dmax} = Maximum elastic deformation force.
$\Delta\ell_1$ = Elastic deformation of anvil
$\Delta\ell_2$ = Elastic deformation of sleeve

A conservative estimate is used and the total material compression is treated as twice that of the anvil ($\Delta\ell_1 = \Delta\ell_2$)

For elastic deformation

$\frac{F_{dmax}}{A} = Y_m \frac{\Delta\ell}{\ell}$

A = Area of impact ($.190 \times .125$ in^2)
Y_m = Young's Modulus (30×10^6)
$\Delta\ell$ = Material Deformation
ℓ = Length of deformed material before compression (1/4 in)

At maximum deformation

$$1/2\, m_{eff}\, V^2 = F_{dmax}\, \Delta\ell \text{ (anvil)} = \frac{\Delta\ell}{AY_m}(F_{dmax})^2$$

$$F_{dmax} = \sqrt{\frac{1/2\, m_{eff}\, V^2\, AY_m}{\ell}} \approx 2{,}700 \text{ lbs (impact)}$$

$$2{,}700 \text{ lbs} \gg \frac{16.1 \text{ in-lbs}}{1.5 \text{ in}} = 10.67 \text{ lbs}$$

Where 10.67 lbs is the force exerted by the rotary actuator at the point of collision if no impact is involved. However, the above calculations are made under the assumption of ideal frictionless elastic conditions. If 10% efficiency is achieved by the locking/unlocking mechanism the antisticktion force is increased by a factor of 80.

$$F_{dmax} = \frac{2{,}700 \text{ lbs}}{\sqrt{10}} = 854 \text{ lbs} \gg 10.67 \text{ lbs.}$$

Thus using kinetic energy techniques can increase available antisticktion force by a factor of 80.

Locking also uses kinetic energy techniques; but these are significantly different from the discussion above. A firm lock, without jamming, over a wide bandwidth of air pressures and mating forces is desired. During locking the robot exerts a downward force on the top of the locking pin. This downward force creates forces and friction between the pin of the cam lock-pin assembly and the top of the slot of the taper pin (Fig. 6). Thus when the rotary actuator is energized by a step function of air and begins its clockwise rotating motion, the cam lock-pin assembly is held in place by frictional forces. The sleeve accelerates the 1/4 inch gap + 1/8 inch additional gap (to be explained later) before it strikes the #10 set screw which is screwed into the lock-pin assembly (Figs. 5, 6). This impacts the cam lock-pin assembly, overcoming the static friction force and the pin starts sliding off the straight section of the taper groove (Fig. 6). After it moves .090 inch (measured), it floats free in space, and the spring of the locking mechanism accelerates the cam lock-pin assembly and shoe anvil faster than the sleeve.

$$a_{sleeve} = \frac{T\text{ (rotary actuator)}}{R_{(impact)} m_{eff(sleeve)}}$$

$$a_{sleeve} = 4.66 \times 10^3 \text{in/sec}^2$$

The "Rapidapt" comes with two springs that measure 11 lbs force each when the mechanism is rotated to open. These were replaced by a single spring, also from the manufacturer, which measured 8 lbs with the mechanism rotated open.

Thus the locking pin assembly with its light mass of 2.838 X 10^{-3} poundals is accelerated by an average spring force of 7 lbs (8 initially and 6 at lock) and has an acceleration of:

$$a_{locking\ pin} = \frac{7\text{ lbs }(12\text{ in/ft})}{2.838 \times 10^{-3}\text{ poundals}} = 2.960 \times 10^4 \text{in/sec}^2$$

$$\text{so } a_{locking\ pin} \approx 6\ a_{sleeve}$$

thus the locking pin accelerates the shoe anvil towards the sleeve hammer. It is perferable that the locking pin strikes the sleeve before it wedges into the slot of the taper pin. This makes the locking force more consistent. The distance in arc length that the shoe anvil of the locking pin must travel from the rear of the slot in the sleeve to the front is 3/8 inch as mentioned above (Fig. 5). Since the process of overcoming the static friction force of the pin (Fig. 6) is essentially an inelastic collision, the sleeve and the cam locking-pin assembly have the same initial velocity V_o when the cam locking-pin assembly begins its acceleration towards the sleeve hammer along the arc of the radius of impact. The equations below use the form $S = V_o t + 1/2\ at^2$ where distances are along the arc and velocities and accelerations are also along the arc (R = 1.564 in).

$$S_c = 3/8 \text{ inch} + S_s$$

S_c = Arc length cam locking-pin must travel to strike the sleeve.

S_s = Arc length sleeve travels

$$1/2\ a_c t^2 + \cancel{V_o t} = 3/8 \text{ in} + 1/2\ a_s t^2 + \cancel{V_o t}$$

a_c = Acceleration of cam locking-pin assembly (2.960×10^4 in/sec^2)

a_s = Acceleration of sleeve (4.66×10^3 in/sec^2)

V_o = Common velocity between sleeve and cam locking-pin assembly immediately after the static friction force described in Fig. 6 is overcome (calculated as 59.1 in/sec worst case using the valve a_s, the 3/8 inch distance the sleeve must travel from rest to begin the locking process and the relationships $S = 1/2\ a_s t^2$, $V_o = a_s t$).

t = The time it takes the cam locking-pin assembly to strike the sleeve hammer. It is calculated to be: $t = 5.48 \times 10^{-3}$ sec.

In this time the sleeve hammer will be located $S_s = 3/8 \text{ in} + V_o t + 1/2\ a_s t^2$ from the locked open position, or <u>.769 in = S_c</u>. Measurements have shown that the sleeve must move approximately 42° or <u>1.144 in + .03 in.</u> arc length (depending on taper variances) to lock.

So the anvil strikes the sleeve hammer slowing the locking pin to the speed of the sleeve. <u>Thus the speed (and energy) with which locking occurs is controlled regardless of mating force between robot and "quick change"</u>. Experiments have confirmed that the anvil does indeed hit the sleeve hammer before locking.

After the sleeve reaches the locking point it continues on for 1/4 inch more where it stops against a hard rubber washer thus saving rotary actuator wear and tear. The #10 set screw (Fig. 5) is set 1/8 inch away from contacting the rear of the sleeve slot so it is not "hammered" or "jammed" into the taper locking groove. A firm; but controlled locking motion results which gives approximately the same locking strength across a wide band of mating forces and air pressures.

Experiments confirm that the locking/unlocking mechanism will perform reliably at any air pressure setting from 45 psi to 90 psi and any robot mating force from 30 lbs to 100 lbs.

IV. <u>REPEATABILITY OF MATING ALIGNMENT</u>

Good repeatability of alignment between "A" and "B" plates permits a sophisticated vision sensor, such as the NBS 3-D vision system, to function at maximum efficiency and accuracy. The NBS system (Fig. 7) uses a flash box to illuminate an object with "structured" light and a camera to interpret the image illuminated by this light. The flash box is based on a periscope principle with the light generating source mounted on the "A" plate (common to all quick change - gripper combinations) and a periscope section (unique to each gripper) mounted on the "B" plate. The light slits in the flash box are .08 in wide so a rotational error in repeatability of ± .02 inch will reduce the light hitting the object by 10%. Tests show the rotational repeatability error is negligible.

Repeatability in tilt between the "A" and "B" plates can also effect the vision system. (Tilt means rotation about X and/or Y axis in Fig. 7). For the particular camera and gripper system used by NBS in its AMRF demonstration, a tilt error of ± .001 inch per 1 inch or greater will affect the camera's calibration. Using three (3) long locating pins (Fig. 8) with .010 inch ± .001 inch clearance, the worst case situation will bring the tilt alignment error to ± .3° before seating in the taper. Since we need a tilt error of ± .001 inch per 1 inch and since we have 1 inch sin .3° = .0052 inch per 1 inch we are off by a factor of 5.2. The taper (advertised to ± .0002 inch in Fig. 4) must make up the difference. (The robot will not help in this instance as its repeatability is only ± .025 inch.)

The following experiment was conducted: 20 matings using a 30 lbs. block as mating force. Repeatability of placing the 30 lb. block in the same position on the back of the "quick change" and thereby providing the same mating force direction is approximately ± 1/32 inch or ± .030 inch (the robot has ± .025 inch repeatability). Three dial gauges were mounted on the circumference of the "B" plate at 120° increments. Gauges 1 and 2 deviated by an average of .00015 in/in, gauges 1 and 3 by .0008839 in/in and gauges 2 and 3 by .00080 in/in therefore the plane of the "B" plate tilts less than .001 in/in as required. Note in Fig. 8b that for the 2nd generation "quick change" we intend to lower the effective height of the long pins and use them as 1/4 inch diameter air conduits while still maintaining their full alignment capabilities.

V. HYDRAULICS/PNEUMATICS

The NBS "quick change" system is capable of interfacing either hydraulic or pneumatic fluid through its modified 1/4 inch hydraulic connectors (see Figs. 2,3,9). Fig. 9b represents a modified 1/4 inch hydraulic connector as used in the prototype "quick change" and 9c shows the same basic 1/4 inch hydraulic connector further modified so as to be substantially shortened. In both cases, the modified hydraulic connectors have machine threads rather than pipe threads so that they can be located accurately on the "A" and "B" plates in all three dimensions. (Pipe threads, with their taper, do not permit the hydraulic fitting to be located accurately with respect to the "A" and "B" plate surfaces whereas straight machine threads do.) To make the location of the modified hydraulic connectors even more accurate a short straight shank section above the thread is added. The location accuracy of the modified hydraulic connectors is ± .001 inch. The "O" rings of the hydraulic connectors are replaced with Buna-N "O" rings reducing their insertion force to a manageable 10 lbs. per connector while lowering their specification from 10,000 psi to 5,000 psi (satisfactory). Clearance measurements between male and female hydraulic components is .007 inch to .010 inch and so the components mate quite easily by robot in all our tests. The double ball configuration of the hydraulic connectors prevents leakage during mating and unmating.

Finally, note that for high repeatability in quick change mating we ensure that the mating contact occurs on the taper surface shown in Fig. 5. Thus the male and female surfaces of the hydraulic connectors are .030 inch ± .005 inch from making full contact. This permits 80% maximum fluid flow rate. In one test hydraulic fluid at 1,750 psi was passed through the quick change without leakage. Two fittings were tested, one at a time. Pressure was limited to 1,750 psi because of the limitations of the hydraulic pump.

VI. ELECTRONICS AND FIBER OPTICS

The NBS electronics/fiber optics interface is based on a self-aligning commercial system shown in Fig. 10. (Multimate drawer connector by AMP, Inc.). AMP makes electronics and fiber optics pins to fit the holes in these connectors. Any combination of electronics and fiber optics is acceptable. On Sept. 26, 1983 these connectors and pins were cycled through 1 hour 45 min. of tests (120 cycles) along with the rest of the quick change with no problem. Their mating requirements are about ± 1/4 inch error in X and Y axis. These connectors are also very forgiving in mating depth (or Z axis) showing continuity along a .24 inch pin length.

VII. HOLSTERS AND GRIPPER STORAGE

Storing grippers around the robot waist is both space and time efficient. Figure 11 shows the configuration and illustrate the space efficiency. The outside faces of the holsters are open so that grippers can be inserted into and taken out with a minimum of motion. Thus, these holsters are also time efficient. Note that the gripper/holster mating surface is slanted at an angle to the horizontal. (This 15° angle is the minimum at which gravity will consistently locate the ears in the same place in the holster each mate/unmate cycle.) Also note (Fig. 12) that the shoes of the holsters which act as receptacles for the "B" plate ears have abbreviated 1/4 inch wide catches to hook the gripper. This 1/4 inch width is more than sufficient and permits easy and accurate programming of the robot in its mating and unmating sequence.

A powerful hydraulic robot has very little compliance when its wrist is held close into the waist, hence compliance was added to the holster (Fig. 13). The design considerations of the compliance system included: a) 1/4 inch or more compliance in displacement along the X,Y, Z axis' and 2° or more in rotation about X,Y,Z axis' to include all 6 vector axis, 3 torques and 3 forces. b) simple, compact, cheap and rugged, design c) repeatability of alignment during extensive use.

A 60 durometer vibration isolator (Fig. 14) was chosen as the element about which a compliance system could be designed and built. Firestone model CA-244, No. 5 vibration isolators were chosen. The bottom set of curves shows the vibration isolator acting as a rubber shear spring. These curves can respond in any direction (360°) in the shear mode. Thus, if we mount them on their side and under the holster we get torques about X, Y and Z plus force in the ± Z direction and force in the + X direction (Fig. 13). All of these forces and torques will be with the rubber acting in its shear mode. Table I, below summarizes the compliance calculations.

TABLE I CALCULATED COMPLIANCE TORQUES AND FORCES

Force Axis	Displacements	Forces
± Z	± .25 in.	± 4(125) = ± 500lbs.
± Y	± .125 in.	no constraint
+ X	+ .25 in.	+4 (125) = + 500 lbs.
- X	- .25 in.	no constraint

Torque Axis	Displacements/Angular Forgiveness	Torque
± Z	+ .35 in. (2.6°)	2(125) 11 = 2750in-lbs.
± X	± .25 in. (2.6°)	4(125) 11 = 5500in-lbs.
± Y	± .25 in. (5.7°)	4(125) 2.6= 1300in-lbs.

During the testing, the compliance system worked well. The first and most obvious advantage was that it was much easier to program the robot through the mating sequence. In teach programming one cannot precisely move the robot; but rather it tends to leap in small increments on the order of 1/16 inch. Thus, without a compliance system, enormous forces can rapidly build up on the locking mechanism without the programmer being aware of them. The compliance system makes this much more manageable. The forces are more easily controlled and the locking/unlocking system is much less likely to jam or be overpowered by the robot.

Secondly, the compliance system permits the robot to correct minor misalignments with ease. In one test the right shoe ear catching tab was misaligned causing the gripper ears to be 1/8 inch or 1.3° (about Z axis) out of alignment from the programmed position. Yet the system continued to mate and unmate for 20 cycles without a problem. One could see the compliance system compensating for the error but no degradation of performance was noted. In other instances the robot gripper was misplaced ± 1/8 inch along the ± Y axis, and 1/2 inch along the -X axis. The compliance and gravity feed of the 15° slant of the holsters easily compensated for these errors.

Thirdly, the compliance system has proven easy to adjust and has stayed in alignment and adjustment throughout extensive use and (in certain instances) abuse. It went thru extensive use before and during a 1 2/3 hour, 120 cycle test, and is still being used today with no evidence of degredation. In some cases during testing, the vibration isolators were stretched to 1/2 inch with no ill effects.

Finally, the design has clearly proved simple, rugged, cheap and compact. A vibration isolator costs only a few dollars and is very easy to replace.

VIII. SENSORS.

The "quick change" system uses two sensors to determine the status of the locking mechanism. 8 states are possible.

TABLE 2 STATES

Sensor 1/2	System Status	Controller Command	OK?
on/off	open	open	ok
off/on	closed	open	No
off/on	closed	closed	ok
off/on	closed	open	No
off/off	in between	open	No
off/off	in between	closed	No
on/on	sensor 1 deficient	closed	No
on/on	sensor 2 deficient	open	No

Subminiature infrared reflective sensors are used (Skan-a-Matic 527 series). These are .4 inch diameter and 1 inch long. They have a range of up to .1 inch and a field of view of .040 inch. They have edge detection repeatability of .001 inch. We imbed them in the "A" plate and attach a section of aluminum ring to the locking mechanism sleeve. We then tune the edge locations of the aluminum ring with a file and let the detectors monitor the ring edges as the locking mechanism moves.

IX. TESTS

Several tests have been conducted on the "quick change" system and described in the text above. These tests and results are summarized below.

TABLE 3 TEST SUMMARY

Test	Results
120 cycle computer controlled docking and undocking	Fault free continuous performance for 1 2/3 hrs and 120 cycles
Mating repeatability using 20 insertions and 30 lbs insertion force	$\pm$ .0009 in/in maximum average tilt between "A" and "B" plates
1,000 psi hydraulic fluid run across quick change interface	no leakage .030 in tilt between "A" and "B" plate - appears easily correctable by reinforcing aluminum "A" plate
6,000 lbs stress test on locking mechanism	did not break
Locking mechanism tests. 30 lbs to 100 lbs mating force. 45 psi to 100 psi air pressure.	will lock & unlock reliably any where in the mating force, actuator psi envelope

X. SUMMARY

The NBS prototype robot "quick change" system has been described to include both the "quick change" itself and the holster system. This description includes both theoretical capabilities and test results. It has been shown that the "quick change" is capable of passing both hydraulics (1,000 psi) and pneumatics and that it mates with high-repeatability - very useful when a sophisticated vision sensor is mounted on it. We have shown that it is capable of passing multiple electronics and fiber optics channels. The holster system was also described. Of particular interest is the compliance system, which has proven useful in allowing grippers to be stored at the robot waist - a region where powerful hydraulic robots are especially rigid.

The "quick change" locking mechanism is described in detail. This locking mechanism features a kinetic energy system which permits a small rotary actuator powered by shop air to reliably secure and release the robot gripper.

A miniaturized version of the prototype system is under development which is 4 inch long vs the 6.5 inch length of the prototype. No capabilities are sacrificed. Modified commercial components are used throughout.

The Robot

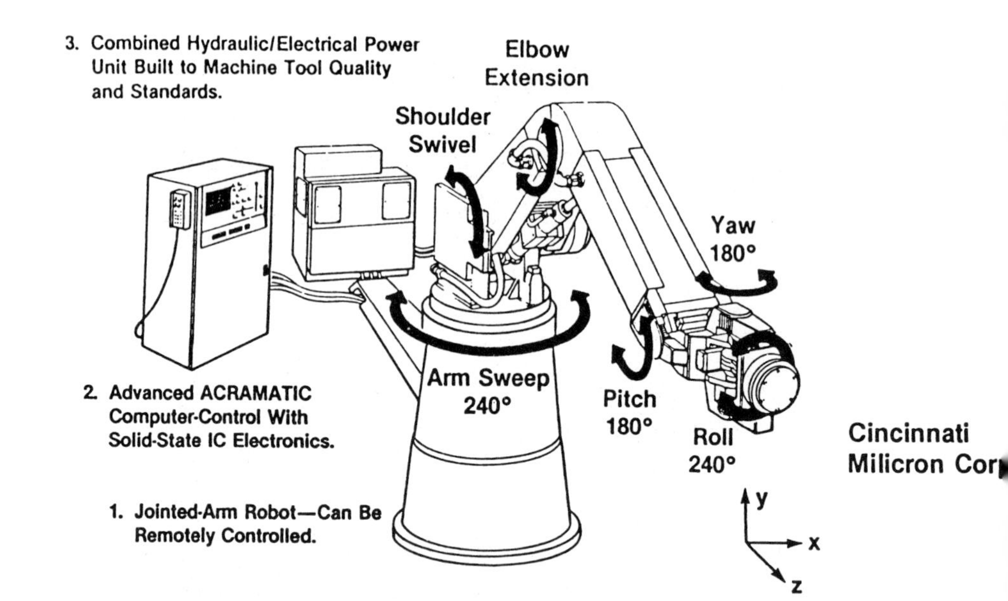

The "Quick Change"

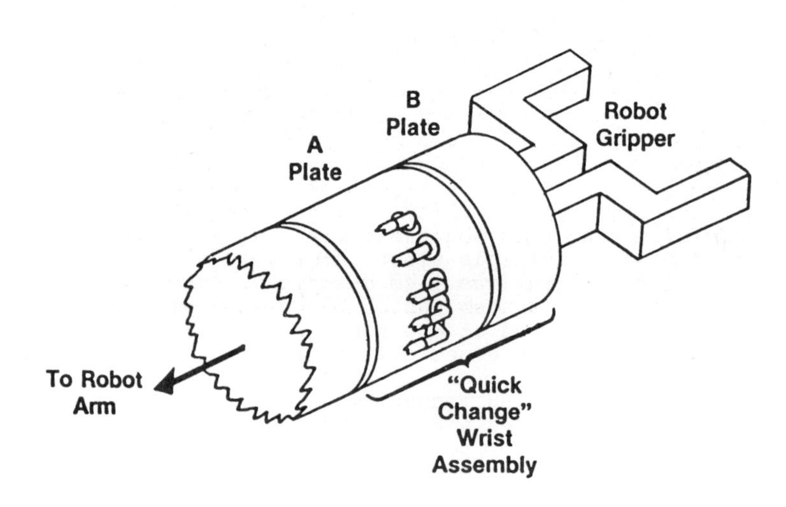

FIGURE 1. ROBOT AND QUICK CHANGE WRIST

FIGURE 2 SYSTEM UNDERGOING TEST

FIGURE 3 THE "A" AND "B" PLATES

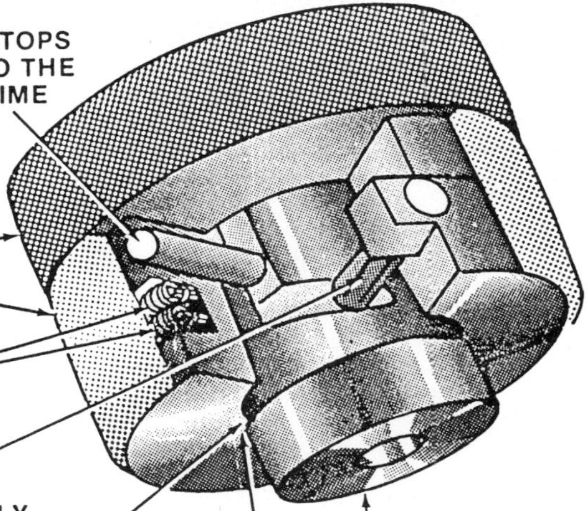

24 OZ. 1.83 IN HIGH, 3 IN DIA.

POSITIVE LOCATING PIN
INSURES SAFE BRAKE STOPS
RELOCATES TOOLING TO THE
SAME POSITION EACH TIME
ALLOWS MACHINING IN
EITHER DIRECTION

KNURLED
RING & BODY
FOR EASY
GRIPPING
NO WRENCHES
REQUIRED

TWO SPRINGS
FOR ADDED SAFETY

SPRING-LOADED
CAM LOCK-PIN ASSEMBLY

CLOSE-TOLERANCE TAPERS
THE PROVEN METHOD FOR MAXIMUM
ACCURACY AND REPEATABILITY

15° ANGLE HELD ±.0001"
IN 1" OR CLOSER ON ALL
RAPIDAPTORS AND ADAPTOOLS

ADAPTOOL—ACCOMMODATES YOUR
STANDARD CUTTING TOOLS

ECONOMICAL—Adaptable to Your Existing Machines and Standard Cutting Tools.

ACCURATE—Repeats Position to .0002 Inch or Better.

SIMPLE DESIGN—Positive Single Cam-Lock Retention, Free of Screws, Wrenches, Threads, or Complex Linkage.

QUICK—Average Tool Change in Less than 5 Seconds.

RIGID—Strong and Rigid Shortest Possible Construction

SAFE—Double-spring Retention Positive Drive in Either Direction Tools Cannot Fall Out.

Younger Tool
Company

FIGURE 4. THE COMMERCIAL LOCKING SYSTEM

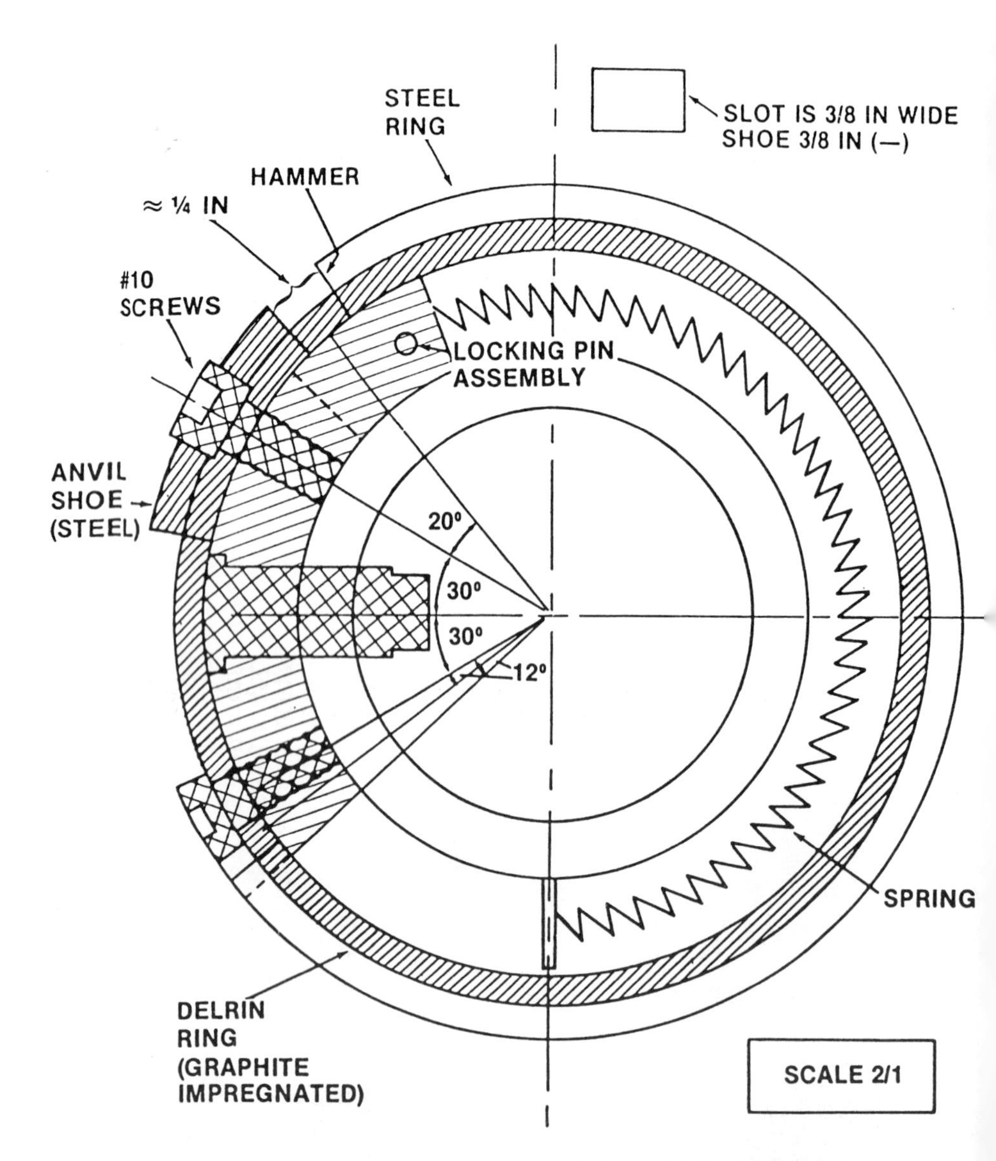

FIGURE 5. IMPACT ACTUATOR SYSTEM

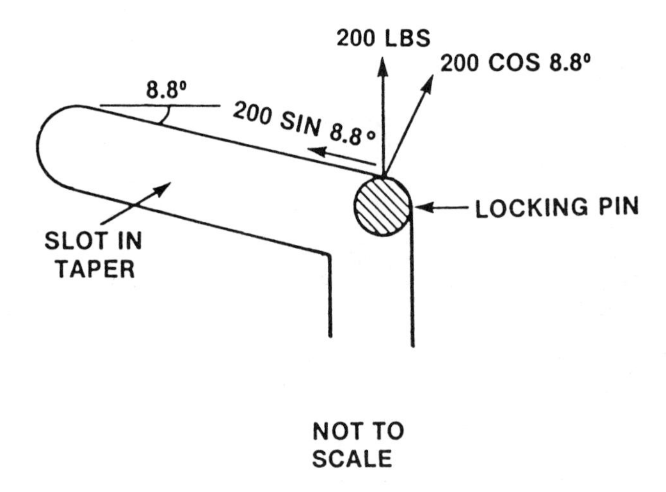

FIGURE 6. FRICTIONAL FORCES BETWEEN LOCKING PIN AND TAPER DURING MATING

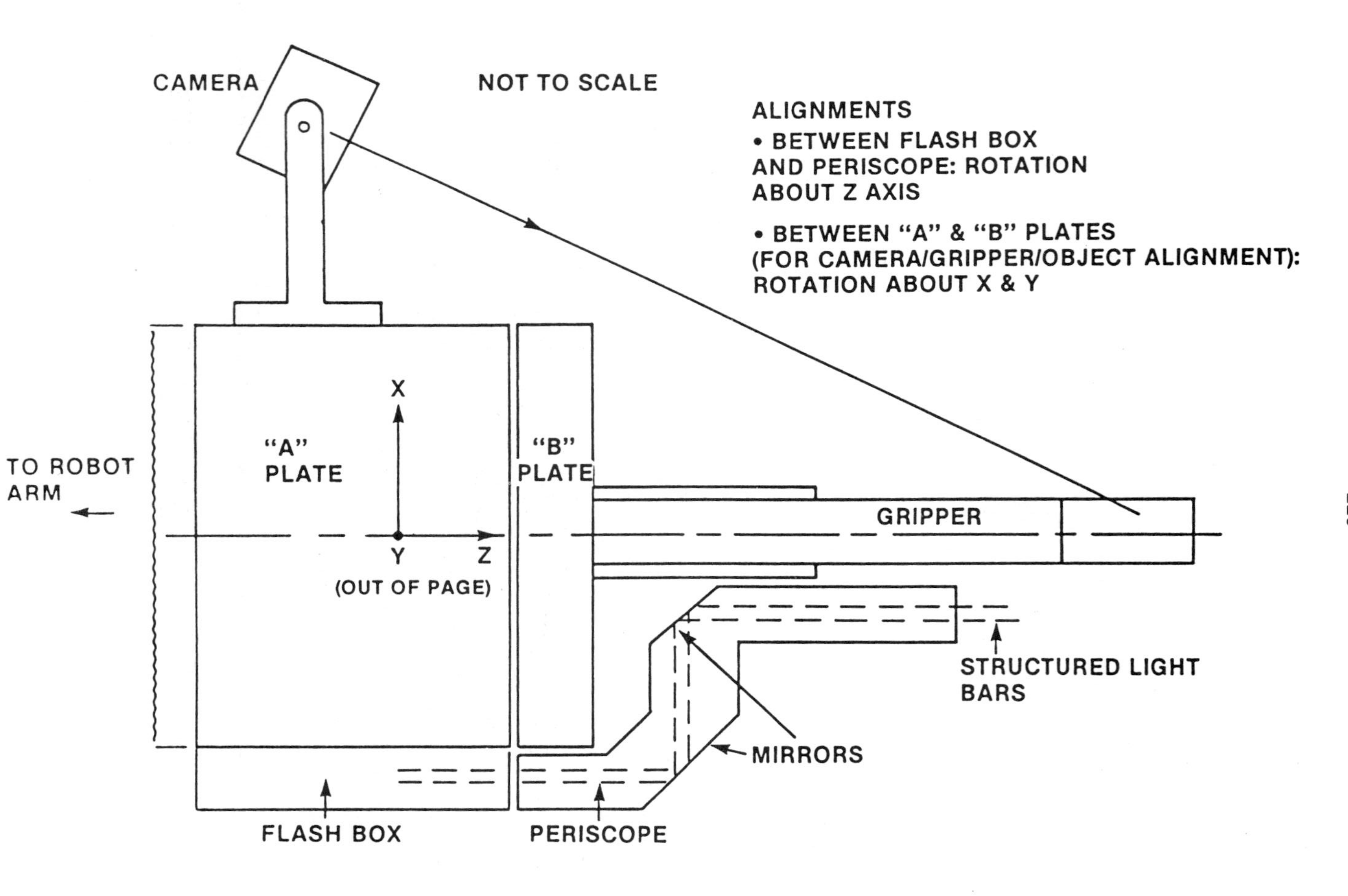

FIGURE 7. NBS VISION SYSTEM RAYS

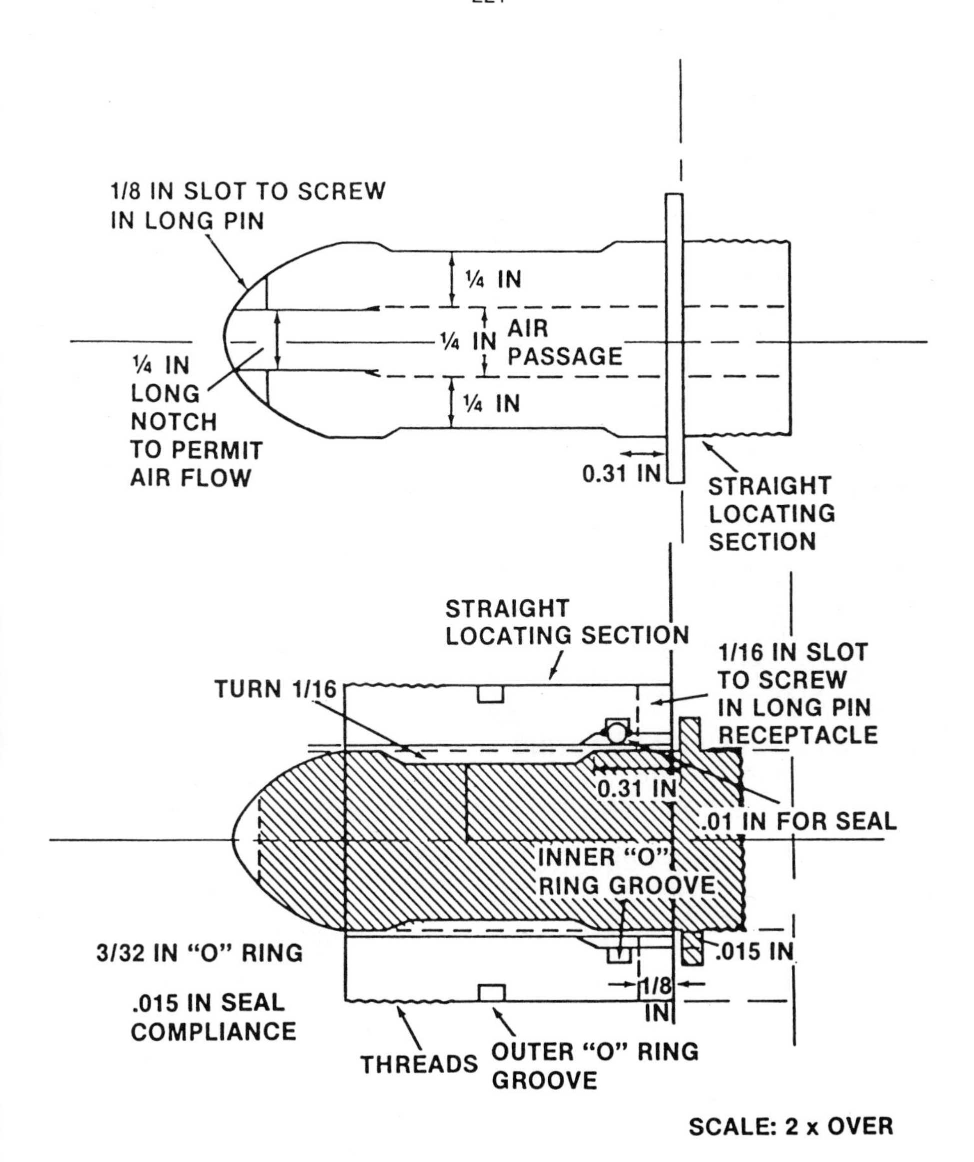

FIGURE 8. ALIGNMENT PINS/PNEUMATIC CONNECTORS

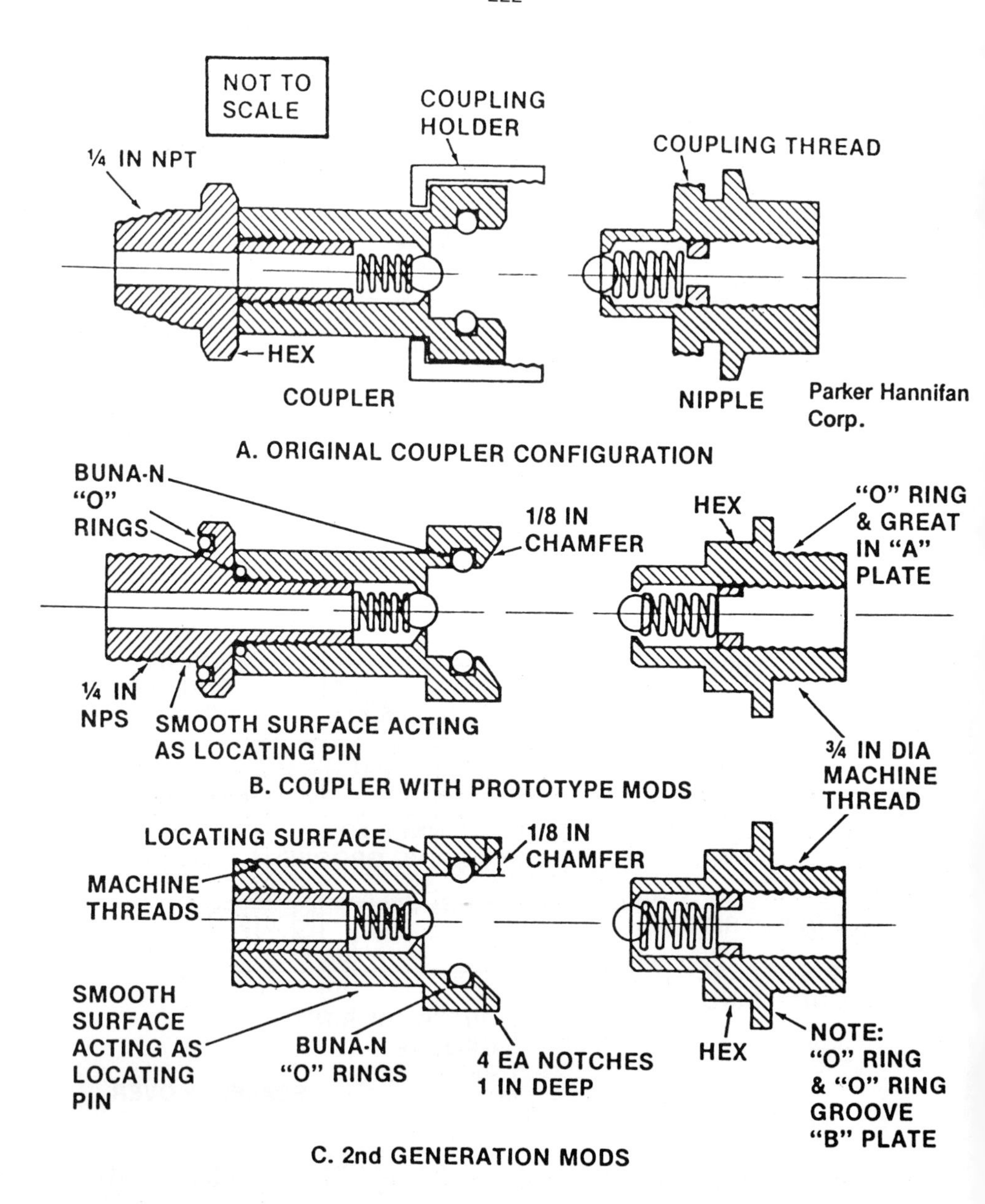

FIGURE 9. HYDRAULIC CONNECTOR MODIFICATIONS

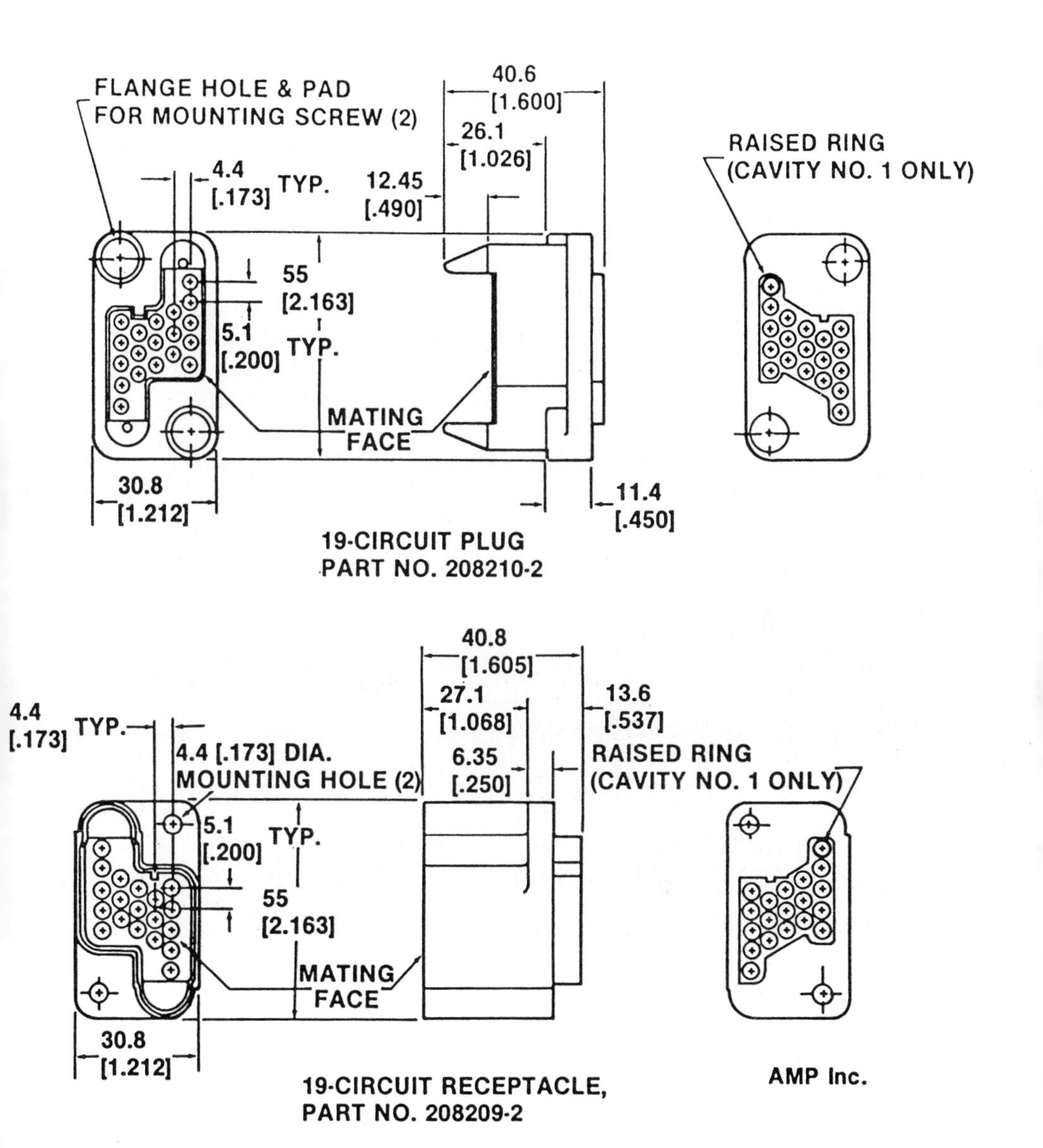

FIGURE 10. ELECTRONICS/FIBER OPTICS CONNECTORS

FIGURE 11   GRIPPER STORAGE

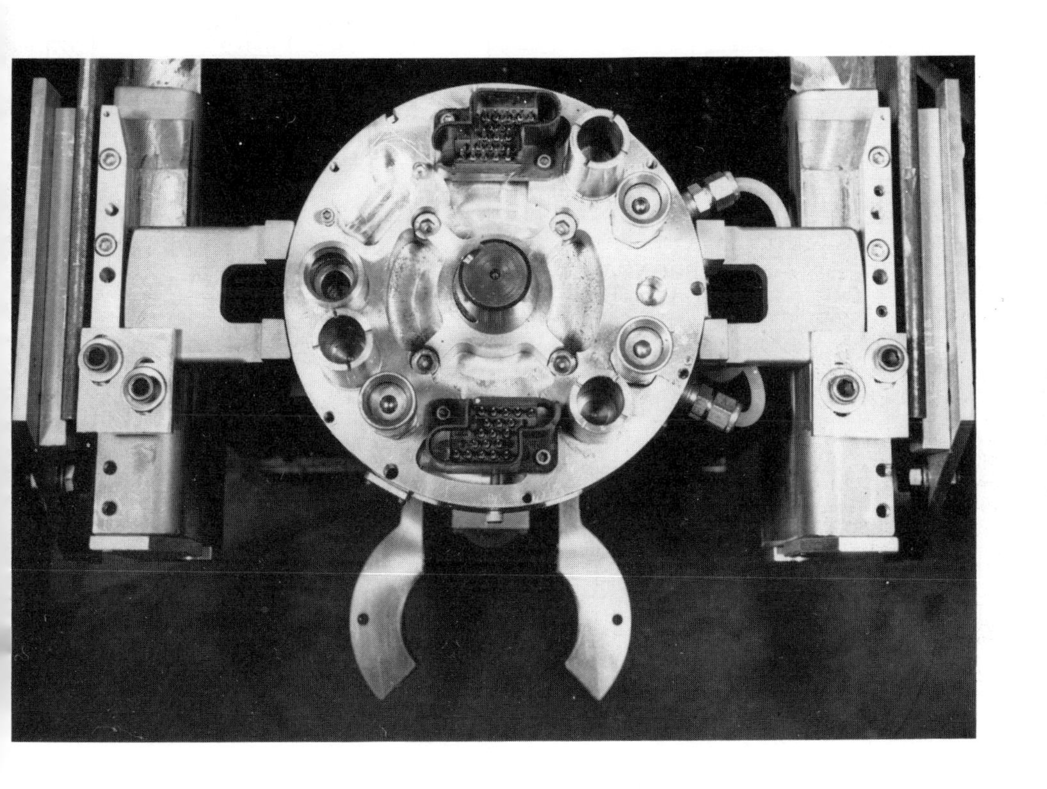

FIGURE 12 GRIPPER EARS AND HOLSTER SHOES

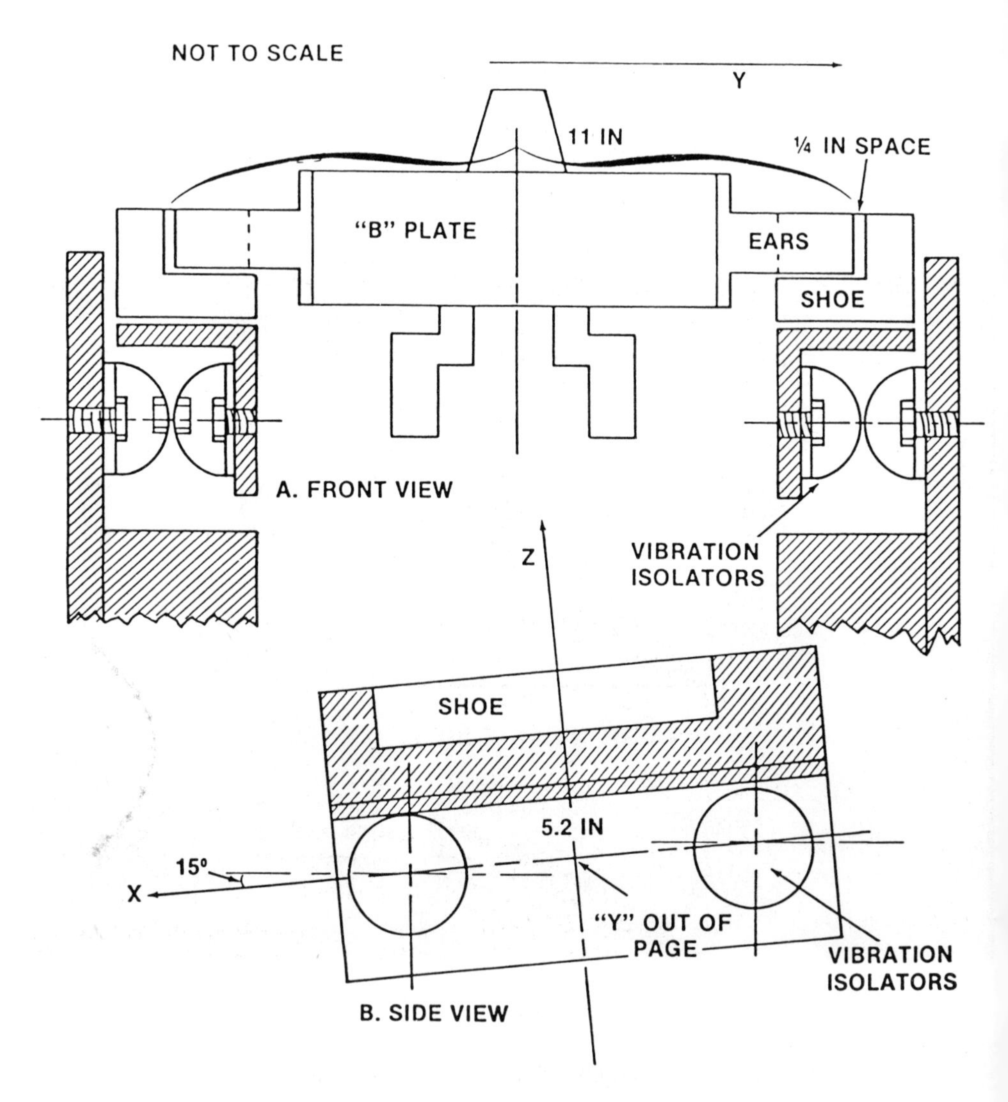

FIGURE 13. THE HOLSTER COMPLIANCE SYSTEM

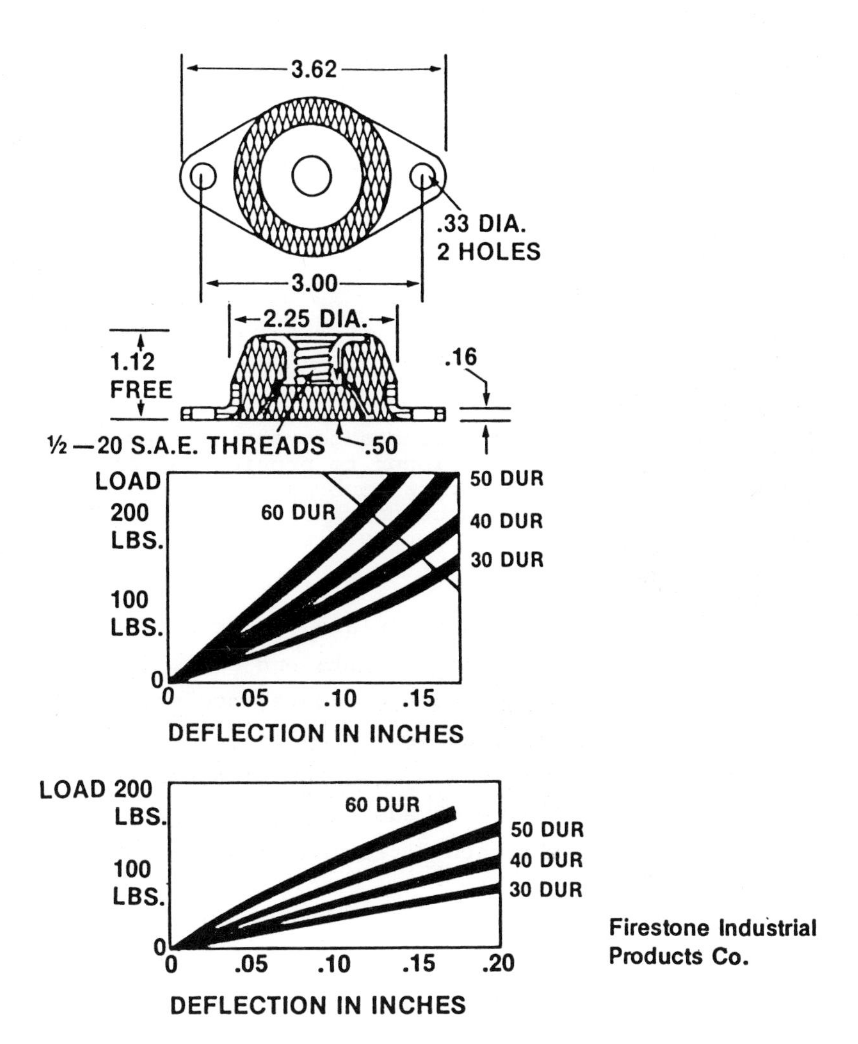

Firestone Industrial
Products Co.

FIGURE 14. VIBRATION ISOLATOR

Local Manipulation Using a N-Link Articulated End Effector

W. Edward Red

College of Engineering Sciences and Technology
Brigham Young University
Provo, Utah

Abstract

An optimization approach selects geometric parameters that permit a two digit articulated gripper to provide balanced strengths during manipulation maneuvers such as grasping, releasing and object conveyance. Reliance upon friction is only required in the slip direction normal to the plane of the fingers.

Introduction

Articulated grippers can manipulate objects of various shapes and sizes. Multiple contact points and shape conforming design features guarantee grasping and conveyance stability for a larger set of objects than can normally be accommodated by the typical single degree of freedom, two link reaction gripper commonly used today [1,2]. In addition less dependence on frictional forces and approach directions for successful grasping is required [3].

In contrast, articulated grippers kinematically are more complicated, more prone to operational failure, require more unobstructed space for grasping and release, and can weigh more when compared to conventional grippers today. To minimize these deficiencies this paper considers a simple design approach that maximizes the grasping strength of a two-digit (two-link), two-fingered gripper while minimizing the geometric features.

Optimization and Manipulation Considerations

Earlier optimization studies [4] of articulated end-effectors having three digit fingers demonstrated a tendency for the last digit to shrink unrealistically in size below the two leading digits. This investigation verifies that a two fingered hand can be optimized for grasping cylindrical shapes to provide balanced strength in the plane of the two fingers and also provide comparable strength in the slip direction perpendicular to the grasping direction.

To minimize the dependence on grasping friction the design neglects the friction of interaction in the hand's grasping plane. This conservative approach avoids the highly nonlinear frictional complexities introduced when a multi-link kinematic device interacts frictionally with a constraint surface such as a cylinder. Friction is incorporated into the analysis to measure the slip strength for forces perpendicular to the grasping plane and to provide a conservative estimate of the capability to grasp cylindrical objects when the center of mass is overhung from the plane of grasp.

Consistent with this conservative design approach it is assumed that all contact is point contact, that the gripper has zero out-of-plane dimensions and thus can only resist bending moments through frictional reaction couples at the contact points.

Furthermore, the pulley design incorporates a constant tension continuous cable that wraps around each joint pulley causing the finger to move as a "rigid body" until contact with the object is made. Contact causes the remaining digits to rotate relative to the digit in contact until they in-turn contact the object. The contact forces increase, thereby providing a stable grasp. It should be noted that the initial contact introduces small contact forces which do not increase significantly until full conformity occurs. This feature enhances the manipulation equilibrium of the object during the grasping/release operation.

<u>Design Equations</u>

Using Figure 1 the appropriate equilibrium equations for design optimization can be expressed nondimensionally by (1) and (2), these resulting from the moment equations about the link pulleys,

$$\overline{F}_2 = \overline{R}_2/\overline{s}_2 \qquad (1a)$$

$$\overline{F}_1 = 1 + (\overline{s}_2 + \overline{L}_1 \cos\theta_2)\overline{F}_2 \qquad (1b)$$

and where $\overline{F}_1 = F_1\, s_1/TR_1$, $\overline{F}_2 = F_2\, s_1/TR_1$ $\overline{R}_2 = R_2/R_1$, $\overline{s}_1 = s_1/R$ and $\overline{s}_2 = s_2/s_1 \leq \overline{L}_2 = L_2/s_1$ define the nondimensional contact forces in terms of:

F_1, F_2 = link contact forces

L_1, L_2 = link lengths

R = cylinder radius

R_1, R_2 = pulley radii

s_1, s_2 = contact distances

T = cable tension

Equations (1) and (2) reflect the grasping assumption that each link will contact the cylinder, thus enabling the angles θ_1 and θ_2 to be determined as

$$\theta_1 = 2\tan^{-1}\left(\frac{\overline{s}_1}{1 + \overline{t}\,\overline{s}_1\overline{R}_1}\right);\ \theta_2 = 2\tan^{-1}\left(\frac{\overline{s}_1\,\overline{s}_2}{1 + \overline{t}\,\overline{s}_1\overline{R}_1}\right) \qquad (2)$$

where $\overline{t} = t/R_1$, the ratio of half the link width to R_1, is chosen such that $t = \max(R_1, R_2)$.

Gripper Optimization

The generalized reduced gradient algorithm within OPTDES. BYU [5] was applied to minimize the objective function Q in (3), the ratio of the finger length to the strength capabilities $\overline{F}_x$ and $\overline{F}_y$, equations (4a) and (4b), respectively:

$$Q = \min\left[(\overline{L}_1 + \overline{L}_2 + 10\,\overline{P}_1)/(\overline{F}_x + \overline{F}_y)\right] \qquad (3)$$

$$\overline{F}_x = 2\left[\overline{F}_1 \sin\theta_1 + \overline{F}_2 \sin(\theta_1 + \theta_2)\right] \qquad (4a)$$

$$\overline{F}_y = -2\left[\overline{F}_1 \cos\theta_1 + \overline{F}_2 \cos(\theta_1 + \theta_2)\right] \qquad (4b)$$

As included in (3), $\overline{P}_1$ will approach zero, thus ensuring moment equilibrium of the contact forces about the first pulley.

$$\overline{P}_1 = 1 - (\overline{s}_2 + \overline{L}_1 \cos\theta_2)\overline{F}_2 - \overline{F}_1 \qquad (5)$$

The following constraint equations reflect the geometric requirements that the link lengths must exceed the pulley radii, each finger must not wrap beyond half the cylinder circumference and the design constraint that $\bar{s}_2 = \bar{L}_2$. In addition, the gripper was designed for $\bar{F}_x = \bar{F}_y$, i.e., balanced strength in the grasping plane.

$$\bar{R}_1 \; \bar{R}_2/\bar{s}_2 \leq 1; \quad \bar{R}_1(1 + \bar{R}_2)/\bar{L}_1 \leq 1 \tag{6}$$

$$(\theta_1 + \theta_2)/\pi \leq 1 \tag{7}$$

$$\bar{s}_2 = \bar{L}_2 = \bar{L}_1 - 1 \tag{8}$$

Typical OPTDES iteration histories are shown in Figure 2.

Friction Strength

Strength in the z or slipping direction, $\bar{F}_z$, is determined as the product of μ, the coefficient of friction, with the normal contact forces of the finger links and the base pad. For stable grasping the base pad force will always be positive.

$$\bar{F}_z = \mu(2\,\bar{F}_1 + \bar{2}F_2 + \bar{F}_y) \tag{9}$$

Capability to withstand bending moments about the x and y axes is a function of non-dimensionalized contact point relative locations $\bar{x}_i$ and $\bar{y}_i$ and the non-dimensionalized overhung distances $\bar{a}_x$ and $\bar{a}_y$ (non-dimensionalized by dividing these distances by s_1). Two worst case situations are examined where the overhung inertial forces are some fraction f_x, f_y of the gripper strengths and perpendicular to $\bar{a}_x$ or $\bar{a}_y$.

The couple strengths about the x and y axes are determined by examining each contact point and evaluating whether its normal contact force is greater than the sum or difference of the contact forces at the remaining contact points (z force equilibrium). If so, this contact point defines the location of a couple axis about which the couple can be determined as the product of the relative locations $(\bar{x}_1, \bar{y}_i)$, μ, and the normal forces, $\overline{F}_i$.

The strength is the minimum couple determined for those contact points having sufficient normal force to satisfy z equilibrium. Because of symmetry, $\overline{M}_x$ couple strength can be determined by examining contact point pairs whereas $\overline{M}_y$ couple strength must be determined by examining each point individually. For $\overline{M}_x$ strength examine the base force $\overline{F}_b = (1 - f_x)\overline{F}_y$ and each link contact force:

$$\textit{Base:} \qquad \overline{F}_b \overset{?}{\geq} 2 \sum_{i=1}^{n} \overline{F}_i \tag{10}$$

$$\textit{Link } j: \qquad 2\overline{F}_j \overset{?}{\geq} - \overline{F}_b - 2 \sum_{i=1}^{j-1} \overline{F}_i + 2 \sum_{i=j+1}^{n} \overline{F}_i \tag{11}$$

If (10) is satisfied the resistance couple in(12) must be compared against resistance couples (13) determined for each link contact force satisfying (11). For the two digit gripper, n = 2

$$\textit{Base:} \qquad \overline{M}_x = 2\mu \sum_{i=1}^{n} \overline{y}_i \overline{F}_x \tag{12}$$

$$\textit{Link } j: \quad M_x = \mu\overline{y}_j \overline{F}_b + 2\mu \sum_{i=1}^{j-1} | \overline{y}_j - \overline{y}_i | \overline{F}_i + 2\mu$$

$$+ 2\mu \sum_{i=j+1}^{n} |\overline{y}_i - \overline{y}_j | \overline{F}_i \tag{13}$$

$\overline{M}_y$ strength, (16) and (17), is determined similarly except that the opposing link forces must be increased on the right finger by $\overline{F}_i^R = (1 + f_y)\overline{F}_i$ and decreased on the left finger by $\overline{F}_i^L = (1 - f_y)\overline{F}_i$, assuming a constant tension "actuator reservoir" (i.e., $T_R + T_L = 2T$ = constant; likewise $\overline{F}_i^R + \overline{F}_i^L = 2\overline{F}_i$ = constant).

$$\textit{Base:} \qquad \overline{F}_y \overset{?}{\geq} 2 \sum_{i=1}^{n} \overline{F}_i \tag{14}$$

$$Link\ j:\quad \overline{F}_j^{\,R} \overset{?}{\geq} -\overline{F}_y - \sum_{i=1}^{n} \overline{F}_i^{\,L} - \sum_{i=1}^{n} \overline{F}_i^{\,R} + \sum_{i=j+1}^{n} \overline{F}^{R} \qquad (15)$$

$$Base:\quad \overline{M}_y = \mu \sum_{i=1}^{n} \overline{x}_i\ \overline{F}_i^{\,L} + \mu \sum_{i=1}^{n} \overline{x}_i\ \overline{F}_i^{\,R} \qquad (16)$$

$$Link\ j:\quad \overline{M}_y = \mu \overline{x}_j\ \overline{F}_y + \mu \sum_{i=1}^{n} |\overline{x}(i) + \overline{x}(j)|\overline{F}_i^{\,L}$$

$$+ \mu \sum_{i=1}^{j-1} |\overline{x}(j) - \overline{x}(i)|\overline{F}_i^{\,R} +$$

$$\mu \sum_{i=j+1}^{n} |\overline{x}(i) - \overline{x}(j)|\overline{F}_i^{\,R} \qquad (17)$$

Non-Optimum Object Manipulation

Grasping of cylinders having radii for which the gripper was not optimized introduces the additional complexity that the first digit, link one, may not necessarily contact the cylinder, thus $\overline{F}_1 \geq 0$. Equations (2) are not applicable, rather a set of constraint equations that permit θ_1 to assume a value less than that determined by (2) must be applied. The brevity of this paper does not permit the presentation of these equations.

In addition, "slip gripping" characterized by the pushing of the cylinder away from the base pad to a suspended equilibrium position must also be considered; these equations are not presented either. Since a robust design will avoid these manipulation difficulties, we skirt the theoretical details but demonstrate graphically one such design in Figure 3 that exhibits both undesirable effects. This gripper was optimized for $\overline{R}_1 = 0.1$ and $\overline{s}_1 = 0.3$.

Design Summary

Optimization studies as conducted for $0.05 \leq \overline{R}_1 \leq 0.5$, $0.2 \leq \overline{s}_1 \leq 2$, and $0.1 \leq \mu \leq 0.7$ demonstrated the optimized designs to be rather insensitive to $\overline{R}_1$ but quite sensitive to $\overline{s}_1$. Figure 4 demonstrates this insensitivity for optimized $\overline{R}_2$ and $\overline{L}_2/\overline{L}_1$ as a function of $\overline{s}_1$ for $\overline{R}_1 = 0.1, 0.5$. It would be expected that including friction in the grasping plane would increase this sensitivity significantly.

From Figure 4 several conclusions can be drawn: 1) grippers having a small base dimension ($2s_1$) require $R_2 > R_1$ and $L_1 \approx L_2$ for balanced design, and 2) for $\bar{s}_1 > 0.5$ the first link dimensions begin to dominate with the ratios L_2/L_1 nearly those of R_2/R_1.

Figure 6 demonstrates that a gripper optimized such that $\bar{F}_x = \bar{F}_y$ provides a slipping strength $\bar{F}_z \approx \bar{F}_x, \bar{F}_y$ for $\mu = 0.4$. This result holds true for grippers designed over large ranges of $\bar{s}_1$. Additionally the strength capability increases with larger $\bar{s}_1$. Unfortunately, grippers optimized for larger $\bar{s}_1$ do not have grasping versatility of grippers designed for intermediate $\bar{s}_1$.

One design optimized for $\bar{R}_1 = 0.1$ and $\bar{s}_1 = 0.7$ demonstrates the desired grasping versatility as shown by Figures 5, 7 and 8. Note in Figure 5 that, in the optimized design, link one applies a zero contact force which increases from zero for $\bar{s}_1 > 0.7$. $\bar{F}_x$ decreases from its balanced value ($\bar{F}_x = \bar{F}_y$ at $\bar{s}_1 = 0.7$) for $\bar{s}_1 > 0.7$ while $\bar{F}_y$ increases. $\bar{F}_2$ is constant for $\bar{s}_1 \geq 0.7$ and only decreases slightly for $\bar{s}_1 < 0.7$.

The overhung capability of two gripper designs optimized for $\bar{s}_1 = 0.7$ and $\bar{R}_1 = 0.1, 0.5$ is compared in Figure 7. At load fractions f_x, f_y below 0.4 the gripper's bending moment strength is greatest about the x axis because $\bar{a}_x > \bar{a}_y$ whereas for $f_x, f_y > 0.4$ the opposite is true; thus balanced bending strength occurs at $f_x = f_y = 0.4$. Note that for $f_x = f_y = 0.4$, $\bar{a}_x \approx \bar{a}_y \approx 2$. This implies that a gripper optimized for $\bar{R}_1 = 0.1$, $\bar{s}_1 = 0.7$, and $\mu = 0.7$ can manipulate cylinders having center of mass offsets up to 1.4 times the radius of the cylinder if the inertial loads do not exceed 40% of the gripper's strength capability.

Figure 8 demonstrates the optimum design discussed gripping

a wide range of $\bar{s}_1$. An articulated hand with coplanar fingers would permit objects of sizes $\bar{s}_1 = 0.5$ to $\bar{s}_1 = 1$ to be successfully gripped (50% size variation) whereas a split-fingered gripper could accommodate a size variation exceeding 100%.

Conclusions

Two digit articulated grippers optimized for balanced strength provide a stable grasp state for end effector manipulation. This brief optimization study indicates that robust grippers can be designed to accommodate a wide range of cylinder sizes and also permit the center of mass to be overhung considerably given sufficient contact friction.

References

1. Salisbury, J.K., and Craig, J.J.: Articulated Hands: Force Control and Kinematic Issues. International Journal of Robotics Research, Vol. 1, No. 1, 1982.

2. Okada, T.: Computer Control of Multijointed Finger System for Precise Object-Handling. IEEE Transactions on Systems, Man, and Cybernetics. Vol SMC-12, No. 3, May/June 1982.

3. Fearing, R.S.: Simplified Grasping and Manipulation with Dextrous Robot Hands. 1984 American Control Conference, San Diego, CA., Vol 1, pp 32-38, 1984.

4. Red, W.E., and Colligan, K.: A Parametric Study of an Articulated End-Effector. Robots West Conference Presentation, Anaheim, CA., November 1984.

5. Parkinson, A.R., Balling, R.J., and Free, J.C.: OPTDES.BYU: A Software System for Optimal Engineering Design. Proceedings of the Int'l Computers in Mechanical Engineering Conference, Vol. 2, Las Vegas, Nevada, August 1984.

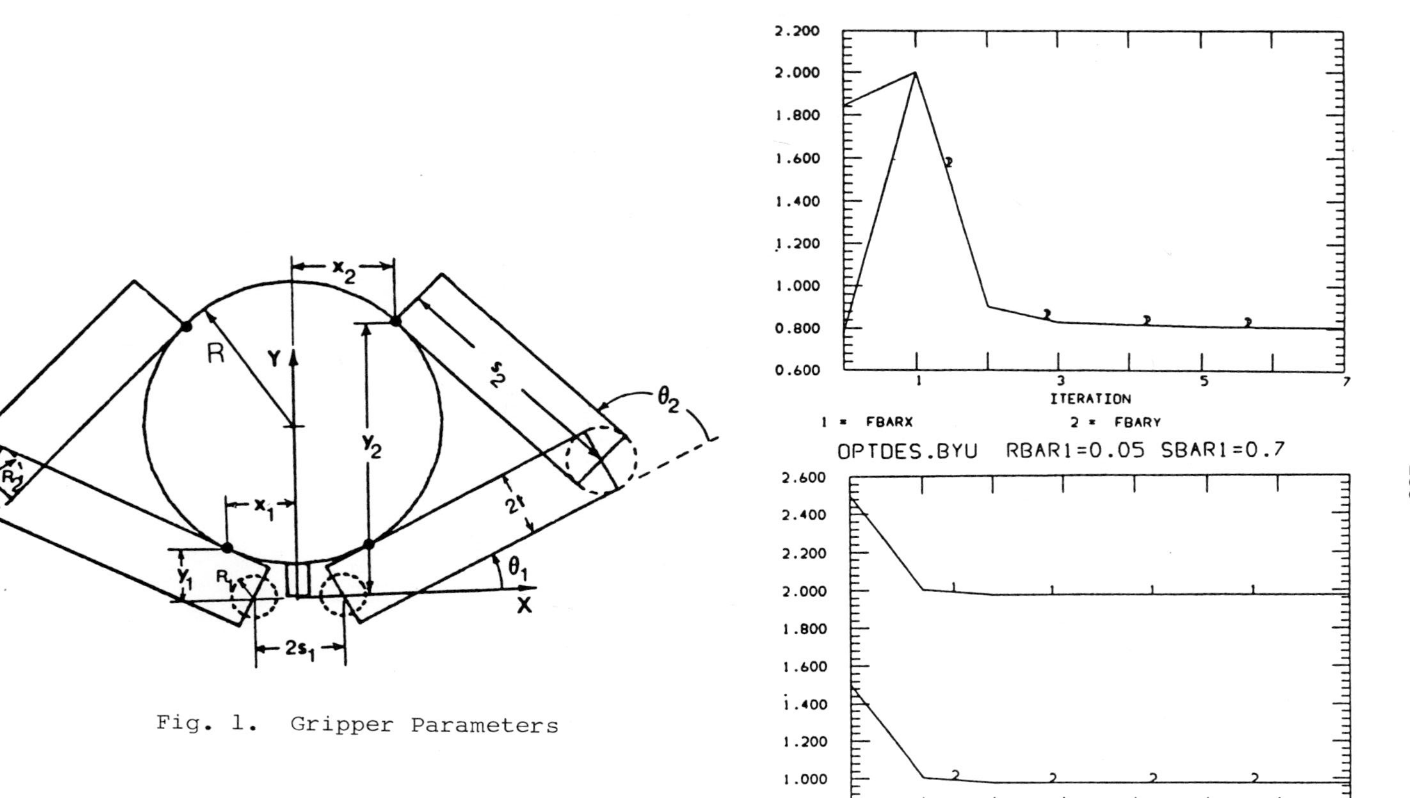

Fig. 1. Gripper Parameters

Fig. 2. OPTDES Iteration Histories

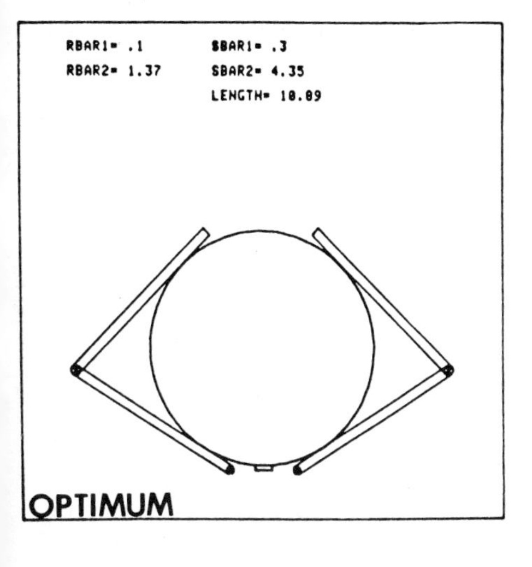

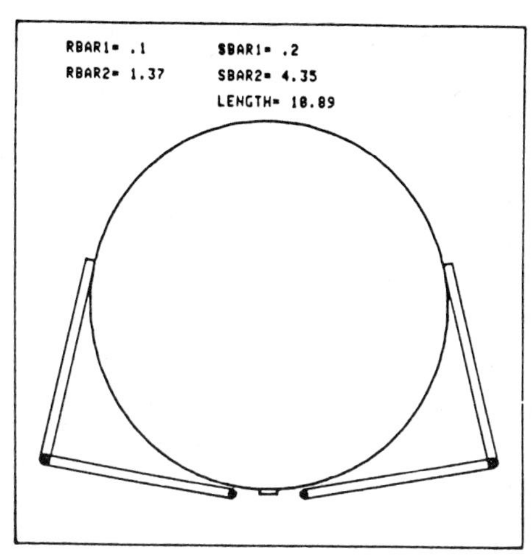

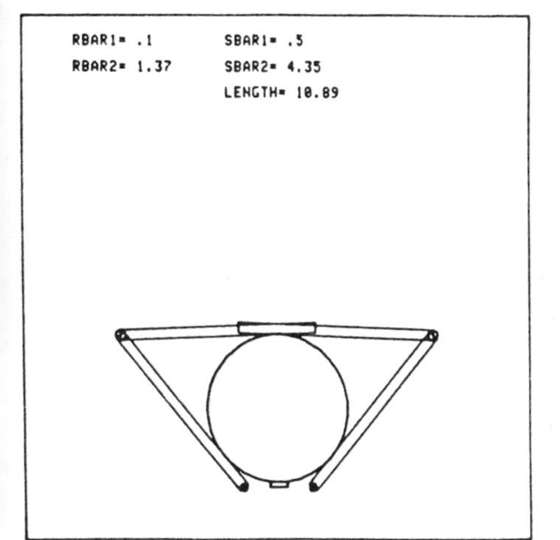

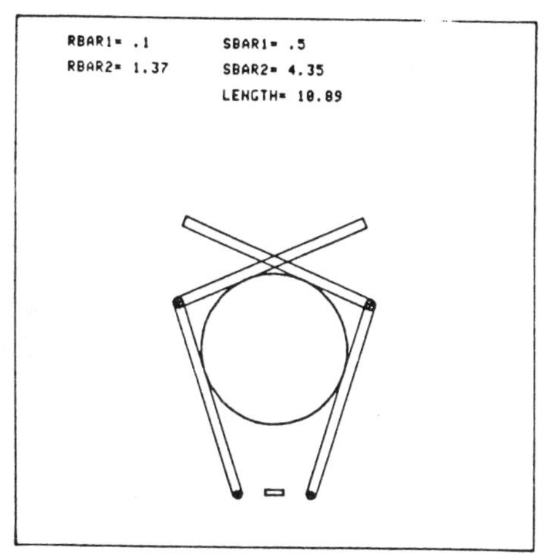

Fig. 3. Grasping Difficulties

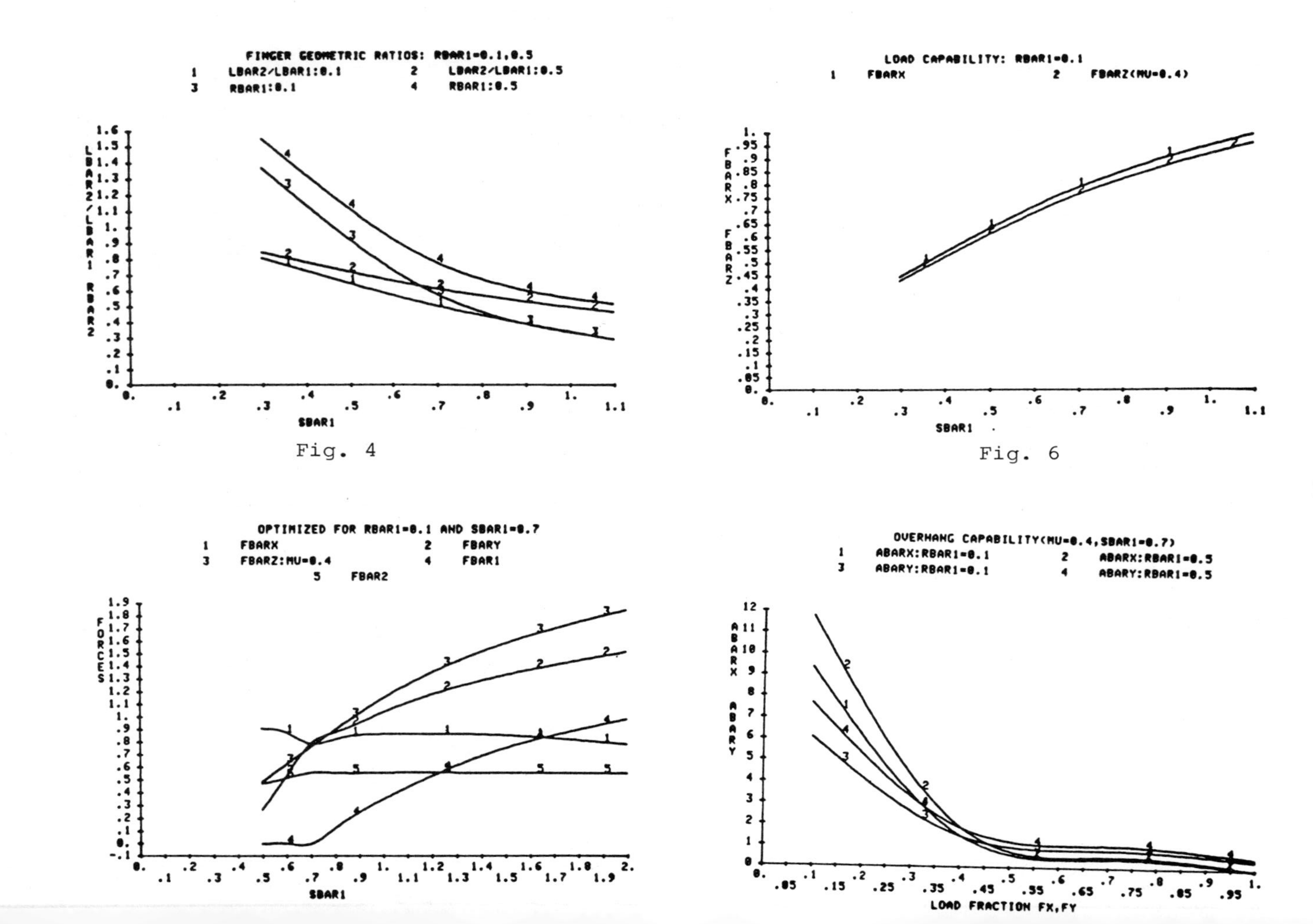
FINGER GEOMETRIC RATIOS: RBAR1=0.1,0.5
1 LBAR2/LBAR1:0.1
2 LBAR2/LBAR1:0.5
3 RBAR1:0.1
4 RBAR1:0.5
LBAR2/LBAR1 RBAR2
SBAR1
LOAD CAPABILITY: RBAR1=0.1
1 FBARX
2 FBARZ(MU=0.4)
FBARX FBARZ
SBAR1
OPTIMIZED FOR RBAR1=0.1 AND SBAR1=0.7
1 FBARX
2 FBARY
3 FBARZ:MU=0.4
4 FBAR1
5 FBAR2
FORCES
SBAR1
OVERHANG CAPABILITY(MU=0.4,SBAR1=0.7)
1 ABARX:RBAR1=0.1
2 ABARX:RBAR1=0.5
3 ABARY:RBAR1=0.1
4 ABARY:RBAR1=0.5
ABARX ABARY
LOAD FRACTION FX,FY

Fig. 4

Fig. 6

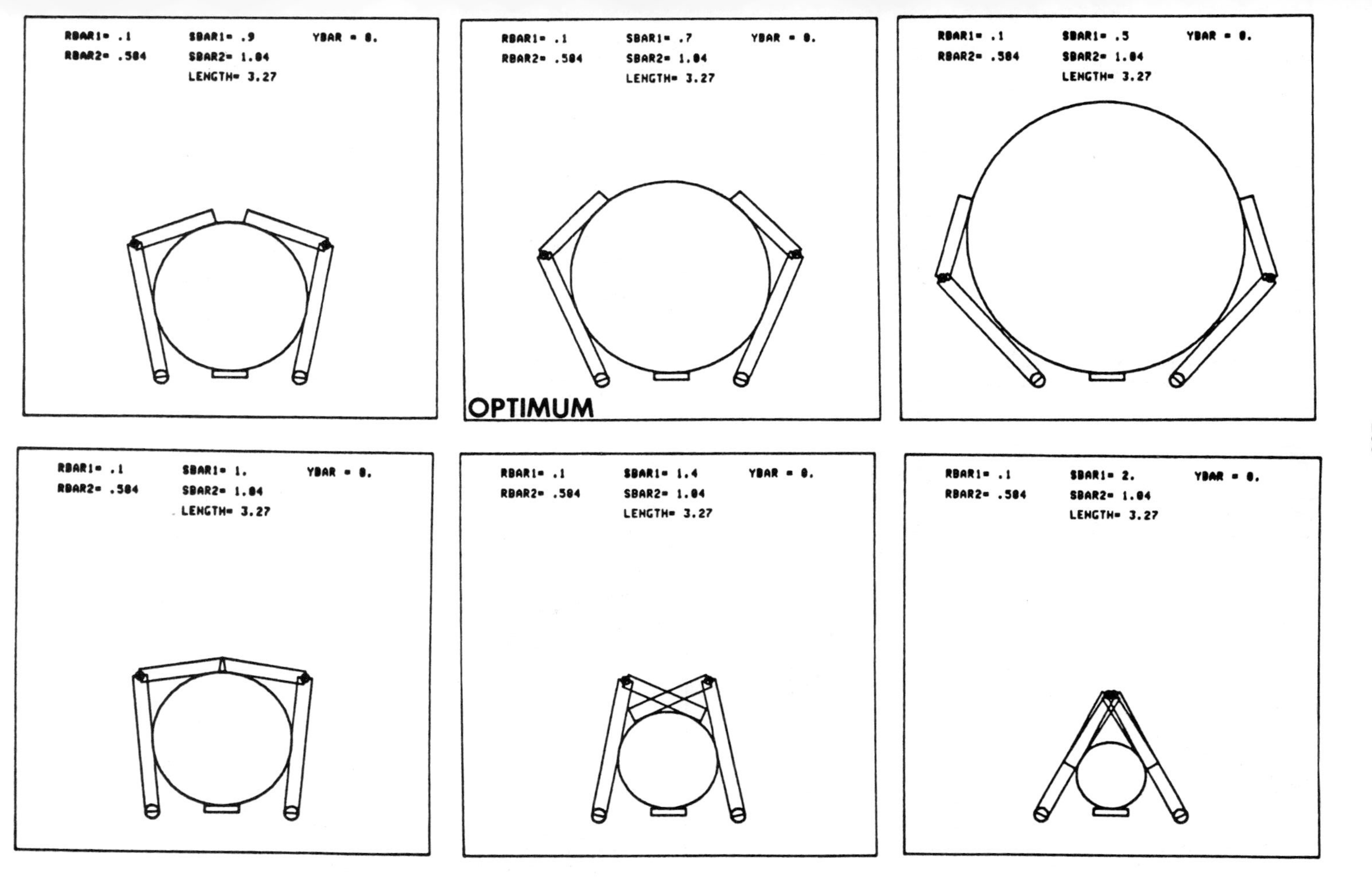

Fig. 8. Robust Gripper Design

Dynamic Synthesis of Manipulator Motion Paths

Dilip Kohli

Associate Professor of Mechanical Engineering
University of Wisconsin--Milwaukee
Milwaukee, Wisconsin 53201

Summary

The current techniques of synthesis of manipulator paths have completely ignored (A) the manipulator linkage dynamics, and (B) the actuator dynamics and the constraints due to bounds on actuator capacities.

In order to synthesize paths which can be executed efficiently and accurately, the interaction of paths with the linkage and actuator dynamics must be studied and the paths designed to account for the dynamics of the system including any constraints imposed by the actuators.

Hybrid Motion Strategy

A preliminary investigation of hybrid motion trajectories for six jointed manipulators with the last three revolute axes intersecting shows a considerable economy in computation time [1]. For such manipulator geometries, the Cartesian path of the wrist center (the point of intersection of the last three revolute axes) is completely determined by the joint values of the first three joints. A computationally efficient hybrid trajectory consists of requiring the wrists center to follow a Cartesian path and joints 4, 5 and 6 to move in joint space motion.

Since the joint values must be found in order to manipulate the arm, the computational costs associated with finding the end point joint solution is included in the total cost. A specific approach would be to consider an example where an arm is to move to a new point with n intermediate points defining the path. First, the function that defines the path must be initialized so that it can define the intermediate points. Next, to generate each intermediate point there is a certain computational cost. Finally, at all n intermediate points, as well as the end point, the joint solution, or inverse kinematics problem, must be solved. So in looking at the total amount of computation there is A) initialization cost B) n intermediate costs and C) n + 1 joint solution costs, where n is the number of intermediate frames. The most reasonable separation of costs is to define the variable cost as the cost associated with the computation of intermediate points. This method of comparing costs for Cartesian and hybrid trajectories is presented in Table 1. The comparison is between hybrid trajectories and Taylor's Method (a), the most efficient method of generating Cartesian trajectories currently published (9).

The following costs are presented using this "shorthand" representation in the Table.

A = Additions or subtractions
M = Multiplications or divisions
S-CP = Sine-cosine pairs
INT = Inverse trigonometric
SQR = Square roots

Table 1

Computational Cost Comparison

OPERATION	TAYLOR [3]	HYBRID APPROXIMATION
Segment Traversal		
Fixed Cost	35a 55M 2S-CP 6INT 4SQR	22A 26M 2S-CP 5INT 2SQR
Variable Cost	36A 60M 3S-CP 5INT 3SQR	9A 12M 2INT 2SQR
Segment Transition		
Fixed Cost	35A 72M 2S-CP 6INT 4SQR	22A 41M 2S-CP 5INT 2SQR
Variable Cost	50A 85M 4S-CP 5INT 3SQR	17A 19M 2INT 2SQR
Pursuit Solution		
Fixed Cost	35A 55M 2S-CP 5INT 4SQR	16A 26M 2S-CP 5INT 2SQR
Variable Cost	37A 60M 3S-CP 5INT 3SQR	16A 12M 2INT 2SQR
Pursuit Traversal		
Fixed Cost	35A 57M 2S-CP 6INT 4SQR	32A 52M 4S-CP 10INT 4SQR
Variable Cost	63A 107M 4S-CP 6INT 4SQR	30A 23M 2INT 2SQR

Significant cost reduction for segment traversal, pursuit formulation, etc., in Table 1 shows the economy that can be achieved using hybrid motion strategy.

Manipulator Dynamics and Motion Paths

Consider a Cartesian trajectory of the end effector denoted by three location coordinates x(t), y(t) and z(t); and three orientation coordinates $\alpha(t)$, $\beta(t)$ and $\gamma(t)$. Let $\underline{x}(t) = \begin{bmatrix} x(t) \\ y(t) \\ z(t) \\ \alpha(t) \\ \beta(t) \\ \gamma(t) \end{bmatrix}$

Let the vector of joint coordinates be denoted by $\{\underline{q}\}$. The vector of torques/forces $\{\underline{\zeta}\}$ at the actuators is given by [25]

$$\zeta = [D(q)]\,\ddot{\underline{q}} + \dot{\underline{q}}^t\,[C]\,\dot{\underline{q}} + \{\tau_g(\underline{q})\} \tag{1}$$

where D(q) is the 6 x 6 system inertia matrix

C(q) is the (6 x 6) coupling matrix containing inertia terms

$\{\tau_g(\underline{q})\}$ is the vector of gravity forces.

The Cartesian coordinates are related to joint coordinates [25], by

$$\dot{x} = [J(\underline{q})]\,\dot{\underline{q}} \tag{2}$$

where [J] is the Jacobian matrix of the system. Differentiating equation (2), one gets

$$\ddot{x} = [\dot{J}]\,\dot{\underline{q}} + [J]\,\ddot{q} \tag{3}$$

from which joint accelerations may be expressed as,

$$\ddot{q} = [J]^{-1}\,[\ddot{x} - \dot{J}\,\dot{\underline{q}}] \tag{4}$$

Let the Cartesian coordinates be a function of independent variable, arc length s, then

$$\dot{x} = \frac{d\underline{x}}{ds}\,\dot{s} \tag{5}$$

and

$$\ddot{x} = \frac{d^2\underline{x}}{ds^2}\,\dot{s}^2 + \frac{d\underline{x}}{ds}\,\ddot{s} \tag{6}$$

Substituting Equations (4) and (6) into Equation (1) yields

$$\{\tau\} = [D(\underline{q})]\ \{[J]^{-1}\ [(\frac{d^2\underline{x}}{ds^2}\ \dot{s}^2 + \frac{d\underline{x}}{ds}\ \ddot{s}) - \dot{J}\ \dot{q}]\}$$

$$+ \dot{\underline{q}}^t\ [c]\ \underline{q} + \{\tau_g(\underline{q})\} \qquad (7)$$

If a Cartesian trajectory is completely specified,

$\underline{x}(s)$, $\frac{d\underline{x}}{ds}$ and $\frac{d^2\underline{x}}{ds^2}$ are known and the manipulator motion is a function of only one unknown variable s. The variable s(t), must be such as to satisfy the actuator bounds. Obviously, the values of s instantaneously may be computed using Eq. (7) for the specified actuator bounds in $\{\tau\}$.

The principle of Pontryagin may be applied [65], and the resulting minimum time motion trajectory s = s(t) may be computed from the solution of coupled nonlinear differential equations.

Joint Motion Strategy

The choice of a polynomial for describing the joint motion or, for that matter, any other constrained function does not assure the compliance of the motion with the actuator bounds, if the time of the motion is predetermined. The constraint functions determining the joint motion may be defined for each joint in terms of independent variables s.

The resultant motion of the manipulator is a function of only one variable s. The variation of s = s(t) may be obtained using minimization of execution time criteria using the Pontryagin principle. Such trajectories will be, in general, of the bang-bang type resulting in high jerks. However, the variation of s as a function time may be obtained using dynamic programming algorithms [65] such that the actuator bounds as well as specified limits on the jerkiness in motion are satisfied.

Actuator Dynamics

There are a variety of actuators used in powering and controlling the motion of links of industrial robots. Generally, the actuator dynamics [42] are formulated by the models of the following form.

$$\dot{\underline{x}}_i = [A_i]\ \underline{x}_i + b_i\ N(u_i) + f_i P_i$$

where

$\underline{x}_i$ is n_i dimensional state vector of the ith actuator.

$[A_i]$ is an ($n_i \times n_i$ - matrix) of the ith actuator model, b_i is an n_i-vector, u^i is the scalar input, $N(u_i)$ is an amplitude saturation type of nonlinearity

$$N(u_i) = \begin{cases} -U_{mi} & \text{for } U_i < -U_{mi} \\ U_i & \text{for } -U_{mi} \leq U_i \leq U_{mi} \\ U_m & \text{for } U_i > U_{mi} \end{cases}$$

f^i is an n_i-vector, P^i is the driving torque acting in the ith degree of freedom. The actuator dynamics [42] may be combined with manipulator dynamics [15]. The resulting dynamics are of very high order and appear to be unsolvable by conventional methods. However, recently developed fast algorithms of recursive dynamics [15] have not yet been applied to the integrated dynamics. The development of such algorithms may render the integrated dynamics solvable in real time.

Pontryagin Maximum Principle

The Pontryagin Maximum Principle [65] has been extensively used in the synthesis of optimal controls. In robotics, Kahn and Roth [2] used it for determining minimum time trajectories in joint space motion for a three link chain consisting of revolute joints where joint angles at the start and the end of the motion as well as the actuator torque bounds were known. The time execution of a motion segment was minimized using the Pontryagin Principle. The joint space trajectories were obtained by solving the resulting boundary value problem involving nonlinear differential equations.

Applying Pontryagin's Maximum Principle for the synthesis of nominal motion trajectories of the manipulator in minimum time yields: Boundary value problems involving coupled nonlinear differential equations. These boundary value problems are difficult and computationally very expensive to solve.

The foremost advantage of using the Pontryagin Principle is in simplicity of formulation and a rather straightforward conversion of a time optimal problem into a local optimum problem in time t.

Consider a Cartesian trajectory segment from 'A' to 'B'. Let the Cartesian path be defined with respect to the length of the curve traversed s. If the total length between segments 'A' and 'B', s_{AB} is very large, the optimal variation of s(t) near A indeed should not depend upon the boundary conditions at B. However, the closer the hand is to station 'B', the greater the influence of the conditions at B will be on the motion.

In general, the time optimal trajectory far away from the end points must be governed by local optimum conditions such as highest possible local speed and/or highest possible local acceleration. The trajectories synthesized on the basis of such local criteria will be very close to the trajectories synthesized on the basis of absolute minimum principles such as Pontryagin's Principle.

Conclusions

In this note, we have presented a brief discussion of the research needed on dynamic synthesis of motion trajectories. General theories of dynamic synthesis are urgently needed for improved efficiency of industrial robots used in a variety of industrial production setups.

References

1. Onan, Lance, and Dilip Kohli: Hybrid motion trajectories for manipulators. Proceedings Robotic Intelligence and Productivity Conference, Wayne State University, Detroit, Nov. 18-19, 1983.

2. Kahn, M. E. and B. Roth: The near minimum-time control of open-loop articulated kinematic chains. J. Dynamic Systems, Measurement, Control, 93 (1971), 164-172.

3. Kahn, M. E.: The near-minimum-time control of open-loop articulated kinematic chains. Stanford Artificial Intelligence Laboratory, AIM 106, December, 1969.

4. Pieper, D. L.: The kinematics of manipulators under computer control. Ph.D. Thesis, Department of Computer Science, Stanford University, 1968.

5. Paul, R. P.: Modelling, trajectory calculation, and serving of a computer controlled arm. Stanford University, Artificial Intelligence Laboratory, AIM 177, November, 1972.

dom are given to allow the end effector (or gripper) to traverse a certain path. Thus, a simple pick and place requirement would necessitate at least two degrees of freedom--one to grip and release, the other to move the gripper from one point to the other. This could be a rotation as shown in Figure 1-1a or a translation as shown below in Figure 1-1f. Most real applications of pick-and-place would require additional degrees of freedom.

The criteria for the selection of the desired degrees of freedom include paths of the end effector in the X, Y, Z axes, as well as the coordinate planes, motions or paths required in the future, the number of intermediate stoppage points, and functions desired at these intermediate points. Current robots are often built with six or more degrees of freedom and are often shown with working axes depicted in Figure 1-1e. Such a robot will supposedly take care of all possible path/stop combinations. There are two important functional drawbacks to this depiction of the robot's capability. The first is that the robot's joints often lock up and interfere with each other before the desired end point is reached. The second is that a moving joint requires an expensive motor and control circuit to generate a movement. Thus, a robot with a multitude of joints would require a complex computer program to synchronize them. A conclusion therefore is that an unnecessary joint is not only expensive but possibly an obstruction.

The modular concept works from a need-based approach. The required path is clearly identified, and then the required joints assembled and controlled to obtain the necessary path. The question arises and as to what might happen in the future; what if future paths are required which are different from the existing path? The modularity concept provides a viable solution--if different paths are required, add new joints or modules if necessary with the necessary software.

Limb Dimensions

Existing robots have the disadvantage of having a fixed working envelope. This necessitates the shifting of existing equipment to suit this envelope, or even worse, designing facilities around the robot. By varying limb dimensions, it is possible to eliminate or minimize this relayout.

Limb dimensions can be changed to accommodate different job configurations as well. This implies that a heavier load would necessitate the use of

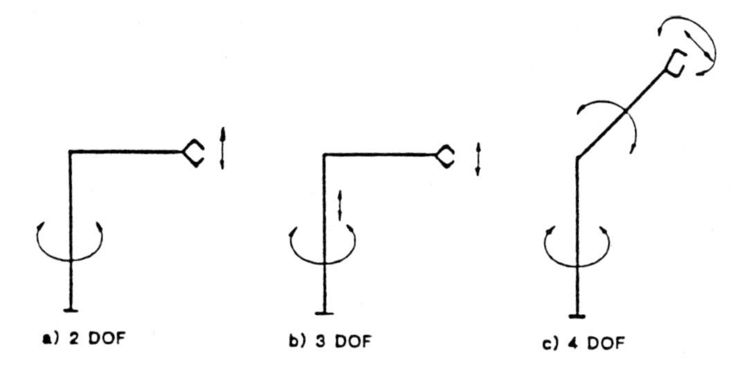

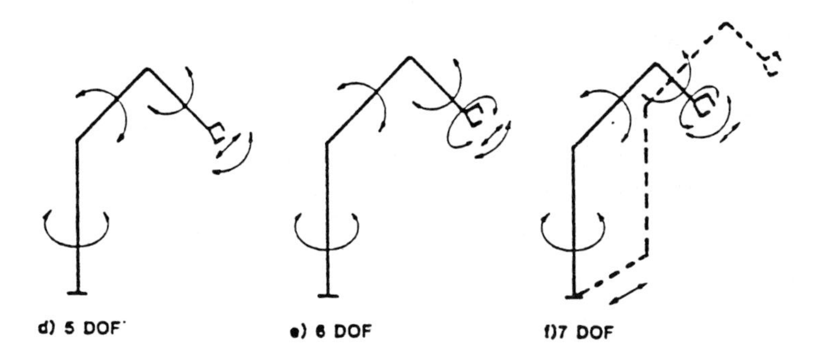

Figure 1-1 Six line diagrams of robots each with a different Configuration and degrees of Freedom (DOF)

heavier, cross-sectional members. Table 1 indicates the variations in the many dimensions of different robot elements. The materials for the robot construction are not mentioned in the table. This is because the variety of materials available lend themselves to wide selection. The most common materials include steel and aluminum. Future growth is expected in the usage of plastics, composites, and graphite fibre structures. The implied trend indicates that lightweight and extremely tough materials will be used in the robots of the future. The point to be made here is that the modular robot permits limb changeability with respect to both dimension and material.

Table 1

Varying Dimensions of Structural Modules

Module (Structural) Member	Module Joint	Varying Dimensions	Degrees of Freedom
Base	-	1. Height	
		2. Cross-section	0 or 1
Body	Waist	1. Height	0 or 1
		2. Cross-section	
Upper arm	Shoulder	1. Length	0, 1 or 2
		2. Cross-section	
Lower arm	Elbow	1. Length	0, 1 or 2
		2. Cross-section	
Swivel	Wrist	1. Configuration	0, 1 or 2
		2. Dimension	

For the modular robot the dimension of a limb can be changed by using many concepts. The first is to stock limbs of different lengths and sections. This has the disadvantage of creating a large inventory. The second involves designing the joints as units and mounting spacers between joints to get different dimensions. Both strategies have the disadvantage that either high inventories must be maintained or long lead times must be quoted to manufacture the required parts. The greatest advantage of varying limb dimensions, of course, is that an efficient robot is manufactured.

The Motive Power Supply System

The three common power supply systems mentioned earlier will now be considered. Here the modularity concept implies using a system which suits a user. For example, a robot feeding a hydraulic press can well be a hydraulic robot as the same hydraulic fluid, pumps, etc., can be used to drive both machines. This would facilitate the bulk purchase of oils, similar maintenance procedures. etc.

Modular robots will offer wider latitudes in the placement of the power drive. Small robots usually have the drive motors mounted on the base to reduce the inertial load of the moving members and so reduce drive torque.

This results in a robot which is always base mounted. A modular robot can possibly have its drive motors near the driven joint. Hence, the mounting of a robot on a building column or roof is also feasible. If a base-mounted modular robot is a must, a wire-cable driven robot offers a higher torque/weight ratio and cost reduction than shaft driven ones, while shaft-driven robots offer more precision. Both, however, could be configured to run off the same computer.

Robot Control

The controller largely determines both the cost of the robot and the effectiveness of its end use. All modern controllers make use of microprocessors technology. The problem with the microprocessors circuits is that the driving programs are as varied as the number of designers of these systems. The modular concept demands a need-based controller, both with respect to control complexity, as well as user compatability. This is important in these days of extremely rapid technological advances and subsequent equally rapid obsolescence. Therefore, the modular robot offers its greatest benefit in this area as one obtains exactly what is required--so a minimum investment is possible with the feasibility of future additions and expansions. Table 2 summarizes the effect a modular robot would have on a robot installation.

Conclusion

Some problems expected with this concept are, first, few machine tool builders are also robot builders; second, programming of the robot for varying degrees of freedom is expected to be more complex; third, the modularity in control technology will have to be experimented with before it can be incorporated; and, finally, the planning of the installation will have to be such that targeted upgrading is possible. These problems are not overriding and initial experimentation has proved the validity of the concept. More work, however, will be required to establish the status of these robots in industry.

Table 2

Effect of Modular Robot Components on Robot Installation

	Degree of Freedom	Limb Dimensions	Power Supply	Control System
Job Characteristics	Product Mix Production Plan	Size, Shape Weight	-	Contouring Characteristics
Operation Characteristics	Process Type	Environment Machine Served	Size of Unit	Acceleration of Characteristics
Supporting Facilities	*	*	*	*
Present Technology Available	*	-	*	*
Safety Standards	-	*	-	-
Existing Plant Layout	*	*	-	-

* indicates some effect

- indicates negligible effect

Dynamic Analysis of Robotic Manipulators with Flexible Links

Om Prakash Agarwal

Dept. of Mechanical Engineering
Temple University
Philadelphia, Pennsylvania

Dept. of Mechanical Engineering
University of Illinois at Chicago
Chicago, Illinois

Abstract

This paper presents a method for dynamic analysis of robotic manipulators consisting of rigid and flexible links. The configuration of each flexible link is represented using two sets of generalized coordinates: reference and elastic generalized coordinates. The elastic generalized coordinates are introduced using a finite element technique. Modal analysis technique is used to reduce the dimension of the problem. Lagrange multiplier technique is used to account for kinematic constraints between two adjacent robotic arms. The equations of motion are integrated using variable order variable step size predictor corrector algorithm. The method is applied to a simple robotic manipulator. Comparission of dynamic response of rigid and flexible robotic manipulator show that the link flexibility can have significant effect on system response.

Introduction

The ever growing demands made for higher industrial productivity have resulted in an increasing use of computer controlled robotic manipulators. Generally these manipulators incorporate various types of driving, sensing, and controlling devices to perform a prespecified task. For most part, the dynamic response of these manipulators is determined assuming that the manipulators are made of rigid links. This is mainly because, the operating speed and performance indices to date, have been relatively low. However, as the demand for higher speed, higher industrial productivity and fine precision work increases, the rigid body assumption will no longer be valid, and a more accurate mathematical model, that account for link flexibility, inertia, and other dynamical effects will be required.

In this paper, an analytical method for modeling the dynamical behavior of robotic manipulators, that consists of rigid and/or flexible links, is presented. The method accounts for the distributed mass and flexibility and their nonlinear dynamic effects. Coupling between gross motion and elastic deformation is considered based upon consistent mass approach. Rigid body translation and large angular rotations are described using a set of Lagrangian coordinates. The deformation of each elastic link is described by elastic generalized coordinates that defines the mode of deformation of links with respect to a reference body axis. Finite element technique is used to account for the complex geometry of the links. It is shown that the reference body axis need not be fixed to a point in the body. A set of reference conditions are used to define the reference axis and the deformation field uniquely.

Background

The general subject of dynamic analysis of flexible multibody system has become a subject of many investigations. An early approach in the analysis of flexible mechanisms is based on a linear theory [1-3]. In this approach the gross motion is first determined, using rigid body analysis and the resulting inertia and reaction forces are introduced in elastic analysis of the components. Linear theory assumptions are no longer accurate enough to represent system dynamics, because coupling between gross motion and elastic deformation can have significant effects.

Sunada and Dubowsky [4-5] recently presented a method for the dynamic analysis of flexible mechanisms. In this method, the existing finite element structural program is combined with a 4x4 matrix dynamic analysis technique. The method has been applied to solve spatial mechanisms and robotic manipulators. In this approach, the mass is assumed to be lumped at finite element grid points and rotary self-inertia at grid points are neglected inorder to avoid the difficulty of using cocsistent mass approach.

Shabana and Wehage [6-7] presented a method for dynamic analysis of large scale inertia-variant flexible systems. In this method the elastic deformation of each flexible component is defined with respect to a reference body-axis that is rigidly attached to a point in the body. The method has been applied to two and three dimensional mechanisms and vehicles. The work presented here is an extension of the method developed in references [6-7] to include different reference and boundary conditions of the inertia-variant flexible links.

Mathematical Formulation

Generalized coordinates and velocities: In order to derive the equations of motion of a flexible body that under go large translations and large rotations, a point on the body must be located with respect to a global coordinate system. As shown in Fig. 1, let XYZ be an inertial coordinate system and $X^iY^iZ^i$ be a coordinate system of body i. Body i is divided into a set of finite elements. The location of a point P on an element ij is given by [8-9],

$$r_p^{ij} = R^i + A^i N^{ij} e^i \tag{1}$$

where R^i is the vector of translational coordinates of reference axis $X^iY^iZ^i$ with respect to global axis XYZ, A^i is an orthogonal transformation matrix from the body $X^iY^iZ^i$ coordinate system to XYZ coordinate system, N^{ij} is the ij^{th} element shape function and e^i is a vector of nodal location of body i with respect to $X^iY^iZ^i$. It should be noted that the nodal variables have been defined with respect to local body-axis. Thus the advantage of local linearity can be taken.

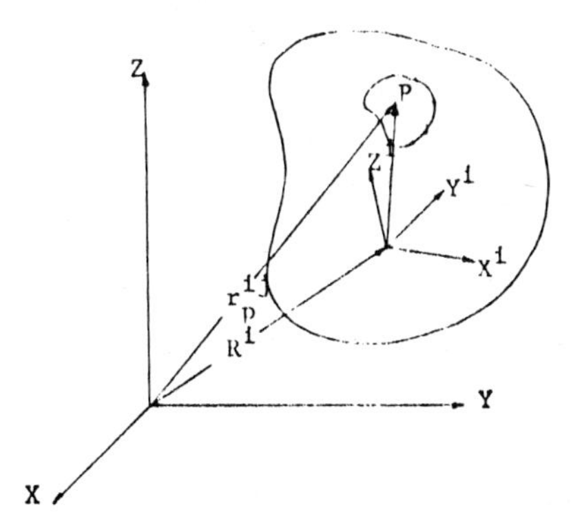

Fig. 1. Generalized Coordinates of Body i

At this point it should be mentioned that the finite element shape function N^{ij} includes rigid body modes that take into account rigid body translation, but is not sufficient to describe large angular rotations. This is mainly because the finite element shape functions assume a small change in rotational variables. These rigid body modes have to be eliminated in order to define a unique displacement field. This is acheived by imposing a set of reference conditions, and the case of zero strain (or rigid body motion) is represented by the set of Lagrangian coordinates. The reference conditions must be selected properly, otherwise it results in improper modes of deformation and even convergence problem. Effects of different reference conditions on mode shapes and on the system response can be found in reference [9]. In general reference conditions are expressed as [9],

$$e^{i} = B_{g}^{i} q_{f}^{i} \tag{2}$$

where q_f^i is a new set of elastic generalized coordinates and B_g^i is a linear transformation matrix whose entries depend on the choice of reference axis. For example, if the reference axis is rigidly attached to a nodal point in the body, then the elastic deformation of the nodal point with respect to the reference axis will be zero. In that case, Eq. 2 is equivalent to eliminati the nodal coordinates of this point.

Substituting Eq. 2 into Eq. 1 and then differentiating Eq. 1 with respect to time, the velocity of point P is given by

$$\dot{r}_{p}^{ij} = R^{i} + \dot{A}^{i} N^{ij} B_{g}^{i} q_{f}^{i} + A^{i} N^{ij} B_{g}^{i} \dot{q}_{f}^{i} \tag{3}$$

Energy expressions: The kinetic energy of body i is given by

$$T^{i} = \tfrac{1}{2} \sum_{j} \int_{V^{ij}} \rho^{ij} \dot{r}^{ij^{T}} \dot{r}^{ij} \, dV^{ij} \tag{4}$$

where ρ^{ij} is the mass density at point P. Using Eq. 3, Eq. 4 can be written in compact form as

$$T^{i} = \tfrac{1}{2} \dot{q}^{i^{T}} M^{i} q^{i} \tag{5}$$

where q^i is a vector of generalized coordinates of body i and M^i is the mass matrix of body i. Matrix M^i can be written as

$$M^i = \begin{bmatrix} m^i_{rr} & m^i_{rf} \\ m^i_{fr} & m^i_{ff} \end{bmatrix} \tag{6}$$

where m^i_{rr} and m^i_{ff} are mass matrices associated with reference and elastic coordinates, and m^i_{rf} and m^i_{fr} represent the inertia coupling between reference and elastic coordinates. It should be mentioned that the matrix m^i_{ff} is a constant conventional mass matrix that arise in finite element analysis. Therefore, a finite element program can be used to generate m^i_{ff}. However, the coupling matrices m^i_{rf} and m^i_{fr} are nonlinear function of reference and elastic coordinates. Therefore, these matrices must be generated and updated at every time step inside the program. These coupling matrices are significantly dependent on the reference conditions and by a proper choice of reference conditions, the calculation of these coupling matrices can be simplified. For example, the inertia coupling can be significantly reduced using mean-axis [9].

Strain energy of body i is written as [6] ,

$$U^i = \tfrac{1}{2}\, q^{i^T} K^i\, q^i \tag{7}$$

where K^i is the stiffness matrix given by

$$K^i = \begin{bmatrix} 0 & 0 \\ 0 & k^i_{ff} \end{bmatrix} \tag{8}$$

Here k^i_{ff} is a stiffness matrix associated with the elastic nodal coordinates. Matrix k^i_{ff} is a positive definite matrix because reference conditions given by Eq. 2 has already been imposed to eliminate the rigid body modes. Notice that there is no stiffness coupling between reference motion and elastic deformation. This is mainly because the elastic coordinates are defined locally with respect to reference axis.

Finally, let Q^i be the generalized force associated with

generalized coordinate q^i. The virtual work of body i is given by

$$\delta W^i = Q^{i^T} \delta q^i \tag{9}$$

where δq^i is the vector of virtual displacement that is consisten with respect to the kinematic constraints. Equation 9 includes all forces except workless constraint forces.

Constraint equations: Compatibility conditions between different links are highly nonlinear because of large relative rotations between individual links. Constraint equations describing the joints can be formulated using a set of nonlinear algebric equations. For example, a revolute joint constraint requires that the joint definition points on two links must coincide, and the axis of the joint on the two links must be collinear at all time 9 . In case of flexible links, the elastic deformations at these joints should also be considered. A set of constraint equations that account for the elastic deformation at the revolut joint are given in reference [9]. Let $\Phi(q,t)$ be the vector of constraints between adjacent links given by

$$\Phi(q,t) = \Phi_1(q,t),\ \Phi_2(q,t),\ \ldots\ ,\ \Phi_m(q,t) = 0 \tag{10}$$

where q is the vector of total generalized coordinates and m is the total number of constraints in the system. These constrain equations are adjoined to the system equations of motion using Lagrange multipliers.

System equation of motion: The variational form of the equations of motion of body i is [6],

$$\frac{d}{dt}\left(T^i_{\dot{q}^i}\right) - T^i_{q^i} + U^i_{q^i} - Q^{i^T}\ \delta q^i = 0 \tag{11}$$

Using Lagrange multiplier technique, and after some manipulations Eq. 11 can be reduced to

$$M^i \ddot{q}^i + K^i q^i = Q^i + F^i(q^i,\dot{q}^i) - \Phi^T_{q^i}\ \lambda = 0 \tag{12}$$

where F^i is a quadratic velocity term that arises from differentiating kinetic energy expression with respect to time and generalized coordinates, λ is the vector of Lagrange multipliers, and Φ_{q^i} is the constraint jacobian matrix.

In general the dimension of Eq. 12 is large because of finite element formulation. Further matrix M^i is a function of generalized coordinates. Therefore, eigenvalues and eigenvectors of this matrix change at every time step. However, the mass and stiffness matrices associated with elastic coordinates are constant matrices. Therefore, a modal analysis is performed 6 assuming the body to be vibrating about its reference axis, and the dimension of the problem is reduced by eliminating higher frequency components. This elimination is justified since, in most practical applications only low frequency modes of vibrations are excited.

Once the dimensions of the problem has been reduced, a variable order, variable step-size predictor corrector algorithm is used to integrate the equations of motion [6].

Numerical Results

The schematic diagram of a robotic manipulator considered here is shown in Fig. 2. In this example arm AB is considered as flexible and arm BC is considered as rigid. The dimensions, material properties, inertia properties, and initial conditions of this robotic manipulator are given in reference [9].

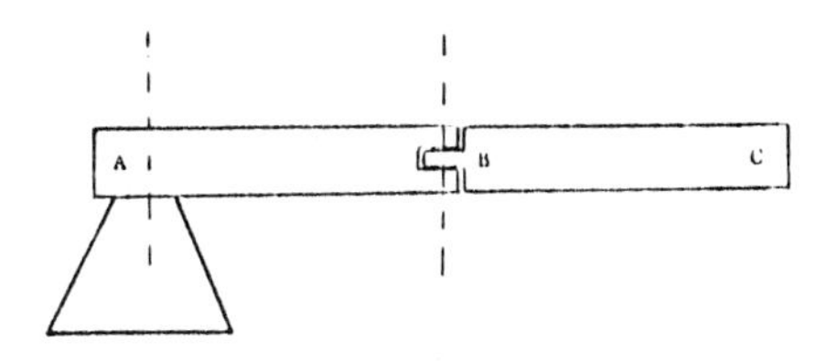

Fig. 2. Robotic Manipulator

In order to reduce the dimension of the problem a set of

elastic deformation modes are identified using reference mean-axis and one out of plane mode is considered. The tip point C is rotated in a circle at an angular velocity of 50 rad/sec. Interest is focused on the out of plane motion of the origin of the reference axis of link AB. The response for rigid and elastic link are given in Fig. 3. Figure 4 shows the difference of the two cases and thus gives the elastic behavior of the flexible link. It can be seen that the elastic displacement can be quite significant.

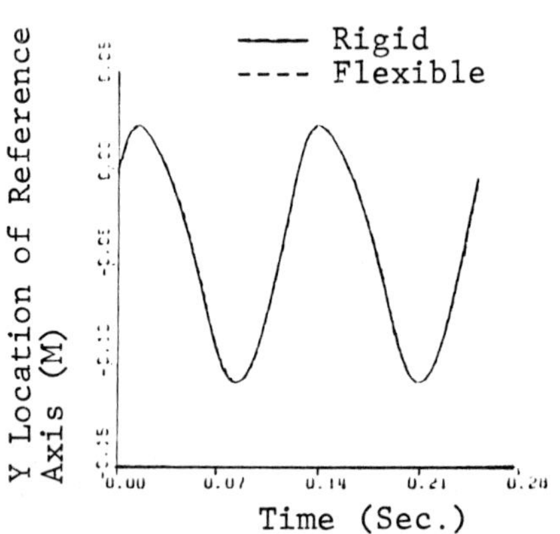

Fig. 3. Comparison of Rigid and Flexible- Arm Response

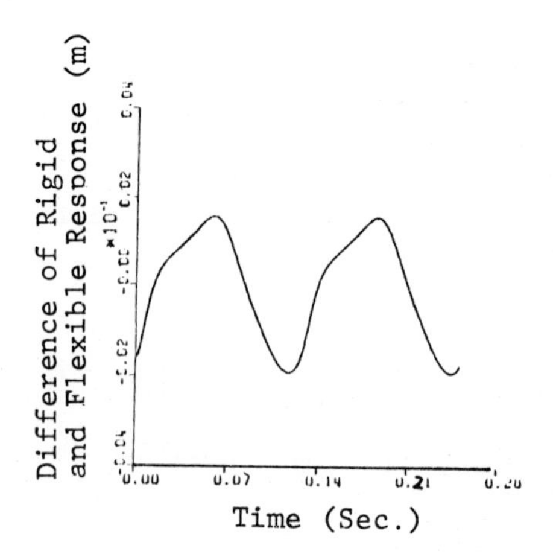

Fig. 4. Difference of Origin of Deformable Mean-Axis and Reference Axis of Rigid Body (Out of Plane Response)

It should be mentioned that the program fails to converge whene an inconsistent inplane mode is considered. This emphasizes the importance of a proper selection of reference axis.

Conclusion

A method for dynamic analysis of flexible robotic manipulators is presented. Configuration of each elastic link is defined using two sets of coordinates: reference and elastic. It is shown that the body axis need not be attacted to a point, rather a set of reference conditions are imposed in order to define the displacement field uniquely. The example considered shows that the elastic displacements can be quite significant.

References

1. Winfrey, R.C.,"Elastic Link Mechanism Dynamics," ASME Journal of Engineering for Industry, Feb. 1971, pp. 268-272.

2. Sadler, J.P., and Sandor, G. N.,"A Lumped Parameter Approach to Vibration and Stress Analysis of Elastic Linkages," ASME Journal of Engineering for Industry, May 1973, pp. 549-557.

3. Erdman, A.C., Sandor, G.N., and Oakberg, R.C.,"A General Method for Kineto-Elastodynamic Analysis and Synthesis," ASME Journal of Engineering for Industry, Nov. 1972, pp. 1193-1205.

4. Sunada, W., and Dubowsky,S.,"The Application of Finite Element Methods to the Dynamic Analysis of Flexible Spatial and Co-planar Linkage Systems," ASME Journal of Mechanical Design, Vol. 103, No. 3, July 1981, pp. 643-651.

5. Sunada, W., and Dubowsky, S.,"On the Dynamic Analysis and Behavior of Industrial Robotic Manipulators with Elastic Members," ASME Journal of Mechanisms, Transmissions, and Automation in Design, Vol.105, No. 1,Mar. 1983, pp. 42-51.

6. Shabana, A.A, and Wehage, R.A,"variable Degree of Freedom Component Mode Analysis of Inertia-Variant Flexible Mechanical Systems," ASME Journal of Mechanisms, Transmissions and Automation in Design,Vol.105,No.3,Sept.1983,pp. 370-378.

7. Shabana, A.A., and Wehage, R.A.,"Spatial Transient Analysis of Inertia-Variant Flexible Mechanical Systems," ASME Journal of Mechanisms, Transmission, and Automation in Design, Vol. 106, No. 2, June 1984, pp. 172-178.

8. Agrawal, O.P., and Shabana, A.A.,"Dynamic Analysis of Multi-Body Systems Using Component Modes," To appear in Computers and Structures.

9. Agrawal, O.P.,"Application of Deformable Body Mean-Axis to Dynamics of Flexible Mechanical Systems," Doctoral Dissertation, Mechanical Engineering Dept., University of Illinois at Chicago, 1984.

Analysis of Structural Compliance for Robot Manipulators

Eugene I. Rivin

Department of Mechanical Engineering
Wayne State University
Detroit, Michigan 48202

Summary

Compliance breakdown is computed for robots with hydraulic and electromechanical drives. It is shown that in all cases the breakdown is dominated by one component whose stiffening would lead to a substantial improvement in overall sructural stiffness and the lowest natural frequency, while stiffening of other components would not cause a significant effect.

The dominating compliance component may represent the drive system (in the hydraulically-driven robot), the transmission, or some structural joints. In all cases considered, link bending contributes less than 20% of the overall compliance. The importance of the results for the structural design and control systems of robots is discussed.

Introduction

Compliance of manipulator structures is an important factor in developing optimal control strategies. However, recent publications on the problem of robot control using compliant models usually consider only the bending compliance of links (e.g., [1]). Since it is clear á priori that transmissions and other components also contribute to compliance, it is very important to identify and quantify the major elements of the compliance breakdown. First, compliance of transmission components, joints, etc., has a lumped spring-like character, as compared to the distributed character of the bending compliance. Accordingly, their treatment in dynamic computations in general and in control system evaluations in particular is different. Another and even more important reason to study compliance breakdown is the need to reduce the compliance of robot structures in order to increase the effective speed, accuracy, and suitability of robots for machining operations [2], etc.

This paper presents the computational results of compliance breakdown for three typical robot systems: a hydraulic robot (similar to the Unimate

2000B); an electromechanical jointed robot with serial links (similar to the PUMA 50); and an electromechanical parallelogram robot similar to the Hitachi Process robot. The word "similar" is used since the initial information employed for the computations (such as types of components, dimensions, etc.) was in many cases taken (or assumed) from available in small-scale public domain drawings and pictures.

A. A Hydraulically Driven Robot (Fig. 1) operates in spherical coordinates. Rotation around the vertical axis is associated with the highest end-of-arm speed and is the most frequently used mode of motion (e.g., for all "pick and place" applications). Accordingly, this mode of motion was selected for analysis of overall compliance/lowest natural frequency. Fig. 1 illustrates design features relevant to this mode. The column carries an aluminum arm carriage with extending steel rods supported by bronze bushings in the carriage and carrying the end block with the wrist and/or the end effector (not shown). The arm carriage is connected to the top surface of the column through the pin-hole pivotal connection. Attached to the lower end of the column is the pinion, which engages with the rack driven by the double cylinder system. Pressurized oil is supplied to the cylinders from the pump through the metal pipes, the hose insert, and the servovalve.

The overall deflection caused by the inertia force P was found to be composed of the following components (all deflections are reduced to the arm end):

a. Bending deflection of the rods, $\delta_a/P = 7.8 \times 10^{-4}$ mm/N;

b. Contact deformation in the rod bushings, $\delta_b/P = 0.034 \times 10^{-4}$ mm/N;

c. Bending deflection of the arm carriage, $\delta_b/P \cong 0$;

d. Contact deformation in the pin-hole connection, $\delta_d/P = 1.6 \times 10^{-4}$ mm/N;

e. Twist of the column, $\delta_e/P = 2.2 \times 10^{-4}$ mm/N;

f. Bending and contact deformation in the rack-and-pinion mesh, $\delta_f/P = 1.8 \times 10^{-4}$ mm/N;

g. Compression of oil in the hydraulic cylinders, $\delta_g/P = 79.8 \times 10^{-4}$ mm/N;

h. Compression of oil in the rigid plumbing, $\delta_h/P = 74.9 \times 10^{-4}$ mm/N;

i. Compression of oil in the hose section, $\delta_i/P = 532 \times 10^{-4}$ mm/N.

In the computational process, contact compliances in the joints were considered according to [3] (item f) and [4] (items b,d). Effective compressibility of the hydraulic oil was modified by deformations of the cylinder/piping/hose walls. Compliance components g, h, i cannot be expected to be very accurate, due to scatter in oil compressibility, hose parameters, boundary conditions, etc. Reduction of a compliance component from its actual value (e_a) to the effective value at the arm end (e_e) was performed using a conventional transformation [3]

$$e_e = e_a \, i^2_{a,e},$$

where $i_{a,e}$ is the ratio of velocities (or displacements) between the points where the component under consideration is located and to where the compliance is reduced.

The total compliance, reduced to the end effector, is a sum of the above identified components, as follows:

$$e = (7.8 + 0.034 + 0 + 1.6 + 2.2 + 1.8 + 79.8 + 74.9 + 532)\ 10^{-4}$$
$$= 699 \times 10^{-4}\ \text{mm/N}$$

Experimental data measured on a Unimate-2000 robot* (with its hydraulic system energized) has shown a wide variation of e in the range of 120-1,050 x 10^{-4} mm/N. Thus the calculated value is in the ballpark -- not a bad correlation considering the indeterminacy and nonlinearity of oil and hose parameters.

Total inertia at the end effector with full payload is $m \cong 90$ kg; thus the fundamental natural frequency f = 2.0 Hz. The range of f, if experimental values of e are used, is f = 4.8 - 1.5 Hz.

This low value for the natural frequency explains why so much time is required for arm-positioning (500 ms for response to a control signal, as indicated in [5]). The compliance breakdown allows us to find simple means to increase stiffness and natural frequency:

α) Replacement of the flexible hose with a state-of-the-art hose <u>or</u> with a metal pipe (compliance 47 x 10^{-4} instead of 532 x 10^{-4} mm/N). The result: $e_\alpha = 21.4 \times 10^{-4}$ mm/N; $f_\alpha = 3.62$ Hz;

β) After α is implemented, additionally shortening the total piping length by 50%. The result: $e_\beta = 15.3 \times 10^{-4}$ mm/N, f_β = 4.28 Hz. This modification implemented without α would not have a noticeable effect.

γ) After α and β are implemented, additional enlargement of the cylinder diameter 1.5 times (with necessary modifications of

---

* Tests were performed together with Mrs. M. Forest.

the hydraulic system) <u>or</u> introduction of a reduction stage between the pinion and the column with 1.5:1 ratio, thus reducing the effective compliance of all hydralic components $1.5^2 = 2.25$ times. The result: $e_\gamma = 75{,}3 \times 10^{-4}$ mm/N, $f_\gamma = 6.1$ Hz. This modification would be effective to some degree even without prior implementation of α and β.

It can be seen that the very modest design modifications α-γ would significantly increase stiffness and natural frequency. However, even after these modifications are implemented, the mechanical compliances, including the most important contributor, bending compliance, play a very small role in the breakdown (less than 2% initially, about 15% after modifications α,β,γ are implemented). This is typical for hydraulically-driven robots.

B. <u>An Electromechanically-Driven Robot With Jointed Structure (Fig. 2)</u>. For a comparison, the same mode of motion is considered: rotation <u>around a vertical (column) axis</u> at maximum outreach of the arm. Unlike the previous case, the compliance components are reduced to torsional compliance around the vertical axis. The identified contributors to the effective compliance are:

a) Bending deflection of the forearm under the inertia force, $e_a = 6.5 \times 10^{-7}$ rad/Nm;

b) Contact deformations in the joint ball bearing between the forearm and the upper arm ("elbow joint"), $e_b = 38.4 \times 10^{-7}$ rad/Nm;

c) Bending deflection of the upper arm, $e_c = 8.1 \times 10^{-7}$ rad/Nm;

d) Contact deformations in the joint ball bearing between the upper arm and the shoulder, $e_d = 13.9 \times 10^{-7}$ rad/Nm;

e) Twisting of the vertical column (waist) inside the trunk, $e_e = 9.3 \times 10^{-7}$ rad/Nm;

f) Angular deformation of the gear train between the driving motor and the column, $e_f = 1.6 \times 10^{-7}$ rad/Nm.

Components b, d, f were computed by [3]. As a result, the breakdown and overall torsional compliance of the robot in the mode considered are

$$e = (6.5+38.4+8.1+12.9+9.3+1.6)10^{-7} \text{ rad/Nm} = 77.8 \times 10^{-7} \text{ rad/Nm}$$

With the full payload of 2.5 kg (5.5 lbs), and considering the structural mass of the linkage, the fundamental natural frequency $f = 26.9$ Hz.

The compliance of this electromechanical robot in this mode of motion is much lower than that of the hydraulically-driven robot considered in Section A, which results in much higher natural frequency (26.9 Hz vs 2.0 Hz) and, accordingly, much better repeatability (± 0.1mm vs 2 mm). Experimental data in [6] for the arm compliance of the PUMA robot in the vertical plane show ten times higher compliance, which is, largely, determined by the compliance of joint motors. The latter can be quite high, as is shown in [3].

The breakdown of compliance in this case is much more uniform, but is still dominated by one component. This component, 38.4×10^{-7} rad/Nm, and the next most important component, 13.9×10^{-7} rad/Nm, are both associated with joints equipped with ball bearings and can be easily improved by not very significant modifications of the joints (e.g, larger spread of the bearings; selection of more rigid bearings). These measures could realistically reduce the respective compliance terms about 50% (in fact, only the elbow joint is critical). Total (bending) compliance of the links is not very significant, 19.0% of the overall compliance without design modifications, and becomes about 26% after the suggested modifications.

C. An Electromechanically-Driven Robot with a Parallelogram Structure and Harmonic Drives. The parallelogram structure shown in Fig. 3 has recently become a popular structural type. For this robot, a purely rotational movement of the payload in the vertical plane is considered. The major contributors to the compliance breakdown reduced to angular compliance of the end link are:

a) Bending of the forearm, $e_a = 12.5 \times 10^{-7}$ rad/Nm.

b) Stretching of the rear upper arm together with bearing deformations, $e_b = 4.5 \times 10^{-7}$ rad/Nm;

c) Stretching of the front upper arm together with bearing deformation, $e_c = 5.4 \times 10^{-7}$ rad/Nm;

d) Angular compliance of the harmonic drive transmission, $e_d = 173 \times 10^{-7}$ rad/Nm.

The overall compliance is, accordingly,

$$e = (12.5 + 4.5 + 5.4 + 173) \times 10^{7} = 195.4 \times 10^{-7} \text{ rad/Nm}$$

Considering the rated payload (10 kg), and the structural mass reduced to the forearm end, the natural frequency is $f = 12.5$ Hz.

The breakdown of compliance in this case is dominated by the harmonic drive compliance (close to 90% of the overall compliance). This situation can be improved either by redesign or by a mechanical insulation of the harmonic drive from the structure through introduction of a reducing stage (e.g., spur or helical gears). The forearm bending compliance constitutes about 6% of the overall compliance. This percentage would rise if the drive stiffness were enhanced.

Discussion

It was shown that the linkage contribution to the overall compliance is small, not exceeding 20% even in the ultimate case. Accordingly, certain suggested effective and sophisticated, but expensive, measures to stiffen links, such as active control of link compliance [7], seem to be unnecessary, at least if the structural design is not radically changed. Similarly, the algorithm of "compliant link control" [1] can be simplified by utilizing the lump parameter "mass-spring" representation of the linkage.

Conclusions

1. Three principal sources of compliance in robot manipulators are found to be: compliance of the drive/power transmission system; structural compliance of linkages; and compliance of structural joints (pivots, bearings, etc.).
2. Overall compliance, as well as the relative importance of these contributors, varies greatly, depending mainly on the type of drive system used (electro-mechanical, hydraulic, pneumatic), and the linkage structure.
3. Computation of static compliance breakdown allows evaluation of the design quality of the manipulator and selection of the most effective means of upgrading it while in the blueprint stage, with relatively little effort.
4. The compliance breakdown of robotic structures is usually dominated by one component, whose identification allows for subsequent significant improvement of the robot's performance.
5. Linkage compliance does not play a noticeable role in the compliance breakdown of the robotic structures considered.
6. In many instances, a reasonably minor design effort aimed at stiffening of the dominating compliant component would lead to a major improvement in the robot's performance (cycle time, responsiveness, accuracy).
7. Accordingly, special efforts are warranted in the development of

basic structural units of low compliance for robotic manipulators, especialy transmissions and joint designs.

8. A mechanical system can be adequately represented as a mass-spring lump parameter system for the purpose of development of an optimized control strategy.

Acknowledgement

This work was supported by the NSF Grant MEA83-14568.

References

1. Book, W.J., Majett, M., "Controller Design of Flexible, Distributed Parameter Mechanical Arms," ASME Journal of Dynamic Systems, Measurement, and Control, 1983, Vol. 105, pp. 245-254.

2. Knight, J.A.G., Chapman, P., "An Assessment of the Potential of an Industrial Robot for Use as a Flexible Drilling Machine," in Proceedings of 23rd International Machine Tool Design and Research Conference, Univ. of Manchester, 1982.

3. Rivin, E.I., "Compilation and Compression of Mathematical Model for a Machine Transmission," ASME Paper 80-DET-104.

4. Levina, Z.M., Reshetov, D.N., "Contact Stiffness of Machines," Mashinostroenie Publish. House, Moscow, 1972 (in Russian).

5. Agin, G., Issues Involving Sensory Control, in "Research Needed to Advance the State of Knowledge in Robotics," Newport, R.I., April 15-17, 1980 (NSF Workshop).

6. Whitney, D.E. Lozinski, C.A., Rourke, J.M., "Industrial Robot Calibration Method and Results, in "Computers in Engineering 1984," ASME, 1984.

7. Zalucky, A., Hardt, D.E., "Active Control of Robot Structure Deflections," ASME Journal of Dynamic Systems, Measurement, and Control, 1984, Vol. 106, pp. 63-69.

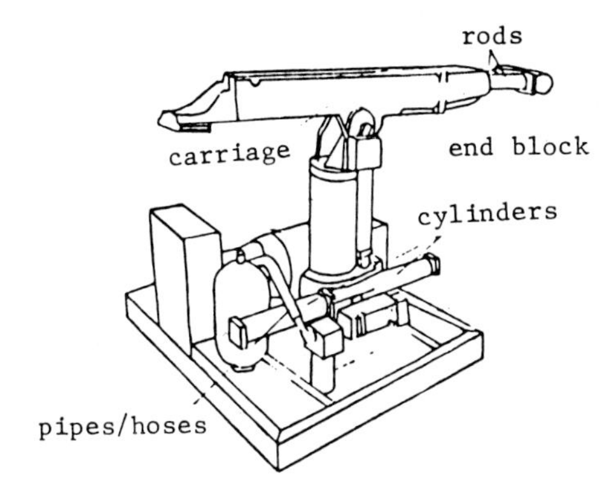

Fig.1. Hydraulically driven robot

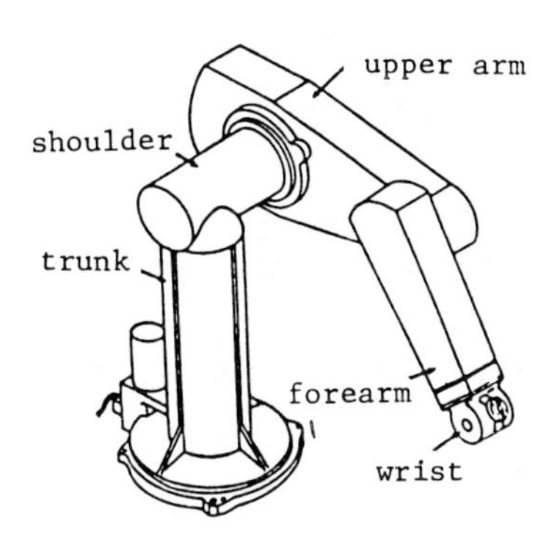

Fig.2. Electromechanical jointed robot

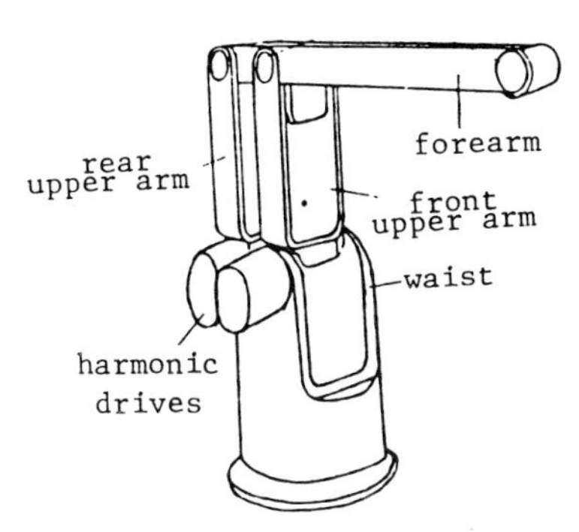

Fig.3. Electromechanical parallelogram robot

Complete Dynamic Model Generation for Articulated Mechanisms

Pierre J. Andre

Centre "Microsystèmes et Robotique" - Unité Associée au C.N.R.S.
E.N.S.M.M. - La Bouloie - 25030 BESANCON CEDEX - FRANCE

ABSTRACT

In order to speed up and make more reliable the setting up of the litteral equations of the geometric, kinematic or dynamic models of a manipulator, we propose a set of programs which, thanks to the simplifications which are performed within the characters strings, leads to compact equations of the complete models, in which not any term is neglected. In addition these programs run under MS/DOS operating system on any 8086/8088 based microcomputer.

1 - INTRODUCTION

The use of matricial representation leads to condensed analytical developments and easy litteral manipulations (1) when long and complicated equations have to be set up. Especially in the field of mechanics applied to robotics.

From a mechanical viewpoint the modeling of an articulated mechanism consists in the setting up of the litteral equations describing the behaviour of a set of articulated bodies submitted to external forces. In case of hand setting up of such equations the risk of mistakes is very important ; however hand computation can be prefered because it allows the operator to control the setting up of the expressions in order to avoid unnecessary lengthening, while a computer usually works juxtaposing terms after only summary comparison and regrouping : it sets the equations without simplifying them as does a man. So, though the expressions obtained by both methods are equivalent, those delivered by automatic generation are much longer. In addition, when the model is used for numerical applications, the terms in excess increase the number of operations and thus the computation time, making difficult a real time control of the manipulator.

Very often, in order to reduce the length of the equations, it is assumed that the upper joints are complete links, the lower being free ones (2). This way leads to an approximate truncated model.

We propose to establish, under a reduced form, the formal equations of the geometric, kinematic or dynamic models of articulated mechanisms by means of algorithms that we have developed and which simulate the simplification logic of a human operator. Although miscellaneous process of symbolic calculus have already been developed (4) we prefered an original method more adapted to mechanics computations.

We remarked that by use of certain wrintings of the derivatives of the frame transformation matrices, it was posible to establish modeling programs involving almost exclusively matrices products even in the case of the dynamic modeling. In the first part of this paper we expose the basic tool consituted by the symbolic matrices product ; among all the modules of the library that we have developed it is by far the most complicated one which needed to be very carefully developed. Its construction has been based upon an original method of formulation of the matricial product. In the last part we compare the results obtained by use of our method to those obtained by RENAUD for the A(i,j) matrix of the simplified dynamic model of the RNUR Vertical 80 robot (2).

2 - EXAMPLE OF MECHANICAL FORMALISM BASED UPON THE MATRICIAL PRODUCT

2-1. Geometric modeling :

An operation currently used in robotics field is to express the position and the orientation of a robot's effector in a fixed frame. To each body S_i constituting the robot a local frame R_i is attached.

So it is possible to represent the coordinate transformation from the R_i to the R_{i-1} frame by use of the following 4 x 4 matrix T_{i-1}^{i} :

$$T_{i-1}^{i} = \begin{vmatrix} \text{3 x 3 frame} & : \alpha \\ \text{transformation} & : \beta \\ \text{---------------} & : \gamma \\ 0 \quad 0 \quad 0 & 1 \end{vmatrix}$$

where α, β, γ represent the coordinates of O_i refered to R_{i-1}.

The T_o^n matrix expressing the coordinates of a point of the effector in the fixed frame is obtained by use of the matrices product :

$$T_o^n = T_o^1 \times T_1^2 \times \ldots \times T_{n-1}^n$$

2-2. Kinematic modeling :

The wrinting of the derivatives of these 4 x 4 matrices under the form of the product of these matrices by the associated differential matricial operator can be efficiently used in order to avoid the creation of symbolic derivation modules which is a very hard programmation work.

Simionescu and Duca (6) translate the iterative relationship :

$$A_i = \overline{A_i} + \left| \frac{\partial A_i}{\partial a_i} \right|_{a_i=\overline{a_i}} da_i + \left| \frac{\partial A_i}{\partial \alpha_i} \right|_{\alpha_i=\overline{\alpha_i}} d\alpha_i + \left| \frac{\partial A_i}{\partial s_i} \right|_{s_i=\overline{s_i}} ds_i + \left| \frac{\partial A_i}{\partial \theta_{in}} \right|_{\theta_{in}=\overline{\theta_{in}}} d\theta_{in}$$

with A_i coordinates transformation matrix from the ith to the (i+1) th body in the closed loop mechanism, $a_i = s_i$ axial distances ; $\theta_i = \alpha_i$ angular differences

into the equation :

$$A_i = \overline{A_i} + \overline{A_i}\Omega_a da_i + \overline{A_i}\Omega_\alpha d\alpha_i + \Omega_s\overline{A_i}ds_i + \Omega\overline{A_i}d\theta_{in}$$

where Ω_a, Ω_α, Ω_s, Ω_θ are the matricial differential operators such that :

$$\left.\frac{\partial A_i}{\partial q_i}\right|_{q_i=\overline{q_i}} = \overline{A_i}\ \Omega_q$$

In our research work we also have used these properties to set up the dynamic model that we present now.

2-3. Dynamic modeling :

To establish the equations we have used a methodology found in a JPL report (7). For a N degrees of freedom manipulator it leads to the following movement equations :

$$\sum_{j=1}^{N}\left\{\sum_{k=1}^{j} \text{Trace } (U_{jk}J_j v_{ji}^t)\ddot{q}_k + \sum_{k=1}^{i}\sum_{p=1}^{j} \text{Trace } (U_{jkp}J_j U_{ji}^t)\dot{q}_k\dot{q}_p - m_j G v_{ji}\overline{\rho_j}\right\} = F_i$$

with : F_i : force or torque actuating the ith joint

q_i : articular variable giving the position of the ith joint

$\dot{q}_i$, $\ddot{q}_i$: speed and acceleration of the ith joint

m_j : mass of the jth body

$\overline{\rho_j}$: vector of the mass centre of the jth body referring to the local frame associated to this body : $\overline{\rho_j} = (x_j, y_j, z_j, 1)^T$

G : gravity acceleration in the fixed frame

$U_{ji} = \frac{\partial T_o^j}{\partial q_i} = T_o^1\ T_1^2 \ldots\ \Omega T_{i-1}^i \ldots\ T_{j-1}^j$

where Ω = matricial differential operator

$U_{jkp} = T^j = T_o^1\ T_1^2 \ldots\ \Omega T_{k-1}^k \ldots\ \Omega T_{i-1}^i \ldots\ T_{j-1}^j$

J_j : inertia pseudo-matrix

If we use the matricial form commonly used in robotics :

$$A\ddot{q} + B\dot{q}\dot{q} = \Gamma + G$$

then the matrices components are (2) :

$$A(i,j) = \sum_{j=\max(i,k)}^{N} \text{Trace } (U_{jk}J_jU_{ji}^t)\ ;\ B(i,j,k) = \sum_{j=\max(i,k,p)}^{N} \text{Trace } (U_{jkp}J_jU_{ji}^t)$$

$$G_i = \sum_{j=1}^{N} m_j \, GU_{ji} \, \overline{\rho_j}$$

So it can be seen that, even for the dynamic model, we avoid the writing of derivation programs if we express U_{jk} and U_{jkp} as products of matrices. So the litteral equations of the dynamic model can be obtained by use of a minimum number of basic modules : this corresponding to the symbolic matricial product and this for the trace computation. We now present these modules.

3 - PRESENTATION OF THE MODULE FOR THE SYMBOLIC MATRICIAL PRODUCT

3-1. Formulation :

We want to multiply the (m,p) A matrix by the (p,q) B matrix. In order to consider the most general case we assume that the terms of both matrices are of the form :

$$a_{ij} = \sum_{k=1}^{\ell_{ij}} a_{ij}^{k} \quad ; \quad b_{ij} = \sum_{r=1}^{s_{ij}} b_{ij}^{r}$$

where ℓ_{ij} and s_{ij} indicate the number of added/substracted terms contained respectively in a_{ij} and b_{ij}.

Let $\ell = \max_{i,j} \ell_{ij}$ and $s = \max_{i,j} s_{ij}$

Under these assuptions A is the sum of ℓ matrices A^k and B is th sum of s matrices B^r :

$$A = \sum_{k=1}^{\ell} A^k \quad ; \quad B = \sum_{r=1}^{s} B^r .$$

So we can write :

$$A^k * B^r = (\sum_{k=1}^{\ell} A^k) * (\sum_{r=1}^{s} B^r) = \sum_{k=1}^{\ell} A^k * (\sum_{r=1}^{s} B^r) = \sum_{k=1}^{\ell} \sum_{r=1}^{s} A^k * B^r$$

If $\overline{A}_j^k$ are the column vectors of A^k and $\underline{B}_i^k$ the line vectors of B^r

Then : $A^k * B^r = \sum_{i=1}^{\ell} \overline{A}_i^k * \underline{B}_i^r$

$\overline{A}_i^k * \underline{B}_i^r$ is a (m,q) matrix, m being the line dimension of A and q the column dimension of B. All the terms of this matrix are monomials. That constitutes the originality of our formulation : due to the distributivity of the multiplication with regard to the addition,the terms distribution is defined at the level of the formula that we establish ; so when programming this formula we "jump" as well above the intermediate stage -commonly used in symbolic calculus programs- needing parentheses, as the use of intermediate variables which does'nt take advantage of the distributivity with regard to the terms included within these variables (1-2).

Sharing the A and B matrices the way we have presented leads to :

$$R = \sum_{k=1}^{\ell} \sum_{r=1}^{s} \sum_{i=1}^{p} \bar{A}_i^k * \underline{B}_i^r \quad ; \quad R \text{ being a } (m,q) \text{ matrix.}$$

Thus the expression of the $r_{ir,jr}$ component of R is as follows :

$$r_{ir,jr} = \sum_{i=1}^{p} \sum_{k=1}^{\ell_{ir,i}} \sum_{r=1}^{s_{i,jr}} a_{ir,i}^k * b_{i,jr}^r \quad ; \quad 1 < ir < m \quad ; \quad 1 < jr < q$$

The most interesting feature of the $a_{ir,i}^k * b_{i,jr}^r$ monomial is to be a product of symbolic and/or numerical variables which can be elevated to any power. We will call such terms "elementary ones.

The use of this formula makes considerably easier the programmation work because it avoids the simplifications of symbolic expressions obtained after the multiplication of non-simple terms.

3-2. Description of the modules :

The elementary modules constituting the modules performing the symbolic matricial multiplication and the trace calculation are listed below (notice that by combination of these elementary modules it is possible to create other modules for matricial computation) :

i - decomposition of a string in order to extract monomials ;

ii - decomposition of a product of factors into a numerical part and a symbolic one ;

iii - decomposition of a symbolic part into its different factors each associated to their power and dissociation of each factor into its symbolic value and its exponent ;

iv - comparison of each factor of a product in order to detect wether or not two products are equal ;

v - multiplication without simplification of two strings ;

vi - analysis of the different factors of a product and eventually rearrangement in order to obtain a compact form ;

vii - addition of two terms applying rules leading to regroup equal values;

viii - addition of n terms with compaction all along the calculation of the result ;

ix - contraction of expressions ;

x - scalar product ;

xi - computation of the trace of a matrix ;

xii - product of N matrices.

4 - APPLICATION

In order to evaluate the efficiency of our method, we will compare our results to these obtained by means of the method proposed by RENAUD (2).

For the term A(1,1), the most complex one of the A(i,j) matrix RENAUD's method gives :

```
A(1,1) = K1 + I2 * S2 * S2 + RO7 * C2 * C2 - 2.0*L3*U3*M3*S2*C2*S3 + 2.0*C2*L3*
         U3*M3*C2*C3 + I3*S2*C3*S2*C3 + I3*S2*C2*S3*C3 + I3*S2*C2*S3*
         C3 + I3*C2*S3*C2*S3 + RO9S2*S3*S2*S3 - RO9*S2*C2*S3*C3 - RO9*
         S2*C2*S3*C3 + RO9*C2*C3*C2*C3
```

Our method leads to :

```
A(1,1) = K1 + I2*S2^2 + RO7*C2^2 - 2*L3*U3*M3*S2*C2*S3 + 2*L3*U3*M3*
         C2^2*C3 + I3*S2^2*C3^2 + 2*I3*S2*C2*S3*C3 + I3*C2^2*S3^2
         +RO9*S2^2*S3^2 - 2*RO9*S2*C2*S3*C3 + RO9*C2^2*C3^2
```

Instead of the 191 characters needed by RENAUD's method the our neds only 147 characters due to the contractions induced by the regrouping of terms.

In this example the expression :

- RO9*S2*C2*S3*C3 - RO9*S2*C2*S3*C3

use 32 characters and needs 8 multiplications and 2 additions for its numerical evaluation ; by use of our method the equivalent term :

- 2RO9*S2*C2*S3*C3

use only 18 characters and only needs 5 multiplications.

For the whole A(1,1) expression the place saving exceeds 23 %, 7 multiplications and 2 additions are saved. Notice that the more complex the expressions are, the greater are the place and computation savings.

5 - CONCLUSION

Our software allows us to get complete dynamic models of various articulated mechanisms under their most compact form.

These programs run under MS/DOS operating system. They have been written in BASIC language and so can be very easily transported.

It has to be noticed that these computations are usually executed offline and so the calculation time that they need is a minor drawback as compared to the advantages they bring : minimal models for simulation purposes, possibility to reliably evaluate the importance of all the terms of the model and justify simplifications, possibility to use i real time the equations since they are the shortest possible and need the smallest number of arithmetic operations.

REFERENCES

(1) ALDON M.J. - Elaboration automatique des modeles dynamiques des robots en vue de leur conception et de leur commande - Thèse d'Etat - MONTPELLIER - 1982

(2) RENAUD M. - Contribution à la modélisation et à la commande dynamique des robots manipulateurs - Thèse d'Etat - TOULOUSE - 1980

(3) MEGAHED S. - RENAUD M. - Minimization of the computation time necessary for the dynamic control of robot manipulators - Proceedings of the 12 th ISIR - PARIS - 1982

(4) LAPLACE A. Un outil de mise au point et d'aide au calcul formel sur ordinateur - Thèse d'Etat - GRENOBLE - 1973

(5) BRAT V. - STEJSKAL V. - OPICKAL F. - The derivation of general kinematic equations of spatial constrained mechanical systems with the aid of a computer - Mechanism and Machine Theory - Vol. 14, 341-347, 1979

(6) SIMIONESCU I. - DUCA C. - On a general method for the synthesis of the path approximation mechanisms - Mechanisms and Machines Theory - Vol.14 289-298 - 1979

(7) BEJCZY A.K. - Robot arm dynamic and control - Technical Memorandum 33-669 - J.P.L. - California Institute of Technology - PASADENA - 1974.

Further Researcch of Lightweight Flexible Robot Arm Design

Ahmad Hemani

Department of Mechanical Engineering
Ecole Polytechnique de Montréal, Montreal

Summary

The results of further research on a new idea towards the design of a light weight flexible robot arm are presented. This string controlled snake-like manipulator will be able to shape itself as necessary in three dimensions through the narrow passages and obstacles. Some more fundamental analysis of the loads leading to the necessary conditions for mechanical design and also further in detail studies of the kinematics of such a system are carried out and discussed.

Introduction

In an earlier paper [1], a novel idea towards the design of a new class of robot arm, that is flexible snake-like arm, was proposed and the results of some research on this subject was presented. This design is attractive from the view point of the fact that such a robot arm will be also light weight. All the movements of the arm are manoeuvered by strings, which are always kept in tension.

A flexible arm is consisting of a number of individually flexible elements. The arm is capable of being figured the required shape, as well as achieving a desired position and orientation at the neck where a tool will be attached. The configurations and the movements of such an arm, of course, are restricted as discussed in [1].

In this paper, some more detailed analysis of the mechanics of the system and its behaviour under different load forces will be studied. This leads to guidelines for later steps of mechanical construction and material selection.

Loads on a flexible element

Figure (1) shows a single flexible element as part of a flexible arm in an arbitrary configuration. This element has to transfer all the loads acting on its plate 2 to the support, partly by its three strings and partly through the elements preceding it. This load, in general, consists of three forces and three torques, respectively along and about three axes of a coordinate system attached to plate 1 of the element, as shown in figure (2).

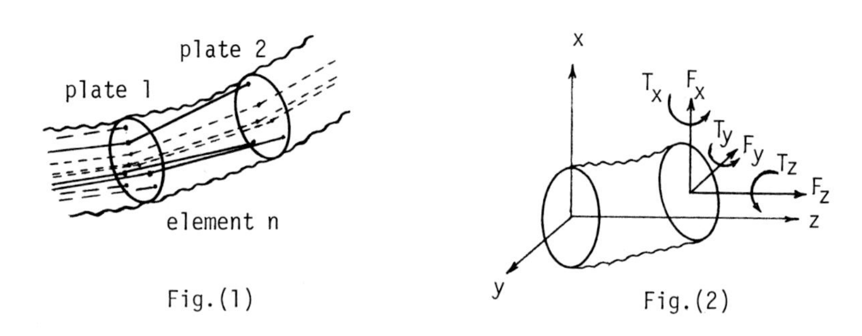

Fig.(1) Fig.(2)

The three strings can only resist the three forces and not the torques. Moreover, since these strings should only partly react to the forces [1], the skin of a flexible element is under the effect of a major portion of the forces and the torques. For the moment, we restrict our discussion only to the torques about ox and oy in figure (2). Furthermore, in the following discussion, for the sake of simplicity and without loss of generality, only the forces along ox or the torques about oy are discussed. This is possible because of symmetry.

First, assuming that we have only forces and no torques, if the force f_0 applied to the centre point P of plate 2 is in the z-direction only (figure 3-a) since the stretch force F_i [1] is also in the z-direction then the shape of the element will remain unchanged. A gradual increase of the force in the x-direction will displace the point P. With no change in the lengths of the three strings, this displacement causes a rotation of plate 2, and the effective force in the x-direction will be the sum of the x-components of both the external and the stretch forces. This causes a further displacement and rotation of plate 2 until an equilibrium is reached (figure 3-b). This further rotation, of course, could be prevented by adjustment of the string lengths. The configuration of an element, defined by the orientation of plate 2 and the displacement of its centre point, therefore, will be governed by the equilibrium of forces for any selected lengths of the three strings inside the element.

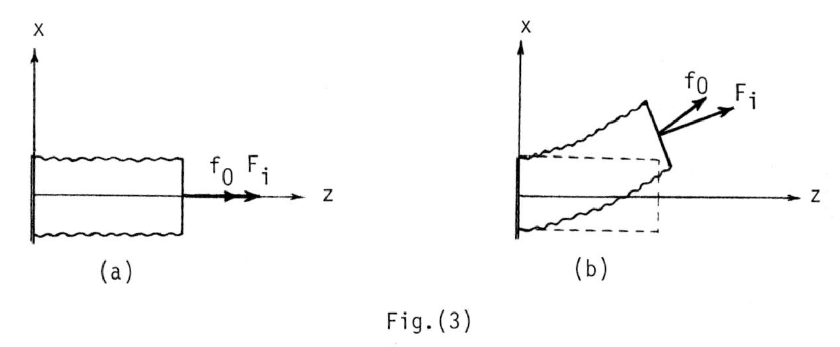

Fig.(3)

In case an element is under the effects of its associated stretch force and its weight only, a unique configuration may be obtained for every possible combination of the three string lengths.

One may assume that a unique displacement (deviation of point p from the z-axis) and a unique string length ratio exists for any given orientation (rotation of plate 2). This can be referred to as the normal configuration for that particular orientation, or that particular length ratio. Any other shape achieved by an element as a result of the effect(s) of other loads when the string lengths have not changed is a distortion from this normal configuration. This distortion can not be determined unless the kinematic equations are solved and a new equilibrium state is found for each case.

In each case, there is a maximum distortion indicating a maximum allowable position for point P, while the strings do not contact the skin; this is a characteristic of an element and is enforced by the physical properties of the element.

For each angle of rotation, or for each set of string length, a family of two curves may be found which indicate the relationships between the variations of the displacement of point P or the orientation of plate 2, respectively, versus the ratio of the lateral force to the stretch force. These relationships are illustrated in figure (4).

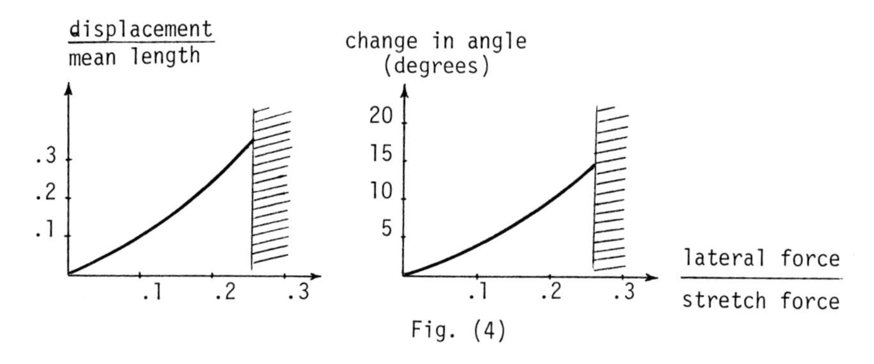

Fig. (4)

Now, if the external load is a torque about oy only, the unshaped element is under the effect of the stretch force and this external torque (figure 5). As a result, plate 2 will rotate and, therefore, the component of the stretch force in the x-direction will pull point P in this direction until an equilibrium state is obtained. The flexible element, hence, may be modelled as a torsional spring where all the resistance against a torque should necessarily be provided by the skin.

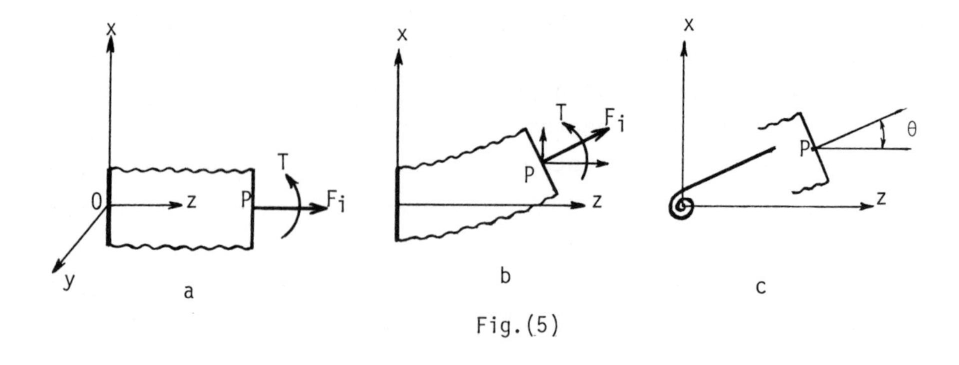

Fig. (5)

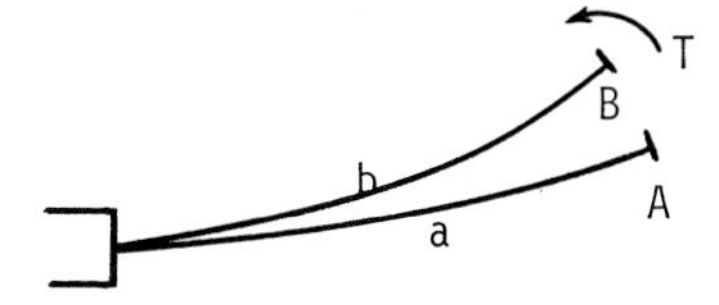

a) Before torque T is applied

b) After torque T is applied

Fig. (10)

From the above discussions, we can also figure out whether it is possible to shape a flexible arm as desired, by dividing this form in segments which are more or less similar to the ones discussed above and by combining the results for presence of both forces and torques.

Concluding remarks

In this paper, further preliminary and fundamental studies are carried out to examine in more depth the feasibility and the limitations of a light-weight design bases on the idea of using strings. It is still too soon to judge the practicability of the idea before some more research is done. However, these results indicate the right directions for further steps. The matters of importance have been pointed out while studying the various loads an arm has to carry. The conclusions are summarized here.

1- The direction of the stretch force must be independant of the orientation of the tool, that is, an end-effector has to be used.

2- With the above option the stretch force direction may be chosen such that, as far as it is possible, it reduces the effect of the load forces. This remarkably narrows the range of variations of the effective force.

3- It is better to keep the magnitude of the effective force constant, rather then the magniture of the stretch force.

4- The flexible elements must gradually become stronger and less flexible as one moves from the arm neck to the bottom.

5- The lengths of the strings cannot be used for measurement of the displacement/orientation as expected in [1]. This is because the string lengths do not represent a unique configuration.

Other important problems that need to be studied in further research, before a final justification on the practicability may be made for this approach, are:

a- The necessary stretch force and material strength for the range of usual loads and their handling as compared with other industrial robots, or for a particular task that a flexible robot may be used only.

b- How to obtain a required position/orientation for a set of two flexible elements; this is a prior problem to how to use the many available degrees of freedom, since the position and orientation of each unit are interrelated but for two units each of these two can be independently selected.

References

1- A. Hemami: «Design of a Light Weight Flexible Robot Arm», proceedings of the Robots 8 Conference, June 1984.

Controls I

Chairman: Robert O. Warrington, Jr., Ruston, Louisiana
Vice Chairman: Y. Hari, University of North Carolina at Charlotte, Charlotte, North Carolina

PID for Profile Control of Rhino

Sudhakar Paidy, Robert Blank, and Michael Buckley

Eastman Kodak Company
Rochester, New York

Rochester Institute of Technology
Rochester, New York

ABSTRACT

The development of an Intel 8085 micropreçessor based system to control profiled movement of a multi-axis robotic manipulator is discussed. The hardware consists of an Intel ISBC 80/24 single board computer, power drive and encoder signal conditioning circuitry, and the Rhino XR-1 six axis educational robot. The firmware is developed to track the robot position and to provide variable pulse-width modulated drive output in a totally digital, closed loop feedback system. The system features concurrent multi axis profiled movement of the manipulator, a serial ASCII operator interface via a CRT terminal, and a serial ASCII host computer interface. The user teaches the robot by specifying points in geometric space or by jogging motors to various positions whcih are then memorized by the system. The user can direct the robot to move from point-to-point, on a path defined by a set of points, or to perform a task of a set of sequences. The host interface facilitates data storage and retrieval and optional robot control remotely.

INTRODUCTION

A microprocessor-based control system is developed as a totally digital multi-axis servo motor driven robot controller. The primary goal of the study was to allow experimentation and demonstration of digital sampled data system characteristics in servo motor positioning and velocity profile control. The result was the transformation of a simple educational tool - the Rhino XR-1 mechanical unit - into an accurate and repeatable jointed arm robot with the feel of much more expensive industrial models. This paper discusses the control concepts, hardware and software used in implementing this powerful robotic controller.

THE ROBOTIC ARM

The Rhino XR-1 robot is a manipulator arm with six degrees of freedom. As compared to a normally configured human being, the six axes mimic:

- The base, side-to-side motion only
- The shoulder joint, bicep
- The elbow joint, forearm
- The wrist with azimuthal capability
- The wrist with rotational capability
- The ability to grasp.

Each of these motions is controlled by an electric servo motor with a D.C. input. The polarity of the D.C. signal determines the direction in which the motor rotates. The actual amount of rotation is conveyed by incremental optical encoders at each of the motors. In order to get maximum resolution, the encoders run at the speed of the motor shaft rather than that of the output axis. This allows extremely accurate positioning of the arm by the controller.

CONTROL CONCEPTS

Motor Position Sensing. The control system receives incremental position information in the form of a dedicated pulsed line for each motor, and a separatae signal indicating the direction of the motor's rotation. Each position pulse is passed on to the software as an interrupt which increments or decrements, based on the direction of rotation, the counter variable for that particular motor.

Motor Speed Control. The axis motors are rated as full power at +12 VDC (forward) and -12 VDC (reverse), and stall at between -6 and +6 VDC. Thus, a pulse-width modulated signal of anywhere between 50% and 100% will drive the motor from zero to full speed. The pulse-width modulation scheme uses the 80/24 system clock to drive a counter and is under total software control. A count is calculated which determines the number of system clock cycles that the motor drive signal will stay on before it times out and turns off.

Closed Loop Motor Control. Speed and position control for each axis motor is accomplished by using desired position (the result of a programmed position versus time curve) as the input, instanteaneous and predicted next position (calculated real time) as the feedback parameters, and system values gain and bias (user selectable) to determine the output motor power level. The calculations are performed at a rate determined by the sampling period set for the system and thus, in conjunction with the pulse-width modulation output routines and position routines constitute a closed-loop, error driven, sampled data control system.

Multi-axis, Profiled Move Scheduling. To execute one discrete robot arm movement, the six axis motors are moved simultaneously through individual position profiles from their starting points to their separately specified end points. The speed of each motor is adjusted so that all motors finish at exactly the same time even when the distance movement of one might be only a fraction of the others. This position versus time profile serves as the source of the "desired position" value in the position error scheme. Desired tasks are performed by instructing the system to execute a sequence of profiled moves through a series of learned endpoints. The motors ramp up to a designated plateau speed, and upon nearing their goal positions will decelerate - smoothing into their endpoints to virtually eliminate overshoot.

User/Host Interface. The user interface allows for the task definition and training the robot. The user can also modify real-time position information, operating parameters, point and sequence data and display system status. An additional communication link allows an external host computer to perform user controls remotely, and provides for off-line storage and retrieval of point and sequence data.

HARDWARE LAYOUT

Control Systems Outline . The heart of the system is an Intel 80/24 single board computer (SBC), called the controller. Through two parallel ports, the SBC pulse-width modulates each motor drive signal, and interprets incoming encoder information, adjusting velocity as needed to accurately produce the desired profile. The SBC is interfaced to the robot through logic and drive hardware which current-boosts the pulse-width modulated motor signal, and decodes a direction signal to move the robot motors in the proper clockwise or counter-clockwise direction. Quadrature encoder signals from the robot are direction discriminated in hardware on the interface board, and forward or reverse going transitions are passed on to the SBC.

2. Position Feedback

The optical "choppers" used on the Rhino robot are guardrature encoders of a high resolution. Comparison of the A and B encoder channels allows for direction discrimination; i.e., an A-channel transition of 1100 would be accompanied by a B-channel pattern of 1001 if the motor is moving CW, or by a B-channel pattern of 0110 if the motor is moving CCW. The circuits used to interface the encoders to the controller do both direction discrimination and the resolution increase.

SOFTWARE DESIGN

Robot Control Software . The control software is organized as three distinct layers: interrups, clocked events and state defined processes which manage all the concurrent activities in a prioritized order.

Interrupts are external stimuli which demand immediate attention. The interrupt hardware/software configuration insures that the system responds to relatively high speed attention seekers; namely, motor incremental position pulses, operator communication transmit/receive signals and a periodic software system timing clock.

Clocked events are activities which the system must perform at periodic intervals, based on the software system timing clock. These events are prioritized. Clocked events include activities such as the pulse modulation drive, motor closed loop power calculation and realtime profile point calculation routines.

State defined processes constitute the remainder of the programs in the system and are executing whenever the CPU is not responding to an interrupt or performing a clocked event. Each state process is in a particular "state" at any one point in time and will by design respond to I/O or memory flag situations. These processes set up execution of point to point profiled moves, manage sequences of moves and homing sequences, and perform the operator and host communication dialogue.

The robot software implements four distinct functions: closed loop position control, position profiling and sequencing algorithms, homing sequences and operator/host dialogue.

Closed loop position control. This consists of a position interrupt routine for each of the six motors plus the clocked events MDRV and SMPL. Each incremental position interrupt routine updates the instantaneous position variable for the associated motor. At full speed, each axis interrupts the system at a rate of approximately 1 kilohertz. At 15ms intervals, the SMPL task samples the instantaneous positions and uses the previous sampled positions to predict the next positions. The difference between these "next" positions and the desired positions (determined by PROE) is multiplied by a gain factor and added to a bias value to determine a "next" power level for each motor, which takes effect at the next iteration of SMPL. The prediction technique eliminates the effect of CPU execution time delays on motor stability; as long as the SMPL task completes its exeuction before the next 15ms interval. Scenarios which take more time to execute such as higher position interrupt rates were found to have no effect on resultant system operation. At 5ms intervals, the MDRV task outputs the power level for each of the motors to the respective programmable interval timers to produce one pulse of the correct pulse-width to each motor power amplifier.

Position profiling and sequencing. Clocked events PROE and the state defined processes MOVR, SEQE and MSEQ together serve these functions. At 40ms intervals the profile execution task PROE determines the desired position for each motor (which SMPL uses) and calculates a "next" desired position for each motor based on a profile acceleration table and on speed scale factors. Like the SMPL task, PROE uses a predictive technique to insure consistent operation regardless of CPU execution scenarios. Note that PROE can be pre-empted by interrupts and also by the MDRV and SMPL tasks. For each profiled move, the mover process MOVR calculates a speed scale factor for each motor and flags the PROE task to begin execution of a profile move. The speed scale factors insure that all motors involved in multi-axis moves start and stop simultaneously and have smooth and predictable concurrent motion. The sequence execution process SEQE directs the MOVR process through a sequence of profiled moves. Likewise, the macro sequence execution process MSEQ directs the SEQE process through a series of sequences. The PROE task increments desired position when the system is in jog mode to provide closed loop jogging.

The homing sequence. The HOMR process directs a two stage homing sequence to place the robot arm in its zero mechanical reference position. In phase one, selected motors are run open loop (SMPL task is turned off) in the selected directions until all pertinent home switches are covered, each motor stopping when its associated home switch closes. In phase two, these motors are run at a lower power level in the forward direction, each until it just uncovers its respective home switch. The two phase feature eliminates the effect of home sensing switch mechanical hysteresis when approached from opposite directions.

User interface. The operator dialogue is performed by the ODLG process and USART transmit and receive interrupts. ODLG is interrupt driven communication driver which processes input command lines and sends required data to the CRT.

When a host link is requested, the HDLG process turns off the system clock and CRT interrupts and places the system in a host link mode. In this mode, the terminal is conncected to the host computer and the robot is transparent to the user. The user may use the host computer as if the terminal is connected directly to the host. However, the host link software intercepts escape sequences which can download data from the host, upload data to the host, and also initiate point and sequence moves. During host initiated profiled moves, the host link is suspended and the terminal returned to "robot mode" until the movement is finished to let the operator abort commands should trouble arise.

RESULTS AND CONCLUSIONS

The system implemented was successful in providing smooth, stable, versatile multi-axis motor control. The 50% utilized 4.84 mHz 8085 CPU supports encoder interrupt at lms intervals for each axis, 5ms PWM drive period, 15ms position error sampling period, 40ms profile calculation period and 9600 Baud operator interface concurrently.

While one of the goals of this implementation was to incorporate an absolute minimum of special hardware, further development could be pursued in the following areas. A more friendly interface such as teach pendant could be used. Feedback sensors in mechanical actuator could permit more adaptive control. An optional "torque mode" could permit handling of fragile objects. Finally, more elaborate profiling techniques (example - erform handwriting) could be employed.

Realization of Threshold Logic for Pattern Classification

K. Yunus and H. M. Razavi

P.T. Super Andalas Steel
Jln. Karo No. 2 Medan
North Sumatra, Indonesia

Electrical Engineering Department
University of North Carolina
Charlotte, Nc 28223

ABSTRACT

In this paper a hardware realization for Threshold Gates and its application to pattern classification is considered. The Gate is programmable so that it can be used for classification,or seperation of different patterns. The classification is Binary, and the gate would only indicate whether an object belongs to a given class or not. The design uses pass transistors and is well suited for VLSI realization.

INTRODUCTION

A Threshold Gate (Element) is defined as:

$$F(X_1, X_2, \ldots\ldots X_n) = Z$$

if $\sum_{i=1}^{n} W_i X_i \geqslant T_1$

or if

$$T_{2j} \geqslant \sum_{i=1}^{n} W_i X_i \geqslant T_{2j+1} \qquad j=1, 2, 3\ldots.$$

and

$$F(X_1, X_2, \ldots\ldots X_n) = \bar{Z} \quad \text{otherwise}$$

where $Z \in (0,1)$

X_i is the ith Binary variable,

W_i is the weight associated with a variable ,

n is the total number of variables, and $\sum_{i=1}^{n} W_i X_i$ is Weighted Sum

In order for the Threshold Gate to be used as a pattern classifier the following conditions must be true.

1- Sensors which measure the properties of an object must have binary outputs. (each sensor would indicate the presence or absence of a property)

2- The objects are seperated in to two classes. Each class with a different set of properties.

Each input to the Gate(X_i) indicates the output of a sensor which measures a given property. Each W_i is an integer associated with each property which emphasizes the importance of a given property for a certain classification. The objects are classified to the two classes based on $\sum_{i=1}^{n} W_i X_i$. i.e. one class of objects has different $\sum_{i=1}^{n} W_i X_i$ than the other class and the two classes are seperated by T_j, Thresholds.

For example if four objects, such as a nut , a bolt, a washer, and a solid bar are to be seperated, the properties that can be measured are the presence or absence of a hole, and presence or absence of an angle.

$X_1 = 1$ the object has a hole

$X_1 = 0$ the object has no hole

$X_2 = 1$ the object has an angle

$X_2 = 0$ the object has no angle

Based on these two properties then the nut would have properties $X_1X_2 = 11$, the bolt $X_1X_2 = 01$, the washer $X_1X_2 = 10$ and the solid bar $X_1X_2 = 00$. Figure 1 shows the four different objects and their properties.

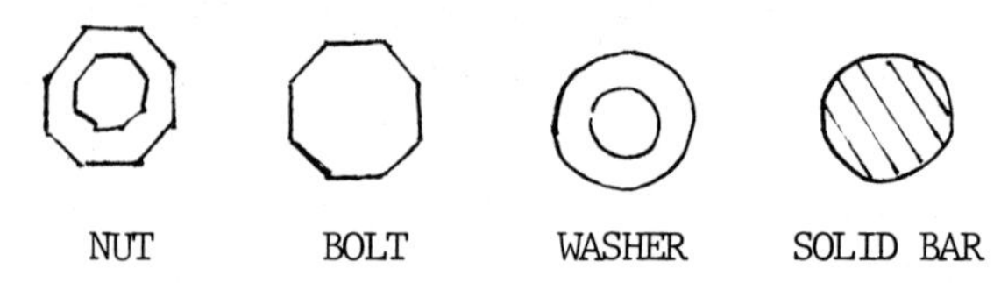

Figure 1. Description of each object based on its properties.

Figure 2 shows a map (Karnaugh) representing the same four objects and the weighted sum for each object. In this case a weight of 1 is assigned to X_1, and a weight of 2 to X_2. This weight assignment should be done so that each of the two classes of the output have different weighted sums. For complete explanation of Weight and Threshold selection see [6,7,8,9,12] .

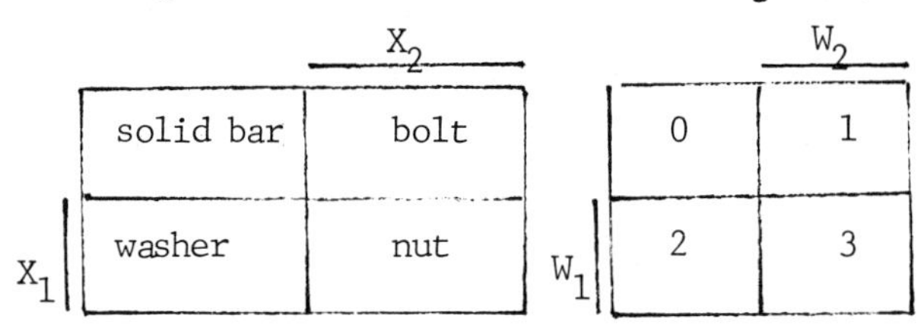

Figure 2 Karnaugh map for the four objects and its corresponding weighted sum map.

Since each object is represented by a different weighted sum, the objects can be grouped in to two classes based on their weighted sums.For example if class one is composed of nuts and bolts and class two consists of washers and solid bars, then the gate has to differenciate between weighted sums 1 and 3 which belong to one class and 0 and two which belong to the second class. It should be noted that the weights selected are not the optimum or the best solution for this example , because an angle is sufficient to seperate these two classes and that it implies that the whole has no significance in this classification and therefore it is possible to assign a weight 0 to the variable X_2 (hole). The smaller the weighted sums are the simpler the circuit would be for classification [8,9,12] .

CIRCUIT REALIZATION

The circuit propose here is composed of two parts. The first part is a digital summer. This is a circuit with n inputs (n measured properties, these inputs are Binary),and n+1 outputs (weighted sums from 0 to n). The circuit assumes that all the weights are equal to 1. However, if a weight other than one is necessary for a variable , then that variable can be applied to more than on input. For example if X_1 is assigned a weight of two then X_1 is connected to two inputs with a weight of one. The summer circuit used here is the " tally circuit described by Mead-COnvey [13]. The basic cell used for the design of the summing logic consists of two pass transistors.

The circuit has a cellular form and can easily be expanded to any number of variables. (see Figure 4).
The second part of the circuit is a selection circuit, which selects the sums belonging to a certain class. The select circuit is also made of the same type of cell that was used in the summing stage. Each pass transistor in the select stage is controlled by a select input which is equivalent to a Threshold. The last stage is an OR gate which signifies whether an object with the given properties belongs to a certain class or not .

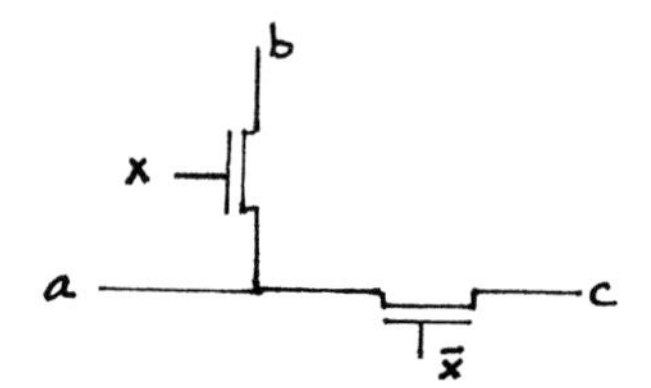

Figure 3 Basic Cell

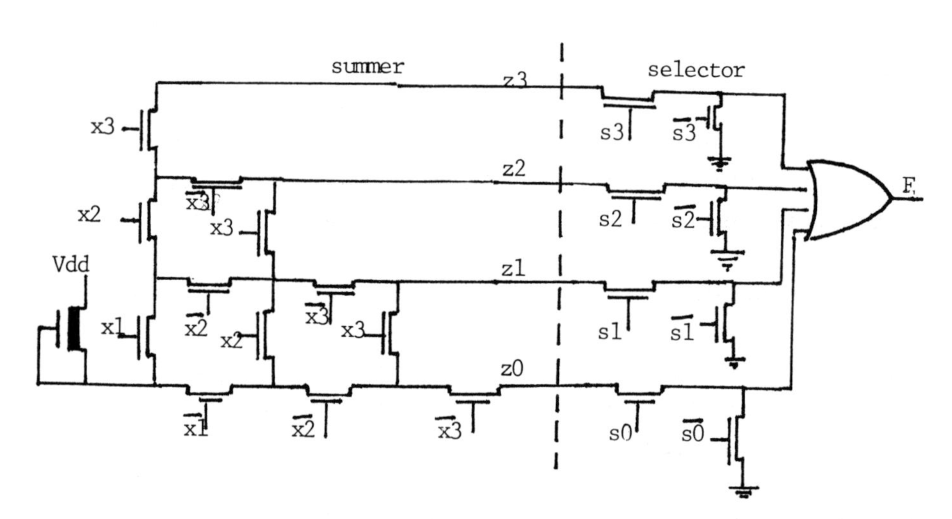

Figure 4 Circuit for Threshold Gate

The circuit shown on Figure 4 can be used for any type of Binary classification as long as the sum of weight is less than or equal to three .

PROGRAMMING THE CIRCUIT

To use the circuit for pattern classification , first the properties that are important to the classification must be measured by a sensor that indicates the presence or absence of that property. Then a set of weights must be chosen for the properties such that the two classes that are seperated or differenciated have different weighted sums. Now all the weighted sums belonging to one class must be activated by the select inputs (applying a logic one to the select input) and all the weighted sums belonging to the second class must be deactivated. (apply a logic 0 to its select input). F= 1 in dicates one class and F= 0 ˙indicates the second class.

EXAMPLE

To use the circuit for the example with nuts and bolts described earlier, the following procedure should be adopted.
apply X_1 to x_1 input of the circuit. Apply X_2 to x_2 and x_3 of the circuit. this gives X_1 a weight of one and X_2 a weigth of two. then z_0 thru z_3 correspond to weighted sums zero to three respectively. Now if the inputs indicate a solid bar z_0 would be at logic one and all the other outputs at logic zero. If the input is a washer then z_2 would be at logic one, If the input is a nut then z_3 would be a one, and others zero. In order to seperate bolts and nuts from washers and solid bars , then weighted sums z_1 and z_2 must be selected. This requires putting a logic one on s_1 and s_2 . Now an F= 1 will indicate nuts and bolts , and F =0 indicates washers and solid bars . For different classifications different select lines must be activated.

RESULTS AND CONCLUSION

The circuit described here was designed and simulated for a gate with weigted sum of up to twelve, using MCNC* tools for VLSI design. The result of the simulation was satisfactory and the total delay time for the circuit was about 45 nsec. The circuit can be expanded to any number of variables, and the only limiting factor are the number of input pins and select pins. As the number of pass transistors that are cascaded reaches more than five transistors, a buffer transistor is needed to restore the voltage level of the signal.

* Microelectronic Center for North Carolina

REFERENCES

1. Baugh, C.R. ; Wooley, B.A. "Statistical Analysis of Threshold Logic Circuits", Proceedings of Twelfth Annual Allerton Conference on Circuits and Systems Theory, 535-541, October 1974.

2. Coates, C.L.; Lewis, P.M. " A Realization Procedure for Threshold Gates Networks", IEEE Transactions on Electronic Computers, EC-12 , 456-461, 1963.

3. Dertouzos, M.L., Threshold Logic: A Synthesis Approach, MIT Press Cambridge Mass. 1965.

4. Gableman, I.J. " An Algorithm for Threshold Element Analysis " IEEE Transactions on Electronic Computers, EC-14, 623-625, 1965.

5. Gosh, S.; Choudhurry, A.K. " Partition of Boolean Functions for Realization with Multi-Threshold Threshold Logic Elements", IEEE Transactions on Computers, C-22, 204-214, February 1973.

6. Haring D.R. " Multi-Threshold Elements", IEEE Transactions on Electronic Computers, EC-15, , 45-65, 1965.

7. Haring D.R. "Multi-Threshold Building Blocks", IEEE Transactions on Electronic Computers, EC-15, 662-663, August 1966.

8. Haring, D.R. ; Diephius, P.J. " A Realization Procedure for Multi- Threshold Threshold Elements", IEEE Transactions on Electronic Computers, EC-16, 828-835, 1967.

9. Haring, D.R. ; Ohori, D. " A Tabular Method for Synthesis of Multi-Threshold Threshold Elements", IEEE Transactions on Electronic Computers (short notes) EC-16, 216-220, 1967.

10. Hurst, S.L. " Digital Summation Threshold Logic Gates, A New Circuit Element", Proc. IEE , 1973, 1301-1307.

11. Lewis, P.M.; Coates, C.L., Threshold Logic, New York, John Wiley and Sons, Inc. 1966.

12. Modaress Razavi, H. " Synthesis of a General Threshold Element", Ph.D. Dissertationfor Electrical Engineering Department of West Virginia University, 1978.

13. Mead, Carver; Convay, Lynn, Introduction to VLSI Systems, Addison - Wesley Publishing Co. October 1980.

14. Sheng, C.L. ; Roy, P.K.S. " An Approach for the Synthesis of Multi- Threshold Threshold Elements", IEEE Transactions on Computers, (short-notes) 913-920, 1972.

Non Linear Control of Uncertain Robotic Systems

S. N. Singh

Vigyan Research Associates, Inc.
28 Research Drive
Hampton, VA 23666

Abstract

We present a state feedback control law for the uncertain robotic systems such that the error between the generalized coordinates and the trajectory of a reference model is uniformly ultimately bounded with respect to any arbitrarily small set of ultimate boundedness.

Introduction

The equations of motion of the robotic systems are highly nonlinear and the parameters of the system are not known in advance since it must handle variable payloads. Modelling errors also exist due to unknown static and dynamic friction forces at the joints and uncertainty in the link parameters.

Recently some attempt has been made for designing control systems for uncertain robotic systems [1-8]. The results of [1-5] are based on linearized time-varying models about the reference trajectories and assume that during the adaptation process the elements of linearized system remain constant. However, for fast motions of the robots, this assymption is not valid. Although, the result presented in [6] treat the nonlinear manipulator dynamics, one must use unbounded feedback gains for the convergence of the error between the responses of the reference model and robotic system to zero. The control laws of [7-8] are based on system invertibility and servocompensation and can not accomplish asymptotic stability for extremely large variations in payload.

We present in this paper, a nonlinear state feedback control law for uncertain robotic systems. We assume that the bounds on the uncertain quantities (variable payload, joint friction force parameters, link parameters, the time-varying reference inputs, etc.) are know. In the closed-loop system the error between the responses of the robotic system and the reference model is shown to be uniformly ultimately bounded with respect to any arbitrarily small set of ultimate boundedness inspite of uncertainties in the system. The derivation of the controller is based on the concept of ultimate boundedness of motion. The controller of this paper differs from those of [9-12]. However the arguments used to establish the result have some similarity to those of [9-12].

Robotic Systems

The dynamics of a n-degree of freedom manipulator is given by

$$D(q)\,\ddot{q}(t) + F(q, \dot{q})\dot{q} + g(q, \dot{q}) = u(t) \tag{1}$$

where q is the vector of n joint coordinates, $D(q)$ is the positive definite symmetric inertia matrix (denoted as $D(q) > 0$), $F(q, \dot{q})\dot{q}$ is the vector grouping the Coriolis and centrifugal forces, and $g(q, \dot{q})$ represents the "parasitic" forces such as force of gravity and friction.

We assume in this paper that the system matrices $D(q)$, $F(q, \dot{q})$ and $g(q, \dot{q})$ in (1) are unknown but satisfy the following assymptions:

(i) For all $q \in R^n$, there exist positive real numbers a_1 and a_2 such that

$$a_1\,||w||^2 \leqq \lambda_{min}(D^{-1}(q))\,||w||^2 \leqq w^T D^{-1}(q)\,w \leqq \lambda_{max}(D^{-1}(q))\,||w||^2 \leqq a_2\,||w||^2 \tag{2}$$

where T denotes transposition and $\lambda_{min}(M)$ [$\lambda_{max}(M)$] denotes minimum [maximum] eigenvalue of a matrix M.

(ii) For all $q, \dot{q} \in R^n$.

$$||F(q, \dot{q})\dot{q}|| \leqq ||\dot{q}||^2\,(b_1 + b_2\;||q||),\; b_i > 0 \tag{3}$$

(iii) $||g(q, \dot{q})|| \leqq c_1 + c_2||q|| + c_3||\dot{q}||,\; c_i > 0$ (4)

Here we assume that the real numbers a_i, b_i and c_i are known. For robotic systems with rotational joints, we can set $b_2 = c_2 = 0$. Furthermore, if no velocity dependent friction forces are present in the system then $c_3 = 0$.

Defining $x = (q^T, \dot{q}^T)^T$, (1) can be written in a state variable form

$$\dot{x} = \begin{bmatrix} \dot{q} \\ D^{-1}(q)\,(-F(q, \dot{q})\dot{q} - g(q, \dot{q})) \end{bmatrix} + \begin{bmatrix} 0 \\ I \end{bmatrix} D^{-1}(q)u \tag{5}$$

It is desired that x follows a desired trajectory $\hat{x}$ of the reference model

$$\frac{d}{dt}\begin{pmatrix} \hat{q} \\ \dot{\hat{q}} \end{pmatrix} = \begin{bmatrix} 0 & I \\ P & Q \end{bmatrix} \begin{bmatrix} \hat{q} \\ \dot{\hat{q}} \end{bmatrix} + \begin{bmatrix} 0 \\ B_1 \end{bmatrix} r \tag{6}$$

$$\triangleq A_M\hat{x} + B_M r$$

where $\hat{x} = (\hat{q}^T, \dot{\hat{q}}^T)^T$, $P = \text{diag}(P_i)$, $Q = \text{diag}(Q_i)$, and $r \in R^n$ is an external input. We assume that A_M is a stable matrix.

Defining $e = (q - \hat{q})$, $z = (e^T, \dot{e}^T)^T$, using (5) and (6) we obtain

$$\dot{z} = A_M z + Bh(q, \dot{q}, r) + BD^{-1}(q)u \tag{7}$$

where $B = (o, I)^T$, and

$$h(q, \dot{q}, r) = -Pq - Q\dot{q} - B_1 r + D^{-1}(q)[-F(q, \dot{q})\dot{q} - g(q, \dot{q})] \tag{8}$$

We are interested in deriving a state feedback control law u such that for any uncertainty in the parameters of the robotic system, the solution of (7) is uniformly ultimately bounded and the set of ultimate boundedness is arbitrarily small.

Remark 1: We may point out that it is possible to derive a controller for the system (5) for ultimate boundedness using the results of [9-11]. For using these results, one writes $BD^{-1}(q)$ in the form $BD^{-1}(q) = BD_*^{-1}(q)[I + E(q)]$, where $E(q)$ includes all the uncertainties of the matrix $D^{-1}(q)$ and $D_*(q)$ is known. Then following [9-10], for satisfying the boundedness condition one requires that $||E(q)|| < 1$. However this holds good only for small uncertainties in the inertia matrix. The controller derived in the next section does not require any restriction on the magnitude of uncertainty in the system.

Control Law

Let us assume that $||\hat{x}(0)|| \leqq d_0$ and $||r(t)|| \leqq d_1 < \infty$ for all t. Then there exists a $d_2 > 0$ such that the solution of (6) satisfies

$$||\hat{x}(t)|| \leqq d_2 \text{ for all } t \geqq 0 \tag{9}$$

In view of (9), we have that for all $t \geqq 0$

$$\begin{aligned} ||q(t)|| &\leqq ||e(t)|| + d_2 \\ ||\dot{q}(t)|| &\leqq ||\dot{e}(t)|| + d_2 \end{aligned} \tag{10}$$

As a consequence of Assymptions (i)-(iii), and (10), we obtain

$$\begin{aligned} &||h(q, \dot{q}, r)|| \leqq ||Pq|| + ||Q\dot{q}|| + ||B_1|| \, d_1 + a_2[(b_1 + b_2||q||) \\ &||\dot{q}||^2 + c_1 + c_2||q|| + c_3||\dot{q}||] \\ &\leqq (||P|| + a_2 \, c_2) \, (||e|| + d_2) + (||Q|| + a_2 \, c_3) \, (||\dot{e}|| + d_2) \\ &+ ||B_1|| \, d_1 + a_2 \, c_1 + a_2 \, \{b_1 + b_2(||e|| + d_2)\} \, (||\dot{e}|| + d_2)^2 \\ &\triangleq \rho(z) \end{aligned} \tag{11}$$

Since A_M is a stable matrix, for any give $S > 0$, there exists a unique $R > 0$ satisfying

$$RA_M + A_M^T R = -S \tag{12}$$

Let us consider the control

$$u(z) = \begin{cases} -\dfrac{B^T R z}{a_1||\mu(z)||}\rho^2(z) & \text{if } ||\mu(z)|| > \varepsilon \\ -\dfrac{B^T R z}{a_1 \varepsilon} \quad \rho^2(z) & \text{if } ||\mu(z)|| \leqq \varepsilon \end{cases} \tag{13}$$

where

$$\mu(z) = B^T R \, z \, \rho(z) \tag{14}$$

Next we consider a closed ball $B(\eta)$, centered at $z = 0$ with radius

$$\eta = [\varepsilon / \{2\lambda_{min}(S)\}]^{1/2} \tag{15}$$

Define ellipsoids

$$Z(k) = \{z \in R^{2n}: \; z^T R z \leq k = \text{constant} > 0\} \tag{16}$$

Now let

$$\underline{k} \triangleq \min \{k \mid Z(k) \supseteq B(\eta)\}$$

Then

$$\underline{k} = \lambda_{max}(R) \, \eta^2 \tag{17}$$

Consider also ellipsoids $Z(\bar{k})$ with $\bar{k} > \underline{k}$ and $Z(k_0)$ with $k_0 = z_0^T R z_0$. Let us define

$$c_0 \triangleq \min \{z^T S z - \tfrac{\varepsilon}{2} \mid z \in Z(k_0) \setminus \mathring{Z}(\bar{k})\} \quad \text{if } z_0 \notin Z(\bar{k}). \tag{18}$$

Now we can state a boundedness theorem.

<u>Theorem 1</u>: Consider system (7) with control (13), satisfying Assumptions (i)-(iii) and (10). Then for every initial condition $(z_0, t_0) \in R^{2n} \times R^1$ there exists a solution of (7) which can be extended over $[t_0, \infty)$. Furthermore, given a $S > 0$ and a $\bar{k} > \underline{k}$, a solution beginning at (z_0, t_0) is uniformly bounded with

$$m(z_0) = \begin{cases} ||z_0|| [\lambda_{max}(R)/\lambda_{min}(R)]^{\frac{1}{2}} \text{ for } z_0 \notin Z(\bar{k}) \\ [\bar{k} / \lambda_{min}(R)]^{\frac{1}{2}} \text{ for } z_0 \in Z(\bar{k}) \end{cases}$$

and is uniformly ultimately bounded with respect to $Z(\bar{k})$ with

$$T(z_0, Z(\bar{k}) = \begin{cases} \dfrac{k_0 - \bar{k}}{c_0} & \text{for } z_0 \notin Z(\bar{k}) \\ 0 & \text{for } z_0 \in Z(\bar{k}) \end{cases}$$

<u>Proof</u>: Consider the Lyapunov function $V(z) = z^T R z$, $z \in R^{2n}$.

The derivative of $V(z)$ along the trajectory of system (7) and (13) is given by

$$\dot{V}(z) = 2z^TR(A_Mz + BD^{-1}(q)u + Bh(q, \dot{q}, r)] \tag{19}$$

On involking (11), and (12), one obtains

$$\dot{V}(z) \leqq -z^TSz + 2z^TR\ BD^{-1}(q)u + 2||\mu(z)|| \tag{20}$$

In view of Assumption (i) and (13), if $||\mu(z)|| > \varepsilon$

$$\begin{aligned}\dot{V}(z) &\leqq -\lambda_{min}(S)||z||^2 - \{2\lambda_{min}[D^{-1}(q)]||B^TRz||\rho(z) / a_1\} + 2||\mu(z)|| \\ &\leqq -\lambda_{min}(S)||z||^2\end{aligned}$$

It can be easily seen that if $||\mu(z)|| \leqq \varepsilon$,

$$\begin{aligned}\dot{V}(z) &\leqq -\lambda_{min}(S)||z||^2 - \{2\lambda_{min}[D^{-1}(q)]||B^TRz||^2\rho^2(z) / \varepsilon\ a_1\} \\ &+ 2||\mu(z)|| \\ &\leqq -\lambda_{min}(S)||z||^2 + \frac{\varepsilon}{2}\end{aligned}$$

Consequently, $\dot{V}(z) < 0$ if $z \notin B(\eta)$ and the theorem can be established using arguments similar to those of [9-10].

We note that $z(t)$ remains in the set $Z(\overline{k})$, $\overline{k} > \underline{k}$ for all $t \geqq t_0 + T(z_0, Z(\overline{k}))$, and the set $Z(\overline{k})$ can be made arbitrarily small by properly choosing S and $\varepsilon > 0$. Thus the error between the responses of the robotic system and the reference model can be made arbitrarily small after a finite interval of time.

Conclusion

Based on the theory of ultimate boundedness, a control law for uncertain robotic systems was derived. In the closed-loop system, the trajectories of the robotic system follow the trajectories of a reference model within arbitrarily small prescribed error bound.

References

[1]. Dubowsky, S.; DesForges, D. T.: The Application of Model Referenced Adaptive Control of Robotic Manipulators, J. Dyn. Syst. Meas. and Contr., Vol. 101, pp. 193-200, Sept. 1979.

[2]. Horowitz, R; Tomizuka, M.: An Adaptive Control Scheme for Mechanical Manipulators: Compensation of Nonlinearity and Decoupling Control, ASME, Paper No. 80, Wa/DSC-6, 1980.

[3]. Koivo, A. J.; Guo, T. H.: Adaptive Linear Controller for Robotic Manipulators, IEEE Trans. Aut. Contr., Vol. AC-28, pp. 162-171, Feb. 1983.

[4]. Lee, C. S. G.; Chung, M. J.: An Adaptive Control Strategy for Mechanical Manipulators, IEEE Trans. Aut. Contr., Vol. AC-29, pp. 837-841, Sept. 1984.

[5]. Kim, B. K.; Shin, K. G.: An Adaptive Model Following Control of Industrial Manipulators, IEEE Trans. on Aerosp. and Elect. Syst., Vol. AES-19, pp. 805-814, Nov. 1983.

[6]. Samson, C.: Robust Nonlinear Control of Robotic Manipulators, Proc. IEEE Conf. on Decision and Contr., pp. 1211-1216, 1983.

[7]. Gilbert, E. G.; Ha, I. J.: An Approach to Nonlinear Feedback Control with Applications to Robotics, Proc. IEEE Conf. on Decision and Control, San Antonio, Texas, Dec. 1983.

[8]. Singh, S. N.; Schy, A. A.: Invertibility and Robust Control of Robotic Systems, IEEE Conf. on Decision and Control, Las Vegas, 1984.

[9]. Leitmann, G.: On the Efficiancy of Nonlinear Control in Uncertain Linear Systems, J. Dyn. Syst. Meas. and Control, Vol. 102, pp. 95-102, 1981.

[10]. Barmish, B. R.; Coreless, M.; Leitmann, G.: A New Class of Stabilizing Controllers for Uncertain Dynamical Systems, SIAM J. Control and Opt., Vol. 21, pp. 246-255, 1983.

[11]. Coreless, M.; Leitmann, G.: Continuous State Feedback Guaranteeing Uniform Ultimate Boundedness for Uncertain Dynamic Systems, IEEE Trans. Aut. Contr., Vol. AC-26, pp. 1139-1144, Oct. 1981.

[12]. Singh, S. N.: Ultimate Boundedness Control of Uncertain Linear Model Following Systems, IEEE Conf. on Decision and Control, Las Vegas, Nevada, 1984.

Interfacing Microcomputer with Numerically Controlled Machine

Z. J. Czajkiewicz and Y. Parikh

DEPARTMENT OF INDUSTRIAL ENGINEERING
WICHITA STATE UNIVERSITY
WICHITA, KANSAS, 67208

ABSTRACT

Rapidly developing microcomputrs are merging with all phases of the manufacturing processes, and therefore it is very conceivable to convert a conventional NC machine to microcomputer controlled machining center or to CNC. In a large manufacturing firms dedicated CNC limits the total flexibility required for manucaturing and therefore DNC (Distributed Numerical Controlled) machining centers are comming in the picture. A low cost microcomputer controlled conversion of conventional NC machine can benefit small to medium size firms as well as large firms in upgrading the facility without large capital investment.

The study involved designing and building an interface. The software was developed to controll the machine. The input of a punched paper tape was replaced by the micro-computer. New system enhanced NC programming in several aspects:

- improved editing capabilities
- milling in straight lines at any angle
- milling circular shape in X-Y plane
- milling any two dimensional curve

In the modified machine, an operator can program his/her part machining sequences on the directly on the microcomputer and do the necessary changes without reprogramming the entire part. All calculations are prepared by the computer. Program can be stored on disk for the future use.

INTRODUCTION

In this research project, an effort was made to convert a conventional open loop Numerical Control machine to Apple II microcomputer controlled machine. The machine under study was a vertical bridgeport milling machine with SLO-SYN point to point controller loacted at the college of engineering, Wichita State University.

The SLO-SYN controller reads a punched paper tape

through a light sensitive tape reading head. The paper tape has 8 channels and the neccessary functions are coded in a standard format, as standrdized by the Electronics Industry Association and Aerospace Industries Association. The main controller of NC machine reads a following sequences:

- TAB function followed by X coordinate, and
- TAB function follwed by Y or Z coordinates, and
- TAB function with any special function code, and
- EOB (End Of Block) statement that terminates the sequence.

The set of instructions before each EOB command is then executed. The process is repeated for each step of X-Y movements. A delete command will delete only last sets of instructions. If the error was detected after punching the tape, then the whole tape has to be repunched to correct the error.

The tape program read by the tape reader is converted in binary codes which are received, interpreted and acted by the SLO-SYN controller. A microcomputer also works with binary numbers and therefore the necessary code can be fed directly from microcomputer to the SLO-SYN controller and thereby eliminating the tape generation and tape reader functions requirements.

PRESENT SYSTEM

The SLO-SYN numerical tape control is three axis positioner with both tape or manual input and ability to

```
SEQUENCE NO. 8.

TAB OR EOB        T
X - INCREMENT     -2000
TAB OR EOB        T
Y - INCREMENT     3750
TAB OR EOB        T
M - FUNCTIONS     02
EOB               E

SEQUENCE NO. 9.

TAB OR EOB END OF INPUT
TAB  X-INCR T/E  Y-INCR  T/E  M-FUN EOB
========================================
 E
----------------------------------------
 T      2000  T     -1875 E
----------------------------------------
 T      2000  E
----------------------------------------
 T      2000  E
----------------------------------------
 T            T     -1875 E
----------------------------------------
 T     -2000  E
----------------------------------------
 T     -2000  E
----------------------------------------
 T     -2000  T      3750  T      02     E
----------------------------------------

PRESS ANY KEY TO START
128 46 2 32 32 32 46 64 1 8 7 5 128 46 2 32 32 32 128 46 2 32 32 32 128 46 46 64
 1 8 7 5 128 46 64 2 32 32 32 128 46 64 2 32 32 32 128 46 64 2 32 32 32 46 3 7 5
 32 46 32 2 128
```

Figure 1: Sample Program Output

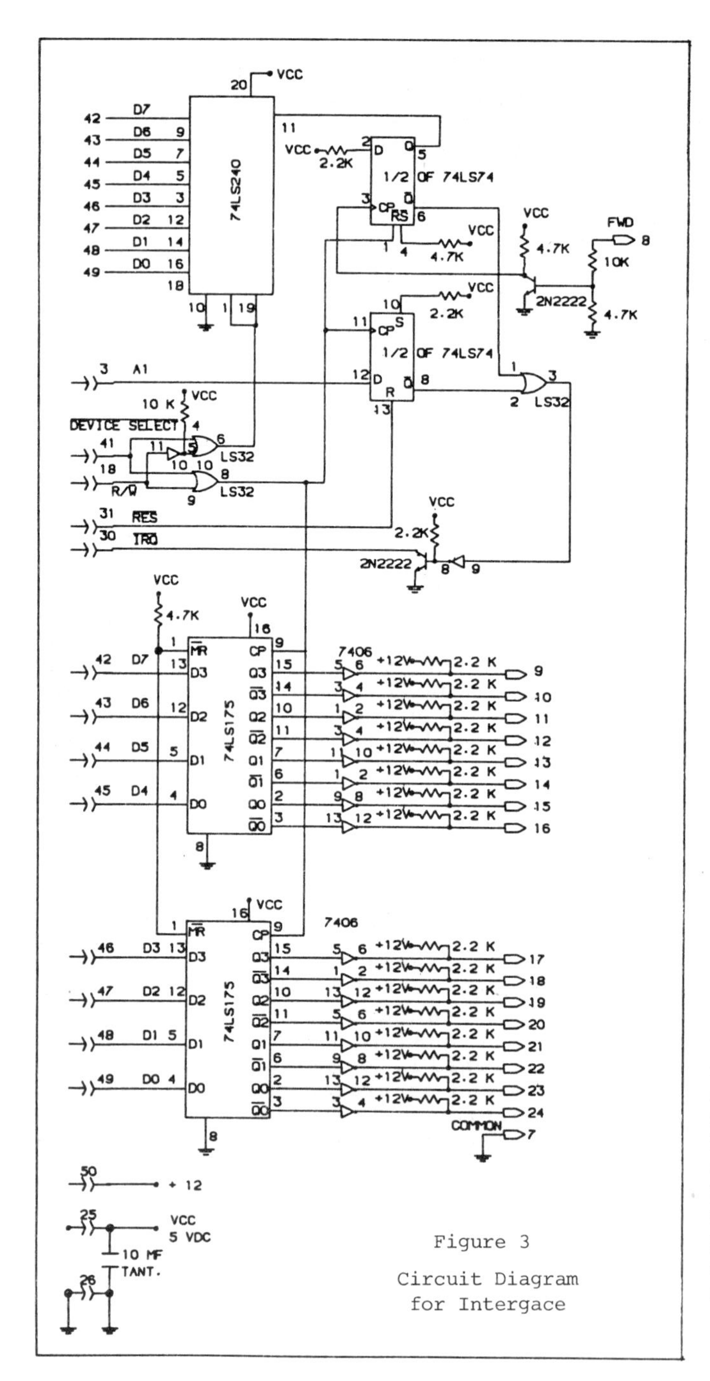

Figure 3

Circuit Diagram for Intergace

NC machining, it is impossible to correct an error after the complete tape has been prepared. Any changes in the program mean repunching the tape all over again. Microcomputer makes programming much easier and can greatly enhance editing capabilities. Programs could be inputed directly from keyboard or stored on disk. Whenever a programmer will detect an error or make even a smallest change in the program, it could be corrected directly on screen of microcomputer. This feature can save hours of work for NC programmers and machinists.

Figure 4

Figure 5

2. Milling in straight lines at any angle.

THE SLO-SYN is equiped with milling at 45 degree angle only. For milling at any other angle the programmers have to calculate necessary coordinates for required tolerance. If the tolerances are tight, i.e. the tolerances are very small, means step size for calculating coordinates is very small and number of required calculations is very large. The great advantage of having microcomputer is let the computer do all the calculation and save the programmers time for doing more productive work.

4. Milling circular shape in X-Y plane.

On SLO-SYN NC machine, in order to mill circular shape other than the diameter of the cutter,the rotary table or indexing table must be used, which means a whole new set up time consumption. Alternatively a cutter can be moved one thousandth of an inch increment in the desired direction and thus obtain the required diameter. A microcomputer is very suitable for this task, whereas punching necessary tape is really cumbersome.

5. Milling any two dimensional curve.

As explained above, if the curve does not fit to the circle, there is no other alternative but to calculate the necessary coordinates and prepare the tape. On the other hand use of microcomputer can solve all the headache of calculation and punching the tape.

6.Three dimensional milling.

It is quite conceivable to develop a routine on micro-computer such that three dimensional or multidimensional (depending upon the available axis of movement on the machine) objects can be milled.

CONCLUSIONS.

Both types of NC machines, closed loop circuit and open loop circuits, can be converted to CNC Machines. Controllers that are commercially available for converting Open Loop circuit NC machines are relatively cheaper than one for closed loop circuit NC machines. The most inexpensive controller has a price tag of $10,000.00 which can not compensate for backlash or generate complex shapes and is dedicated to a given machine only. Commercial CNC controller for converting NC to CNC with much needed functions, such as complex shape generation, backlash compensations and multiaxes control has price tags of $25,000.00 to $35,000.00.

Having or developing a proper interface between a micro computer and the NC machine, microcomputer is the most inexpensive, most versatile, extremely portable and tremendously powerful controller for converting NC machine to CNC machine.With further research in this area, it is possible to convert other more traditional machines to CNC machines. Furthermore, user freindly software can be generated that can ease the operators in programming and operating the CNC machine without having operators to go through the specialized training.

REFERENCES

1. Bernhart W.D., "Research and Instruction in Computer Aided Design and Manufacturing", Research Report for Boeing Company, not published.

2. Thornhill R.B., "Engineering Graphics and Numerical Control", McGraw-Hill Company, 1967.

3. "Apple II - Reference Manual", Apple Computers Inc.,1981.

4. Slo-Syn Operators Manual for NC Bridgeport Milling Machine.

Dynamic Control for a Tentacle Manipulator

Mircea Ivanescu and Ion Badea

Dr. Mircea Ivanescu
Automatic and Computer Department
University of Craiova
1100 Craiova, Romania

Ion Badea
Craiova Computer Centre
1100 Craiova, Romania

Abstract

A class of manipulator arms with an infinite number of degrees-of-freedom (a tentacle) is studied.A distributed parameter model is determined using the generalised form of Lagrange's equation.This model is characterised by an integral-differential equation which generates the dynamic of the manipulator.

First,the unconstrained control problem is solved and an optimal algorithm for the minimum energy control is determined.The constrained problem for state,control and final time is also developed and necessary conditions of these problems are obtained.

INTRODUCTION

During recent years the problem of manipulator action control has been attracting considerable attention.There are a great number of publications and achievments in the field of manipulator control.Out of a great number of them we notice papers in which the control problem of motion is a problem of controlling multivariable systems possessing redundant degrees-of-freedom in a finite number.

This paper presents a new manipulator with ideal flexible elements and distributed mass and torques,a tentacle model.This model is equivalent to a mechanism with infinite number of degrees-of-freedom.It is shown that the dynamic of the system is defined by a set of integral-differential equations obtained by using the generalised Lagrange's principle.

Such systems offer a great possibility of control in some points in the space,a good flexibility for a large number of applications etc.

The control of this system is an important problem.It requires to find the control law of distributed torques so that the general trajectory in space should become the imposed trajectory.Several difficulties arise associated with the complex form of dynamic equations and with the highly non-linear,interactive nature of the system.

In this paper it is shown that by applying the variational method for the integral-differential model of the tentacle manipulator it is possible to obtain the necessary conditions for optimal control for unconstrained and constrained problems.

The analysis of the control methods for a planar tentacle is described but this system possesses all the important features so that the application of the methods to more complex systems in a 3-dimensional space may be readily inferred.

DYNAMIC MODEL

We shallconsider the tentacle arm of Fig.1.The system can operate in a 2-dimensional vertical space,XOY.

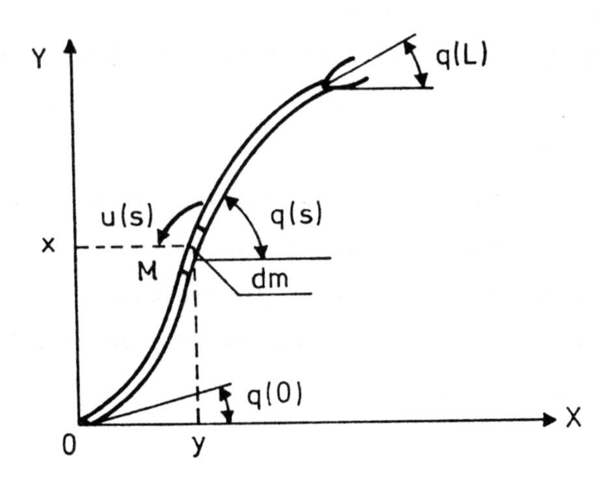

Fig.1

We assume the arm as perfectly flexible,without any viscous friction.Its great flexibility determines an analysis similar to that used in infinite-dimensional systems,the distributed parameter systems.

Technologically such systems can be obtained by using a cellular structure for every element of the arm.The control can be produced using an electrohydraulic or pneumatic action which determines the contraction or dilatation of the peripheral cells.In this paper we shall discuss only the dynamic of the control without dealing with these technological problems.

Let us take the arm in Fig.1.We assume a uniform distributed mass,with a linear density ρ [kg/m].Also,we shall neglect the effects of the section in dynamic motion of the arm.We denote by s the spatial variable on the length of the arm,$s \in [o,L]$.

The position of some point M of the arm can be defined in Cartesian coordinates (x,y) or in Lagrange generalised coordinates q , s,where q represents the absolute angle.

$$x = \int_0^s \cos q' \, ds' \tag{1}$$

$$y = \int_0^s \sin q' \, ds' \tag{2}$$

where q' represents

$$q' = q(s') \qquad s' \in [o,s] \tag{3}$$

In a dynamic analysis,the generalised coordinate q will be a function of time t ,

$$q = q(s,t) \tag{4}$$

The dynamic behaviour of the tentacle arm is determined by the effects of a distributed torque on overall length of the arm u(s,t) and by the potential energy of the system.

Let v_x ,v_y be the velocity components with respect to X and Y axes.From (1) and (2) results,

$$v_x = \dot{x} = -\int_0^s \sin q' \frac{\partial q'}{\partial t} \, ds' \quad (5)$$

$$v_y = \dot{y} = \int_0^s \cos q' \frac{\partial q'}{\partial t} \, ds' \quad (6)$$

For an element dm, the kinetic energy will be,

$$dT_c = \frac{1}{2} \, dm \, v^2 \quad (7)$$

Using (5),(6), we have

$$T_c = \frac{1}{2} \int_0^L \rho \left(\left(\int_0^s \frac{\partial q'}{\partial t} \sin q' ds' \right)^2 + \left(\int_0^s \frac{\partial q'}{\partial t} \cos q' \, ds' \right)^2 \right) ds \quad (8)$$

where T_c represents the kinetic energy for the overall arm. Obviously, this energy is a functional by the form

$$T_c = T_c(q, \frac{\partial q}{\partial t}, s) \quad (9)$$

Similarly we obtain the potential energy,

$$dT_p = dm \, g \, y \quad (10)$$

and from (2) results

$$T_p = \int_0^L \rho g \int_0^s \sin q' \, ds' ds \quad (11)$$

where

$$T_p = T_p \, (q, s) \quad (12)$$

The dynamic behaviour of the arm will be obtained using Lagrange principle developed for infinite-dimensional systems [3].

$$\frac{\partial}{\partial t} \cdot \frac{\delta T_c}{\delta \left(\frac{\partial q}{\partial t} \right)} - \frac{\delta T_c}{\delta q} - \frac{\delta T_c}{\delta s} - \frac{\delta T_p}{\delta q} - \frac{\delta T_p}{\delta s} = Q \quad (13)$$

where $\delta T / \delta q$ denotes a functional partial (variational) derivative [1], which is defined as the variation of the functional T with respect to the function q at a point $s \in [0, L]$, and Q is the generalised input of the system,

$$Q = \int_0^L u(s,t) \, ds \quad (14)$$

Substituting the relations (8) and (12) into (13) and expanding them we obtain the general form for the system motion,

The adjoint equations are the same as the equations obtained in (21)-(23).

APPROXIMATE SYSTEM

We shall use an approximate model based on a spatial discretization which has the advantage to be strongly connected to the physical structure of the system. Let $s_o, \ldots, s_N$ be the spatial discrete variable and Δ the size of the space increment. We denote by

$$q_i(t) = q(t, s_i) \quad , \quad i = o,1,\ldots,(N-1) \tag{36}$$

We select N and Δ so that

$$|q_i(t) - q_{i-1}(t)| < \varepsilon \qquad \begin{matrix} i = 1,\ldots,(N-1) \\ t \in [o, t_F] \end{matrix} \tag{37}$$

where ε is a positive constant sufficiently small.

By considering different values of i, from (15),(16) we obtain the following set of differential equations

$$\sum_{i=o}^{m} \left[\frac{d^2 q_i}{dt^2} \cos(q_m - q_i) - \frac{dq_m}{dt} \cdot \frac{dq_i}{dt} \cos(q_m - q_i) - \left(\frac{dq_i}{dt}\right)^2 \sin(q_m - q_i) \right] -$$

$$- g(\sin q_m + \cos q_m) = \frac{1}{\varsigma \cdot \Delta} u_m \qquad m = o,1,\ldots,(N-1) \tag{38}$$

with initial conditions

$$q_i(o) = q_i^o \tag{39}$$

$$\frac{dq_i}{dt}(o) = q_{1i}^o \tag{4o}$$

where $u_m(t) = u(t, s_m)$.

A new approximate model of the system can be obtained by considering the specifical form of (38) and using the inequality (37). This is

$$\frac{d^2 q_i}{dt^2} - \left(\frac{dq_i}{dt}\right)^2 = \frac{1}{\varsigma \cdot \Delta}(u_i - u_{i-1}) \quad , \quad i = 1,2,\ldots,N-1 \tag{41}$$

where u_i, u_{i-1} denotes two consecutive sequences of the control.

A similar procedure can be developed for the adjoint equations, (19)-(23).

$$-\frac{d\lambda_{1i}}{dt}+\lambda_{2i}\,q_{2i}^{2}\,\Delta+\frac{\lambda_{2i}}{\lambda_{2,i-1}}\left(\frac{d\lambda_{1,i-1}}{dt}-K_{i-1}\right)+K_{i}=0 \qquad (42)$$

$$-\frac{d\lambda_{2i}}{dt}-\lambda_{1,i}-q_{2i}(1+\Delta)+\left(\frac{d\lambda_{2,i-1}}{dt}+\lambda_{1,i-1}+q_{2i-1}\right)=0 \qquad (43)$$

with final conditions

$$(x_F-\sum_{0}^{i}\cos q_{1j}\;\Delta)\sin q_{1j}-(y_F-\sum_{0}^{i}\sin q_{1j})\cos q_{1i}+\lambda_{1i}=0 \qquad (44)$$

$$\lambda_{2i}(t_F)=0 \qquad (45)$$

where

$$\lambda_{i}=\lambda(t,s_i)$$

$$K_i=-q_{2i}^{2}-g\cos q_{1i}+g\sin q_{1i}$$

The equations (36)-(45) represent an approximate model for unconstrained problem.A similar technique can be used for the problems with state,input or final time constraints.

CONCLUSION

The paper presents the dynamic of a new manipulator,a tentacle model.A parameter distributed system described by an integral differential equation is derived using the generalised form of Lagrange's equation.

The optimal control problems are then studied.First,the unconstrained control problem and a minimum energy control is developed.Then,the constraints of the state,control and final time are introduced and the integral-differential equations associated to these problems are determined.

References

1. Wang,P.K.C.,Optimum Control of Distributed Parameter Systems, IEEE Trans.Automatic Control,vol.AC-9,pp 13-22,1964.
2. Hewit,J.R.,Padovan,J.,Decoupled Feedback Control of Robot and Manipulator Arms,3rd Symp.on Theory of Robots and Manipulators Udine,Sept.,1978,pp 251-266.

Manipulation II

Chairman: Dilip Kohli, University of Wisconsin-Milwaukee, Milwaukee, Wisconsin
Vice Chairman: V. Singh, Banaras Hindu University, Varanasi India

A Method of Grasping Unoriented Objects Using Force Sensing

Yoshihide Nishida, Yoja Hirata, and Mitsuhito Watanabe

Manufacturing Development Laboratory
Mitsubishi Electric Corp.
1-1 Tsukaguchi-Honmachi 8-Chome,
Amagasaki, 661 Japan

Abstract

A new method for grasping unoriented objects using force information has been developed. It is based on the estimate of the position and orientation of a workpiece by detecting three components of contact force between a workpiece and a finger-tip. The feasibility of the method has been demonstrated by a sensory hand to pick up an unoriented bolt on the horizontal plane.

1. Introduction

It is necessary to detect the position and orientation of workpieces, if a robot has to grasp unoriented objects. Vision sensors and tactile sensors are thought to be effective to obtain these information, and many researches have been conducted using visual means [1][2]. However there are little reports on tactile sensors.
We think tactile sensors are more immune to environmental influence, and more economical than visual ones. Of course, it is difficult to recognize a complicated shape using tactile information, but there are many cases in which we want to detect the position and orientation of simple-shaped workpieces in actual situation.
In this study, we set a target : to develop a method using force sensing to detect the position and orientation of an unoriented cylindrical workpiece.

2. Method of detection

This chapter describes that the position and orientation of a cylindrical workpiece unoriented in three demensional space can be obtained by detecting three components of contact force between a workpiece and a finger-tip.

Suppose that a hand moves in the Z direction in the force sensor coordinate system and the finger touches the workpiece on its curved surface (See Fig. 1)
We denote the unit vector $\underline{a}$ along the central axis of the cylindrical workpiece

$$\underline{a} = ax \cdot \underline{i} + ay \cdot \underline{j} + az \cdot \underline{k} \tag{1}$$

where ax, ay, az are the scalar components of the vector $\underline{a}$, and $\underline{i}$, $\underline{j}$, $\underline{k}$ the unit vectors corresponding respectively to the x, y, and z axes, also we denote the normal vector $\underline{n_1}$ at the first contact point, and $\underline{n_2}$ at the second contact point respectively

$$\underline{n_1} = n_1x \cdot \underline{i} + n_1y \cdot \underline{j} + n_1z \cdot \underline{k} \tag{2}$$

$$\underline{n_2} = n_2x \cdot \underline{i} + n_2y \cdot \underline{j} + n_2z \cdot \underline{k} \tag{3}$$

where n_1x, n_1y, n_1z are the scalar components of $\underline{n_1}$, and n_2x, n_2y, n_2z the scalar components of $\underline{n_2}$. Since $\underline{a}$ is orthogonal to $\underline{n_1}$ and to $\underline{n_2}$, we obtain the following equations.

$$n_1x \cdot ax + n_1y \cdot ay + n_1z \cdot az = 0 \tag{4}$$

$$n_2x \cdot ax + n_2y \cdot ay + n_2z \cdot az = 0 \tag{5}$$

The vector $\underline{a}$ is a unit vector. Hence, using Eq. (2) and (3), the components of $\underline{a}$ are expressed in terms of the components of $\underline{n_1}$ and $\underline{n_2}$.

$$ax = \frac{1}{1+C_1{}^2+C_2{}^2}, \quad ay = C_1 \cdot \frac{1}{1+C_1{}^2+C_2{}^2},$$

$$az = C_2 \cdot \frac{1}{1+C_1{}^2+C_2{}^2} \tag{6}$$

where

$$C = - \frac{n_2z \cdot n_1x - n_1z \cdot n_2x}{n_2x \cdot n_1y - n_1z \cdot n_2y}, \quad C = - \frac{n_1x + n_1y \cdot C_1}{n_1z}$$

Therefore, if we obtain two normal vectors at contact points in the curved surface, we can determine the axial vector representing the orientation of the cylindrical workpiece.
The normal vector at the contact point is obtained as follows. The components of the first contact force vector $\underline{F_1}$ detected by force sensor are

$$F_1x = (|\underline{N_1}|\cdot\sin\alpha_1 - \mu\cdot|\underline{N_1}|\cdot\cos\alpha_1)\ \cos\beta_1$$
$$F_1y = (|\underline{N_1}|\cdot\sin\alpha_1 - \mu\cdot|\underline{N_1}|\cdot\cos\alpha_1)\ \sin\beta_1 \qquad (7)$$
$$F_1z = (|\underline{N_1}|\cdot\cos\alpha_1 + \mu\cdot|\underline{N_1}|\cdot\sin\alpha_1)$$

where

$\underline{N_1}$ = normal force vector at the first contact point.
α_1 = angle that $\underline{N_1}$ forms with the Z axis.
β_1 = angle that the vertical plane containing $\underline{N_1}$ forms with the X-Z plane.
μ = the coefficient of friction.

Solving Eq. (7) for α_1 and β_1, we have

$$\alpha_1 = \theta f + \cos^{-1}\left(\frac{F_1z}{\sqrt{F_1x^2+F_1y^2+F_1z^2}}\cdot\cos\theta f\right) \qquad (8)$$
$$\beta_1 = \tan^{-1}(F_1y\ /\ F_1x)$$

where θf is the friction angle and $\theta f = \tan^{-1}\mu$. Expressing the components of the normal vector $\underline{n_1}$ in terms of α_1 and β_1, we have

$$n_1x = \sin\alpha_1\cdot\cos\beta_1,\ n_1y = \sin\alpha_1\cdot\sin\beta_1,\ n_1z = \cos\alpha_1\cdot \qquad (9)$$

Hence, using Eq. (6), we obtain the axial vector, that is, the orientation of the cylindrical workpiece.
Let the Y' axis be chosen in the direction of $\underline{a}$, the Z' axis be chosen in the direction of $\underline{b}$ which is in the plane containing $\underline{a}$ and the Z axis, the X' axis be chosen in the direction perpendicular to $\underline{a}$ and to $\underline{b}$. The coordinates X' and Z' of the first contact point are

$$X' = r\cos\theta\ ,\quad Z' = r\sin\theta \qquad (10)$$

where r = radius of the cylindrical workpiece

θ = angle between $\underline{n_1}$ and $\underline{b}$.

The angle θ is determined by comparing Eq.(9) with the following equation (11) obtained by rotating $\underline{b}$ through θ about $\underline{a}$.

$$\underline{n_1} = \left(\sin\theta\frac{ay}{\sqrt{ax^2+ay^2}} - \cos\theta\frac{az\cdot ax}{\sqrt{ax^2+ay^2}}\right)\underline{i} + \left(-\sin\theta\frac{ax}{\sqrt{ax^2+ay^2}} - \cos\theta\frac{ax\cdot ay}{\sqrt{ax^2+ay^2}}\right)\underline{j} + \cos\theta\frac{ax^2+ay^2}{\sqrt{ax^2+ay^2}}\underline{k} \qquad (11)$$

The contact point coordinates in the X,Y,Z coordinate system are determined by the transformation of the coordinates system.
In many cases, cylindrical workpieces in factory are on the horizontal plane. In this case the detection of orientation and position may be remarkably simplified. As shown in Fig. 2, the orientation angle β_1 is obtained from Eq. (8). Using the angle α_1 obtained from Eq. (8), the contact point is represented in the coordinates X' and Z.

$$X' = r\sin\alpha_1 \ , \ Z = r\cos\alpha_1 \tag{12}$$

or, in the coordinates X and Z

$$X = r\cdot\sin\alpha_1 \ / \ \cos\beta_1 \ , \ Z = r\sin\alpha_1 \cdot \tag{13}$$

However, in the case where the angle α_1 is smaller than the friction angle θf, the forces Fx, Fy can not be sensed by the force sensor. In this case, it is necessary to bring the finger into contact with the region where α_1 is larger than θf.

3. Application to a bolt on the horizontal plane

3.1 Experimental system

An experimental system was developed in order to examine the feasibility of the proposed method. The system consists of a sensory hand, a X,Z,θ positioning table, and a controller. As shown in Fig. 3, the sensory hand has two fingers. These fingers are positioned by a stepping motor. The grasping force is generated by a spring. The upper section of the finger is a three degree of freedom force sensor, and the lower section is a nail with a detent preventing a grasped workpiece from tilting. The finger-mounted force sensor consists of 12 semiconductor strain gauges bounded on a T-shaped elastic body made of chromium-molybdenum-steel. Groups of the strain gauges X_1 - X_4, Y_1 - Y_4, and Z_1 - Z_4 form wheatstone bridges to detect forces in the X, Y and Z directions respectively.

3.2 Experiment

The system was applied to the acquisition of an unoriented

bolt on the horizontal plane.

To check the ability of the finger-mounted force sensor to detect the position and orientation of a bolt, the angle of circumference α and the orientation angle β at the contact point was measured by it. The experimental results are shown in Fig. 4. The error of measured α is not influenced by the angle α and β at the contact point, but the error of measured β is influenced fairly by α. The error of β is excessive in the region where the angle α at the contact point is smaller than 30°. Therefore the finger should be brought into contact with the region where the angle α is larger than 30° in order to detect accurately.

Considering the above results, the step in the grasping process is as follows.

(1) The hand goes downward detecting the forces acting on two fingers until the proper contact force is obtained.

(2) The position and orientation of the bolt are calculated. If the finger touches in the region where α is smaller than 30°, or if the outside of the finger touches, the hand goes upward and moves in the X direction.

(3) Repeat (1) - (2) until the inside of the finger touches in the region where α is larger than 30°.

(4) The hand is positioned and oriented according to the calculated value, and goes to grasp it.

Fig. 5 shows a sequence of grasping motion. The hand adapts to the position and orientation of a bolt, and grasps it. The experimental results demonstrate the feasibility of the proposed method.

4. Conclusion

A new method for grasping an unoriented cylindrical workpiece using force information has been discussed. At the same time, on the basis of the method, a sensory hand has been developed which can detect the position and orientation of workpieces. The most significant feature of the hand is that high sensitive three degree of freedom force sensors are mounted on the fingers. The feasibility of the proposed method has been demonstrated by the hand. The method will be useful in solving bin-picking problem.

References

1. Kelley, R. B.; Birk, J. R.; Martins, H. A. S.; Tella, R.: A robot system which aquires cylindrical workpieces from bins. IEEE Trans. Syst., Man, Cybern., vol. SMC-12, No.2, pp. 204 ∿ 213, Mach 1982.

2. Hermann, J. P. : Pattern recognition in the factory : An example, Proc. 12th Int. Symp. Ind. Robots, pp. 271 ∿ 280, 1982.

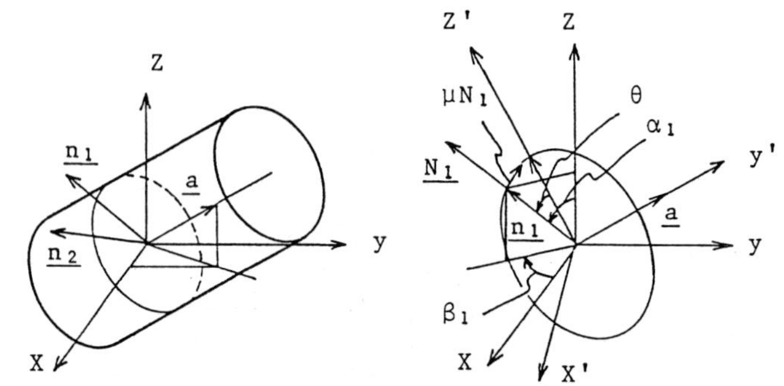

Fig. 1 Forces applied to the cylindrical workpiece in three dimensional space.

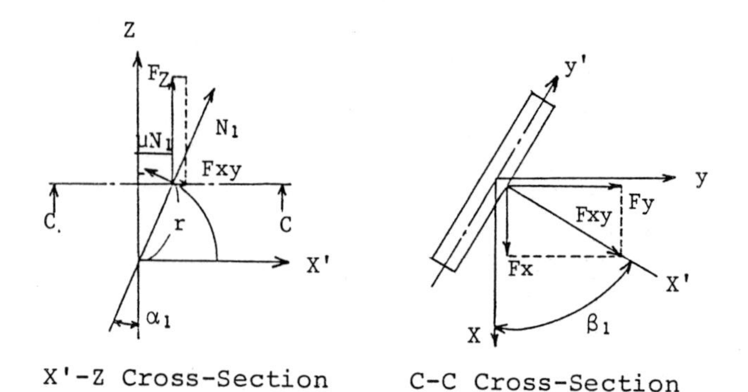

X'-Z Cross-Section   C-C Cross-Section

Fig. 2 Forces applied to the cylindrical workpiece on the horizontal plane.

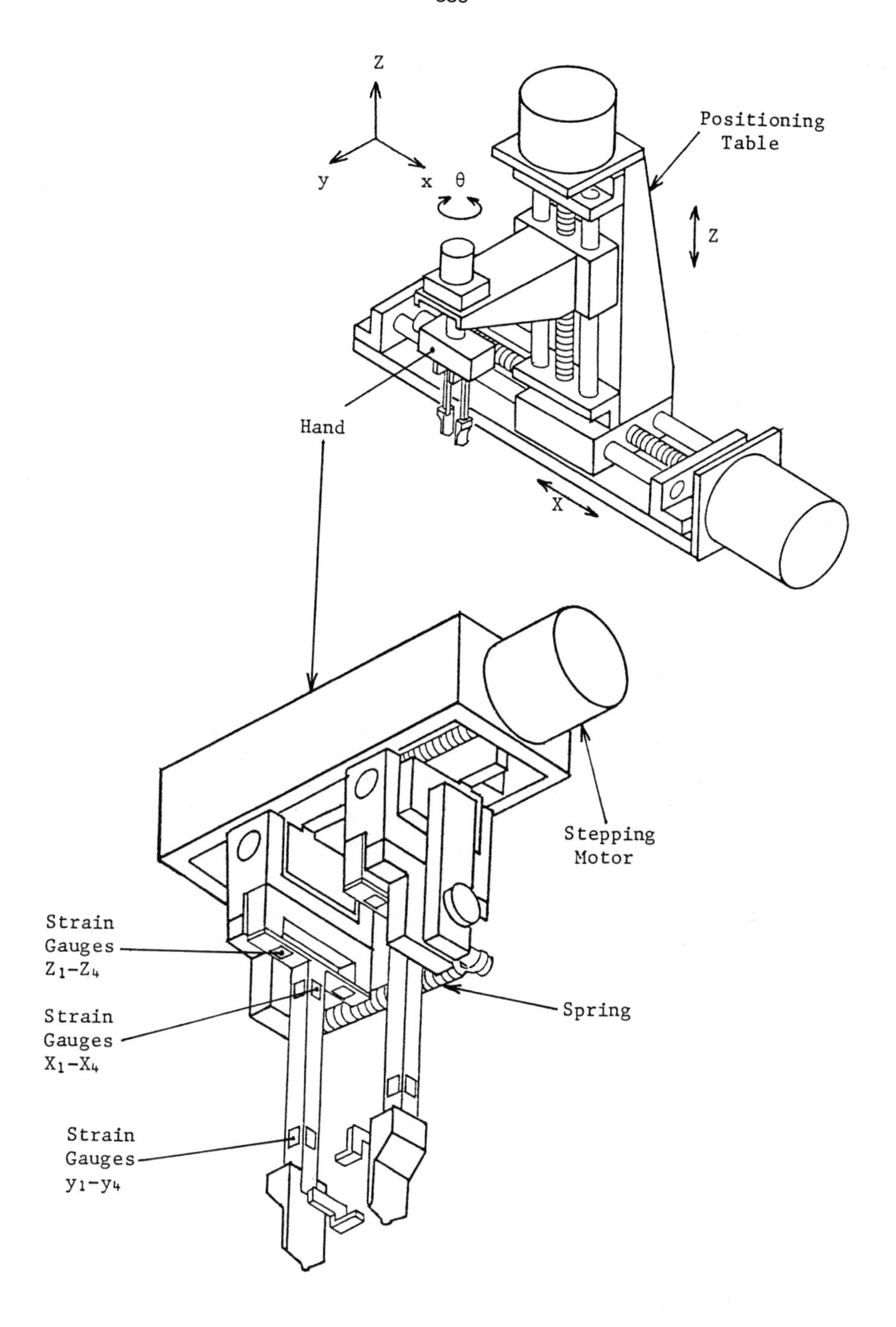

Fig. 3 Mechanical System

Set Angle β

○ - 0° ▽ - 10° □ - 20° ◇ - 30° △ - 40° ● - 60°

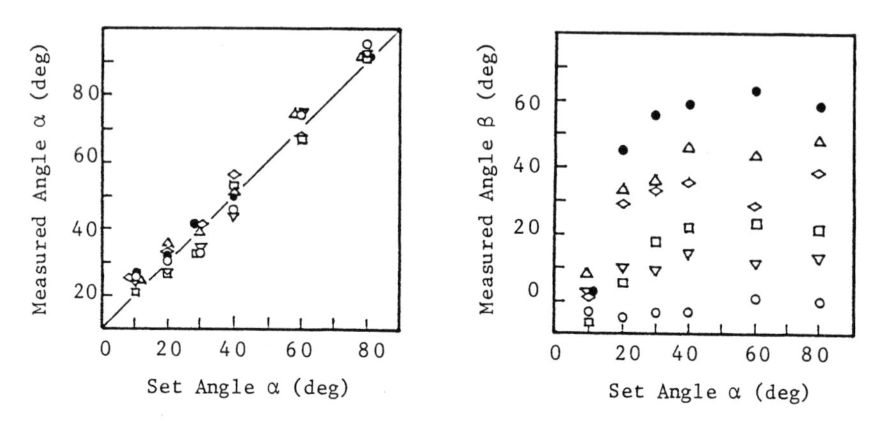

Fig. 4 Measurement of the angle α and β

(1) Touching (2) Positioning and Orienting (3) Picking up

Fig. 5 Grasping Motion

Mapping of Positioning Error of Robot Manipulators

C. O. Huey, Jr. and Joseph Anand

Quality Assurance Engineer
Calma Corporation
San Diego, CA

Associate Professor of Mechanical Engineering
Clemson University
Clemson, SC

Abstract

The output error for open kinematic chains is predicted based on values for the errors present in the fixed and controlled geometric parameters and on the effect of joint clearances. The output error is computed over a specified region and a mapping utility program is used to generate contour maps of this error. The statistical nature of the error is considered. The method has been applied to R-R-R, R-P-R, and R-P-P mechanisms. Examples for R-R-R mechanisms are included in the paper.

Introduction

It is clear that, for any position of a mechanism, the relative locations of the components of the mechanism are influenced by the presence of error in the geometric parameters that define the mechanism and its position. To simplify the view here, consider the simple open chain shown in Figure 1 and assume a specific interest in the location of point P relative to the x,y coordinate system. It is easily seen that the departure from the nominal position induced by errors in the geometric parameters, ℓ_1, ℓ_2, θ_1, and θ_2, will be influenced by the nominal values of these same parameters (note the effect of changing θ_1 and θ_2). The extension of this notion to robot manipulators is obvious: The error in position of the end effector of a robot is influenced by errors in the geometric parameters that define the mechanism itself and by errors in the controlled parameters that define its position and by the location of the end effector in the workspace. Given a convenient means of quantifying the error induced, by these factors it would be possible to efficiently match robots to tasks on a rational basis when accuracy is an issue. For example, the task might be organized so that operations requiring high precision could be executed in the region where accuracy is high while less demanding operations occur elsewhere. In other cases

[1]Former graduate student, Clemson University

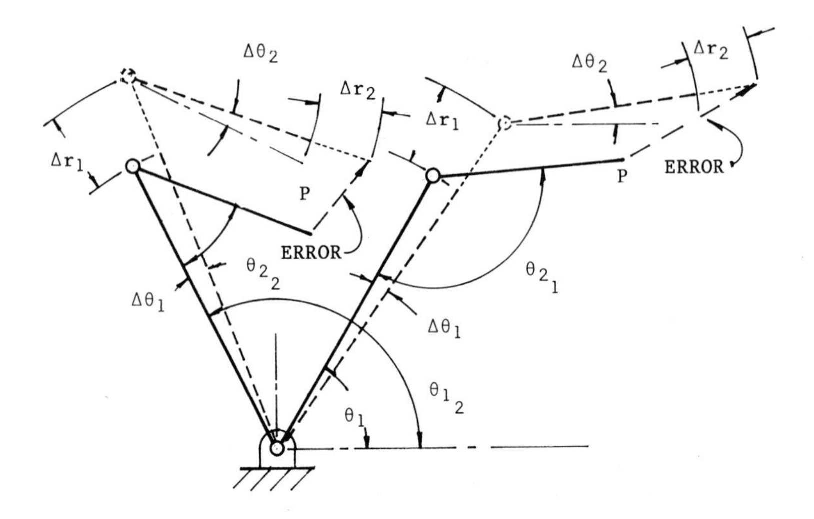

Fig.1. Error in the output position of an open kinematic chain as a function of error in both the defining parameters and the position of the chain.

it might be desirable to simply define the region of the workspace where the error would be below a threshold value.

The development that follows considers a general open kinematic chain of three degrees of freedom. In the original study the effects of joint clearances were also considered, however, owing to the constraints of space, they will be but briefly mentioned here. For the same reason the examples that follow involve only R-R-R mechanisms even though the method has been applied to R-P-R and R-P-P mechanisms as well.

Methodology

Figure 2 shows a R-R-R mechanism arranged in a fashion typical of robot applications. Using the familiar Hartenberg-Denavit matrix representation of the links, the end point, shown as point P in the figure, is located in the fixed reference frame by the following [1,2][1]:

$$[P]_1 = [{}_2A_1]\ [{}_3A_2]\ [{}_4A_3]\ [P]_4 \tag{1}$$

[1]Numbers in brackets refer to entries in the bibliography

where:

$[_{i+1}A_i]$ is the transformation from coordinate system i+1 to system i

$[P]_1$ locates point P in the base coordinate system

$[P]_4$ locates point P in the coordinate system attached to the final link

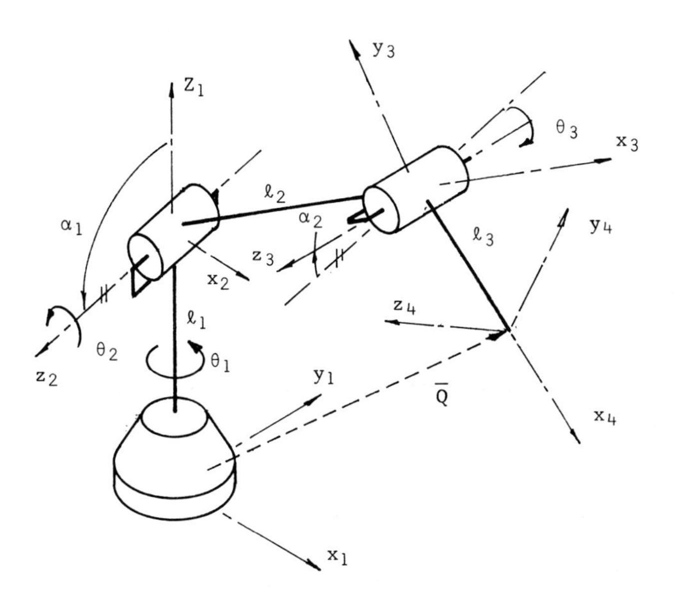

Fig.2. Schematic of an R-R-R mechanism as frequently employed in robots.

If $\overline{Q}$ is the position vector locating the output point in the fixed reference system a general component of $\overline{Q}$ can be written as

$$q = q(v_1, v_2, v_3, \ldots v_n) \tag{2}$$

where:

v_i is either a controlled variable or a fixed mechanism parameter

Expanding the function, q, in a Taylor series about the nominal values of the variables, v_{im}

$$q = q_m + \sum_{i=1}^{n} \left.\frac{\partial q}{\partial v_i}\right|_{v_{im}} (\Delta v_i) + \frac{1}{2!} \sum_{i=1}^{n} \sum_{j=1}^{n} \left.\frac{\partial q}{\partial v_i \partial v_j}\right|_{v_{im}, v_{jm}} (\Delta v_i)(\Delta v_j) + \ldots \tag{3}$$

is obtained. Assuming small values for the variations in the in parameters, v_i, the variation in q becomes

$$\Delta q = \sum_{i=1}^{n} \frac{\partial q}{\partial v_i}\bigg|_{v_{im}} (\Delta v_i) \quad . \tag{4}$$

If the errors in these variables are independent, uncorrelated, and normally distributed with zero mean, then the error in q will also be normally distributed with zero mean [3]. In such cases, the standard deviation of the output error distribution can be written as

$$\sigma_q = \sqrt{\sum_{i=1}^{n} \left(\frac{\partial q}{\partial v_i}\bigg|_{v_{im}} (\sigma_{v_i})\right)^2} \quad . \tag{5}$$

This analysis can be applied separately to determine the effects of error in the fixed mechanism parameters and in the controlled parameters. Assuming linear theory due to the small effects being considered the separate results can be superimposed.

Considering the controlled parameters alone and applying the notions above, the deviation of the output position vector from the nominal position can be found as

$$[\Delta_q] = [J_1][\Delta\Gamma_1] \quad . \tag{6}$$

where:

$[\Delta\Gamma_1]$ is the matrix of errors in the actively controlled parameters

$[J_1]$ is the Jacobian matrix for the set of parameters $[\Gamma_1]$

Extending the analysis to geometric parameters we can write the output error as

$$[\Delta_q] = [J_2][\Delta\Gamma_2] \tag{7}$$

where:

$[\Delta\Gamma_2]$ is the matrix of errors in the fixed mechanism parameters

$[J_2]$ is the Jacobian matrix for the set of parameters $[\Gamma_2]$

Tribo-System in Robot Design

J. P. Sharma and S. N. Dwivedi

College of Engineering, University of North Carolina at Charlotte, USA

Abstract

Robots are an assembly of mechanical systems, computers, instrumentation, controls and transducers. The deterioration of Electro-Mechanical components first starts at surface and subsurface of interacting material pairs and contact points. The surrounding environment also contributes severity to the degradation of these components. These are tribolgical factors which should be incorporated in the design of robots for applications in work space having mild to hostile environment. The accuracy and sensitivity of motion, position and movement to perform a predetermined task will depend on tribo-design of interacting electro-mechanical components in robots. Since the first cost of deployment of robot is very high, a careful tribo-design consideration will help in evaluating the performance and life cycle casting of robot. In this paper, the authors have discussed the critical tribo-failure problems of robot components.

A 'robot' in general gives an impression of an electro-mechanical human being which will have a life-style similar to that of an intelligent person. But, for technologist the robots are nothing but assembly of the following sub-systems:

(a) Mechanical Sub-systems

(b) Opto/Electro/Mechanical or combination transducers to pick up signals related to operating parameters

(c) Instrumentation and controls for processing signals received from transducers and energize mechanical systems (articulate end effectors, etc) based on command received from computers

(d) Microprocessor to process stored data from the memory based on specified instructions as an input provided by the computer command language and/or the signals received from work space

(e) Computers act as a source of information in the form of programmed language which controls and defines the specific task and retrieves the data from memory to microprocessor

The interface of these sub-systems and synchronization of their work with respect to time in the form of robot is shown in figure 1:

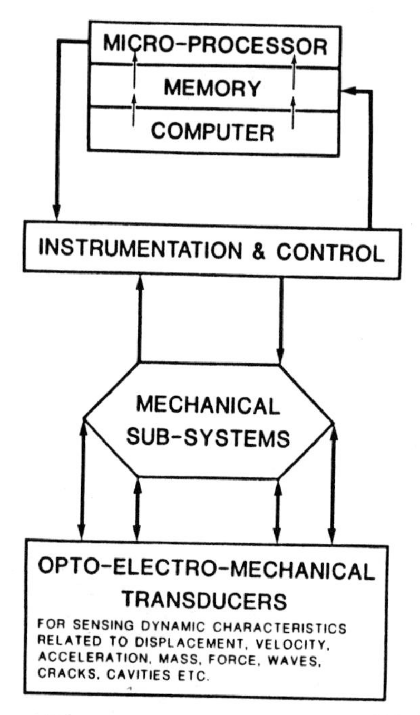

FIGURE 1: INTERACTION OF VARIOUS SUB-SYSTEMS IN A ROBOT

The interfacing of these systems, synchronization of communications, analysis, and operational work function with respect to time on the basis of logical command allow the whole system to act as a 'robot'.

It is a common practice now a days to make a statement that robots can work indefinitely without showing any sign of fatigue in any environment. This is not true. Robots do get tired. They show signs of stress and ageing if continuously allowed to operate without rest over a period of time in difficult and hostile environment. Every component of which the robot is made of undergo through stress and strain cycle resulting in fretting wear and localized loss of strength in the mechanical components due to thermal loading in the operational environment. The degradation of these components will appear in the form of thermal distortion, increase in tolerances, change in resisitivity and inpedence. This will affect the performances, sensitivity, accuracy of doing work, operational life and efficiency of robots.

This will lead to heavy expenditure for its maintenance and up keep in sound condition. In general the degradation of robot components occur first at the contact surfaces between contacting material pair. The dynamic interaction of these surfaces depends upon the contacting materials and its geometry. The surface - dynamics in the form of Tribology at the micro-level becomes extremely complicated in presence of difficult environment. The surface and sub-surface loss appears in the form of wear, tribo-corrosion, noise ageing, friction, heat, pitting, etc. These are known as tribological losses. A system diagram showing interaction of tribological systems in robot components belonging to mechanical and electro-mechanical sub-systems are shown in figure 2:

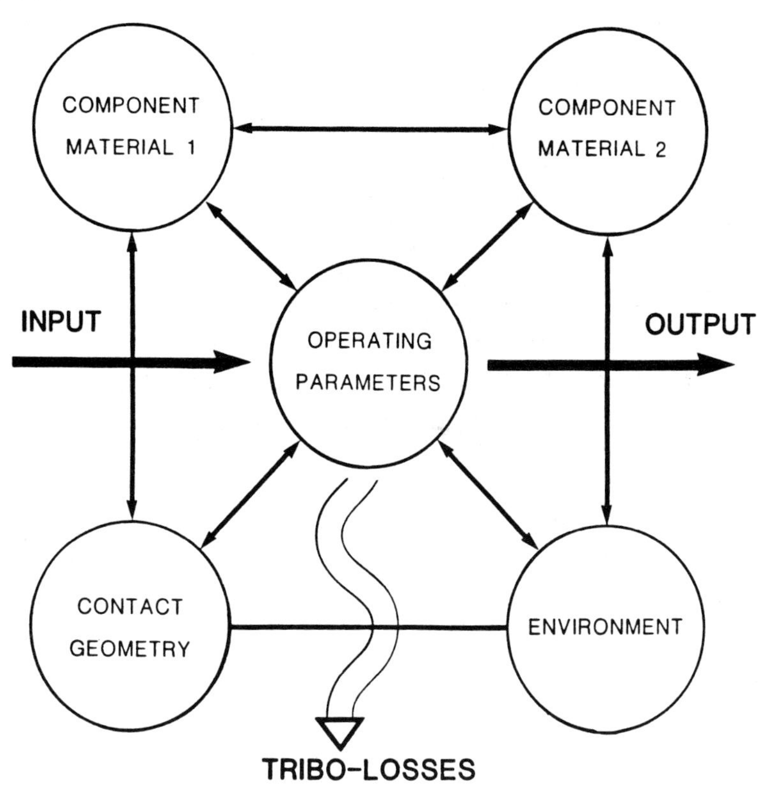

FIGURE 2: SYSTEMS DIAGRAM SHOWING INTERACTION OF TRIBOLOGICAL SYSTEMS IN ROBOTS

The details of various tribological losses in combination with operational parameters and environments is shown in figure 3. These points are emphasized because robots of today and that of the future will not only be seen in manufacturing but also in deep mines, underwater expedition for minerals and oil drilling and space tasks.

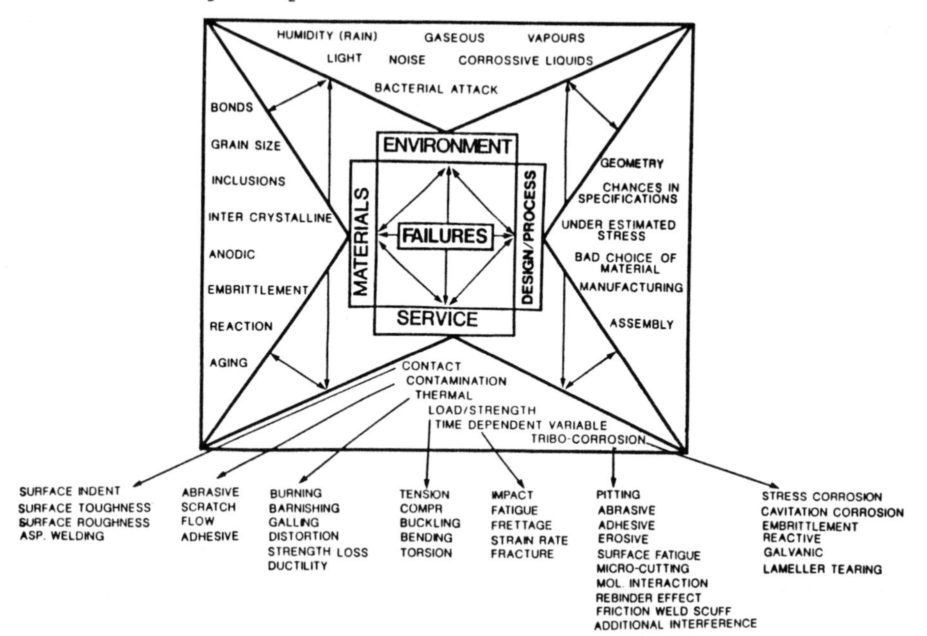

FIGURE 3: TRIBOLOGICAL LOSSES IN ROBOT COMPONENTS OPERATING IN HOSTILE ENVIRONMENT

In order to design a sound, a perfect and an intelligent robot, it is necessary to asses and synthesize the functions of these components, its expected life-cycle and possible performance, malfunction and failure of each system. This is not only necessary for design and manufacturing of robot but to predict life cycle costing of this artificial intelligent man.

This paper highlights the tribological losses of the components in various sub-systems of which the robot is made of to evolve tribo-design considerations for designing the robots.

Mechanical Sub-Systems

Mechanical Sub-Systems basically is the inversion of mechanism. The kinematics deals mostly with the precise and controlled motion of the connecting components. In addition, these components should show considerable strength to perform a function without any deformation and yielding over a number of cycles of operation.

Most common mechanical sub-system organ of robot is classified as manipulator which incorporates segments, articulations, convertors, actuators, and transmission units. The components which interact with the work space are generally known as end effectors, robot arms and fingers as grippers. They get the energy through synchronous motor, and by fluidic drive system. These drives are controlled by computers which provides instructions to microprocessor. The signal energizes the feed back control or automotive control system based on signal received from sensors performing a particular function in the work space. The main function of the robot arm is to move, to extend, manipulate and to perform a task in a predetermined position with high sensitivity and accuracy. The robot manipulator and end effectors consist of several mechanical components. The assembly of these components in the form of extension arm, elbow joints, fingers, etc either singly or in group performs a desired function. These functions are performed at a predetermined and predefined positions with respect to time. Thus accuracy and sensitivity of 'motion', 'position', 'velocity', 'acceleration', and 'force' at any instant are extremely important. Any deviation in the tolerance of these components and its assembly will alter the end effectors task related to motion and position at work space. Any slight shift in the position will lead to process loss in productivity or accuracy of work. The motion or movement of end effector or robot arm or gripper are generally controlled by drive and transmission system. Drives are very often electric or fluid motors connected to gears, bearings, guideways, through mechanical couplings. These components are to be carefully designed to prevent tribological losses to keep them in specified tolerance limit.

The gears in robot should be light weight to avoid inertia effect. It should exhibit good self lubricating properties to control friction and wear. Wear on gear teeth will increase backlash which in turn affect both the position and motion or movement of robot arms or organ because a slip will be introduced due to vibration.

If the friction value of these gears are not kept within permissible limit during performance will bring in thermal distortion on the tooth profile. This will induce noise and vibration. This may subsequently result in interference in control signals. The excess friction and wear of end effector joints, elbows and grippers will affect their performance because of variation of slide/roll ratio and clearance. This will not be acceptable for accuracy of performance of the task at work space. The transducers at grippers may not be able to communicate their correct observations in the form of analog signal to instruments and control unit. The synchronization will be lost.

Similarly, any deviation in the clearance of bearings and its fits and tolerance limits may result in unbalance of mechanical components and misalignment due to wear. This will affect the accuracy of 'position' and "controlled motion" of the mechanical components at the work space. Corrosion over a period of time will also introduce further misalignment of the system. Change in friction value due to ineffective self lubrication properties may result in loss of surface properties of the contacting materials and thermal distortion in bearings.

The stick-slip friction phenomena may also occur at joints and guideways when end effectors are to perform certain functions like loading, lifting or positioning of some jobs. This will lead to uneven distribution of stress and strain which may cause yielding, wear and heat build-up in localized area. Systems which generate a combination of reciprocating and rotating motion consist of seals, bearings, gears, chain or belt drives. Their performance, sensitivity and accuracy, generally deteriorates due to friction, wear, lubrication and surface failure which depend upon interaction at localized contact surface.

To summarize the tribological problems which are responsible for the reduced performance life of end effectors or manipulators in robot which need careful design attention are:

(1) Surface properties of materials to stand surface fretting
(2) Stick-slip friction
(3) Heat build-up and dissipation at contact
(4) Surface rigidity and PV factor for wear
(5) Coating for self lubrication in difficult environment
(6) Tribo-corrosion in areas where the robots have to work in humid and dusty environment
(7) Friction welding problems in vacuum
(8) Stress distribution at contact points to avoid tribo-welding and yielding

Instrumentation & Controls:

These are sets of instruments which convert analog signals into digital signals which can be read and then fed to computer system through combination of electronic instruments to provide predetermined desired signal for operation of mechanical system. This sub-system consists of optical system, electrical system, mechanical system, fluidics or combination of all these systems. The signals received from transducers goes through control unit which operates control switches, energizes electronic circuits and operates actuaters. These systems then stimulate mechanical organ of robot. The important tribological losses in this system are more often seen at 'contact' points in switches and circuit network. Any change of resisitivity in these components are to be attributed due to environmental effect and physio-chemo adsorption or excessive heat. This will result in formation of tribo-layer known as non-conducting layer on the contact points. This will stop the flow of current and will change its resistivity. The control circuit may become ineffective for accurate response. Care should be taken for designing of these contact points to avoid the formation of tribo-layer. Due to several cycles of make and break of contact points will result in adhesion, metal transfer and wear. A careful study of these tribo-problems will help in selecting suitable material pair, coating, surface treatment or alloying of the contact points for accurate transmission of voltage and amperage signals. The surface fatigue due to several make and break contact and heat build-up should also be carefully examined as this may lead to fretting and loss of physical, chemical and mechanical properties of the materials used for electrical contacts. The surrounding atmosphere and environment in which the robot organ or component have to function is very vital. The excess humidity, reactive gases, radio active radiation, thermal fluctuation and corrosive environment will fast deteriorate the surface properties of the circuit connections and contact points. This will lead to component degradation. A proper control signal will then not synchronize with time and will lead to delayed communication. This will affect the work performance and accuracy in job function.

The important tribo - effect in instrumentation and controls are as follows:

(1) Tribolayer in electrical components at contact points due to adsorption and reaction from environment causing change of resistivity and surface properties.

(2) Problems related adhesion of contact switches, and contact points

(3) Surface fatigue or fretting due to repetative contacts, heat build-up, errosion due to discharge and tribo-corrosion

(4) Coating, surface treatment, diffusion and alloying of materials to prevent wear, adhesion and metal transfer

Transducers

Transducers in the form of sensors are most commonly used to measure displacement, velocity, acceleration, pressure, temperature, defects, stress - strain, wave propogation signals, without showing any sign of hysterisis loss. These measurements should conform to linear relationship with high degree of resolution, sensitivity and accuracy. These transducers are mostly non contact and contact type. From tribological considerations, the difference in optical transducers and electro mechanical transducers are to be carefully observed for defect initiation and growth.

(1) Optical transducers: These types of transducers use light as a source with different frequencies Any change in their intensity or frequencies will be calibrated with the quantity to be measured. These signals are then processed to give direct information in the form of measurable quantities. The lenses or fiber glass are usually coated to produce predetermined reflection, refraction and diffraction. These coatings on optical transducers should have good adhesion in difficult environment if the transducer has to work effectively over a long period of time. Adhesion is a tribological consideration and should be carefully evaluated.

(2) Electromechanical transducers: These types of tranducers are quite sensitive to heat and corrosive environment. Any change in their resistivity or stiffness due to tribochemical reaction and change in its physical and chemical properties during continuous operation in heated atmosphere may result in deviation of predetermined signal. This will affect its sensing capability. The sensitivity and accuracy of the transducer will be lost. These transducers should incorporate tribo-design aspect for effective sensing of measurable quantities useful for robot control and performance.

The transducers mostly pick-up the response at the interface. Continuous use of these transducers over a period of time in varying environment will lead to their degradation due to tribo-chemical reactions, diffusion, and thermal loading resulting in change of resistivity and surface properties.

Computers

This unit consists of some mechanical components like keyboards, contact points. These keyboards make contact with electronic circuit in the computer. The "contact" points may go bad if tribo-design aspects are not incorporated. This may result in wear, metal transfer, errosion and corrosion at the contact point. Similarly, in tape or disc drive and micro processor the "motion" occurs at the magnetic head for reading and processing data and are controlled by electro-mechanical components. Any deviation in their movement due to friction and wear may lead to computer going erratic and so the computer controlled robot.

Conclusion:

As such, the role of tribology in design of Robot will help make robot efficient and economical for applications in difficult environment. The important problems related to tribo-design are as follows:
Stick-slip friction, dynamic friction, wear, metal transfer, adhesion, coating, self lubrication, contact surface stresses, adsorption, deterioration of surface properties, heat build-up, thermal distortion, fretting, pitting, erosion. These should be carefully taken into consideration to maintain predetermined tolerances in all contact points and moving joints to avoid excessive wear over a long period of use. The Table I gives the summary of various robot components commonly used and the tribo problems faced by them.

Table I

Mechanical Sub-System	Tribo Problems
Concentric Columns and arms	Friction stick slip friction, wear tolerance, lubrication
Elbow bends	Dynamic friction, slip, tolerance, wear lubrication
Lead screws & ball screws	Stick slip friction, wear, lubrication, tolerance
Belt Drive	Friction & wear
Chain Drive	Friction, wear, stick-slip link breakage, adhesion lubrication
Gears	Friction, wear, clearance, surface durability, fretting adhesion, noise
Bearings	Metal transfer friction, wear, clearance, stiffness, fretting, noise, thermal distortion, lubrication
Slideways	Tolerances, stick slip friction, lubrication, wear, thermal distortion lubrication
Actuators	Seals, stick-slip wear
Drives	Friction, wear lubrication, slip, etc.

Instrumentation, Sensors, Controls & Computers

Contact Switches	Adsorption, Corrosion, Contact wear, metal transfer, erosion
Components	change in resistivity due to heat and tribo-reactions, ageing, adhesion, coating degradation
Sensors	loss of stiffness in resistivity due to physio chemical reaction from the environment, tribo-corrosion, interferance with surrounding signals.

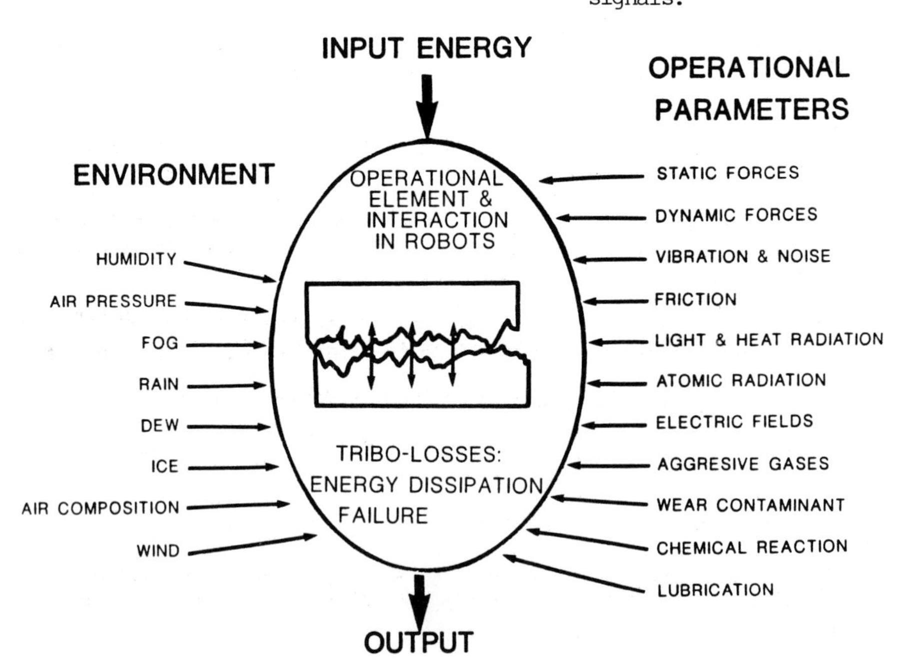

FIGURE 4: COMBINED EFFECT OF ENVIRONMENT AND OPERATIONAL PARAMETER AT INTERFACE OF MECHANICAL ELEMENTS

Finally the life of the robot as a system has to be determined in terms of the combined effect of environment and operational parameters as shown in figure 4. The life cycle costing will be one of the important criteria and will be regulated by these factors. This is essential because the deployment of the effective robots in the factory of future or its service to humanity for exploration of resources in difficult environments will be assesed on the basis of its performance and life cycle costing.

References

1. Dwivedi, S.N., "Custom Design of Robots for Specific Industrial Applications" International Symposium On Design and Synthesis, Tokyo, July 11-13, 1984

2. Dwivedi, S.N., "Practical Problems in Application of Industrial Robots" ASTM Conference on Automated Manufacturing Systems, San Diego, Ca April 4-7, 1983

3. Dwivedi, S.N. and Davis, D., "Concept of Product Design for Robotic Assembly", 8th Applied Mechanism Conference, St. Louis, Sept. 19-21, 1983

4. Dwivedi, S.N. and Kohli, Dilip, "Kinematic Analysis of Regional Structure of Robots Containing One and Two Prismatic Pairs". Proc. Academic National De Ingenieria, Mexico, 1982

5. H. Czichos: Tribology - a systems approach to the science and technology of friction, lubrication and wear, Series 1, Elsevier, 1978

6. Philippe Coiffet: Robot Components, Vol. 4 Prentice Hall, 1983

7. E. L. Safford, Jr.: Handbook of advanced robotics, Tab Books Inc. USA, 1982

8. P. A. Engel: Impact Wear of Materials, Elsevier, N.Y. 1978

9. Gupta, B.K., Sharma, J.P., Bhushan, B & Christensen H: "Friction and Wear Control Through Coatings and Surface Treatments" - Manuscript ready for publication.

End Effector for Manipulation of Orbital Construction Components

J. H. Batton, R. L. Hudson, J. P. Sharma, and Suren N. Dwivedi

Department of Mechanical Engineering Science
University of North Carolina at Charlotte, Charlotte, NC 28223

Summary

The utilization of large truss structures as orbital assembly components will necessitate that a means be devised to manipulate these structures. The Space/Shuttle is expected to serve as a base of operations in the early phases of orbital construction. Current Shuttle capabilities must be enhanced to accommodate the special needs of employing such structures.

The Shuttle's RMS will provide a means of working with truss elements. What will be required is an end-effector which will allow the RMS to secure and manipulate the structures. Due to the structural limitations of the trusses, the end-effector will be required to distribute as well as regulate the inertial loading experienced by the trusses.

Introduction

In the next decade or so, the National Aeronautics and Space Administration plans to make great strides in the development and deployment of large space structures. These structures will be able to accommodate multiple payloads, which will provide cost savings to users through the sharing of spacecraft utilities and the simplifying of maintenance. A particularly beneficial use of these space platforms will be the clustering of geosynchronous communication satellites and the subsequent reduction in crowding of this location.

Present designs center around the structures being deployed from the Space Shuttle cargo bay. In these designs the structures will be assembled in space, since preassembled units would be too large to fit within the Shuttle bay. NASA is leaning toward the implementation of collapsible trusses which can be automatically erected and deployed without the use of a construction fixture or extravehicular activity (EVA). These trusses could then be assembled together, in space, into the required space structure shape.

The proposed trusses are to be constructed of lightweight materials - probably aluminum alloys or graphite epoxys to minimize the loading of the Shuttle cargo bay.

The truss shapes can vary,according to use, though square and triangular designs are the most likely candidates for widespread use. Typical dimensions of these trusses will be 40 meters long with cross-sectional dimensions of 2 meters x 2 meters.

Because of the given dimensions and material composition, these trusses are relatively flexible by Earth standards, particularly in the direction perpendicular to the principal truss members. This flexibility necessitates the need for exercising great care in the movement and deployment of the trusses.

Need Assessment

Once a truss has been erected out of the cargo bay, it must then be moved and positioned into its place as part of a space structure. In the NASA concept, the remote manipulation system (RMS) of the Space Shuttle will be used for this positioning activity. Controlled by an operator within the Shuttle, the RMS, or Shuttle "arm", can grapple an extended truss to position and orient it properly.

The existing end-effector permantly connected to the Shuttle arm would require that truss designs be modified to allow for grappling pins to protrude from certain members. Due to the need for trusses to be folded into the smallest units possible, protruding pins could cause extreme packaging problems. Also, the point loading which the RMS would create is totally unacceptable. As mentioned, the trusses to be used for space structures are relatively flexible, especially in the lateral directions. Due to the inertial forces encountered in positioning, single point attachment to the truss is highly likely to deform the member and possible distort the entire truss.

These two shortcomings of the existing RMS end-effector create the need for a specialized end-effector to grapple trusses for orbital assembly. The optimal design would be a robotic device, interfaced to the Shuttle computers, that could be held to the Shuttle arm by the permanent RMS end-effector. This device, dedicated to the manipulation of space structure trusses, would meet all of the following efficiency requirements:

1) The ability to attach and adequately hold the trusses for RMS positioning.

2) The ability to distribute the inertial loads across the trusses and thus minimize the chance of distortions.

3) The ability to monitor and regulate, via computer interfacing, inertial loading on the trusses.

The manipulator/gripper designs common in robotics applications today could not be used on the end-effector, though modifications of these designs may prove beneficial. Existing manipulator designs are generally rejected because they do not address the special conditions of space. For example, vacuum and pneumatic systems are totally unacceptable for space applications. Also, the required end-effector manipulator must completely surround or dock with a truss member before any contact is made, because in the environment of space a slight premature bump would send the truss floating from within the gripper's grasp. Existing manipulator designs do not provide adequate sensing capability or feedback to attach to the relatively weak truss member without causing distortion.

End-Effector

With the establishment of a need for a specialized end-effector to attach to the Shuttle RMS, attention must now be focused on the design considerations. Three main areas must be addressed: 1) mechanical design, 2) control and load regulation, 3) material limitations. Mechanical issues include the actual design and dimensions of end-effector components, the strength of these components, and their interaction. End-effector control includes the attachment to truss members, sensory perception of truss loading conditions, and computer monitoring of these conditions. Material limitations take into account the vacuum, low temperatures, and radiation effects of the space environment on the materials used in constructing the end-effector.

Mechanical Considerations

As part of the end-effector's mechanical considerations, two of the previously mentioned requirements must be addressed: 1) The end-effector must be able to attach itself securely to a truss, allowing the Shuttle RMS to impart both translational and rotational motion.

2) The end-effector should, through the existence of multiple contact points with the truss, be able to distribute loading on the truss during manipulation and enhance the ability to control the motions of the truss. Other mechanical considerations necessary in the design of an orbital assembly end-effector may be summarized as follows:

.The end-effector should be able to configure for several different truss configurations each with its own set of dimensions.

.The end-effector should be capable of being locked into a certain configuration.

.All components of the end-effector should be strong enough to handle any reasonable loads that may be applied. Within the preceding constraints, a general design of the end-effector can be formulated. basically, the device will be designed to attach to the Space Shuttle's RMS gripper. It will depend on the RMS for (1) its capability to roll about the axis perpendicular to the end-effector-truss contact plane, and (2) on the RMS and Shuttle for translation.

Designing the end-effector to engage various sized and configurations of trusses increases the versatility and usefulness of the end-effector, but also increases its complexity. To allow for this versatility, the end-effector is constructed with four arms arranged in a square pattern with each arm located at a corner of the square (Fig. 1&2).

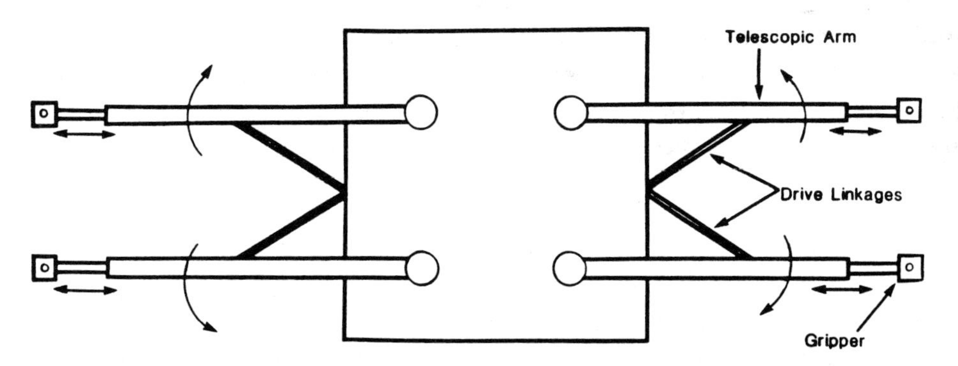

FIGURE 1: PARALLEL ORIENTATION OF END-EFFECTOR ARMS

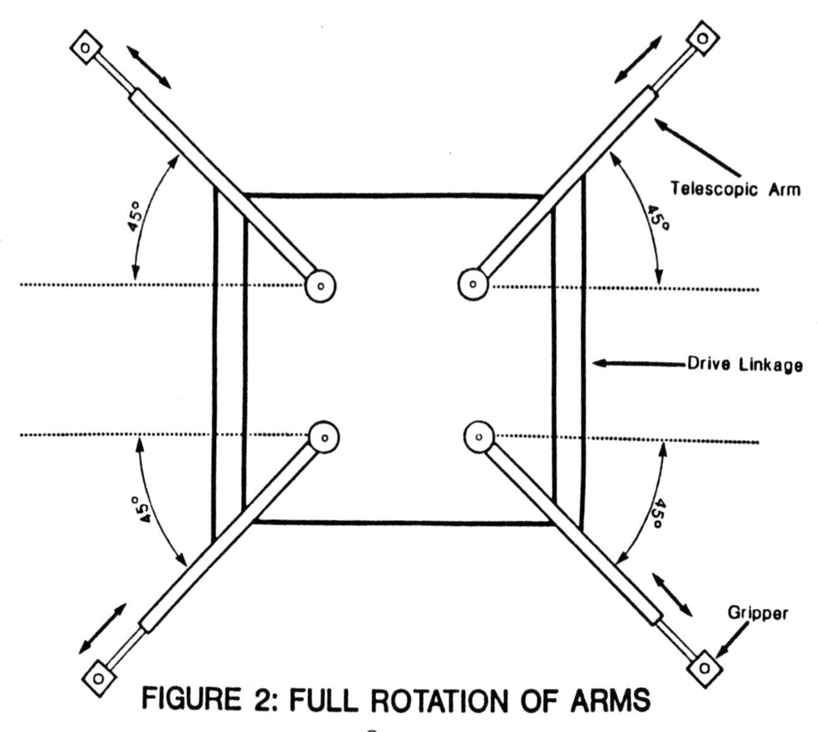

FIGURE 2: FULL ROTATION OF ARMS

These arms can be rotated at a 45° angle from a parallel position so as to separate the contact points with the truss. In addition to the rotation motion, the arms may be extended out longitudinally to nearly twice their original length. This variability of rotation angle and extension also provides for a more effective distribution of the truss loads on the end-effector.

Due to the symmetry of the structures to be handled, the individual arms of the end-effector need not be manipulated independently of each other. When one arm is in a rotated or extended position, all four must behave in a like manner. This feature greatly simplifies the control requirement of the device and facilitates the easy incorporation of system redundancy.

The rotation of the arms is carried out by means of a motor-driven lead screw attached to a linkage. The angular orientation of the arms is photoelectricly recorded through an optical disk encoder attached to the motor's drive shaft. The encoder sends information regarding the number of turns of the motor to the end-effector processor unit. The extension or retraction of the arms is affected in a similar manner.

The motor which controls the telescoping of the arms is connected by means of a gearbox to four flexible-shaft cables. These cables are, in turn, connected to a power lead screw located in the base element of each arm. These screws drive the telescoping sections in and out. The lead screws used for both the angular and extension positioning mechanisms also function as locking elements to hold the arms in their desired locations. The screws must therefore withstand any outside force which may tend to rotate the arms or force the arms along their telescoping direction.

From this point arises both the mechanical strength and tribo-design considerations for the end-effector. Due to the inertial loadings produced by various combinations of rotational and translational motions of the entire end-effector, the arms, arm joints, and contact points with the truss must withstand a reasonable range of force and magnitudes and tribological influences . Particularly crucial is the design of the effective arm joints for bending moment created by the truss load on the end of each arm and friction, wear control at joints for positioning accuracy and to avoid tribowelding. The arms themselves must be designed with sufficient strength and rigidity to avoid any possible deflections caused by forces perpendicular or parallel to the contact plane of the truss. Again, the lead screws will be required to handle both compressive and tensile loading, and therefore must be designed accordingly to prevent thread deformation.

To adequately meet the needs for the end-effector to connect and to manipulate truss structures, special consideration must be given to the tribodesign of "gripper" that will attach to the end of each arm. At present, two designs approach meeting the specified needs. The first is a somewhat conventional gripper, composed of two mechanical "fingers" which would completely surround a truss and then enclose on it. The second is a probe that would enter into holes in plates mounted on members of the truss and latch into place. From all preliminary indications, the probe design best qualifies as the end-effector's connecting device. The probe on each of the four arms would enter a hole in truss members. Once in place, latching mechanisms from each probe would extend and secure the truss to the probe's baseplate. Once docking is accomplished, Piezo-electric sensor elements attached to the baseplate would be held firmly against the surface of the truss members. A side and conceptual view of the end-effector are shown in figures 3 and 4.

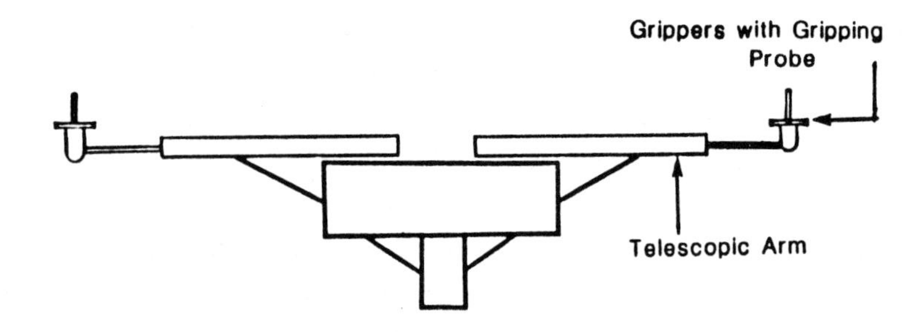

FIGURE 3: SIDE VIEW OF END-EFFECTOR

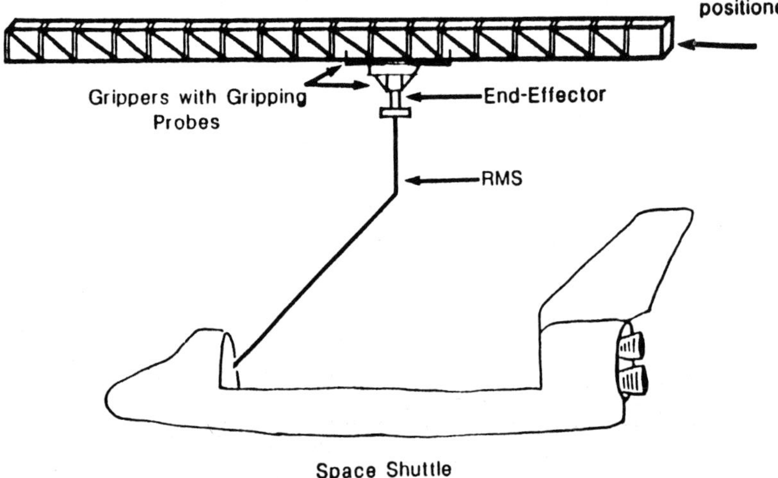

FIGURE 4: CONCEPTUAL VIEW OF SHUTTLE RMS AND END-EFFECTOR GRAPPLING A TRUSS

Control and Load Regulation

It is these Piezo-electric sensors that will be used to determine and to regulate truss loading conditions. The sensor elements are mounted on the circular baseplate which concentrically backs the docking probe. There are four individual sensors mounted at 90° intervals on each probe baseplate. By virtue of the probe locking mechanism, these sensors are kept in firm contact with members of the truss at all times during the period when the truss and end-effector are attached. By comparing the outputs from each of the four contact locations (from a total of 16 sensor elements), information can be obtained describing the condition of the truss and any loading irregularities, structural deflections, or resonance development which may be occurring. Additional research is needed in the area of space environmental and tribological effects on Piezo-electric sensors interacting with gripper and truss to be lifted. The end-effector is initially intended to be used in conjunction with the operator-controlled Shuttle manipulator arm. While there is also a provision for the manual operation of the end-effector, the sensory inputs can be linked to the Shuttle computer, controlling movement by both the arm and the Shuttle's reaction control system (RCS). Feeding the sensory information through the end-effector processor to the controller in the Space Shuttle (with or without a human element in the loop) will enable the loading to be adjusted - whether through RMS or RCS action - so as to limit any inertial loading damage to the truss structure. This sensory feedback thus facilitates the active regulation of the loading on the truss during manipulation (Fig. 5). The control interface between the end-effector Processor and Shuttle can be accomplished via a radio-telemetry link.

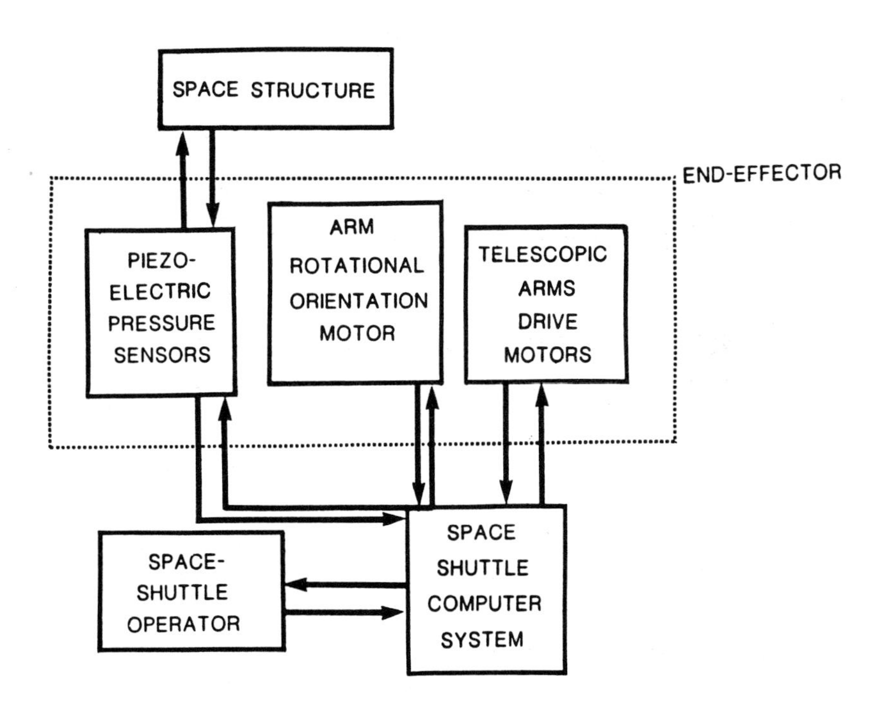

FIGURE 5 : SENSORY CONTROL SYSTEM

Material Considerations

Material considerations are extremely important in the design of an orbital end-effector. The materials used must be lightweight and of minimal dimensions to fit easily into the Shuttle's cargo bay. All components of the end-effector must be strong and exhibit excellent tribological properties at contacting surface, with very low chance of failure. The materials must be capable of withstanding the environmental conditions of space. Surface properties of the material should be such that it should be capable to withstand erosion due to cyclic impact of space particles and surface fatigue. The surface of materials should have considerable endurance strength and interact with other surface with controlled friction without causing any localized friction welding problems.

In the low-earth orbit where the Space Shuttle operates, temperature conditions can vary greatly. The peak temperature of a structure can range from 100°C to 80°C depending on whether the surface of the structure faces the sun. Solar radiation must be considered when selecting materials. The materials used must not have their molecular composition altered due to radiation bombardment. With a factor of safety included, space materials should have a useable lifespan of 30 years.

An additional consideration in choosing materials is that of vacuum-welding created when different components, made of the same material, come in contact with one another. Minimizing the number and area of metal-to-metal contact points is one partial solution to this problem. The use of dissimilar materials for components is another possible solution. At joints and surfaces requiring bearings, the bearings themselves could be constructed of a non-welding material to separate similar material components. Coatings, such as teflons, could be used to prevent the vacuum-welding of parts as well as function as a lubricant. Since vacuum-welding is most likely to occur at the interface of smooth similar surfaces, the intentional defacing of a compoment surface could be sufficient to eliminate the potential for welding.

Conclusion

Among the many concerns which develop in an enormous project such as shuttle base orbital construction, the design of the RMS end-effector is a small and vital one. Use of existing gripper configurations for the end-effector is not possible. The design of the entire end-effector, particularly the grippers must be governed by the mechanical, tribology and material limitations imposed by the environment and the structures. Likewise, control systems to supply accurate positioning, load regulation and system redundancy must be incorporated.

References

1. "Development of Deployable Structures for Large Space Platforms", interium report by Rockwell International for NASA George C. Marshall Space Flight Center, vol. 1, Contract NAS 8-34677, August 1982.

2. Covault, Craig, "NASA Shifts Large Structures Planning", Aviation Week and Space Technology, April 9, 1979, pp. 43-48.

3. Smith, Bruce A., "Large Space Structure Work Advances", Aviation Week and Space Technology, October 27, 1980, p. 49.

4. Dwivedi, S.N. and Kohli, Dilip, "Kinematic Analysis of Regional Structure of Robots Containing One and Two Prismatic Pairs". Proc Academic National De Ingenieria, Mexico, 1982

5. Dwivedi, S. N., "Custom Design of Robots for Specific Industrial Applications" International Symposium on Design and Synthesis, Tokyo, July 11-13, 1984.

Computer Vision and Sensors

Chairman: Lawrence A. Goshorn, International Robomation/Intelligence, Carlsbad, California
Vice Chairman: Murali Varanasi, University of South Florida, Tampa, Florida

A Laboratory Environment for Robotics and Vision Research

R. Peres and M. Varanasi

Department of Computer Science and Engineering
University of South Florida
Tampa, Florida 33620

Introduction

Robotic systems offer tremendous promise for a diversity of applications in manufacturing. Sensory integration and flexibility for multiple applications through programming can provide these systems tremendous capabilities that can be channeled into several application areas. Integration of manipulation and sensory capabilities into working systems requires a hierarchial approach in control, architecture, programming and knowledge representation.

There are many useful and technologically exciting applications for robots, such as medium batch manufacturing, space exploration, deep-sea mining, fire-fighting, etc., that require that a robot function in an unstructured environment by making decisions based on sensory information gathered from the operating environment [1]. Vision appears to be the most useful sense for intelligent robots because it offers the largest amount of information in the widest range of applications [2] and is supported by a wealth of knowledge. Other factors that make vision sensing attractive are its non-intrusive nature and the availability of a wide range of vision sensing and processing equipment that can be interfaced to different types of equipment at different levels of sophistication and cost, depending on the application. Even though there has been considerable research in the past few years in the area of computer vision, it is not yet clear what combination of equipment and algorithms best supports various specific applications [3]. Specifically, for robot control some of the questions that need to be addressed are: when should black and white image acquisition be used instead of color; what is the most effective resolution to use when digitizing the image; when stereovision is required where should the cameras be situated. Other considerations are the trade-off between the accuracy in determining position and size of the objects in the work space versus the added computational complexity to achieve it and the different types of collision avoidance strategies and trajectory determination when more than one robot arm is in the work area.

The main components in our laboratory have been selected to create a modular and flexible environment where the different aspects of robotics and vision questions can be considered without major reorganization of the system components [4]. An example of this is our image processing and display subsystem that can handle either black and white or color images; it has its own video processor and image memory that can be controlled by our VAX 11/750 using either low level pixel commands or high level commands that implement operations on the entire image. Figure 1 shows the essential subsystems in the laboratory and Figure 2 the more detailed overall equipment configuration. Each of the three subsystems can be individually controlled by the central processor. After all the individual subsystems have been developed and tested for a particular application, they can then be linked together and integrated in the central processor. This type of laboratory environment provides not only flexibility to one researcher when working on a particular application, but it also allows for multiplicity of investigations to be carried out in parallel fashion.

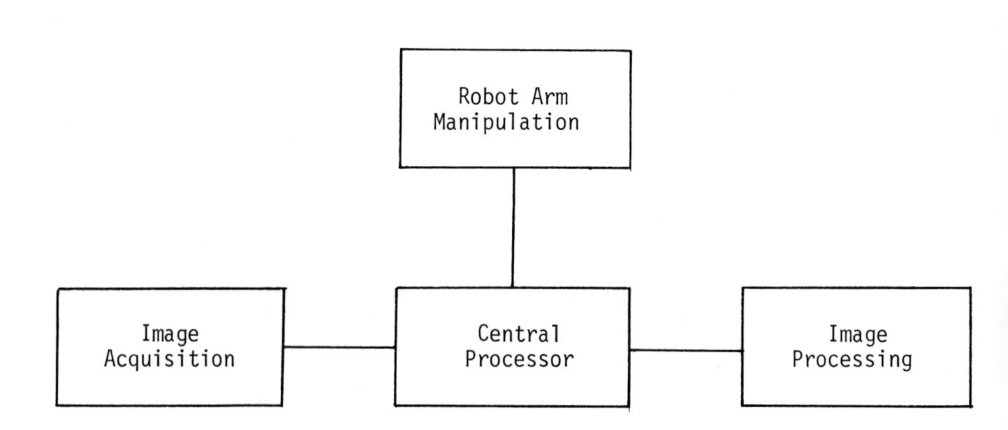

Figure 1. Basic Modular Components

Robotic Environment

At the present time, three robot arms are interfaced to the central processor module in the laboratory. Two of these arms are of the type Microbot Alpha and the third arm is a Mitshubishi RM501. These three arms are placed in a

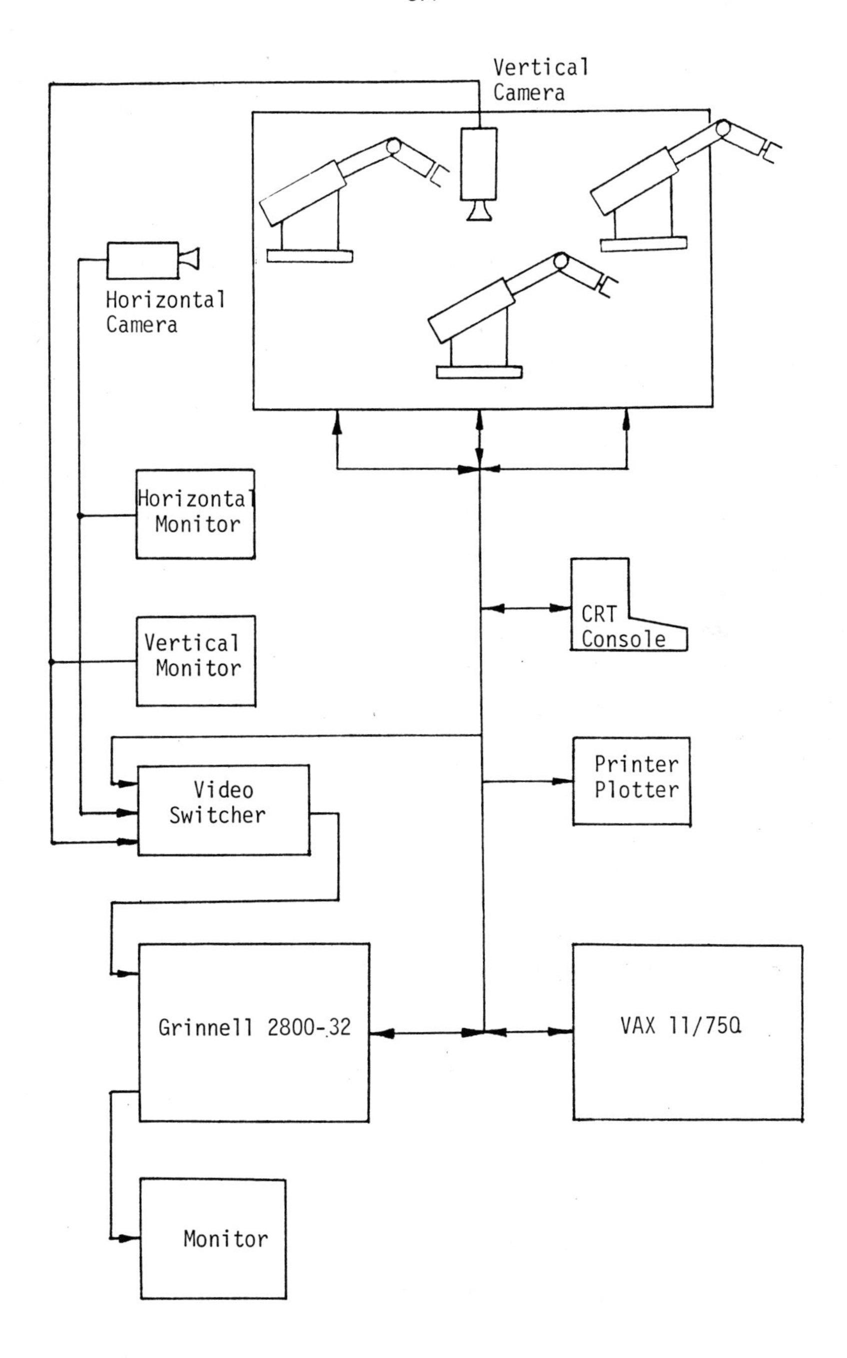

Figure 2. Overall System Configuration

triangular pattern in a common work space that can be reached by all three. Coordinating the movements of the arms then is a function that is implemented in the central processor module which has control of all three arms and also receives the information from the image processing system.

The robot arms can lift a minimum of 1.5 pounds and have a top speed of approximately 1.5 ft/sec. The base of the arms is fixed in place but each arm has freedom to move in 6 different ways: waist, shoulder, elbow, wrist pitch, wrist roll, and gripper open-close. Each one of these movements is controlled by a motor which is in turn controlled by a local microprocessor control system in each robot arm. One basic difference between the Alpha and the RM501 robot arm is that the joints of the Alpha are controlled by stepper motors in an open loop mode of control while the RM501 arm joints are controlled by D-C servo motors with position encoding used as feedback. This provides the opportunity to study the role of greater accuracy of the closed loop servo motors would play in an environment with visual feedback.

Both types of robot arms can be programmed in one of two ways: through the Teach-Control box of each robot or from a host computer. Even though the Teach-Control method is a useful programming capability used for pre-structured and often-used sequence of arm movements, it is the host computer control method which is the key in using vision information for robot arm control. The central processor of Figure 1 is a VAX 11/750 computer which communicates with each robot microprocessor control system through an RS232 asynchronous serial interface at 9600 baud rate. The two Alphas can be cascaded on one serial line but a different serial line must be used for the RM501. The robot arms respond to ASCII characters which represent the commands sent from the central processor via the serial interface. Each robot arm also communicates to the central processor with respect to the position register information for each joint motor as well as acknowledgement after each command is received from the central processor.

Algorithms to accomplish effective control of the robot arms are at the present time implemented in the central processor. These control algorithms revolve around transformations between arm joint angles and the cartesian coordinate system. The forward arm solution determines the x, y, and z coordinates of the end point of the robot arm as well as the pitch and roll of the wrist from a measurement of the joint angles θ_1, θ_2, θ_3, θ_4, θ_5, as shown in Figure 3. These joint angles are determined by reading the position

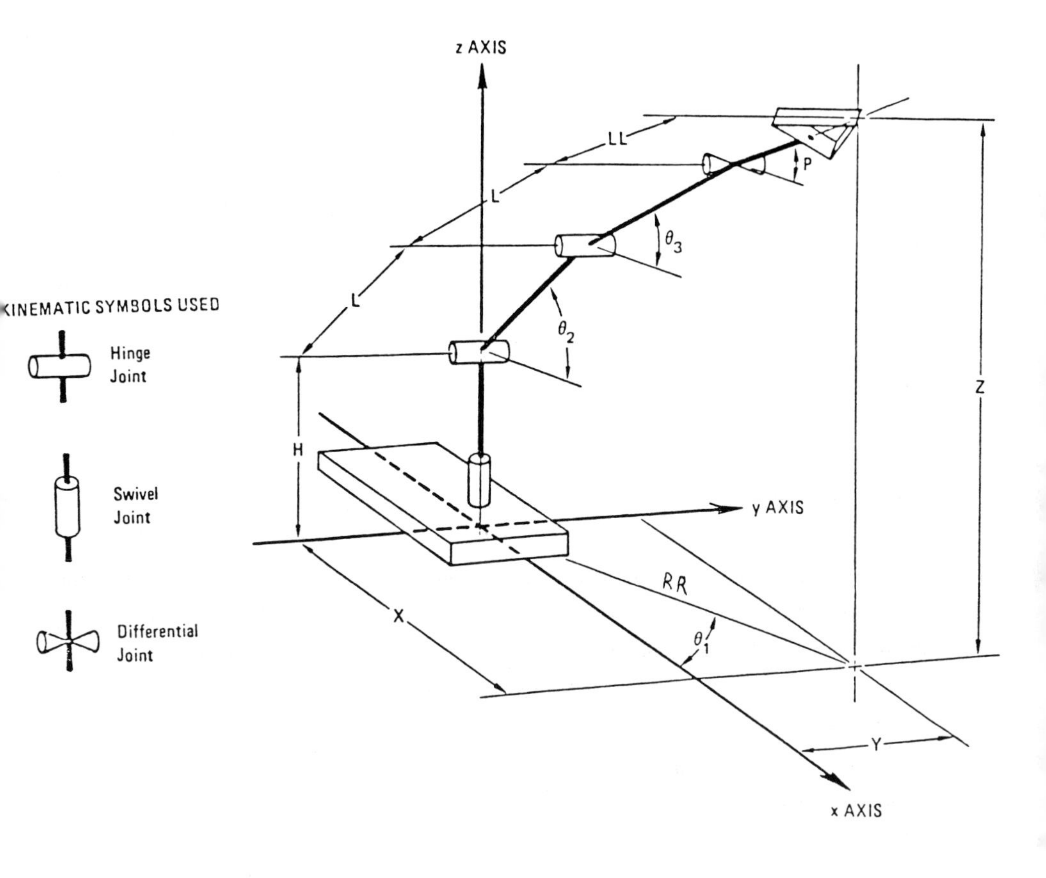

Figure 3. Kinematic Model of the Alpha Arm

registers of the joint motors. The forward arm solution is obtained using the following equations [5]

$$P = (\theta_5 + \theta_4)/2 \quad , \quad \text{Roll} = (\theta_5 - \theta_4)/2$$

$$RR = L \cos \theta_2 + L \cos \theta_3 + LL \cos P$$

$$X = RR \cos \theta_1 \; ; Y = RR \sin \theta_1$$

$$Z = H + L \sin \theta_2 + L \sin \theta_3 + LL \sin P$$

On the other hand, the backward-arm solution is used when the end point of the robot arm needs to be moved to a specified x, y, z position and orientation of the wrist. Transformation equations shown below [5] are used to determine the joint angles θ_1, θ_2, θ_3, θ_4, θ_5 and from these the proper commands can be sent to each of the joint motors to drive the arm to the desired position.

$$\theta_1 = \tan^{-1}(Y/X) \quad ; \quad RR = \sqrt{X^2 + Y^2}$$

$$\theta_5 = P + Roll \quad ; \quad \theta_4 = P - Roll \quad ; \quad R_0 = RR - LL \cos P$$

$$Z_0 = Z - LL \sin P - H \quad ; \quad \beta = \tan^{-1}(Z_0/R_0)$$

$$\alpha = \tan^{-1}\sqrt{4L^2/(R_0^2 + Z_0^2) - 1} \quad ; \quad \theta_2 = \alpha + \beta \quad ; \quad \theta_3 = \beta - \alpha$$

The work space where the three robot arms function resembles a cube of approximately 2.5 feet on each side and is shown on Figure 4. The position of the different objects in the work space is not known before-hand to the robot control algorithms. Therefore a task requiring coordination of the robot arms to manipulate these objects in this work space must make use of fairly accurate sensory information such as vision. Two television cameras are used to gather visual information from the work space and they are discussed in the following section.

Image Acquisition

Two GE2505 charge injection device (CID) cameras with a resolution of 388 x 244 are presently used for image acquisition. These CID cameras offer advantages such as small size, tolerance to processing defects, avoidance of charge transfer loss and lack of blooming. As shown on Figure 4, one of the cameras is mounted vertically over the work area and the visual information sensed is used to determine the orientation and location of objects in the work area. The second camera is mounted horizontally and the sensed information together with that of the vertical camera, is used to determine the height of the object. The video outputs from the two cameras are connected through a video switcher, to the central processor to enable selection of one of the two camera signals to be digitized for processing by the image processing system. In this manner, the central processor can dynamically select either vertical or horizontal position information for use in the image recognition process.

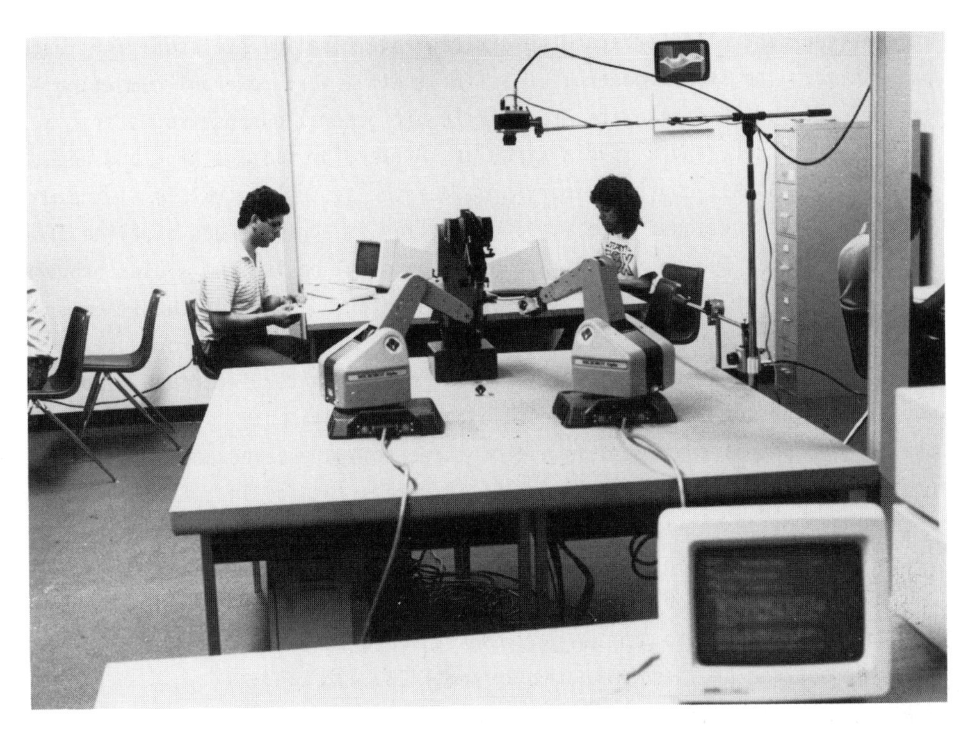

Figure 4. Robot Work Space with Cameras

The video signal from the switcher is digitized, stored, and manipulated under control of the central processor. Imaging functions are performed in a special purpose processor, the Grinnell System 2800-32, a modular special purpose system for black and white, as well as RGB image processing. It is capable of digitizing a video signal with 512x512x8 bits resolution which can then be stored in memory. The system offers the flexibility of configuring the memory under software control so that the stored information can be addressed as a single bank of 480 lines x 640 pixels x 8 bits. The contents of the memory can be accessed for display on a television monitor through a digital to analog converter dedicated to each of the memory boards. This allows the users visual display of intermediate results of the different image processing and robot arm control algorithms being developed.

The system architecture of this image processing system is organized around a system controller utilizing a high speed bit-sliced arithmetic logic unit and a video controller arranged as a video rate pipeline processor for image processing. The system controller is in itself a very powerful computing system. It has a 16 bit wide ALU with 16-bit general purpose registers, a quotient register, and a 16-bit circular shifter for fast packing and unpacking. A 4k x 64 bit control memory provides the system controller a wide array of graphic instructions. The video controller is a high speed pipeline processor which can be controlled on a pixel by pixel basis from a user programmable 1k x 32 bit wide control memory. The interface between the image processing system and the central processor, which as mentioned previously is a VAX 11/750, is done through an intelligent host computer interface board mounted in the VAX 11/750. The communication of data and instructions between the two systems is performed in one of three software selectable modes: DMA, programmed I/O, and memory mapped I/O. Programs running in the VAX 11/750 can access many different types of pixel operations performed by the image processor. A repertoire of standard pixel operations is provided by the Grinnell 2800-32 from which high level operation can be synthesized. Further, the system enables users to define many non-standard functions of interest through FORTRAN routines provided with the Grinnell 2800-32.

Image Recognition

Image recognition is the information processing task of understanding a scene from its projected images. After the images of the scene have been digitized and stored in memory, it involves implementing algorithms to provide a useful interpretation of the scene to execute a given task [6].

Edge detection is among the first basic steps required for image recognition. It is based upon a gradient map showing the difference in brightness between an object and its background. The end result of the edge detection algorithm is a binary map consisting of ones for all pixels lying on the outline or edge of an object and zeroes for all other pixels in the image. Different windows containing a pixel and its neighbors can be used to obtain a gradient map. Since the size of the window and the calculations performed with the pixels in the window affect the quality of edge detection as well as the computation time required for it, we are presently evaluating the Sobel and the Roberts Cross gradient operators [2] to determine how effective they are in our operating environment.

The Sobel Gradient Operator uses a 3 x 3 window centered around the primary pixel. The intensity of each pixel is represented by a through i which follows.

$$\begin{matrix} a & b & c \\ d & e & f \\ g & h & i \end{matrix}$$

The gradient magnitude is then obtained by using the function:

$$S(c) = \left[[(a + 2b + c) - (g + 2h + i)]^2 + [(a + 2d + g) - (c + 2f + i)]^2 \right]^{1/2}$$

The Roberts Cross operator uses a 2 x 2 window with the primary pixel being the upper lefthand element:

$$\begin{matrix} a & b \\ c & d \end{matrix}$$

The gradient magnitude is given by

$$R(a) = [(a - d)^2 + (b - c)^2]^{1/2}$$

There are also several methods for building the binary map from the gradient map. A method employing a global threshold will compare one global value to each element of the gradient map. If the gradient value is greater than the threshold value, then the pixel is part of an edge and a one is placed in the corresponding position of the binary map. A local threshold method calculates a threshold value for each element of the gradient map. This threshold value depends on some local property, typically the average intensity. This method is useful when there is not a clear difference between object and background. If the threshold value also depends on the coordinates of the pixel, then it is called a dynamic threshold and it is useful with pixels that share some property that can be used to separate the objects from the background. Even though a local or a dynamic threshold method of obtaining binary images can be computationally slow, they are more accurate and we are using them to evaluate the effectiveness of a global threshold.

Since most of the pixels in a binary edge map have zero intensity, manipulating the entire edge map wastes memory and computation time. Chain encoding [7] provides a simple and efficient method to store the edge information and it is presently the method being implemented in our central processor. Chain encoding of a binary edge map involves three steps: scanning the image for starting a chain, selecting all edges that do not belong to a closed loop, from which closed boundaries are obtained in the last step. At the end of the chain encoding procedure each closed outline of an object is represented by a list containing the address of a node pixel and a list of numbers representing the direction between all other edge pixels in the outline of the object.

The result of the chain-encoding procedure of the visual information from the vertical camera is used by feature extraction algorithms implemented in the central processor. Through simple calculations of the direction vectors in the chain code, basic shape and size characteristics such as object perimeter, center of mass, and region area, which remain constant regardless of rotation and translation of the object, are used to discriminate between objects. A(p + q)th order moment m_{pq} for a uniformly bright region R is given by 8

$$m_{pq} = \iint_R x^p y^q dxdy$$

The first six moments of a region can be approximated using the information derived from the chain code [8]. After these features have been calculated for an object, they can be compared to the features that have been stored for model objects. Image recognition for simple objects quickly follows.

For robot manipulation of objects, object recognition is not sufficient. Information on the location and orientation of the object in world coordinates is also required. This information can also be derived from the moments above. The second and third terms of the region's moments give the center of mass of an image relative to its node point. Since the coordinates of the node point are known, and these in turn can be transformed to world coordinates, the location of the center of the object information can be transmitted to the robot control algorithm. The orientation of the object can then be found by calculating the angle of a region's axis of minimum moment of inertia which can also be found from the moments [8]. This angle is then used to calculate the wrist pitch and roll with which the robot arm would best grab the object.

Programming efforts are being directed to use the visual information from the horizontal camera, together with the position information derived from the vertical camera, to determine the height of the different objects in the work space. This would lead to a more complete understanding of the scene and to more effective arm coordination algorithms.

Acknowledgements

Several individuals have provided valuable contributions to the development of the robotics laboratory environment that has become the pride and joy of the department. We owe special thanks to Professors O. N. Garcia and K. H. Kim who have helped in the creation of the laboratory, and Paul Andre, Mathias Merx, and Morgan Paul who helped directly in supporting the activities of the laboratory. Without their unselfish support, progress of the research activities would not have been possible.

References

1. An Overview of Artificial Intelligence and Robotics. Volume II-Robotics, NBSIR 82-2479, March 1982.

2. Gonzalez, R. C. and Safabakhsh, R., "Computer Vision Techniques for Industrial Inspection and Robot Control: A Tutorial Overview," IEEE Computer Society Tutorial on Robotics, 1983.

3. "An Overview of Computer Vision," NATIONAL BUREAU OF STANDARDS, NBSIR 82-2582, September 1982.

4. Jarvis, R. A., "A Computer Vision and Robotics Laboratory," Computer, Volume 15, Number 6, June 1982.

5. Microbot Inc., Alpha Reference Guide, Volume 1, December 1982.

6. Horn, B. K. P., "Artificial Intelligence and the Science of Image Understanding," in Computer Vision and Sensor-Based Robots, G. G. Dodd and L. Rossol (Eds.), N.Y.: Plenum Press, 1979, pp. 69-77.

7. Freeman, H., "Computer Processing of Line-Drawing Images," Computing Surveys, Vol. 6, No. 1, March 1974, pp. 57-97.

8. "The Mathematician's Corner", ROBOTICS AGE, March/April 1981, pp 17,18.

Video Vision Robot Guidance

James W. Mooere, Eugene S. McVey, and Rafael M. Inigo

School of Engineering and Applied Science
University of Virginia
Charlottesville, Virginia 22901

Abstract

Fundamentals for the autonomous and semiautonomous control of vehicles using 3-D machine vision is presented. Major problems for the automated guidance of vehicles are: pathway identification and description; obstacle avoidance; real time decision execution; closed loop vehicle control.

Introduction

Research at the University of Virginia on the guidance and control of vehicles that makes use of machine vision is reported here. This is an example of real time image processing that has significant industrial applications such as for the transportation of manufacturing materials, parts and tools from inventory to the factory floor by line guided and free ranging robots. Neff (1981) has reported [1] that a Japanese wire guided robot is making significant economic improvements in manufacturing. Another application would be for guidance of robots working in a confined space where materials must be transported short distances from one work station to another or for a robot that must move to different parts of a work station to weld, drill, paint, etc. Such mobility could enhance the performance and reduce the number of robots required to do a job.

Another application that has motivated part of our work is the semiautonomous operation of automobiles. Although autonomous operation may eventually be feasible in controlled conditions such as on U.S. interstate highways, semiautonomous applications such as collision avoidance and other safety related uses have near term feasibility.

Apparently, until recently little work has been done on automated vehicle operation work in which the vehicle is to have the capability to access the operating environment in terms of such factors as "finding the roadway,"

The research on which this paper is based was supported by the National Science Foundation Grant ECS-82-15443 and the Department of Transportation Contract No. DTRS5680-C-00033.

locating and avoiding obstacles, and making navigation decisions. Related literature starting about a decade ago, concerned with the application of radar to the automobile collision avoidance problem is available which would be expected to be useful but it contains little useful information other than providing assurance that high level visual information processing will be necessary to reliably solve even the simplest problems. A summary of the radar and related work is available in [2].

However, useful mobile robot research which is directly related to our work has been in progress for at least two decades [3]. This Stanford Research Institute work using a TV camera and a telemetry link to a minicomputer [4] enabled a vehicle to follow a white line painted on a flat surface. Moravec, also at SRI picked up the idea with the objective of applying it to a rover for planetary exploration [5]. Research by the Tokyo Institute of Technology [6] and automotive work by Nozaki [7] were reported in 1977. Propulsion rover work [8] and the ETL system [9] were reported on in 1979.

Other recent publications further demonstrate the international interest in free ranging mobile robots. Iijima, et al. [10] (1981) suggested that their interest includes applications in streets, fires, under water, in nuclear reactors and in mines for the robot "Yamabico 3," which uses four supersonic sensors. Larcombe [11] of England contributed an analysis of the nonlinear steering problem for wire and optical line guided robots. While his results are applicable to differential optical, tracking including off-center tracking, they should be extendable to "offset" tracking using machine vision as we propose. The use of machine vision should make possible the removal of limitations he reports due to line fouling and limited offsets. The "HILARE" mobile robot project [12] was started in 1977 to provide experimental support for several French teams. They are working on theory of obstacle avoidance.

Fijii, et al. [13] are working with a robot arm mounted on a mobile cart guided by vision. They refer to this overall robotic system as a "Locomotive Robot." Cook [14] defends the use of free ranging robots as opposed to wire guided or mere vision line following robots. The obstacle detection problem appears to be critical to obtain good results in general industrial applications. Stanford University investigators [15] appear to be close to a working system for following a navigation line for relatively simple applications. Their system uses a 16 phototransistor linear sensor to determine on a binary basis the location of a reflective tape navigation line. The work by Komo [16] has elements that are similar to our theoretical work, but at the present stage of development their robot moves relatively slowly in a limited area, dragging power and other cables to which it is connected.

We do not yet have a moving system but we do have contributions to the literature. In [17] the possibility of using fuzzy set theory for segmentation is considered. An overview of our automotive work is contained in [18] and [19]. Road edge detection is discussed in [20] and pattern coding for guidance is contained in [21]. Stereovision distance measurement problems are addressed in [22] and [23] and camera control problems for roadway observation are presented in [2]. Stereovision system error analysis [24] indicates that this method of distance measurement looks promising for industrial mobile robots. A new method of machine vision tracking which serves as a background for motion detection is contained in [25]. A summary of our recent work may be found in [26] and two papers [27,28] have been submitted recently on navigation line guidance problems. Although vision based mobile robot research is active in several countries, judging from the literature the total effort is relatively low. This is not surprising because of the many robotic problems yet to be solved that do not have the added complexity of motion.

The general system configuration of interest is a platform supported by wheels. It can be configured with different options. For example, it could have stereoscopic machine vision and onboard computing and data processing capability to act on instructions for travel to nearby or remote locations. Or, it could use a contrasting navigation line painted on the floor, wall or overhead. This system could use a one dimensional optical sensor array, which is more economical and requires less data processing than a two dimensional sensor. Fundamental problems include:

1. Identification and location of the pathway.
2. Detection, location and avoidance of obstacles.
3. Recognition and management of navigation information.
4. Vehicle position control.
5. Human safety.

Pathway Idenfitication

Identification of the space in which a robot can operate is a fundamental problem for mobile robots. For example, a robot that operates between a factory floor and an inventory area will have to be able to navigate in passageways with walls, move along a wall, travel in open space, travel in areas restricted by contrasting lines, etc. A robot confined to a work station will have some of these problems too and, in addition, may have to maneuver among obstacles consisting of machines, materials, etc.

Research on solution of this general problem will undoubtedly continue into the indefinite future, especially for unstructured environments because this is a difficult pattern recognition problem. However, many applications have suitably structured work spaces such that the problem solution is practical now. Our general approach has been to identify texture and changes in texture given some a priori information. For specific reasonably well defined situations this reduces to an edge detection problem. For an analytical treatment of navigation line errors and optimum line width refer to [27].

Edge Detection and Location

High speed capability is of paramount importance in the selection of an edge detection algorithm. The second order discrete difference algorithm

$$[\Delta^2(x,y)]^2 = [I(x,y-1)-2I(x,y)+I(x,y+1)]^2 \qquad (1)$$

has been found satisfactory working in conjunction with the Hough transform [29,30] for edge location determination. The intensity value is I, the horizontal direction is y and the value is squared to make the result positive and monotocally increasing as the rate of grey level value change increases.

The Hough transform uses the edge detection algorithm output to estimate the edge location in the robot's coordinate system and acts as a noise filter. It provides an equation for the boundary of regions with different intensities. The passageway edge will be modeled as a polynomial. In the case of a straight line the representation could be

$$y = mx + b \qquad (2)$$

Points on the edge represented by this equation would be mapped by the transformation

$$T:(x_i,y_i)\ (m,b) \qquad m = \frac{y-b}{x} \qquad b = y-mx \qquad (3)$$

which gives an estimation of (2). That is, the points (x_i,y_i) are reapplied into the straight line (2). This scheme of passageway detection has been laboratory tested with lines on the laboratory floor using a PDP/03 computer.

Range Detection

Stereopsis analysis [22] is one method of range determination or free ranging robots. We have proposed [28] a ranging method which makes use of navigation line width for the line following class of robots. There is research activity involving laser ranging and structured light which may eventually solve the range determination problem.

Obstacle Avoidance

Obstacle avoidance is a fundamental problem for all types of mobile robots. We are working with a two camera method which detects patterns above the roadway to distinguish obstacles from shadows [18,26].

Guidance and Control

Control involves the two major areas of guidance and navigation. To date navigation, the ability to follow a prescribed path, has not been addressed except for short time navigation as part of collision avoidance. Guidance can also be subdivided into two major areas: longitudinal control and lateral control. Under longitudinal control would be put acceleration schedules, collision avoidance, and station keeping, i.e., maintaining a fixed and safe position in relation to other vehicles in moving traffic. Lateral control includes maintaining lateral position in a highway lane and also includes collision avoidance.

In considering both longitudinal and lateral control research work reported by Fenton is most useful.

Longitudinal Control

Fenton [31,32] reported a successful model which has been experimentally verified. This is shown in Fig. 1 in block diagram form. This model represents the automobile in a form for velocity control. It includes two major nonlinearities. One is the engine torque representation. This shows a variation in gain of approximately 8 to 1 and a variation in a first order lag time constant of approximately 10 to 1. The second nonlinearity represents slip of the driving wheels with respect to the roadway. This shows up as a difference between actual vehicle speed and speed as represented by the speedometer. This also has the form of a first order lag with a gain of one and a variation of time constant from 1 second at low speeds to 0.1 second at high speed. Since the stability in a station keeping mode is not affected by wheel slip it may be eliminated from the model. The torque nonlinearity is handled by an approximate cancellation scheme involving an inverse of the nonlinearity. For computer control design Fenton's modified model was put into discrete state variable form. We have done simulation with the model at a sampling rate of 10 samples per second using two approaches with assumed vision input. These were a Minimal Response Time Digital Controller (MRTC) and a Proportional, Integral, Derivative controller (PID). For minimum settling time and for steady state error of zero the controller used was

$$D(z) = \frac{2z^3 - 3.478z^2 : 1.971z - .366}{0.377z^3 - 0.0502z^2 - 0.0127z + 0.2525} \qquad (4)$$

The basic model is available in [31,32]. The response of the system for this controller is shown in Fig. 2. Responses for the other form of controller were also satisfactory.

Lateral Control

Here, as in longitudinal control, a model described by Courmier and Fenton is used. A block diagram is shown in Fig. 3. Values of the parameters are those of a 1965 Plymouth sedan weighing 2168 kg.

The basic lateral control task is to maintain a path in a highway lane. Thus, the lateral control system is to be primarily a regulator. It can then also operate in a navigation mode such as lane changing or collision avoidance by simply following a path defined by the control computer. Thus its task is still essentially that of a regulator.

The general approach we have used is to minimize a quadratic performance index using the Lagrange multiplier.

A control law was derived and a plot of controller sequence and lateral deviation versus time is given in Fig. 4.

It has been determined in our earlier work [33] that the sampling period need not exceed 0.1 second for stable control. During this time a vehicle traveling at 30 m/sec (67 mph) would travel 3 meters or about 10 feet. This is the time available for analyzing a video scene.

Bibliography

1. Neff, Robert, "Japanese Work to Robot Production," Electronics, V54, No 2, Oct. 6, 1981, pp. 87-88.
2. Walden, K.Y. and McVey, E.S., "Distance Considerations for Vehicle Control Using Machine Vision," accept. Proc. IEEE Southeastcon '82, Gainesville, FL, April 1982.
3. Nilsson, N. J., Proc. IICAEI, 1979, pp. 509-520.
4. Schnidt, Jr., R.A., "A Study of the Real Time Control of a Computer Driven," Stanford Artificial Intel. Proj. Memo No. AIM-49, Aug. 1971.
5. Moravec, H.P., "Obstacle Avoidance and Navigation in the Real World by a Seeing Robot Rover," Carnegie Mellon Univ. Robotics Inst., Rept. CMU-RI-TR3, 1980.
6. Matsushima, K., et al., Proc. 21st IPS, Japan, 1977, pp. 569-576.
7. Nozoki, T., Trans. S.I.C.I., 13-5, 1977, pp. 51-57.
8. Delrotin, B., Proc. 5th IICAI, 1979, pp. 723-732.
9. Tsukwsa, H., ETL Bulletin, 1979, pp. 43-728.
10. Iijima, J., Yuta, S. and Kanazama, Y., "Elementary Functions of a Self-Contained Role at 'YAMABICO 3.1'," Proc. 11th Int. Symp. on Indus. Robots, Tokyo, 1981.
11. Larcombe, M.H.E., "Tracking Stability of Wire Guided Vehicles," Proc. 1st Intl. Conf. on Automated Guided Veh. Sys., Stratford Upon Avon, June 1981, pp. 137-144.
12. Geralt, G., Sahek, R. and Chatila, R., "A Multi-Level Planning Navigation System for a Mobile Robot: A first Approach to Hilare," Proc. 6th Al Conf., Aug. 1979, pp. 335-337.
13. Fijii, S., Ohtsuki, H. Matsumoto, H. Yoshimoto, K. Kihui, H. and Yamada, K, "Computer Control of a Locomotive Robot with Visual Feedback," Proc. 11th

Intl. Symp. on Indust. Robots, Tokyo, 1981, pp. 219-226.
14. Cooke, R.A., "Microcomputer Control of Free Ranging Robots," 13th Intl. Symp. on Indust. Robots and Robot 7, V2 (April 1983), pp. 13-109 -13-120.
15. Weber, D.M., "Smart Carts Tackle In-Plant Tasks," Electronics Week, Aug. 1984, pp. 37-41.
16. Komo, J., "Moving Robot Applied Work Tracing," Proc. IFAC, Budapest, Hungary, July 1984.
17. Inigo, R.M., Hinkey, M. and Ruest, C., "The Use of Fuzzy Set Theory in Image Segmentation," Proc. IEEE Southeastcon, Huntsville, AL, April 1981, pp. 669-673.
18. McVey, E.S., Inigo, R. M., Moore, J.W. and Wirtz, M.J., "Application of Machine Vision to Vehicle Guidance and Safety," Proc. of the 15th Asilomar Conf. on Circ. Sys. & Comp, Pacific Grove, CA, Nov. 9-11, 1981.
19. Inigo, R. M. and McVey, E.S., "Machine Vision Applied to Vehicle Guidance and Safety," inv. paper 32nd IEEE Vehicular Tech. Conf., San Diego, CA CA, May 1982.
20. Wirtz, M.J. and McVey, E. S., "A Pattern Recognition System for Measuring Vehicle to Road Edge Distances on a Moving Vehicle," 14th Southeastern Symp. on Sys. Theory, VPI, Blacksburg, VA, April 1981.
21. Berger, B.J. and Inigo, R.M., "Block Stored Pattern Coding of Video Signals," accept. Proc. IEEE Southeastcon '82, Gainesville, FL, April 1982.
22. Lee. J.S., McVey, E.S. and Inigo, R.M., "Distance Measurement Using Deconvolution in a Stereo Vision System," Proc. 13th Southeastern Symp. on Sys. Theory, Orlando, FL, March 1981.
23. Moon, C.W. and McVey, E.S., "Computer Vision Distance Measurement Error Analysis for Imperfect Camera Alignment," accept. Proc. 14th Southeastern Symp. on Sys. Theory, Blacksburg, VA April 1982.
24. McVey, E.S. and Lee, J.W., "Stereovision Accuracy and Resolution Associated with Sensor Size," accept. IEEE Trans. PAMI, Feb. 1982.
25. Schalkoff, R.J. and McVey, E.S., "A Model and Tracking Algorithm for a Class of Video Targets," IEEE PAMI, V. PAMI-4, No. 1, Jan 1982, pp. 2-10.
26. Inigo, R.M., McVey, E.S., Berger, B.J. and Wirtz, M.J., "Machine Vision Applied to Vehicle Guidance," accept. IEEE PAMI.
27. Drake. K. S., McVey, E.S. and Inigo, R.M., "Sensing Error for a Mobile Robot Using Line Navigation," sub. IEEE PAMI, July 1984.
28. McVey, E.S. Drake. K. C. and Inigo, R.M., "Range Measurements by a Mobile Robot Using a Navigation Line," subm. IEEE PAMI, July 1984
29. Duda, R.O. and Hart, P.E., "Use of the Hough Transform to Detect Lines and Curves in Pictures," Comm. of the ACM, V15, N1, Jan 1972, pp. 11-15.
30. Ianniono, A. and Shapiro, S.D., "A Survey of the Hough Transform and its Extensions for Curve Detection," Proc. 1978 Intl. Conf. on Pattern Recognition, pp. 32-38.
31. Cormier, W.H. and Fenton, R.E., "On Steering of Automated Vehicles-A Velocity Adaptive Controller," IEEE Trans Veh. Tech., V. VT-29, N.4, Nov. 1980.
32. Fenton, R. E. and Chu, P.M., "On Vehicle Longitudinal Stability," Trans. Science, V. 11, No. 1, Jan 1977.
33. Inigo, R.M., McVey, E. S. and Moore, J.W., "Application of Image Processing to Vehicle Guidance and Safety," Phase I Report, DOT Contract No. DTRS5680-C-00033, 1981.

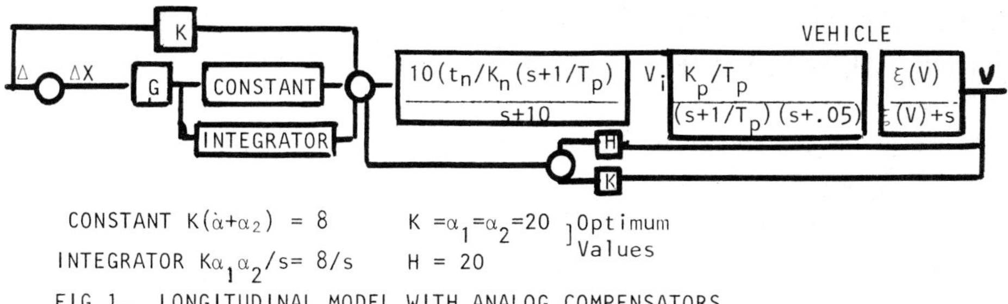

CONSTANT $K(\dot{\alpha}+\alpha_2) = 8$ $\quad$ $K = \alpha_1 = \alpha_2 = 20$] Optimum Values

INTEGRATOR $K\alpha_1\alpha_2/s = 8/s$ $\quad$ $H = 20$

FIG 1. LONGITUDINAL MODEL WITH ANALOG COMPENSATORS

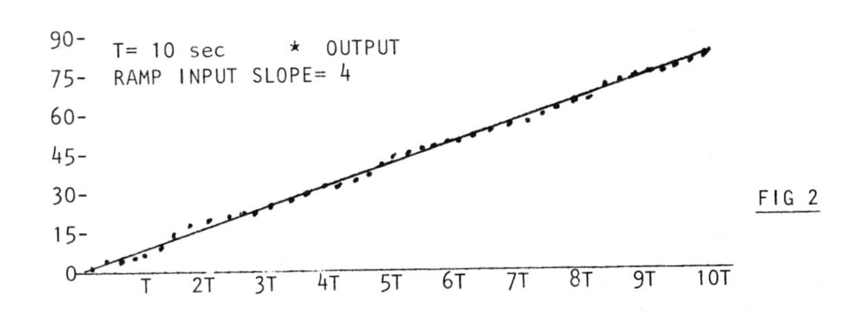

FIG 2

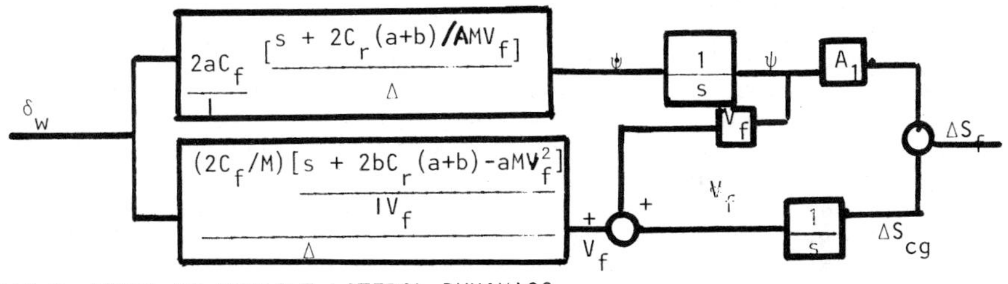

FIG 3. MODEL OF VEHICLE LATERAL DYNAMICS

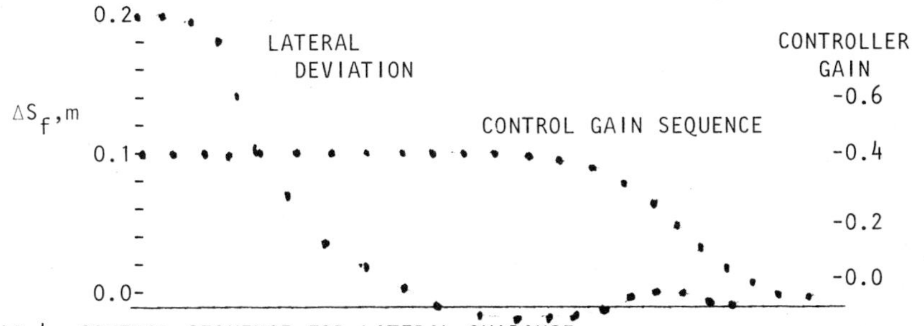

FIG 4. CONTROL SEQUENCE FOR LATERAL GUIDANCE

General Knowledge Representations of Road Maps and Their Applications

Mingfa Zhu and Pepe Siy

Department of Electrical and Computer Engineering
Wayne State Univesity
Detroit, MI 48202

Summary

This paper discusses knowledge representation methods for a class of quite general road maps (with NS, EW, and diagonal roads). Several methods are suggested to model intersection points (nodes) of roads. Instead of bit-matrices, general matrices are used to represent knowledge of entire road maps. Based on this knowledge-based matrix, a rule based production system is generated for goal-searching. Road obstructions can be easily handled in these representations. Some algorithm of goal-searching is suggested, which is quite efficient because of searching by roads rather than nodes.

I. Introduction

Goal-achieving or trip-planning has been a very important research area in artificial intelligence especially in mobile robot and automous vehicle intelligence. Shortest path in general graphs could be found by using well-known algorithms. But these algorithms are not necessary efficient because the graphs in which shortest paths are searched are too "general". On the other hand, road maps, in terms of which mobile robots or automous vehicles travel, are usually not so general that quite simple models of them could be used in trip-planning. Roughly speaking, a road map is a connected, usually not complete graph, that is, from one node one can go to any other node along a sequence of edges (road segments), and the minumum number of edges travelled usually is not one. Usually, in a road map, there are some main roads which have uniform orientations, for example, in New York city there is one set of "avenues" between uptown and downtown and one set of "streets" perpendicular to "avenues". There are some "road-switching" points at intersections of a "avenue" and "streets" at which one can switch roads between an "avenue" and a "street". Several authors achieved representations of knowledge of road maps and efficient procedures for trip-planning. The main point of their work is the emphasis of the role of roads instead of that of graph nodes, that is, the only concern is that which roads a vehicle can switch to along the current road on which it is moving, rather than that from a given node, which nodes can be reached directly. The authors of this paper suggested a bit-matrix method for knowledge representation of roadmaps of only east-west (EW) roads and north-south (NS) roads and a production system for goal-achieving. It is based on this simple road map model that efficiency of search process is achieved.

	B1	B2	B3	B4	B5	B6
A1	20AA	0088	20AA	20AA	38AE	00AA
A2	20A8	268B	20AA	38EE	00AA	20AA
A3	0289	20AA	3FFF	00AA	20AA	20AA
A4	0	3FF3	00A8	3EBA	202A	0022
A5	0088	20AA	0298	00AA	20AA	00AA

Fig.2(c)

	B1	B2	B3	B4	B5	B6
A1	1	0	1	1	1	0
A2	1	1	1	1	0	1
A3	0	1	1	0	0	1
A4	0	1	0	1	1	0
A5	0	1	0	0	1	0

Fig.3(a)

d_{ij} 0 1 2 3 4 5 6 7 8

NE NW SW SE
RA mark bits

NE(SW)--NW(SE)
NW(SE)--EW
NW(SE)--NS
NE(SW)--EW
NE(SW)--NS
RS mark bits

Fig.3(b)

	B1	B2	B3	B4	B5	B6
A1	0	0	0	0	182	0
A2	0	061	0	18A	0	0
A3	021	0	1FF	0	0	182
A4	0	1FD	0	1FD	0	0
A5	0	0	024	0	064	0

Fig.3(c)

	B1	B2	B3	Bn
A1	X11	X12	X13......	X1n
A2	X21	X22	X23......	X2n
.	.	.	.	.
Am	Xm1	Xm2	Xm3......	Xmn

Fig.4(a)

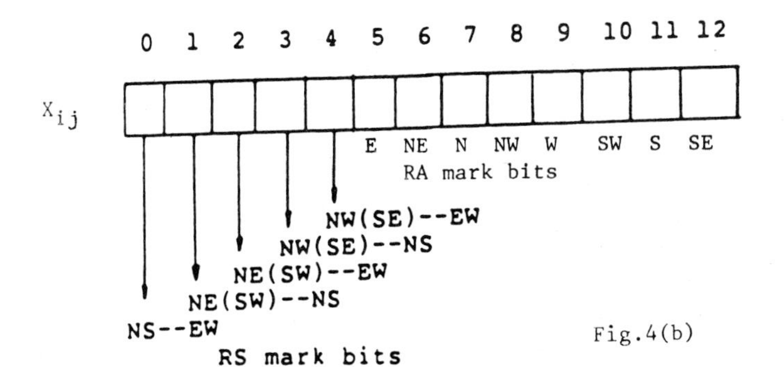

Fig.4(b)

Algorithms for Touch Sensing in Real Time Parts Recognition Systems

James K. Blundell and Donald W. Thompson

James K. Blundell — Dept. of Mechanical Engineering, University of Missouri-Columbia at Kansas City

Donald W. Thompson — United States Air Force

ABSTRACT

Sensory research for industrial robotic applications is an emerging science. The underlying motive for this research is to provide a robot with the ability to interact with its environment. This study concentrates on the development of a versatile touch sensory system for recognizing and retrieving various objects from a conveyor. The development of a linear array of touch sensors and the corresponding interface requirements for implementation on a microcomputer are detailed.

1.0 INTRODUCTION

Tactile sensing can be defined as the ability of a robot to "feel" objects in its working space using force or pressure transducers. A distinction is made between tactile sensing and touch sensing. Tactile sensing is described as "the continuous-variable sensing of forces . . ." (1), i.e., the output of the transducer is an analog signal proportional to the amount of force or pressure applied. Touch sensing, however, often referred to as simple touch or contact sensing, is restricted to binary states. When the threshold value of either the input (force) or output (voltage) variable of a transducer is exceeded, a binary system is created which lends itself to detecting only the presence of an object. To eliminate the need for making many measurements to detect the shape of an object, numerous sensors generally are placed in an array for both tactile and touch sensing.

An extensive survey was performed by Harmon (1) to assess the needs, uses, and requirements of tactile and touch sensing in the industrial environment. The capability to recognize the position, size, contour, and orientation of an object was emphasized. Tactile sensing was noted to be useful in almost every facet of manufacturing, although the need for a resolution finer than 10 * 10 elements per square inch was rarely indicated. However, sensor requirements such as rapid response time (1-10 msec), low threshold sensitivity (1 gram), ability to endure industrial environment, and above all, low cost were stressed. Interestingly, the need for a compliant "skin-like" material containing a matrix of sensors was often cited for tactile sensing. Also, to process the vast amount of sensory data, the development of new algorithms and data processing strategies were indicated.

As a result of Harmon's survey, a massive research program was initiated worldwide to develop a tactile sensing material suitable for industrial use. Many different types of materials and applications have been implemented with some success (2)-(12). Regardless, these materials have not been enthusiastically adopted by industry, and their utilization will be limited to research activities until the 1990's at which time major advances in sensory technology should have taken place (13). Also, the premise that the sensors presently available are dedicated to a particular application is cited by Knight (13) as a reason for their less than enthusiastic acceptance in industry.

Since the thrust of tactile research is directed toward the development of skin-like materials to be applied to the end effector (or gripper) of a robot, another problem has arisen. The general location of an object must be known for it to be grasped and recognized by the manipulator. Search techniques for static objects using "whisker" sensors mounted to a robot gripper were explored by Russell (14). Using a sweeping motion, the locations of several objects were discerned to within 1 mm. Likewise, the successful use of simple sensors interfaced to a microcomputer was demonstrated by the research.

In the industrial environment, however, static situations are seldomly encountered. Most processes involving robot manipulators are characterized by conveyors and dynamic workpart handling systems. The position of an object must be tracked for proper interaction with the manipulators; this tracking function is predominated by vision systems.

Vision systems have been plagued with visual noise, the need for high contrast between the object and background, and the necessity for large computers to process the immense amount of data (15). CONSIGHT-I (15), a linear-array-vision-based robot system developed by the General Motors Research Laboratories, was shown to remove many of these constraints while providing two-dimensional object tracking and sizing capability in addition to determining the orientation of an object. However, special lighting configurations to alleviate shadows were noted as essential to its operation.

The concept of the CONSIGHT-I program can be equally applied to touch sensing technology. A nondedicated system employing simple touch sensing can be developed to provide the preliminary recognition and orientation determination of an object. Once an object is recognized, a robot arm can be used to remove the object from the conveyor and place it in a specific position. Special lighting requirements can be eliminated, and data processing can be accomplished with a microcomputer.

2.0 THEORY

Sensory systems, citing visual, tactile, and simple touch as examples, are utilized by "intelligent" machines to provide the data essential for their interaction with the environment. Feedforward systems, primarily vision, are used to relay preliminary information about a process prior to the action of a manipulator. Tracking, sizing, and the determination of the orientation of an object are cited as basic functions of these types of systems. On the other hand, feedback systems are characterized as elements which provide data relevant to the operation in progress. Tactile and touch sensing systems are used primarily as feedback devices; the presence of a force is indicated by touch sensors, while the magnitude of a force is revealed by tactile sensors.

Tactile, touch, and vision sensory systems, although different in hardware design, are related by their similarity of application and operation. The concepts of visual pattern recognition are equally applied to tactile and touch sensing. The sensing process is divided into two functions: (1) image sensing and (2) analysis.

2.1 Image Sensing

Image sensing is provided by transducers which typically are arranged in an array. The location of a transducer in an array is designated by its columnar and row vectors. The output value of a transducer, in digital form, along with its columnar and row vectors, is stored in the memory of a computer. Thus, a two-dimensional image of an object is contained in the computer.

2.2 Image Analysis

Once the data is accumulated in the memory of the computer, it is analyzed according to the algorithm of the particular application. If needed, some type of segmentation process is performed. Segmentation is used to separate an object from its background and to classify the characteristic property of each pixel into a more workable form. Segmentation is primarily applied to vision and tactile sensing systems. In most cases of touch sensing, segmentation is provided by the transducer, therefore, segmentation algorithms are not required. A wide variety of segmentation algorithms are cited in the literature (16)-(21).

Provisions for identifying an object or image are also included in the analysis process. Strategies such as Pixel-by-pixel comparison with a master image (22), shape analysis from object outlines (23)-(25), and feature extraction (16), (22), (26) are cited in the literature. Pixel-by-pixel comparison techniques are limited to applications employing large computers because a minimum of three image arrays are required (22). One array is allocated for the master image; a second array is needed for the sensed object, and a third array is designated for containing the difference relationships between the master and sensed image. The orientation of the sensed image is also required for proper alignment with the master or mask image. The orientation generally is determined by global feature extraction.

In shape analysis techniques, the image is broken down into constituent geometric elements (16) which are identified by boundaries. The boundaries are determined by boundary tracing techniques (23). These routines progress in a predescribed manner to check for discontinuities in pixel patterns. When a discontinuity is discovered between a pixel and one of its adjacent elements, the pixel is flagged as a boundary point.

Pavlidis (23) classifies boundary tracers as local or global. Local boundary followers, such as sequential scanning, move from point-to-point along the periphery of an image searching for discontinuities. Typically, only the external boundary is found with these techniques, so they are often termed external boundary followers.

On the other hand, global boundary tracers lend themselves to determine the interior boundaries of an image if they exist (23). By scanning all pixels in the image, tracers such as "skip across" followers can determine if internal or peripheral voids exist and can discriminate the borders of these voids. In this technique, the pixels are scanned along a vector (row or column), and each discontinuity on the vector is flagged as it is encountered. This process is illustrated in Figure 1. Although termed global, this technique is also used to flag only the extreme discontinuities along a vector which represent the external boundary. This process is also illustrated in Figure 1. However, for accurate border determination, this case is limited to images with no peripheral voids. Once the outline of an object is determined by global or local methods, global features such as perimeter, centroid, and included area are extracted from the image (24). In addition, local properties such as concavity and convexity are derived from the outline (24). The combination of these features characterize an image and provide a means to recognize the object.

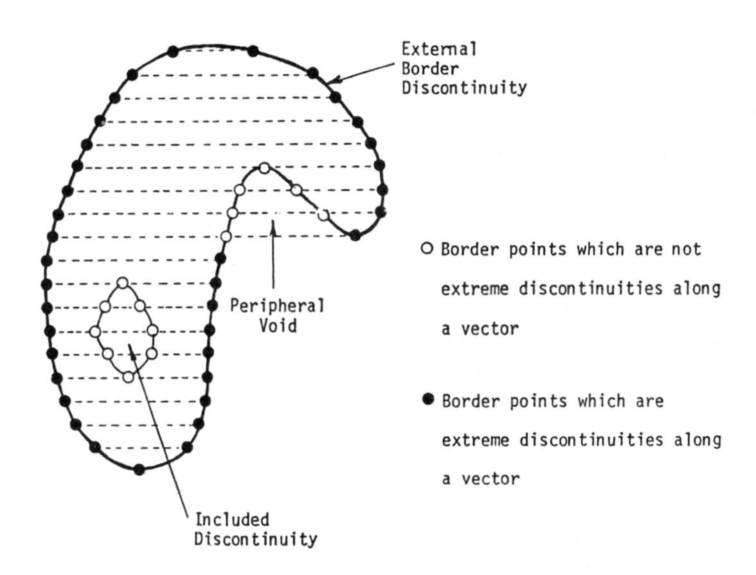

Figure 1 Image Characterization

2.3 Classical Moment Techniques

Object features such as centroid, area, area moments of inertia, and orientation of the principal axes are derived using classical methods.

The major properties of an object are determined in this manner, and in the cases of simple geometric shapes, additional descriptive features are not warranted (26). These shapes are recognized by comparing their properties with those of a master image. For more complex object geometries, these properties are used as a basis from which additional descriptive feature algorithms can be implemented if necessary. Commonly, local features such as concavity, convexity, and average radii are determined with these supplemental techniques (24). Thus, through the use of feedback and feedforward elements, a computer image of an object is obtained, and this image is analyzed and recognized to provide input for a manufacturing or material handling process.

3.0 HARDWARE REQUIREMENTS

The equipment required for an industrial application of a robot system is divided into the following basic categories: (1) sensing elements, (2) interface and input/output hardware, (3) the controller, and (4) the manipulator.

Limited by the provision of a single bit of data, microswitches and simple contact switches are most commonly used for digital transducers. Actually, the resistance of these devices varies in the same manner as analog transducers, but due to force amplification and surface conductivities of the contacts, the resistance is altered almost instantaneously to provide a full-scale step change in voltage. In other cases, an analog output voltage is created proportional to the surface conductivity, and the digital switching is provided by the input voltage sensitivity of the digital interface circuitry.

The touch transducer is depicted in Figure 2. The initial streamwise location of the object on the conveyor, as determined by the first instance of a triggered switch, would be varied with the height of the object. However, due to the inclined position of the arms, this error for

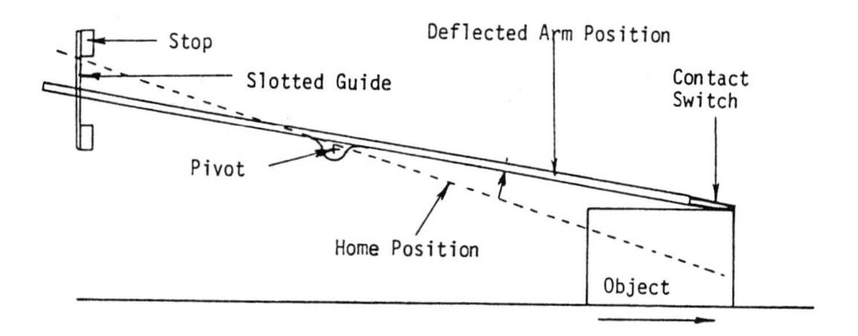

Figure 2 Touch Transducer Schematic

tracking initiation could be minimized. Since relatively small tracking forces would be induced, the limitation of transverse displacement of the arm would be provided by a smaller arm cross-section than previous designs. Likewise, additional transverse displacement restraints could be provided by slotted guides mounted to the arm stops.

From design analysis, the required overall length of the arm was found to be approximately 12.5 in. and the distance from the pivot point to the sensing tip was determined to be approximately 7.5 in. Tracking forces of approximately $3.0 * 10^{-4}$ lb were calculated.

Forty of these sensing arms were assembled on a 0.25 in diameter pivot shaft which in turn was mounted to the support structure.

The second basic category of equipment, interface hardware, is comprised of electronic circuits and connections to link the transducers to the computer. To transfer sensory data to the computer, a means of accessing the data, address, and control buses of the computer is needed.

The transducer interface required the scanning of 40 separate digital signals but because eight bit logic was used parallel processing of the data was accomplished using an input buffer. After the transducer data is loaded into the buffer five eight-bit data blocks are generated and access is provided by a tri-state octal line driver. Non-zero values on the data bus indicate an activated sensor. The hardware is readily expanded and its performance is only limited by software flexibility.

The controlling function of a robot system is performed by the computer. In addition to receiving the data from the sensory equipment, the movement of the robot arm is controlled by this unit.

Finally, the last category of equipment in a robotic system is comprised of the manipulator.

In combining these components to yield the sensory system for this study, many of the hardware configuration concepts of the CONSIGHT-I program (15) would be instituted. Since an object would be traveling along a conveyor, only the provision for a linear array of touch sensors mounted transversely to the conveyor would be warranted. The second dimension of the image array would be provided by an incremental encoder mounted to the conveyor; the object would be scanned at each increment of the encoder. Thus, a composite image would be generated by combining each of the linear image scans into a two-dimensional array. Likewise, the use of a nondedicated image generation system would be implemented. Object tracking would be provided by the encoder and corresponding software. Also, in accordance with the requisites of Harmon (1), the equipment would be designed to provide a resolution of approximately 10 * 10 elements per square inch.

4.0 SOFTWARE CONCEPTS

Software provides the means to integrate theory and equipment into a working robotic sensory system. Coordination of all activities of the system is accomplished by software. Although the equipment may be designed to be flexible, this effort may be in vain if flexibility is not also designed into the software. The ultimate success of a robotic sensory system lies equally on equipment and software development.

Modularity of subroutines is one method of providing software flexibility. In most cases, subroutines may be added, deleted, or changed as necessary with minimal impact on the other subroutines. In the software developed for this study, most of the actions are accomplished in subroutines. These subroutines are called by the main program and other subroutines. In many instances, the use of a subroutine depends on the state of a flag.

For prototype software development, a high-level language is desirable. By using such a language, a programmer can concentrate on the task without involving himself with the intricacies of machine-coded manipulations associated with lower level languages. Eventually, a low-level or machine program may be written once the concepts are proven valid by a high-level language. In this study, the software is written in BASIC. An extension of the vocabulary of this version of BASIC is provided by ARMBASIC. This enhancement allows the robot to respond to commands similar to those of BASIC. The addition of this vocabulary is transparent to the user.

4.1 Software Architecture

The software architecture for this study consists of the main program and subroutines assigned according to function. Each subroutine is given a nonoperational name which characterizes its function. This architecture is shown in Figure 3.

As shown in Figure 3, the software is comprised of a main program and four categories of subroutines. A detailed description of each of these portions of software follows.

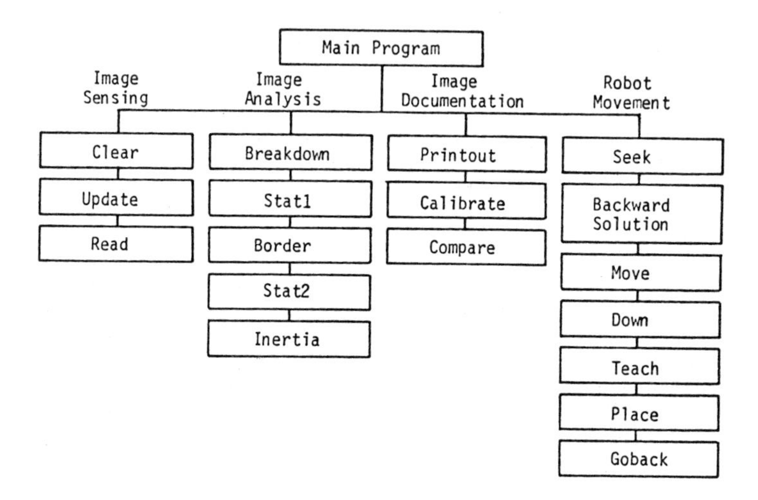

Figure 3 Software Architecture

4.2 Main Program

The main program sets up the variables and provides the control function for the logical access of subroutines for data generation and manipulation. Much of the interaction with the operator is accomplished in the main program. The operator provides inputs that determine the maximum size of an object to be recognized, selects whether a printout of the object image is to be made and whether the general model or line-area model, referred to as the "fill" model, will be used in analyzing the image. A tolerance for comparing a sensed image to a calibration image is also set by the operator in the main program.

During the initialization of the main program, the operator chooses the calibration mode which sets the relevant flag. The operator then places the robot arm in a previously determined home position. Once the home position of the robot is identified, the common image-sensing portion of the software is invoked. The image array is initially set to zero by the "clear" subroutine. The "read" subroutine is then entered in which image data is collected as an object passes under the linear array of touch sensors.

Upon completion of data collection, the main program calls the subroutines for image analysis. The same qualified statistics are generated by both of these models. In addition, the sensed image is displayed on the monitor.

This subroutine compares the statistics of a sensed image with those of a master or calibration image. If a discrepancy between the two sets of statistics exceeds a tolerance entered by the operator, a "no-match" flag, FQ, is set. The operator can then opt to alter the comparison tolerance and save the sensed image statistics as the calibration model.

The subsequent portion of the main program pertains to robot movement. Subroutines are called to seek and grasp the object. If the system is in the calibration mode, the "teach" subroutine is called and the pick-and-place movements are described by operator inputs. If the system is in the normal mode, the predetermined pick-and-place actions are performed by calling the "place" subroutine. The main program then returns to the image sensing portion of the software to prepare the system for recognition of another object.

4.3 Image Sensing

Explanation of the image sensing routines requires an understanding of the coordinate systems used in this study. Two coordinate systems are implemented in the image sensing and analysis routines. A global coordinate system is used to track the leading edge of an object in reference to the linear array of touch sensors. The axes of this system are perpendicular and parallel to the direction of conveyor motion with the origin established by the first sensor in the sensor array. This coordinate system is shown in Figure 4.

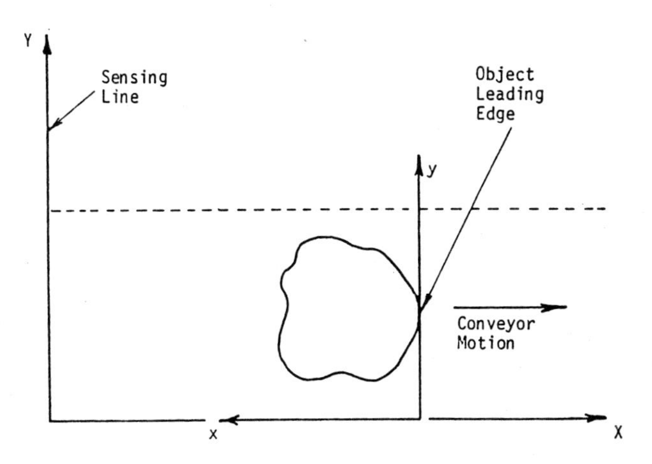

Figure 4 Image Sensing Coordinate System

A nonstationary local coordinate system is related to the object. This system is originated by the leading edge of the object. Unlike the global coordinate system, the local system follows the nonstandard left-hand convention with axes perpendicular and parallel to the global axes. The local coordinate system is the principal reference for image sensing and analysis.

Image sensing is then performed by the "read" subroutine. The "read" subroutine inputs sensory data at sequential scans of the x coordinate. During each data input cycle, the data is loaded into the parallel input buffer and stored in the computer memory by reading each sensory gate sequentially.

When the "read" subroutine is departed, a two-dimensional image array consistent with the local coordinate system is stored in memory. This image array is comprised of a series of five data blocks. Each data block contains an integer value related to the state of eight touch sensors at a particular displacement count or x coordinate. In addition, a linear array of displacement counts, corresponding to each image array, is stored in memory. These arrays are then used by the image analysis subroutines for object recognition.

4.4 Image Analysis

Image analysis is performed according to the theory previously discussed. To implement this theory, five subroutines are utilized. Analysis can proceed according to the general or "fill" model. The method used is determined by the "fill" flag which is set by the operator.

Image analysis begins with entry into the "breakdown" subroutine.

The boundary points are specified as the extrema of each successive y-coordinate scan (Figure 1).

The "breakdown" subroutine then proceeds in a manner similar to the "read" subroutine. Each of the five data blocks corresponding to a scan count is then analyzed, and the analysis continues by incrementing the x-coordinate scan count up to its maximum value.

The "border" subroutine basically compares the x and y coordinate counters with the boundary pointers previously stored.

The statistical subroutines, "stat1" and "stat2" collect the data for the finite element images. The total area of the image is summed accordingly. The first area moment of the image about the X axis is summed and the first area moment of the image about the Y axis is summed. The moments of inertia about the X and Y axes, WX and WY respectively, and the product of inertia, WZ, are determined. These subroutines also display the sensed image on the monitor by setting the particular pixels corresponding to the scan, word, and bit counts.

Following the statistical analysis of the image, the "inertia" subroutine is called. This subroutine calculates the object centroid coordinates, the second area moments with respect to the object centroid, the maximum and minimum moments of inertia, and the angle of orientation of the principal axes in reference to the local coordinate system.

The analysis of all images is performed in this manner. The identification of the area, and the maximum and minimum moments of inertia of an object allows the software to qualify the object for recognition. Recognition of an object involves comparing the statistics of a sensed image with those of a master or calibration image. This process is accomplished by image documentation.

4.5 Image Documentation

Image documentation for this study is performed with three subroutines. The "calibrate" subroutine merely stores the area and maximum and minimum moments of inertia of an image for future reference. The "compare" subroutine compares the area and maximum and minimum moments of inertia of a sensed image with those of the calibration image. A tolerance for the comparisons is set by the operator. If the statistics do no match within this tolerance, the image is not recognized, and the "no-match" flag is set. The tolerance is required due to the lack of a data base for image description. Over a long period, a data base for a particular image can be established to derive the calibration statistics and the match tolerance for an image. Therefore, to develop a data base, a means of documenting the sensed image and its statistics is needed. This function is provided by the "printout" subroutine.

4.6 Robot Movement

Once an image is analyzed and recognized, the robot must retrieve the object from the conveyor and perform predescribed movements relating to the application. In this study, a simple pick-and-place operation is performed.

In summary, the software is written to interact with the robotic sensory equipment. An image is sensed by the image sensing software as an object passes under the touch sensor array. The array of data collected during image sensing is then analyzed by moment techniques, and statistics are generated to qualify the object for recognition. Once recognized, the object is manipulated by the robot according to the robot movement software. Thus, the entire robotic sensory system is coordinated by software.

5.0 OPERATION AND PERFORMANCE

Testing of the touch sensory system was accomplished using five different shapes. These were a rectangle, an ellipse, an isosceles triangle, an I-section, and a thick ring. Trial runs were made with each of these objects to determine the effectiveness of the touch sensory system in collecting, analyzing, and recognizing image patterns. Also, the ability of the robot to successfully orient itself for retrieval of the objects was evaluated.

The effectiveness of the touch sensors and the image sensing software is realized from the printout of the sensed image shown in Figure 5. Sensing problems evidenced by these images include voids and stray pixel activations which are not representative of the objects.

A void is caused by the loss of adequate contact between the sensing filaments of a tracking arm during a particular read scan of an object. A void may be extended over many scans which indicates that contact is not

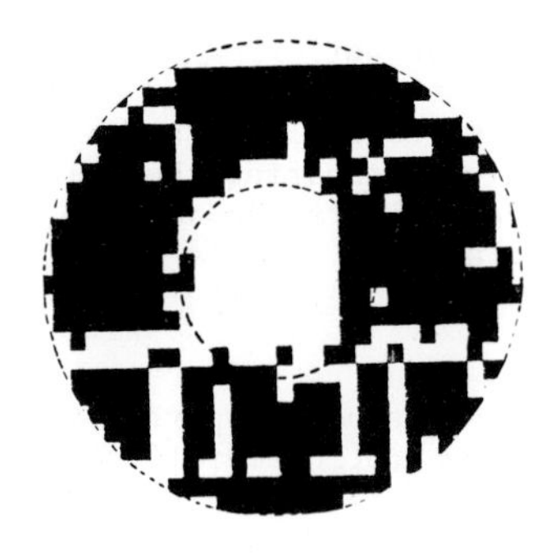

AREA: 3.35347 IMAX: 1.67932 IMIN: 1.58181 THETA:-4.00399

Figure 5 Example of Sensed Image

made between the filaments, or a void may be isolated to a single occurrence which indicates bouncing of the switch or lack of adequate current draw through the switch. Switch instability is identified by the occasional making and breaking of contact between the filaments. In most cases, contact of the filaments is made, as witnessed by the illumination of the light emitting diode, but poor contact causes most of the reference voltage to be dropped across the switch rather than the pull-up resistor and diode. Consequently, the desired low voltage reference at the input to th integrated circuit is not made available, so the threshold voltage for recognition of a set switch is not realized. Therefore, a principal problem with the switches appears to be the relatively high surface resistivity of the contact filaments.

Stray pixel activation is also noted in the sensed images. These stray marks are the vertical excursions shown in many of the image printouts. These marks may be caused by extraneous Radio Frequency Interference (RFI) induced on the sensor filaments and the data paths between the sensors and the computer. It should be noted that RFI was not accounted for in the design of the system. It is possible that the voids are also caused by Radio Frequency Interference.

The effect of voids is diminished by the line-area model. Since the area between the two extrema of a scan level is assumed to be continuous, voids in this region are inconsequential. Thus, relating to its function, the line-area model is termed the "fill" model because voids are filled during the process. On the other hand, the effects of stray marks are accented by the "fill" model. The stray marks cause areas which should be voids to be filled in by the "fill" model.

Despite the sensing errors, a suitable image for object recognition may be generated by the hardware. The generation and recognition process typically take 0.6-0.8 seconds.

6.0 CONCLUSIONS AND RECOMMENDATIONS

The touch sensory system developed for this study was shown to operate adequately despite the errors associated with data collection of the image. General representations of objects were analyzed and recognized, and the robot successfully retrieved the object from the conveyor in most cases. However, inconsistent sensory switch activation degraded the operation of the system. Voids were evidenced in all sensed images, so accurate descriptions of the objects were not achieved by analysis. No significant software problems were identified; the analysis proceeded according to the theory. The lack of accuracy of the system was directly attributed to the inconsistency of the sensing process. Thus, to optimize the operational characteristics of the system, further research in the area of image sensing, particularly switch design, is needed.

7.0 REFERENCES

(1) Harmon, Leon D., Automated Tactile Sensing. Society of Manufacturing Engineers Technical Report No. MSR82-02, Dearborn, MI, 1982; p. 57.

(2) Larcombe, M. H. E., "Carbon Fibre Tactile Sensors," Proceedings of the First International Conference on Robot Vision and Sensory Controls, April 1-3, 1981, Stratford-upon-Avon, UK; pp. 273-275.

(3) Hills, W. D., "A High Resolution Imaging Touch Sensor," Proceedings of the ASME Winter Annual Meeting, Nov. 14-19, 1982, Phoenix, AZ.

(4) Robertson, B. E., and Walkden, A. J., "Tactile Sensor System for Robotics," Proceedings of the 2nd International Conference on Robot Vision and Sensory Controls, Nov. 2-4, 1982; pp. 572-577.

(5) Allen, Peter, "Visually Driven Tactile Recognition and Acquisition," Proceedings of the IEEE Computer Science Conference on Computer Vision and Pattern Recognition, 1983; pp. 280-282.

(6) Dillman, R., "A sensor Controlled Gripper with Tactile and Non-Tactile Sensor Environment," Proceedings of the 2nd International Conference on Robot Vision and Sensory Control, Nov. 2-4, 1982; pp. 159-165.

(7) Minuro, Ueda, and Kazuhide, Iwata, "Tactile Sensors for an Industrial Robot to Detect a Slip," Proceedings of the 2nd International Symposium on Industrial Robots, May 16-18, 1982.

(8) Dixon, J. K., et al, "Research on Tactile Sensors for an Intelligent Naval Robot," Proceedings of the 9th International Symposium on Industrial Robots, March 13-15, 1979; pp. 507-518.

(9) Stute, G., and Erne, H., "The Control Design of an Industrial Robot with Advanced Tactile Sensitivity," Proceedings of the 9th International Symposium on Industrial Robots, March 13-15, 1970; pp. 519-528.

(10) Briot, M., "The Utilization of an 'Artificial Skin' Sensor for the Identification of Solid Objects," Proceedings of the 9th International Symposium on Industrial Robots, March 13-15, 1979; pp. 529-548.

(11) Brown, T. D., and Muratori, D. R., "Miniature Piezoresistive

Transducers for Transient Soft-body Contact-stress Problems," Experimental Mechanics, Volume 19, No. 1, Jan. 1979; pp. 214-219.

(12) Purbrick, John A., "A Force Transducer Employing Conductive Silicone Rubber," Proceedings of the First International Conference on Robot Vision and Sensory Controls, April 1-3, 1981, Stratford-upon-Avon, UK; pp. 73-77.

(13) Knight, J. A. G., "Sensors for Robots: The State of the Art," Proceedings of the 2nd European Conference on Automated Manufacturing, May 16-19, 1983, Birmingham, UK; pp. 127-131.

(14) Russell, R. A., "Closing the Sensor-Computer-Robot Control Loop," Robotics Age, Volume 6, No. 4, April, 1984; pp. 15-20.

(15) Holland, S. W., Rossol, L. and Ward, M. R., "CONSIGHT-I: A Vision-controlled Robot System for Transferring Parts from Belt Conveyors," Computer Vision and Sensor Based Robots, Edited by Dodd, G., and Rossol, L., Plenum Press, New York-London, 1979; pp. 81-99.

(16) Gonzalez, Rafael C., and Safabakhsh, Reza, "Computer Vision Techniques for Industrial Applications and Robot Control," Computer, December 1982; pp. 17-32.

(17) Hord, R. M., Digital Imaging Processing of Romotely Sensed Data, Academic Press: New York, 1982; pp. 65-127.

(18) Zuker, Steven, "Algorithms for Image Segmentation," Digital Image Processing and Analysis, Edited by Rosenfeld, A., and Simon, Jean C., Noordhoff Leyden, 1977; pp. 169-183.

(19) Pratt, William K., Digital Image Processing, John Wiley & Sons, 1978; pp. 201-279.

(20) Rosenfeld, Azriel, Picture Processing by Computer, Academic Press: New York, 1969.

(21) Roselfeld, Azriel, and Kak, Avinash C., Digital Picture Processing, Volume 2, 2nd Ed., Academic Press: New York, 1982.

(22) Fu, King-Sun, "Pattern Recognition for Automatic Visual Inspection," Computer, December, 1982; pp. 34-40.

(23) Pavlidis, Theodosios, "A Review of Algorithms for Shape Analysis," Computer Graphics and Image Processing, Volume 7, Academic Press: New York, 1978; pp. 243-258.

(24) Horaud, R., Olympieff, S., and Charras, J. P., "Shape and Position Recognition of Mechanical Parts from Their Outlines," Proceedings of the First International Conference on Robot Vision and Sensory Controls, April 1-3, 1981, Stratford-upon-Avon, UK; pp. 125-131.

(25) Schachter, Bruce J., "A Matching Algorithm for Robot Vision," Proceedings of the IEEE Computer Science Conference on Computer Vision and Pattern Recognition, 1983; pp. 490-491.

(26) Lucas, Dean, "Moment Techniques in Picture Analysis," Proceedings of the IEEE Computer Science Conference on Computer Vision and Pattern Recognition, 1983; pp. 178-187.

Direct Contact Sensors Based on Carbon Fibre

A. Pruski and B. Mutel

L. A.E.I. University of METZ . FRANCE

SUMMARY

Industrial products made with carbon fibre present interesting properties for the development of sensors. The concept is based on the modification of the electrical resistance of the material under pressure. The main characteristics are low hysteresis a high sensitivity and no noise.

We describe in this paper three applications of carbon fibre felt in the field of robotic.

First, we propose a matricial tactile sensor that allows to measure the force repartition on a plane or a curved surface. It is able to estimate an interaction between a robot and a elementary pattern recognition.

The second sensor, that we describe is a slip sensor. Fixed on a robot gripper, it allows to realise an adaptative grasping.

The third type of sensor derived from the preceding are is able to detect a contact on a large surface. This characteristic has permitted to develop a security sensor to detect an obstacle on the trajectory of a robot in order to stop its motion.

For all the sensors we present the principles, the experimental developments, the associated electronics and applications.

I. INTRODUCTION

The faculty of adaptation of a robot to a duty is linked to sensors which implement it.

The sensors used in robotic are decomposed in two parts:

- Sensors without contact with the environment. They deliver a big flow of informations but their use involve the respect of particular constraints (light, shadow, material...).

- The sensor in contact with the environment.

The are mainly used in operations which require the knowledge of the interactions between robot and handled objects.

The technics used in the design of direct contact sensors are numerous and various (piezoresistivity /4/, piezoelectricity /5/, strain gange, optoelectronic /6/,...). The object of this paper is to present principles, characteristics and applications of three types of direct contact sensors mode by carbon fibres.

II. THE MATERIALS

The carbon fibres are filaments of several micron tickness with resistivity between 500 to 1500 cm.
There are several types of industrial products with carbon fibres :

- independant fibres
- texture fibres
- tufted felt with long fibres (several cm)
- short fibres felt.

The principle of the proposed sensors lays on the electrical resistivity variation of a lot of carbon fibres subjected to pression variation. The studies reported in /1/, /2/ show the specific characteristics of the previous type products. The short fibre felts presents a large sensibility, a no noised output and a neglectable hysteresis. The other products generate important mesure noise.

III. TACTILE SENSOR

Some applications in robotic require the use of devices measuring the pressure distribution which appears during the contact of handled objects. These informations contribute to the recognition of the object it self or to the object position depending of the impress. During assembly operations, the knowledge of interaction variation between robot and handled object allows to control the robot path so the contact forces are minimized. This is the active compliance.

A tactile sensor is, in general, based on a matrix ordering of elementary tactile cells. (tactels)

a) current well sensors

The sensor with current well is constituted with a conductor plate in which plugs are isolated (fig. 1). A carbon felt, with metallized face is glued on the plate by a silver conductor past. A voltage generation supplies the felt on a homogeneous manner from the metallized face. The conductor area with the surrounded current plugs (guard ring) cancels the intercorrelation between the sensors. So we get a good resolution of the spacial informations.

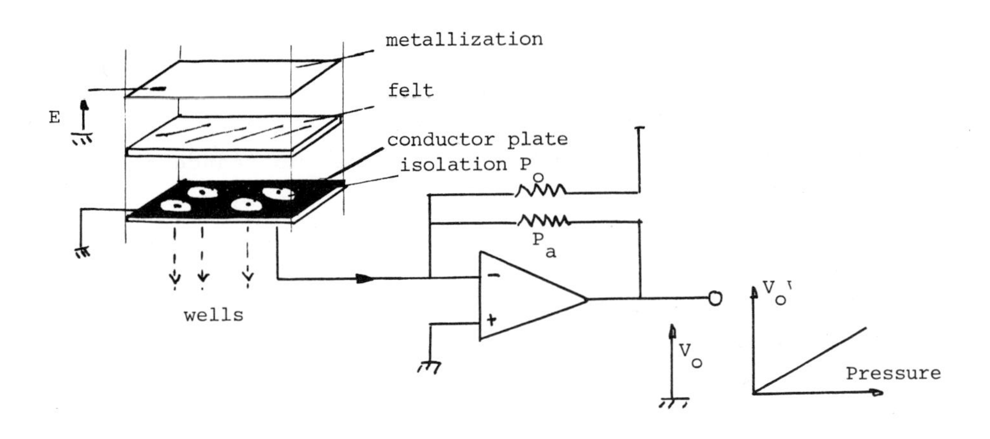

Figure 1. Sensor and its interface

The intercorrelation between sensors is shown on figure 2. For distances less than 5 mm the resolution is not more good. It cames from the isolation ring around the plug and from the mechanical material behaviour. The metallized face makes the Sandwich stiffer which limits, the resolution of the matrix sensor.

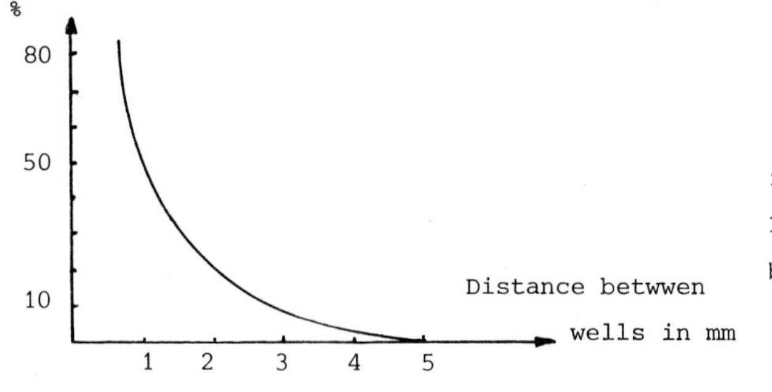

figure 2
Intercorrelation
between wells

The interface between the plug current and the data logger is made by a current voltage converter. The electric signal out put by the plugs presents a linear response in the used pression range (0 to 200 gr/cm). The non-uniform sensitivity of the felt implies the adjustement of the slope and offset sensor by the potentiometers P_o, P_a.

b) <u>sensor with matrix power</u>

The previous described matrix sensor is simple but it requires on important electronic board. A current voltage converter is associated to each well. In order to reduce the interface volume we propose some modification to plug current sensor. First the felt power is realized by parallel conductor strips. The wells along the lines perpendicular to the power strips are connected together. The supply of the conductor strips is made in sequences one strip at a time is supplied while the others are connected to the earth. The result signal on the well line measures the pression applied at the cross between the considered power strip and well line. Such power structure introduce a distribution of line current in the felt which modifies the behaviour of the guard ring. The isolation between wells on the same line is not performed. To guarantee the independance between the wells we realize a band felt sensor parallel to the power strips (fig. 3).

Address
decoder
metallization
felt
V_o

fig. 3 Matrix power sensor

After a sequential multiplexer acquisition the sensor signals are sent to the data logger. In this case, the un-homogeneous behaviour of the felt is software computed.

The used material is tough, soft and can be easily taillored. So this carbon fibre sensors are suitable for plane or curve areas. The matrix sensor has been implemented on a parallel jaws gripper of a robot. So we can set functions as the control of tightness strain, the position recognition of the object in the jaws and the measure of pression variation in the jaws during assembly. This tactile gripper allow us to realize a sofware for the control of a prismatic part assembly where the position parts to be assembled are not well specified.

IV. SLIPPING SENSOR

Following the human model the grasp of an object is related to an autoadaptative grasping and not to a fixed value of the applied effort. In fact is the slipping of the object in the hand which defines a correct grasping.

The principle of the proposed sensor lays on the measure of the noise generated by the tufted felt with long carbon fibres. This noise cames from the fuzzy contact variations of the fibres under the slipping effect between the surface of the felt and the object. The sensor is supplied by an electric current with two electrods glued on one face. The figure 4 shows the processing system of the sensor signal when a slipping appears the processing system delivers a logical signal to control the grasping. Experiences showed that the sensor is working when the applied pressures of a slipping is less than 300 gr/cm. Over this value the sensibility decreases, the fibres are no more free to generate noise.

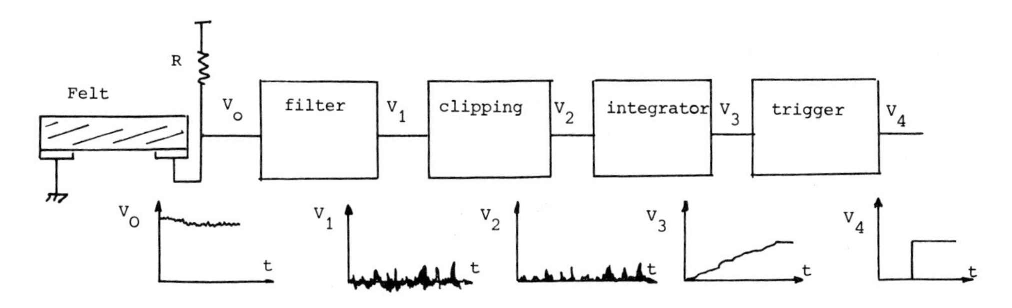

Fig. 4 : Processing system for slipping measurement.

V. SECURITY SENSOR

The introduction of robots in workshops implies a particular warning about the security of the staff and machines. The used security systems concern the prevention of any approach near the working station (shelter, fence) the proximity sensor on the working robot area (optical systems, air stream, transceiver,...) on the detection of contact on the moving arm. Here we propose the use of carbon fibre texture in the design of a direct contact security sensor. The principle is derived from the slipping sensor. A light contact applied on the device generates a noise. The electronic circuits associated to this security sensor are decomposed in two parts :

- Processing the noise signal to detect a contact and set on an interruption
- continuous control of the good working of the sensor and send this information to the robot control board by a binary key word (signature) to avoid falsification (Figure 5).

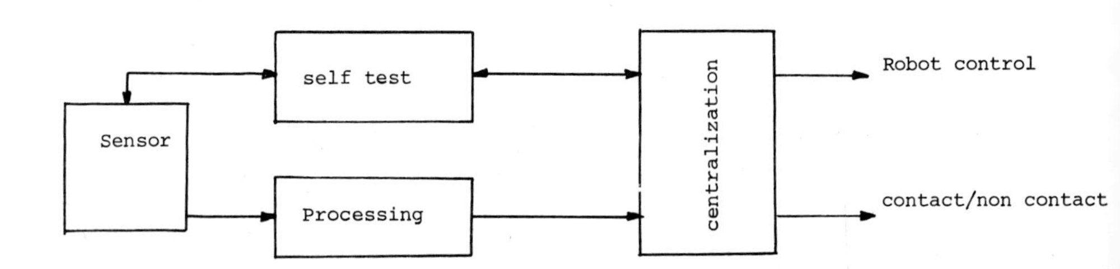

Figure 5. Security sensor

Tests mode on large sensor surface have shown a good sensibility of the device whatever the contact points on the surface. The used material can be easily taillored in order to adapt the sensor to any shopes. For instance, the sensor may cover the robot arm, its working area or some other moving devices.

VI. CONCLUSION

This paper presents the use of textures base on carbon fibres in the design of direct contact sensors. Two materials particularly used : felts with short fibres and felts with long fibres. For the first one we have a linear relation between pressure and resistance variation. The output signal of these sensors is non noised with good sensitivity and neglectable hysteresis.
On the opposite the second one generates big noised signal. This behaviour is used to detect, with great sensitivity, a slipping or a point contact. This study is supported by the A.R.A. project from CNRS. The slipping an security sensors are patented.

BIBLIOGRAPHY

/1/ PRUSKI A., MUTEL B. - Capteur sensoriel pour robot industriel.
Congrès AFCET Productique et Robotique Intelligente, Nov. 83 - Besançon - France

/2/ PRUSKI A. - Etude de capteurs tactiles à fibres de carbone. Application à l'assemblage de pièces mécaniques par robot.
Thèse de 3è cycle - Juin 84 METZ Université.

/3/ LARCOMBE MNE - Carbon fibre tactile sensors.
1st ROVISEC - Conferences April 81.

/4/ PURBRICK J.
A force transducer employing conductive silicone rubber.
1st ROVISEC 1981.

/5/ DARIO P.
Touch sensitive polymere skin uses piezoelectric propertises to recognize orientation of objects.

/6/ REBMAN J.
A tactile sensor with electrooptical transduction.
3st ROVISEC Cambridge 1983

Vision Systems Using Microprocessor Based Microcomputers

Andrew Weilert

Graduate Research Assistant
University of Kansas Center for Research, Inc.
&
Department of Mechanical Engineering
Industrial Innovation Lab
Lawrence, Kansas

Abstract:

This paper discusses the possibilities and advantages of using a microcomputer based vision system for manufacturing and assembly task control. Topics covered include the limitations of current generation microcomputers as applied to vision analysis, conditions under which a microprocessor based system could function successfully and examples of applications meeting the functionality conditions of the small scale vision system. Additionally, a description of the development and application of an I.B.M. Personal Computer based vision system illustrates the major stumbling blocks of microcomputer based vision along with some of its advantages over large systems.

Introduction:

The current drive for flexible, intelligent automation has sparked two parallel trends in the areas of computer vision systems and microprocessor based microcomputers. Since the demand for flexibility requires more intelligence from the machine tools involved, microcomputers and microprocessors have taken a larger role in the area of machine control giving the previously dumb machine tools a new level of sophistication. Likewise, since more extensive automation in turn requires more extensive testing, process monitoring and quality control, vision systems have grown more useful for inspection, sorting and positioning tasks. Unfortunately, as the vision systems have become more useful, they have also become exceedingly expensive thus limiting their use, usually to a single alignment or inspection station. A system for automation using inexpensive, distributed computer vision systems would provide better control and monitoring capabilities and in turn better flexibility and intelligence. The obvious solution to providing this low level vision ability is to merge the microprocessor based microcomputer with a camera system to produce a small scale vision system with the ability to directly control the machine or process what it is monitoring.

The key points to be addressed are the same as those facing any new technology or system. First, is the system technologically feasible? Can a system be developed using the current state of the art? Next, can the system perform useful work? Finally, can the system be justified economically? Will the system pay for itself? The following sections cover these topics point by point and lead into a description of a vision system based on a garden variety personal computer capable of rudimentary vision tasks.

Technological Feasibility:

The question of technological feasibility can be approached from the standpoint of functionality. A system can be deemed feasible if it can perform the tasks required. For a vision system the basic tasks boil down into three main categories: separation, computation and comparison, although all three may not be required for a given application.

Image separation is the process of separating the overall image input from a camera into individual objects of interest. For instance, if an image of several nuts and bolts is presented to the system it should be able to separate and label each individual part with a code number or name. Since this process tends to involve a global scan of the image pixels, the segmentation stage of image analysis is usually the most time consuming. Traditional vision systems have relied on brute computing power to process image arrays stored in memory. Obviously, for a micro based system to achieve any level of performance, a new method must be used which accents the attributes of the microcomputer while deemphasizing its faults. One method for performing image segmentation of binary images on a micro involves a merging of hardware shift register methods used on some large systems with the string handling abilities of the current generation of microprocessors. The details of this method are discussed in the description of the system developed for this project and demonstrate the technical feasibility of using a micro for this stage of image processing with the binary images typical in industrial situations.

The next major task confronting any vision system is calculation. Once the objects in the field of view have been separated physical parameters must be calculated for use in analysis, testing or recognition. The techniques used in this area are generally simple mathematical relationships and summations. Specifically, moment factors can be easily calculated on-the-fly by summation and provide a wealth of information including the area and centroids of each object and the factors required to compute the axis angle of each object. For typical industrial applications using binary images, the calculations involved generally don't require a great deal of precision and can be handled readily by a microcomputer's single precision arithmetic. Additionally, the mathematical coprocessors available for many of today's micros provide an added speed boost for the calculation phase.

The final task required of a vision system is some sort of comparison step which allows for testing against inspection go/nogo limits and recognition of objects using comparison against previously calculated factors. The key problem is that certain parameters vary for different locations and orientations so that care must be taken to choose position invariant parameters. For inspection applications the comparison step is simple, compare the calculated value for some invariant physical parameter for the current object to the preset acceptance range using an if/then test. Recognition testing is slightly more complex. A recognition system easily adapted to a microcomputer is the sieve test for SRI (Stanford Research Institute) algorithm. The process is simple: calculate a number of parameters for each object then successively compare the objects with a series of acceptance ranges. Again, the comparison stage is readily adaptable to micros if invariant parameters are chosen to alleviate any complex normalization calculations.

Overall, as the micro based vision system developed for this project demonstrates, the current technology can be put to use in a small scale vision system if care is taken to avoid the areas where the microprocessor based computers are lacking and emphasize those areas where the micros shine.

Performing Useful Work:

Beyond the simple level of basis system functionality, the task of performing useful work with a microprocessor based vision system stands as the major stumbling block to the real world feasibility of such a system. Of course, useful work is hard to define, but a simple categorization might be that useful work is any task that needs to be performed to produce a desired end result. As such, any number of the jobs performed by current industrial vision systems, inspection, sorting, positioning and control, qualify. As is true for any type of equipment, the usefulness of a micro based vision system depends upon matching system abilities to application demands.

Probable Applications:

At the present time the abilities of most microcomputers are rather severely limited which in turn limits the areas where a small scale system could be applied. Specifically, until advances in microprocessor technology increase the capabilities and speed of the typical microcomputer, the tasks to which a microcomputer based vision system could be applied would have to satisfy the following constraints.

First, unless the camera system used allows for some automatic functions, the application must not be time critical since the current generation of micros generally requires a few seconds to perform a scan of even a simple image. While this time frame may be fine for some low volume applications it is probably unacceptable for producing two or three parts per second.

Additionally, since speed is at a premium on a microcomputer, the application must limit the number and complexity of the calculations required. Generally, areas, centroids and moments of inertia can be quickly and easily computed and used for positioning and simple recognition, but power spectra and Fourier transforms are probably beyond the capabilities of most currently available micros.

Lastly, the limited memory available on most microcomputers precludes any use where the storage of an entire image in a byte array is required. Typically, the memory limits on a micro will necessitate the use of a simple bit mapped binary image.

In spite of these limits, a number of tasks could be performed with the current generation of microprocessor based systems. For instance:

Simple sorting--separating large parts from small, only a minimal amount of analysis is required. Examples would include sorting fragile or odd shaped parts which couldn't survive a sorting screen or some similar method.

Simple inspection--checking for presence or absence, size limits, and similar tasks using simple area summations and calculations. An example would be checking to see if a required number of holes had been properly drilled; if the total area of all holes is below some preset limit, a hole has been missed.

Positioning for hand or slow automatic work--time is not critical so the relatively slow speed of the current microcomputers would not be a problem. Examples would include positioning integrated circuit substrates under test equipment probes, typically performed manually with a video camera and a joystock controlled positioning table.

In general, the current generation of microcomputers should be capable of handling a number of useful tasks if the problems of speed, calculation and memory limits can be overcome.

Economic Justification:

Of course, given that a microcomputer based vision system could be used in a situation, the next question broached is why such a system should be used. In a typical automated industrial situation there are always tasks which must be performed and/or monitored in some way. In general a vision system may perform the monitor and control functions better than a human operator; however, large scale vision systems can't be used to monitor every process in the plant due to their high cost. Thus, a microcomputer based vision system fills a gap; the small system allows a greater number of processes to be carefully monitored since numerous micro based systems could be implemented with the capital required for a single large system. Additionally, the micro based systems could serve as a supplementary vision system to a large system; while the large system performs the time critical, calculation intensive tasks like exhaustive inspection at one station, the micro based systems could be pre-inspecting for gross flaws, thus lessening the load on the main system. Also, since two key applications for vision are robot sensing and inspection in harsh environments, micro based systems provide the added advantage of small size and weight for robot and tight quartered use and easy encapsulation for use in harsh environments. Finally, of course, the main reason for employing a micro based vision system for an application is economic. If only minimal analysis is required and speed is not a factor, why spend the money for an underutilized, full scale vision system when an inexpensive micro based system will suffice.

System Implementation Example:

Perhaps the best way to show the feasibility and future of microprocessor based vision systems is to describe the development and application of one such system.

This project was sponsored by DIT-MCO International, Kansas City, Kansas, and supervised for the Center for Research by Professor B. G. Barr. DIT-MCO manufactures printed circuit board and integrated circuit substrate test equipment. Their effort to automate their product identified the need for a vision system capable of positioning parts without doubling the cost of the testing fixture.

This application fit perfectly with the constraints of a small scale vision system: low cost was vital, no complex calculations were required and due to the nature of the test procedure time was not critical.

The system was based on an I.B.M. Personal Computer and a Micron Technology Microneye camera using a low cost dynamic ram chip as the image sensor. The cost limits were met with the $2200 computer and $300 camera.

For ease of programming, compiled BASIC was used for the menu driven analysis program which used connected component analysis to separate the image and comparisons against area and moment values for recognition. Although the compiled basic program required an average of 36 seconds to scan the 128 by 64 image, recoding in Intel 8088 assembly language should reduce the cycle time to a respectable five seconds.

Reiterating, the key to using a microcomputer in vision is to emphasize the machine's good points while avoiding the flaws. Specifically, in this case the faults are memory and speed limits while the high points were the built in string handling capabilities of the 8088. The image segmentation routine provides the prime example since it employs string manipulations to eliminate the need to store the image as an array variable.

Specifically, the connected component algorithm uses a string variable to store a single line of the image in shift register form. For a given pixel, the pixels to the left, above and above on each diagonal must be checked to see if they belong to a previously labeled object. Using a shift register running from the upper left location across the previous line and down to the last pixel tested reduces the labeling tests to checking the first three and the last locations in the shift register. The string manipulations facilitate this by allowing the old values to be shifted left by one space and the new value entered on the right thus forming a continuously updated list of the previous scan line's object labels without requiring variable array storage of the entire image simultaneously and taking full advantage of the 8088's superior string handling capabilities.

Finally, the system was demonstrated by sorting integrated circuit packages, with the vision software calculating the positional data needed to drive a set of positioning tables (with gripper) directly from the P.C. The system performed flawlessly, consistently sorting large chips from small and giving accurate positions and orientations.

Conclusion:

There is no doubt that as microcomputer technology continues its incredible advance, vision systems based on micros will become commonplace. However, as the project described above points out, with careful programming, resource allocation and application selection there is no need to wait for the new technology since the current generation of microcomputers can already provide a low cost capability for handling a fair range of rudimentary vision tasks.

Design and Implementation of Hand-Based Tactile Sensors for Industrial Robots

Ren-Chyuan Luo

Department of Electrical and Computer Engineering
North Carolina State University
P.O. Box 7911
Raleigh, NC 27695-7911

Abstract

A capacitive slip sensor has been constructed and tested, utilizing an original design, which enables the slip of an object to be measured directly. Incorporated with this is a novel capacitive tactile force sensor. This highly sensitive slip sensor can be used to correct the grasping force adaptively.

The effective use of a robot in performing many manufacturing tasks is hampered if objects become badly deformed, damaged or dropped due to improper grasping force during the work process. Furthermore, dropped objects may injure people while the robot arm moves up and down exerting an accelerating or decelerating force on the object. Therefore, to successfully hold the object with the least necessary force, the slippage of the object must be detected, and the grasping force controlled. The design and implementation described in this paper meets realistic industrial needs.

In addition to improving the capability of robotics systems for application in industry, the investigations provide an advancement of the state of knowledge in the application of the intelligent sensor-based robotic hands in flexible automation.

1. Introduction

Industrial robot systems have been found to be tremendously effective tools for replacing humans in dull repetitive tasks or hazardous environments in manufacturing and material handling industries. To accomplish many automatic manufacturing processes by using industrial robots, requires a considerable improvement in the capability of the robots to perceive and interact with the surrounding environment. In particular, it is desirable to develop sensor-based, computer-controlled interactive systems that can emulate human capabilities.

Based upon sensor signals and stored computer programs, a control computer can automatically perform with minimum assistance by human operators. Sensors for general automation applications have been presented elsewhere [1,2,3].

Without sensory capabilities, the robots will cause some difficulties for grasping both hard, heavy objects and fragile objects, since the robot hand is usually constructed as an open-loop control system. Therefore, the objects sometimes become badly damaged or deformed due to the excessively strong grasping force, or they are dropped due to the excessively weak grasping force.

In view of the recent trend toward the extensive use of industrial robots for handling objects of unknown characteristics, it is very important to determine the optimum grasping force correctly by some adaptive means so that the objects are not damaged or dropped. One of the most effective means to adjust the optimum grasping force correctly is to use a slip sensor [4,5,6].

The slip of an object can be measured in various ways [7]: for example, estimate the slip from the pressure exerted on the finger tip [8]; detect the slip using a piezo-electric crystal which senses object slide relative to it [9]. Other techniques, including the use of semi-conductor, electromagnetic transducers, photoelectric devices and acoustic microphone system were reported [10,11]. All of the slip sensors cited so far are qualitative devices [12]. These slip sensors are used mainly as input signal detectors of a robot sequential controller. They are not suitable for generating the optimum grasping force. Two recent studies described designs of tactile sensing array for recognizing the shape, position and orientation of the object in the hand [13,14]. These methods might be used to sense the slip signal, however, it may not succeed due to the poor sensitivity in tangential slippage direction.

An original design of a capacitive slip sensor based on phase detection principle, which enables the slip of an object to be measured directly has been built and successfully tested. A novel capacitive tactile force sensor which is used to incorporate with slip sensor for compensating the grasping force is also described in detail.

2. Capacitive Slip Sensor

2.1 Working Principle of Slip Sensor

The slippage of an object in a robot finger has been defined as a relative motion of the object with respect to the finger in the direction perpendicular to the grasping finger force [6]. The proposed slip sensor detects the change of phase angle while the change of the capacitance is sensed through the rotating motion of rollers contacting the object. The amount of detected phase shift is a measure quantity of the slip displacement.

2.2 Mechanical Design and Characteristics

The slip sensor was designed with five major parts: rollers, a roller recess, an inner casing, an outer casing and linear bearings, as shown in a cutaway view in Figure 1.
The rollers are composed of two materials, conductive and nonconductive, which were machined as half cylinders and fastened together as full cylindrical rollers. These rollers were mounted in the nonconductive inner casing with small bearings which enable them to rotate. The roller recess is an aluminum block with semicircular grooves cut from it.
This block was placed very precisely beneath the rollers with clearance less than 0.01 inches. In combination, both rollers and roller recess constitute a variable capacitor. The less the clearance between the rollers and roller recess, the greater the total maximum capacitance.

The rollers were mounted on the inner casing, protruding 0.04 inches above the edges of the inner casing. The inner casing fits into the outer casing and is separated from it by linear bearings on its side and by a reset spring on its bottom. When an object is grasped by the robot gripper, the rollers to touch the object first while the inner casing sinks into the outer casing against the reset spring, until the edges of the outer casing grab the object. The rollers then return to the

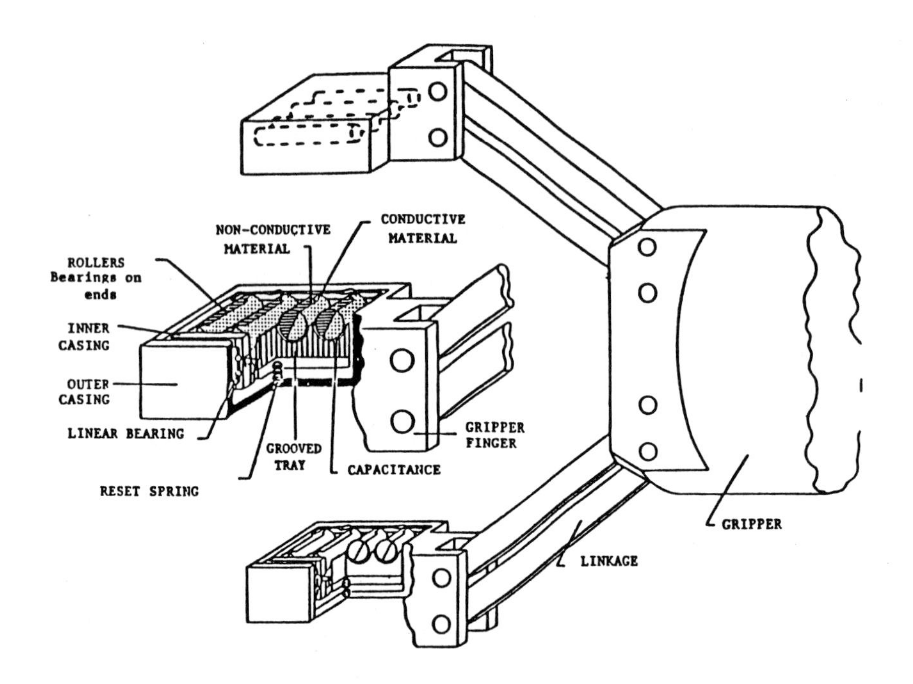

Fig. 1 Sketch of slip sensor for robot gripper.

finger surface pushed by the reset spring. As a result, the object is supported not only by the rollers but also by the whole finger surface.

2.3 Principle of Phase Shift Detection

For detecting the phase shift by the change of capacitance resulting from a rotation of the roller in contact with the object, a phase shift detection circuit was designed as shown in Figure 2.

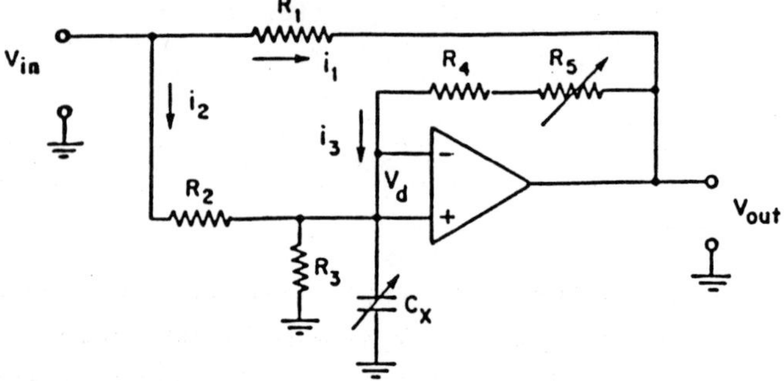

Fig. 2 Phase shift detection circuit.

By giving an a.c. voltage through an ideal capacitance circuit, the phase difference between voltage and current is $-\pi/2$, i.e. the phase of the current will lead the voltage by 90°. The transfer function can be derived as follows:

$$i_1 = \frac{V_{in} - V_{out}}{R_1} \qquad (1)$$

$$i_2 = \frac{V_{in}}{R_2 + R_3} = \frac{V_d}{R_3} \qquad (2)$$

$$i_3 = \frac{V_{out}}{R_4 + R_5 + 1/_{j\omega C_x}} = j\omega C_x V_d \qquad (3)$$

$$i_1 = i_3 \qquad (4)$$

From eqs. (2) and (3) we get the transfer function:

$$M(j\omega) = \frac{V_{out}}{V_{in}} = = \frac{1 + j\omega C_x (R_4 + R_5)}{1 + R_2 / R_3} \qquad (5)$$

The phase shift angle

$$\phi = \tan^{-1} \omega C_x (R_4 + R_5) \qquad (6)$$

By giving constant resistance (R_4, R_5) and constant input frequency, the change of the capacitance C_x is a linear measure quantity of phase shift angle as well as the displacement of the object slippage. To detect the phase shift angle, a proper design of signal conditioning circuit will be needed. A control circuit is also required, which would manipulate the operation of the start-stop oscillator and enable the counting of the phase shift time period. The number of the pulses which are counted during the start-stop period is proportional to the phase shift. This phase shift can be converted to the d.c. voltage for further processing.

Fig. 3 illustrates an experimentasl test result for the capacitive slip sensor. Between the detected phase angle and the count output demonstrates a good linear relationship.

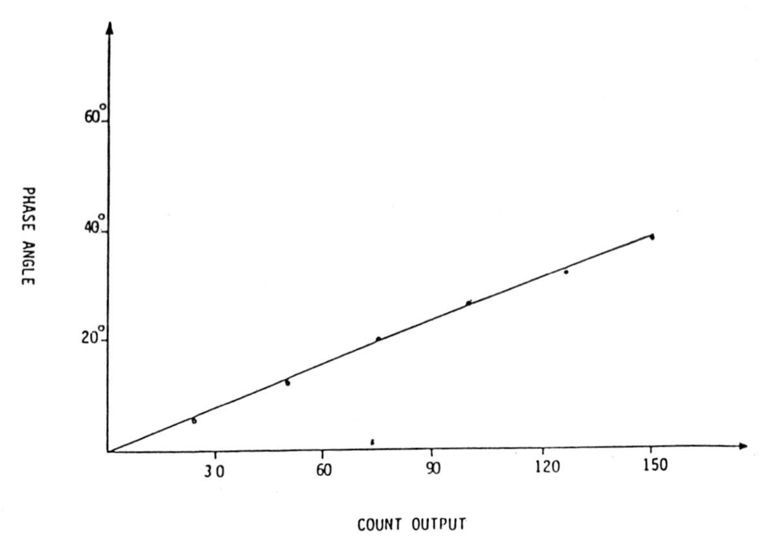

Fig. 3. Phase angle vs. count output for the slip sensor.

3. Capacitive Tactile Force Sensor

The capacitive transduction sensing method can sense the force with great sensitivity and accuracy. Capacitive transduction elements convert a change in displacement/force into a change of capacitance. This changeable capacitance will be mounted on the fingertips, connected into a Wheatstone-bridge circuit, across which a voltage is applied, as shown in Fig. 4.

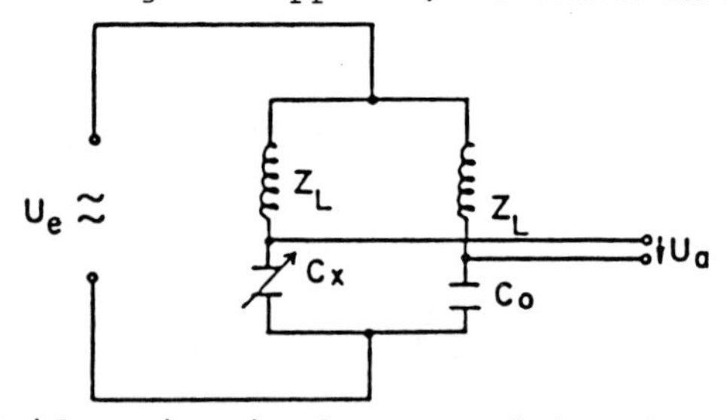

Fig. 4. Bridge circuit for capacitive force sensor.

The relationship between output voltage Ua and input voltage Ue which corresponds to Fig. 4 can be derived as follows:

$$\frac{U_a}{U_e} = \frac{Z_{C_0}}{Z_{C_0} + Z_L} - \frac{Z_{C_X}}{Z_{C_X} + Z_L} \qquad (7)$$

$$Z_{C_0} = \frac{1}{j\omega C_0} = -j \frac{d_0}{\omega \varepsilon_x \varepsilon_0 A_e} = -jkd'_0 \tag{8}$$

Where

ε_r = Dielectrical constant

ε_0 = Electrical field constant

A_e = Plate area of the capacitance

$$K = \frac{1}{\omega \varepsilon_r \varepsilon_0 A_e}$$

$$Z_{C_X} = jK\,(d_0 + \Delta d)$$

$$Z_L = j\omega L \tag{9}$$

$$\frac{U_a}{U_e} = \frac{-jkd_0}{jkd_0 + j\omega L} - \frac{-jk(d_0 + \Delta d)}{-jk(d_0 + \Delta d) + j\omega L} \tag{10}$$

After mathematical operations, we obtain

$$\frac{U_a}{U_e} = \frac{K\omega L \Delta d/d_0}{-K^2 d_0 + \omega^2 L^2 - 2K\omega L + (K^2 d_0 - K\omega L)\,\frac{\Delta d}{d_0}} \tag{11}$$

The linearity and sensitivity are obviously then maximal if the term $\Delta d/d_0$ in denominator is zero, i.e., $Kd_0 = \omega L$ we obtain

$$\frac{U_a}{U_e} = K_1 \frac{\Delta d}{d_0} \tag{12}$$

Where K_1 is a constant.

When the robot gripper retrieves the objects and applies the force on the fingertip, it will cause the change of the capacitance in one side of the bridge branch shown in Fig. 4.

As a result, the relationship between capacitance and plate distance d_0 will be defined as follows:

$$C_0 = \varepsilon_0 \varepsilon_r A_e / d_0 \tag{13}$$

Applying for a force,

$$C_0 + \Delta C = \varepsilon_0 \varepsilon_r A_e / d_0 - \Delta d \tag{14}$$

Dividing (15) by (16), we obtain:

$$\frac{C_0}{C_0 + \Delta C} = \frac{1/d_0}{1/d_0 - \Delta d} ; \frac{1}{1 + \frac{\Delta C}{C_0}} = \frac{d_0 - \Delta d}{d_0}$$

then we get

$$1 - \frac{\Delta C}{C_0} = 1 - \frac{\Delta d}{d_0} ; \frac{\Delta C}{C_0} = \frac{\Delta d}{d_0} \tag{15}$$

By combining eq. 12 and eq. 15, we obtain a linear relationship between output voltage U_a and the change of the capacitance ΔC, as well as the change of the plate distance Δd, i.e., the force $F \sim \Delta d \sim \Delta C \sim Ua$.

The electronic measurement circuit should transmit the movement of the small distance Δd of the capacitive sensor into d.c. analog voltage. Figure 5 shows the schematic diagram of the signal processing.

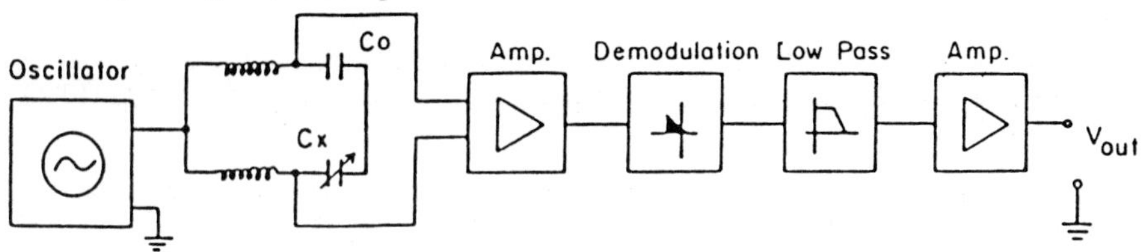

Fig. 5. Schematic block diagram of capacitive force sensor signal processing.

Once the forces are applied to the fingertips, one side of capacitance will be changed. It will cause a phase shifting diagonal voltage and input voltage. The diagonal voltage will then be amplified, demodulated, and, through low-pass generate an analog d.c. voltage. This output d.c. signal is proportional to the force which is applied to the fingertips.

4. Grasping Force Compensation Using Combined Slip and Force Sensor

In many industrial robot appplications, for example, in the area of forging, assembly, material handling, it is always necessary to have grasping force compensation during the operation. This situation will occur when the finger is about to lift the object from the bin, the object is going to transfer from hand to hand or the arm moves up and down exerting an accelerating or decelerating force on the object [6]. The object will begin to slip when the grasping force becomes smaller than the minimum grasping force. The slip can be stopped by increasing the grasping force of the finger. It is helpful to indicate the grasping force exerted on the object.

The slippage of the object and the grasping force exerted on the object require a pertinent compromise, which can be reached by employing optimal control theory, to succeed in holding the object with the least necessary force. In combination of the capacitive slip and tactile force sensor described can be implemented to fulfill such industrial needs.

References

1. A. K. Bejczy, "Effect of Hand-based Sensors on Manipulator Control Performance," Mechanism and Machine Theory, Vol. 12, 1977.

2. A. K. Bejczy, "Sensors, Controls, and Man-machine Interface for Advanced Teleoperation," Science, No. 208, June 1980.

3. G. G. Dodd and L. Rossol, Computer Vision and Sensor-based Robots, Plenum Press, 1979.

4. M. Ueds, T. Shimizu, "Sensors and Systems Necessary for Industrial Robots in the Future," Memoris of the Faculty of Engineering, Nagoya Univ., Vol. 27, No. 2, 1975.

5. R. Masuda, K. Hasegawa, "Control Method of Industrial Robot Hand by Force Signal Feedback," Trans. SICE (Japan) Vol. 10, No. 4, 1974.

6. R. Masuda, K. Hasegawa, "Slip Sensor of Industrial Robot and Its Application," Electrical Engineering in Japan, Vol. 96, No. 5, 1976.

7. M. Ueda, K. Iwata, "Tactile Sensors for an Industrial Robot to Detect a Slip," Proceedings of the Second International Symposium on Industrial Robots, May, 1972.

8. M. Ueda, K. Iwata, "Adaptive Grasping Operation of an Industrial Robot," Proceedings of the Third International Symposium on Industrial Robots, May, 1973.

9. T. J. Armstrong, "Manual Performance and Industrial Safety," report prepared for Dir. Safety Research of Nat. Industry for Occup. Safety and Health, Morgantown, WV, Task Order No. 78-10433, 1978.

10. R. Tomovic, Z. Stojiljkovic, "Multifunctional Terminal Device with Adaptive Grasping Force," Automatica, Vol. II, 1975.

11. R. Dillmann, "A Sensor Controlled Gripper with Tactile and Non-Tactile Sensor Environment," Proceedings of the 12th International Symposium on Industrial Robots, June 1982, France.

12. L. Harmon, "Tactile Sensing for Robots," Proceedings of the NATO Advanced Study Institute on Robotics and Artificial Intelligence, Lucca, Italy, 1983.

13. M. Raibert, J. Tanner, "Design and Implementation of a VLSI Tactile Sensing Computer," Journal of Robotics Research, Vol. 1, No. 3, 1982.

14. J. Rebman, "Tactile Sensor with Multimode Capabilities," 3rd International Conference on Robot Vision and Sensory Controls, Nov. 1983.

Robot Cell and Mobile Robot

Chairman: James B. Canner, Sterling Detroit Company, Detroit, Michigan
Vice Chairman: Jim Hansen, Borg Warner Research, Des Plaines, Illinois

Industrial Robots and Flexible Manufacturing

W. R. Tanner

Productivity Systems, Inc.
1210 E. Maple Road
Troy, MI 48083

Summary

Industrial robots and flexible manufacturing systems (FMS) share a common characteristic: flexibility. Robots can be an element of flexible machining systems, performing a number of workpiece, fixture and pallet handling functions. In addition, robots can perform inspection, cleaning, tool changing and deburring operations within the flexible machining system. Robots can also be used in flexible assembly systems, performing insertion, inspection, dispensing, fastening and machining tasks.

The Robotic Industries Association (RIA) defines an industrial robot as, "a reprogrammable multifunctional manipulator designed to move material, parts, tools, or specialized devices through variable programmed motions for the performance of a variety of tasks." Modern Machine Shop, in their "1984 NC/CAM Guidebook" issue, describes a flexible manufacturing system (FMS) as, "a computer-integrated group or cluster of multiple NC machines or work stations linked together by work-transfer devices, for the complete automatic processing of differing product parts or the assembly of parts into differing units." There is a characteristic common to both robots and FMS and that is flexibility. The two technologies are not, however, interchangeable; robots can be an element of FMS but are not, of themselves, flexible manufacturing systems.

Typically, in an FMS, workpieces are placed into palletized fixtures, one per fixture, at a load station. The pallets are then transported from machine to machine for processing of the part, with the pallet automatically shuttled into, registered and clamped, and shuttled out of each machine. FMS processing of smaller parts may be accomplished either by locating and clamping a number of parts in a single palletized fixture for transport through the system or by transporting loose parts through the system with locating and clamping performed at each machine. There are a number of workpiece, fixture and pallet handling functions in an FMS which could be accomplished with industrial robots.

Fixture Loading and Unloading

In the staging area, where parts, fixtures, pallets and transporters come together, a robot could be used to load and unload fixtures. This function would be limited to parts weighing no more than 75 to 100 kg, because workpiece positioning into fixtures would probably require at least two wrist axes on the robot, and this generally limits payload capacity. In a loop FMS configuration, a single robot could be used both to load and to unload parts. If inspection of finished parts was performed in the staging area, the robot might also load and unload a coordinate measuring machine (CMM) or other inspection/measuring device. A gantry robot configuration or a cylindrical coordinate or anthropomorphic robot on a one-axis transport table might be required for access to load, unload and inspection stations.

Pallet Loading and Unloading

Also in the staging area, a robot might be used to load fixtures to and unload fixtures from the pallets which are registered and clamped on the machines. Payload limitations would, again, have to be considered; however, fixture handling might not require any wrist motions and there are several robots capable of handling loads in excess of 500 kg, without a multiple-axis wrist. As with fixture loading and unloading, a loop FMS would permit a single robot to load and unload pallets. Considering load capacity requirements, a gantry or cylindrical robot should be used.

Pallet or Workpiece Transport

Although they do not resemble the typical industrial robot arm, self-powered computer-controlled vehicles (automatic guided vehicles, or AGV's) can also be considered as robots. An AGV might incorporate elevator and shuttle mechanisms which would enable it to be automatically loaded with a pallet in the staging area and to not only transport the pallet from machine to machine in the FMS, but also to transfer the pallet to and from each machine. Where loose parts are moved through the FMS, the AGV might transport quantities of parts arrayed in racks or pallets to and from a staging area and from machine to machine and might also transfer these racks or pallets to and from load stations at each machine. The AGV system could also be employed to transport cutting tools to and from machines. Unlike tow carts, conveyors and pallet flotation systems, AGV's provide the flexibility for queuing and for alternative processing of different parts or for bypassing machines or stations which are out of service.

Inspection

Handling operations, other than fixture loading and unloading, might also be done by robots. As previously mentioned, a robot in the staging area could load and unload a CMM or other inspection/measuring device. At work stations within the FMS, robots could also load smaller workpieces into automatic gages for in-process inspection. As an alternative, with fixtured/pallet-mounted parts, the robot might function as the inspection device. Although robots lack the accuracy and repeatability for direct, precise measurement, they could be used to handle probes and similar self-contained measuring devices. Robots could also perform visual inspections by handling solid-state imaging devices such as digital cameras or fiber optics detectors. One potential advantage is that the robot could manipulate such a device into almost inaccessible locations for close-range visual inspection. The robot's versatility would enable it to handle more than one inspection device and its repeatability would assure that all desired inspection operations were performed as programmed.

Part, Fixture and Pallet Cleaning

Another robot function in the FMS might be part, fixture and pallet cleaning. With smaller parts, the robot might simply transfer workpieces to an automatic washer, perhaps as a secondary step to fixture unloading. For larger parts and for fixtures and pallets, the robot might perform the cleaning functions by manipulating spray devices mounted on the end of its arm. This approach would effectively utilize the positioning and orienting capabilities, the continuous-path motions and the programmability of the robot. It would be fast, energy-efficient and make economical use of cleaning fluids. It would be effective because the robot could be programmed to direct fluids into holes and passages and to flush chips away from blind areas and corners. It also would remove people from contact with potentially irritating or toxic materials and from an unpleasant environment.

Tool Changing

Robots might be used to change cutting tools on NC machines. A robot could be dedicated to this task at a single machine, where it might also load and unload parts as well as change tools, or a traveling robot might change tools on several machines. For example, in an FMS at General Electric's Erie, Pennsylvania, locomotive works, a GE robot on a traverse base changes cutting tools on two special vertical milling machines. As mentioned previously, an AGV could be used to transport racks of cutting tools between robot tool changers and the tool room, thereby automating completely the tool handling functions.

Machine Load and Unload

Where parts are not fixtured in a staging area for transport through the FMS, that is, where "loose" parts are transported, robots could be used to load and unload the machine tools or work stations. Several approaches might be taken: an array of parts, in some type of tray or dunnage might be transported by AGV to a machine for robotic loading/unloading; parts might be presented, one at a time by means of a conveyor, to a machine for robotic loading/unloading; or, an array of parts (perhaps stacked in trays several layers deep) might be transported by AGV, conveyor or two cart to a marshalling area, for robotic transfer to individual machines, machine loading/unloading and transfer to another marshalling area and repalletizing. The type of robots used and the "packaging" of the parts would be different in each of these situations.

For parts delivered in quantity to an individual machine or to a group of machines served by a single robot, the parts should be separated and roughly positioned and oriented in racks, trays or dunnage. The robot would probably be a large, servo-controlled unit, floor-mounted in a fixed location. It could load and unload more than one machine and could process parts sequentially through all machines in the cell or through individual machines, perhaps in random order. A machine vision system might be incorporated to facilitate part acquisition and in-process gaging of parts might also be included.

For parts delivered one at a time to an individual machine, a small non-servo or servo-controlled robot could be used. This robot could be floor-mounted or it could be mounted on the frame or bed of the machine which it serviced. The robot's function would usually be limited to loading and unloading the machine tool and, if it were free-standing rather than machine-mounted, perhaps to loading and unloading an automatic gage, as well. The same conveyor which delivered the parts to a machine would generally be used to transport the parts from the machine (probably to the next machine in the system). Initial placement of parts on the conveyor might be done manually, or a bulk container unloader could be combined with automatic, mechanical part orienters for a conveyor loading system.

Where parts were delivered, in quantity, to a marshalling area for processing through a series of machines, large gantry robots, single-rail overhead units or cylindrical or anthropomorphic arms on traversing bases would be suitable. More than one robot arm might be used, with the first robot depalletizing parts and loading the first machine, the second robot unloading, transferring and loading the next machines in the sequence, and perhaps a third robot unloading the last machine, gaging and repalletizing the parts. In this example, the second robot would probably be fitted with dual end-effectors, for quick interchange of parts at each machine.

Deburring

Another potential robot application in an FMS would be for

deburring of machined parts. Similar operations, such as chamfering or polishing might also be considered. The robot's ability to perform a number of different tasks, in random order, coupled with a relatively large work envelope and up to 6 degrees of freedom, plus a capability to exert uniform, programmed, controlled force, make it especially well suited to deburring and similar operations. The incorporation of machine vision into the robotic deburring cell would provide a capability for inspection and adaptive control. Like the robot part/fixture/pallet washer, robotic deburring would become a discrete operational segment of an FMS, rather than an enhancement of another FMS element.

Flexible Assembly Systems

Thus far, this discussion of robotics and FMS has been centered on flexible machining systems. With wider use of robots, FMS could go well beyond machining, particularly into flexible assembly. Japan's use of robots in assembly operations has already reached 19%, according to the "1982 RIA World Wide Robotics Survey and Directory", compared to 1% utilization in assembly in the United States. In addition to the usual parts handling functions, robots in flexible assembly systems could accomplish a number of tasks.

Fitted with passive compliance end-effectors (remote-center compliance devices, or RCC's) or with wrist force sensors, robots could perform close-tolerance insertion operations in mechanical assembly. Combined with high-resolution machine vision systems, robots could accomplish precise placement of surface-mounted components in electronic assembly. Robots could dispense adhesives, sealants and lubricants in complex patterns and controlled quantities. Robots could drive screws, rivets and other mechanical fasteners. They could join metal parts by resistance welding or ultrasonically weld plastic components. They could perform precise, controlled soldering of critical components or joints in electronic devices. They could perform simple machining operations, such as drilling, tapping, spot facing or chamfering.

As in machining systems, robots, in flexible assembly systems, could also accomplish a number of inspection tasks. Using vision, they could inspect for missing components, verify that correct components were installed, and assure the correct orientation of components. They could also perform critical measurements, with resolution of better than 0.05mm. The robots could check torque on rotating devices and could perform some functional tests during assembly operations. Finally, robots could be used in the packaging of the finished products at the end of the flexible assembly system.

Potential

The long-term evolution of FMS is toward the automatic factory, but in the meantime, the need is to improve the efficiency of batch manufacturing, which represents between 60% and 80% of metal cutting and related processing in the U.S. Today, metal cutting occurs during only 6% to 8% of available machine time in batch processing, compared to 45% to 55% in dedicated transfer lines. The potential of an FMS, especially one making maximum use of robotics, is to drive batch manufacturing efficiencies to nearly the level of transfer line automation.

Use of Simulation in Planning of Robot Work Cell

Jim Hansen

Research Engineer

Borg-Warner Research Center
Des Plaines, Illinois

Abstract

This paper addresses some of the steps involved in planning and layout of a manufacturing work cell involving robots. The use of computer simulation systems for robot work cell planning is discussed along with their benefits. Some general capabilities of the current simulation systems are brought out.

Introduction

Planning and layout out of a manufacturing work cell involving robots is an involved process. The necessary range of motions, lifting capacity, operating speed, controls, and interfaces needed for the robot must be considered. Additional equipment such as end effectors, fixtures, or material handling setups must be designed or selected from outside vendors. Before purchasing the robot(s) and other equipment, you want to be as confident as possible that the cell will work as expected. A careful planning and layout process that addresses such potential problems as inadequate cycle times, collisions or interferences, and exceeding robot motions/strength limits/or accuracy will help increase the odds for success.

The first part of this paper will deal with the steps involved in planning, initial layout, and mockup of the work cell. The second part of this paper will deal with using simulation of the robot work cell to aid in the planning and layout process. A number of currently available systems for robot work cell simulation and generic capabilities of the systems will be discussed.

Planning Layout, and Mock Up of Cell

You have selected an operation in your factory to be adapted for using a robot or robot(s). The reasons could be that you are looking for higher productivity, better quality, reduced exposure of workers to hazardous conditions, or others. A first step should be to detail out a sequence of operations that should be going on with the robot(s) and associated machinery. An example is;

Step 1 - Grab part A at conveyor station 1

Step 2 - Lift and transport to test station

Step 3 - Put part A into test station and release

Step 4 - Retract arm and wait for test to finish

Each step may be associated with some specification or design constraint for the robot(s) and other machinery. Step 1 involves the design of the end effector in grabbing part A. Step 2 involves the weight capacity and range of motion of the robot. Step 3 may involve some repeatability constraint on the robot in putting the part in the test station. Step 4 implies a signal to pass between the test station and robot for the end of the test. Time constraints on certain steps would involve the speed capacity of the robot.

The necessary specifications for the robot(s) can now be written out from consideration of the sequence of operations. If the specifications are too restrictive or require a more sophisticated and expensive robot than you had anticipated, the operations have to be examined to see what changes will affect the necessary robot(s) requirements. Design changes in parts or auxiliary equipment may help in affecting the robot specifications also. A specialized device to present a part to the robot may eliminate a motion so that you only need 5 axes of motion instead of 6 axes. Chamfers on corresponding pins and holes with selective compliance built into the end effector may drastically affect the accuracy/repeatability requirement for the robot.

After the robot(s) has been selected and the necessary auxiliary equipment has been decided upon, detailed layouts can be done of the work cell. Top and side views (Fig. 1) showing the robot(s), the robot(s) workspace, and other machinery can be done in 2-D or a 3-D CAD system could

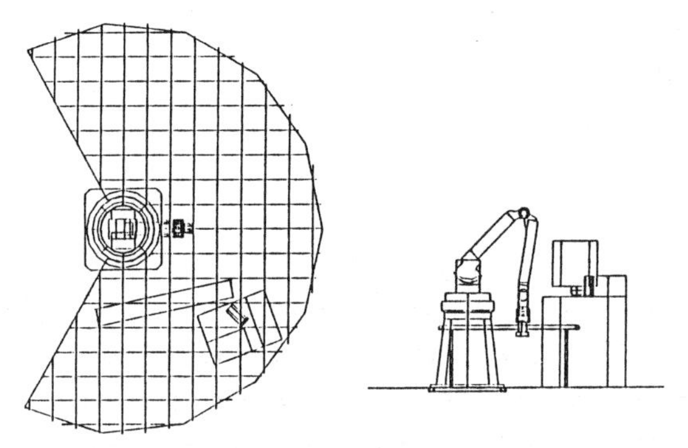

Fig. 1. Top and Side Views of Work Cell

also be used. The layouts allow you to visualize better what is going on in the work cell. Problems of service accessibility, interferences, and operator safety can be seen with the layouts. Better estimates of cycle time for the operation can be done using the distances measured off the detailed layout and the robot speeds you have settled on. At this point you can make your financial calculations and see if your original reasons for using robots are still being met.

A final step before buying and installing the robot(s) and other machinery in your factory should be a hardware mock up of the cell. This may involve setting up the complete working cell in a lab or having the vendor demonstrate to you the critical working aspects of your cell using the robot you plan to purchase. This hardware mock up can allow you to train your people on the new equipment, find various electrical or interface problems, check out your calculations of cycle time, and find any other problems without the pressure of production.

This process of planning, layout, and mock up has been extremely helpful in putting together successful robot installations in the factory. Previous efforts involved significant amounts of time in manual layout of the cell and calculations of cycle time. It took less time and effort to just mock up the cell and correct problems there. With more involved robot work cells being planned and with the wide range of robots available, the hardware mock up becomes a very expensive way to find your problems.

Simulation of Robot Work Cells

A number of companies are now offering computer hardware and software for simulating robot work cells 1 . An animated display of the work cell is shown on a graphics terminal in a wireframe or solid image. Some currently available systems are PLACE from McDonnell Douglas Automation Co., Robot-Sim from GE-Calma, and Robographix from Computervision. The auxiliary machines, end effectors, and robot(s) are set up on a 3-D CAD system. Enough detail is needed to show a realistic picture and to try to predict collision problems. The robot work cell simulation software could run on the same 3-D CAD system hardware or on another graphics terminal. Figure 2 shows an early version of PLACE running on an Evans & Sutherland MPS terminal. Machine and end effector designs were done on a Unigraphics CAD system. A number of different robot models are stored in a library with the system.

by the integration of many computer controlled machines under the control of a central computer. Most manufacturers agree that the integrated systems do not have many of the characteristics listed above, and that results in serious maintenance problems for local technicians when they take over the maintenance responsibilities. The Automated Manufacturing technology program at Piedmont Technical College is designed to train technicians for this type of system maintenance.

SYSTEM TRAINING CONCEPT

The curriculum will train technicians to install, maintain and service the high technology systems which are currently being used in new automated manufacturing facilities and to prepare technicians for the systems which are planned for the future. The graduates of the program will be prepared to install, operate, analyze, troubleshoot and repair the following automated systems:

- Advanced manufacturing systems used in the shaping, forming and processing of raw materials into finished parts and products.

- Advanced assembly systems using automation, vision and robots to assemble parts into finished products.

- Automatic material handling systmes which include conveyors, manless parts vehicles, robots and transporters.

- Automated warehousing and parts storage systems.

- Data communication networks of computers which include manhine controllers, robot controllers, work cell computers, inventory control computers.

LABORATORY DESIGN GUIDELINES

The Robotics Laboratory must provide hands-on training for participants and must possess the following capabilities:

- Permit training on large and small robot systems, on electric, hydraulic and pneumatic drive systems, on servo and non-servo models, on stop-to-stop, point-to-point and continuous path machines, on the three major arm geometries and on low and high technology systems.

- Permit training in basic hydraulics and pneumatics.

- Include some work cells shich are designed to provide robotics training at as low a cost per station as possible.

- Permit training which imitates actual industrial work cells.

- Permit training in the application of programmable controllers in automated work cells.

- Permit training in the application of work cell sensors using limit switches, photoelectric sensors, proximity sensors, pressure pads and vision.

- Permit training in interfacing the robot with sensors, vision systems and other computer-controlled devices.

- Include equipment for robot applications in welding, machine tending, palletizing, inspection, assembly and coating.

- Permit training in system and robot troubleshooting.

- Permit laboratory class sizes of twelve students.

- Permit training in the programming of industral robots and programmable controllers.

- Demonstrate the use of safety devices to protect both the robot and the operator.

WORK CELLS IMPLEMENTED

The robotics laboratory has 14 work cells for training in automated manufacturing concepts. The cells offer training in low and high technology robot systems with a variety of power sources and types of path control. The work cells currently implemented in the laboratory include four classified as high technology, four classified as low technology and five designed for basic concept training. The basic work cell and the simulated production processes in each are described below:

HIGH TECHNOLOGY GROUP

Large work cell with a Cincinnati Milacron T3 hydraulic robot with a Modicon programmable controller, conveyor sustem, welding system, machine mock-up, parts feeder and contact and non-contact sensors.

Medium work cell with a Cincinnati Milacron T3 - 726 small electric robot with a T1 programmable controller, parts assembly mock-up and system sensors for interfacing.

Small assembly work cell with an IBM 7535 electric robot, parts feeder, assembly fuxtyre and system sensors.

A similar assembly work cell with a Seiko TR-3000 electric robot and IBM PC computer.

LOW TECHNOLOGY GROUP

Pneumatic work cell with a Seiko 700 pick-and-place robot, Allenair rotary index table, parts feeder, conveyor, T1 programmable controller, system sensors and system control panel. Machine loading and unloading is simulated.

Two work cells, each equipped with a pneumatic modular robot from Mack, Inc., T1 programmable controller, parts feeder, system sensors and system control panel. Material handling is simulated at these work cells.

A work cell with a Platt Saco Lowell RC-4 pneumatic robot.

TRAINING GROUP

Two work cells each equipped with a Sandhu Machine Rhino electric robot, conveyor system, parts feeder, amchine mock-up and Apple computer system. Training in basic system concepts is emphasized at these work cells.

Two more work cells, similar but not identical to the two preceding stations, contain an additional linear base. Also, a rotary table replaces the conveyor.

A work cell with a Microbot Minimover 5 robot and automated factory simulation hardware.

SUPPORT HARDWARE

A Machine Intelligence Corporation VS-110 vision system for use in any of the work cells where vision would enhance the manufacturing operation being simulated.

A Digital Equipment Corporation PDP 11/23 minicomputer networked to a VAX 750 for use with any of the work cells where host computer control is necessary.

LABORATORY TRAINING LEVELS

Each of the work cells will support many levels of training from basic machine operation to troubleshooting a complete automated work cell. Projects developed for each cell fall into one of six progressive training levels. Completion of each lower level is a prerequisite for the next higher level. A general description of each level follows:

LEVEL ONE - In the first level projects and activities are designed to train participants in the operation of the basic equipment which is included in the work cell. For example, learning the operation of the robot would be an objective for this level, along with programming both the robot and any programmable controller which is present in the work cell. In addition, the operation of all support and peripheral equipment in the cell would be covered in a level one project.

LEVEL TWO - Level two projects are designed to train participants in the integration of the robot and the basic work cell support equipment necessary to perform a manufacturing function. A typical project would include the integration of the robot with a gripper and parts feeder to permit the development of a palletizing program. In a low technology cell the project would include assembling a modular system, connecting the pneumatic lines, wiring the valves to the controlling device and writing the necessary program.

LEVEL THREE- Level three projects are built directly on the level two projects which preceded them. The work cell must be upgraded to include the sensors necessary for continuous untended operation. These devices may be contact sensors for some projects, proximity and photoelectric sensors for others and vision for still others. Also included at this level is the interfacing of the sensors to the work cell control system and the integration of the sensor information into the program driving the cell.

LEVEL FOUR - This level builds on previous project work in level three. The operating work cell with sensors interfaced to the controller does not have any control panel to display work cell conditions for an operator to use. Rather, this level will require the interfacing of controller conditions to a control display panel to indicate work cell conditions. For example, a robot may be halted because the part supply is low in the parts feeder.

LEVEL FIVE - Projects are included in level five which require interfacing of robot controllers to other control devices, such as programmable controllers and hose computers.

LEVEL SIX - The last level includes projects which require troubleshooting of the work cell to identify the location of malfunctions in the total system.

EDUCATIONAL ROBOT SYSTEMS

All of the high technology and low technology work cells present in our automation laboratory are capable of supporting the six levels of training identified above. This is a result of the language present in the robot controller and the hardware assembled in the work cell. Educational robots will not support a system implementation which will allow all six levels to be tested. This results from several educational robot system shortcomings which are:

No input and output ports for receiving sensor inputs and for driving external loads. When inputs and outputs are present they are not integrated into the robot programming language and are not opto-isolated.

No high level language which permits conditional branching based on the condition of input ports and which supports teaching of robot moves with a portable teach pendant.

In addition, the operating systems of educational robots do not resemble those of the industrial robors currently in use.

PIEDMONT INDUSTRIAL ROBOT SIMULATOR

The system described in Figure 1 will permit a Rhino educational robot to implement the six levels of training required for the work cells in Piedmont's Automation Laboratory. Also, the system will permit the educational robots to be used for introductory training on the Cincinnati Milacron T3 robot systems. The industrial robot simulator includes the following hardware:

Rhino robot with deluxe gripper and standard controller

Apple II, II+ or IIe computer with a language card and 80 column card.

Rhino special serial card which permits the Apple to send the Rhino controller the desired move in hole increments compatible with the controller limitations.

A teach pendant designed to simulate the Cincinnati Milacron pendant and to operate all the functions available in the Rhino System.

A Bell parallel port VIA card to intergace the teach pendant to the Apple computer.

An Opto 22 input and output board and opto-isolated modules for interfacing sensor inputs and control outputs.

A California Computer Systems parallel card to interface the Opto 22 I/O modules to the Apple computer.

System software to simulate the operation of the Cincinnati Milicron T3 on the Apple computer and Rhino robot.

The simulator will duplicate the monitor output screens of the Cincinnati Milacron T3 version 4.0 controller on the Apple monitor. The teach terminal keyboard on the Cincinnati is simulated with the Apple keyboard, and the T3 teach pendant is simulated with the 24 pushbutton teach pendant made at Piedmont for the Rhino system. The input and output modules and interface cards for the Apple are supplied by vendors.

The software is a compiled Pascal code and will permit the use of the following T3 language commands:

Delay
NOP
Output
Wait
Perform
Execute

The last two are conditional and unconditional branch commnads.

The system level commands supported by the simulator include:

Close Path
Erase
Function
Home
Goto
Menu

The teach pendant operation will be most like the operation of the version 3.0 controller teach pendant used on the T3 Hydraulic robots by Cincinnati Milacron.

CONCLUSION

With the industrial robot simulator the Rhino robots will support types of training not currently possible. First, all six training levels required for work cells in the laboratory will be possible with the Rhino system, and second, the Rhino robots can be used for basic training on the Cincinnati Milacron T3 series of industrial robots.

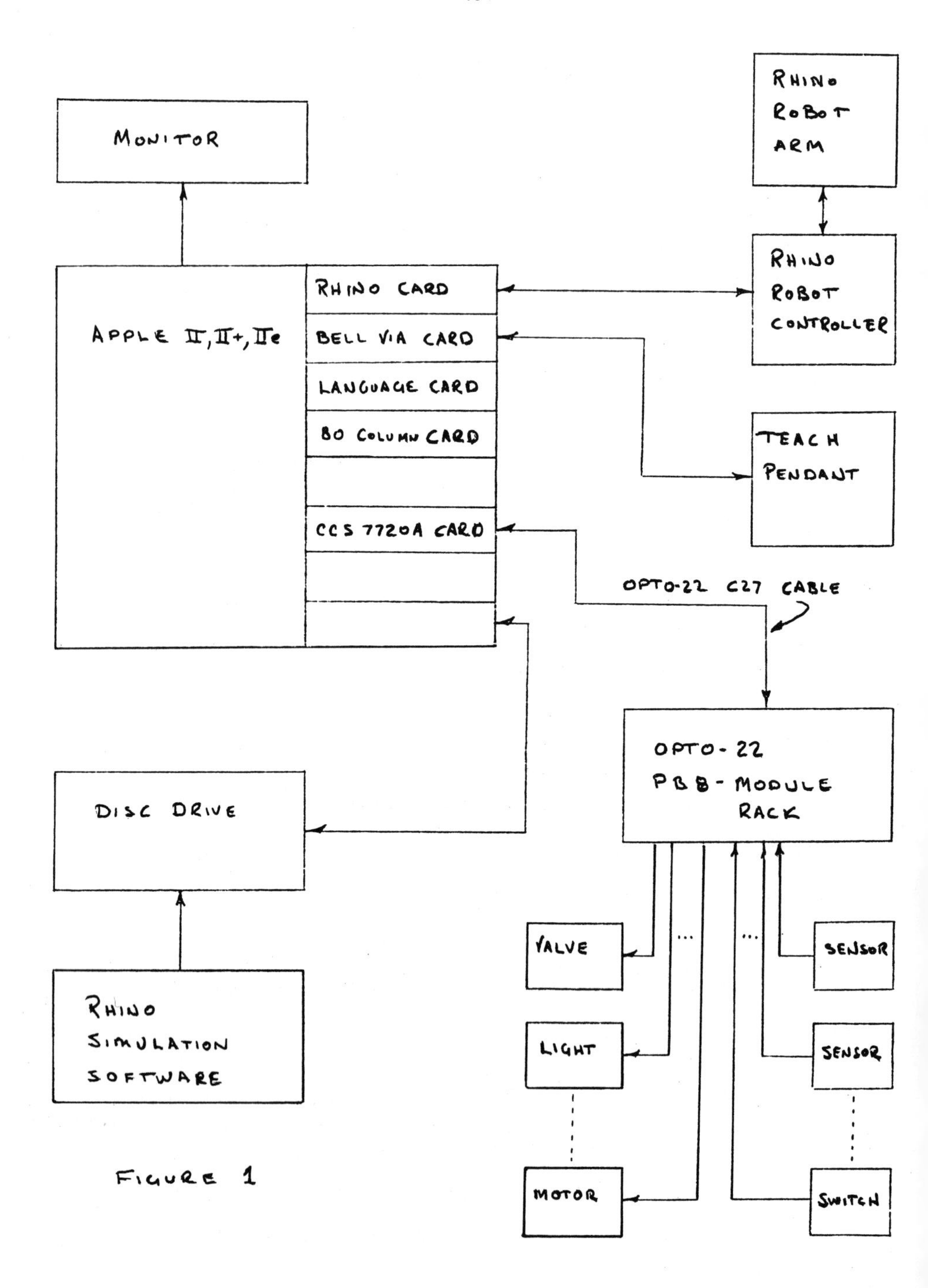
MONITOR
RHINO ROBOT ARM
APPLE II, II+, IIe
RHINO CARD
BELL VIA CARD
LANGUAGE CARD
80 COLUMN CARD
CCS 7720A CARD
RHINO ROBOT CONTROLLER
TEACH PENDANT
OPTO-22 C27 CABLE
OPTO-22 PB8-MODULE RACK
DISC DRIVE
RHINO SIMULATION SOFTWARE
VALVE
LIGHT
MOTOR
SENSOR
SENSOR
SWITCH

FIGURE 1

An MTTS-Robot Work Cell Model

Leon Nguyen and Suren N. Dwivedi

Test Equipment Engg.
IBM Corporation
Charlotte, NC

College of Engineering
Univ. of N.C. at Charlotte
Charlotte, NC

Summary:

A tester-robot work-cell consisting of an IBM Mixed Technology Tester System (MTTS) and an Unimation PUMA 760 robot was successfully demonstrated in the IBM Charlotte Test Equipment laboratory. The model was very simple and required little engineering effort to build. The model made use of the existing handling equipment from the testing floor and required little modification to the tester fixture.

This paper presents the design concepts and the performance of this MTTS-Robot work-cell model.

Introduction:

In the electronic card manufacturing process, the cards are assembled, then soldered, then tested before being shipped. During the testing function of this process, the cards are sorted on a tester to separate the good cards from the defective cards. The defective cards are then analyzed and repaired.

Among the labor involved in the testing function, the sorting operation is the most repetitive. Our objective was to automate this sorting function in order to free the operators for other parts of the job that require more of their technical skills.

This paper describes a robotic application to the IBM MTTS tester (Mixed Technology Tester System). It provides a quick implementation of a work-cell that is dedicated to the sorting operation and capable of meeting the rapid increase in the work load of the testing floor.

I.- The Present Testing Routine:

Cards to be tested are delivered to the test floor in "black trays," which are stacked on skids. Then, the skids are hand trucked to the tester area.

A typical job of 300 cards size about 110 mm x 170 mm (4.5 x 7 in.) would be delivered on one skid with 4 stacks of 19 trays each.

After completing a set-up routine at the tester, the operator performs the sorting operation. During a typical cycle, the operator grasps a card from a tray then inserts the card into the tester fixture. When the testing is complete, cards that tested "good" are marked and replaced onto the "black trays." The defective cards are then analyzed for defects either on-line or at the work station. In this process, the operator makes use of his or her technical skill to identify the components or circuits to be repaired. The cards are then re-tested.

Studies have shown that 20% to 25% of the operator's time is spent during the sorting mode. The balance is spent for the analysis and repair work. On-line analysis can account for up to 5% of the tester's utilization time.

II.- The Design Concepts:

The design of the MTTS-Robot work-cell model was based on the following concepts:

a/- To make full use of the capability of the Puma 760 robot and the existing cards-handling system.

b/- To minimize the modification to the tester and to the cards-handling system so that the transition from the human operation to the robotic operation of the tester remains smooth and simple.

III.- The Hardware:

To implement the above concept, the following modifications to the existing equipment were needed to integrate the work-cell.

(a) **Tray modification:** As shown in Fig.(1), two holes were punched on the bottom of the trays to provide for the gripping locations by the robot "hand". The holes were spaced at a distance equal to the length of the cards to permit the use of one set of grippers for the handling of both the cards and the trays.

Spacers were attached to the bottom of the trays to provide for the coarse location of the cards in the trays.

(b) **Skid modification:** As shown in Fig.(2), square brackets were attached to the sides of the skids to provide for positive location of the stacks of trays on the skids.

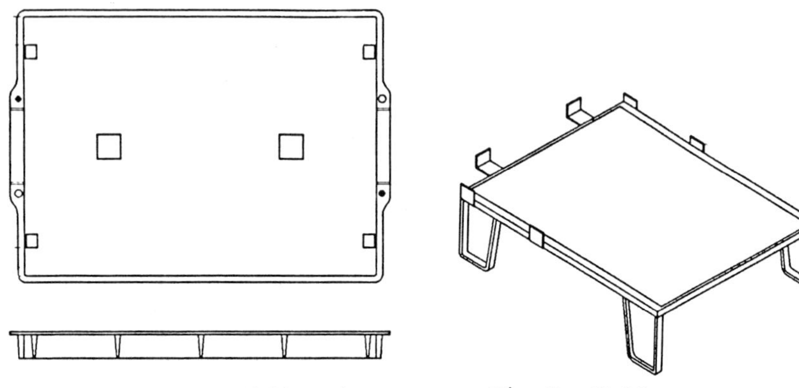

Fig 1 Tray Modification

Fig 2 Skid Modification

(c) **Tester fixture modification:** Front loading and unloading of card to the tester was chosen to take advantage of the robot's capability for moving in the Z direction of the tool coordinates.

The upper and lower rails of the tester were stripped off the front guides and guide plates. A new set of sectioned guides were installed in place to provide for an openning suitable for the loading and unloading operation.

IV.- The Card Handling Scheme:

Three skids were positioned at the work-cell: one skid with the incoming job, one skid to receive the outgoing good cards, and one skid to receive the outgoing bad cards.

Because the skids were located coarsely by square brackets fastened to the floor and because the trays were also located coarsely by the square brackets attached to the skids, the locations of the cards were not precise enough for a direct move from the tray to the tester. A card registration tool was installed to provide for the needed precision.

This registration tool was a simple, shallow, open box-like fixture with chute-guides in a funnel shape. Gravity aligned the card at the register (see Fig 3).

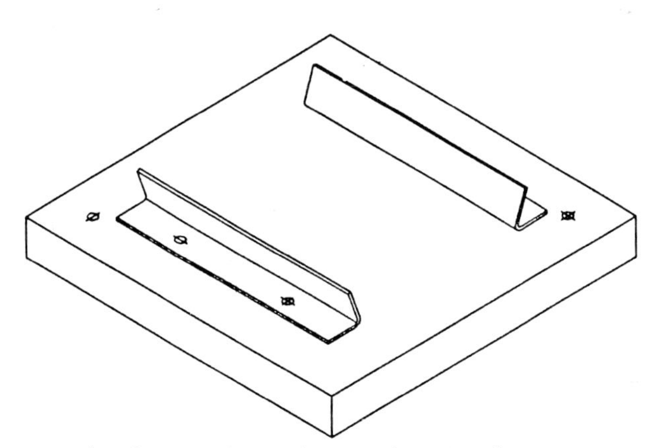

Fig 3 Card Registration Tool

V.- The Gripper Design:

Several gripper designs were tried. A succesfull set of grippers featured horizontal grooves at the gripping surfaces to compensate for the random warpage of the cards.

As the experimentation continued in the laboratory, the most satisfactory gripper was found to be one manufactured by PHD Inc. and equipped with a set of "fingers" having pinching capabilities in the Z direction of the tool coordinates (see Fig 4).

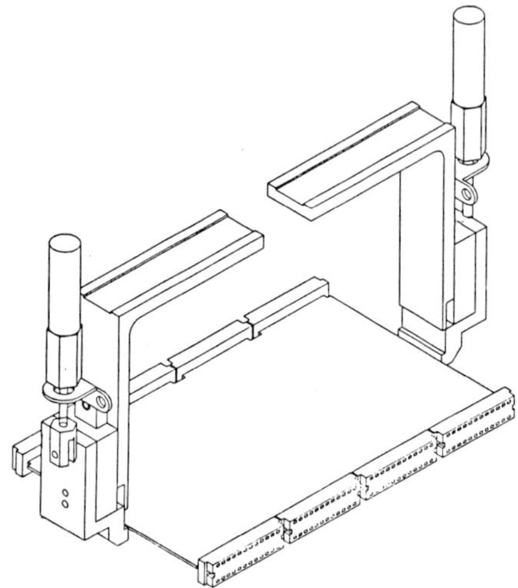

Fig 4 The Effective Gripper

With this design, the fingers were adjusted to close at the width of the card without squeezing the card. A set of "finger nails" activated by air cylinders provided the grasping forces to hold the cards by the edge.

This gripper design accomodated a wide range of card warpage and provided a positive grasp on the card edges without bowing the card.

VI.- The Robot Programming:

The program was written in modular form. A main routine calls different subroutines to perform different tasks, depending on the status of the system as controlled by internal counters or by digital input/output.

The complete cycle starts with a reset of internal counters. Then a full tray is moved from the incoming skid to the service location. A card is then picked-up from the service tray and moved to the registration tool where it is aligned and picked-up again. The card is then inserted into the tester fixture where it is held for 2 seconds to simulate the testing socket time. An manual input signal to the robot controller simulates the testing results (good or bad). Depending on this signal, the robot removes the card from the tester and moves it on to the appropriate "good" or "bad" outgoing tray.

After every pick or place operation, the controller keeps track of the location, then indexes to the next pick or place location for the next time around.

Indexing is used in the X-Y plane for the picking placing of cards in the trays; indexing in the vertical plane, as well, is used for the picking and placing of trays.
The sort cycle is stopped when all cards are tested.

VII.- System Performance:

The MTTS-Robot work-cell model was tested with a sample job of 48 cards in 12 trays (2 stacks). The system performed repeatly without any failure for an extended period of about 1 month. At a speed setting of 80% of maximum, a thruput of 120 cards/hour was obtained.

The above was the direct result of the general layout of the work-cell. There was room for improvement by rearranging the layout or by redesigning the sequence of the robot moves.
The work-cell was built in a laboratory environment. In order to implement this design to the testing floor, some refinement would be needed. Especially, the following actions should be taken:

(a) To redesign the mode of locating the cards in the trays. The spacers we used did not provide adequate precision for the location of the cards. The use of a molded liner would be more effective.

(b) To add some minor modifications to the tester socket control software in order to transmit the test result signal from the tester to the robot controller.

VIII.- Conclusions:

The MTTS-PUMA work-cell model was a simple robotic application to the tester. It required very little engineering effort to design and build and it made use of the existing handling equipment. It also made use of the full capabilities of the robot. The implementation of this model would result in an effective automation of the testers in a minimum of time and without introducing excessive auxiliary equipment normally found in a "hard automation" set up.

References:

1. Unimate Puma Mark II Robot 700 series Equipment and Programming Manual, Unimation, Danbury, CT. August 1983.

2. MTTS Reference Manual, General Technology Division, IBM Corporation, Endicott, NY. 1983 (IBM Internal Use only.)

3. Altamuro, V.M., How to Achieve Employee Support, Safety and Success in Your First Robot Installation", Proceedings of Robot 8, June 4-7/1984, Detroit, Michigan.

4. Dwivedi, S.N., "Custom Design of Robot for Specific Industrial Applications", International Symposium on Design and Synthesis, Tokyo, July 11-13/1984.

5. Kerstetter, Keith L., "Flexible Material Handling for Industrial Robots", Proceedings of Robot 6, March 2-4/1984, Detroit, Michigan.

Designing Robotic Workcells for PCB Assembly

Dennis Miller **General Manager of Workcells**

Control Automation
P.O. BOX 2304
Princeton NJ 08540

A b s t r a c t

Flexible, accurate, low cost automation is a concept whose time has come on the PCB manufacturing floor. Workcells designed specifically for flexible PCB assembly, offer a practical and turnkey alternative to hard automation, semi-automatic, and manual assembly. Automation is one of the few tangible means of keeping up in today's competitive markets, and turnkey applications of assembly robots are leading the way.

I n t r o d u c t i o n

Increasing labor, material, and inventory costs and a competitive marketplace are migrating more and more manufacturers to designing and constructing continuous inline automatic manufacturing lines. This is particularly true in the areas of electronic and printed circuit board assembly. This trend seems to be independent of whether the manufacturer is high volume with low product mix or low volume with high product mix. The only difference in manufacturers appears to be the percentage of the assembly line that is dedicated, flexible, and manual. The important fact is that while no one that I am aware, has a totally automated printed circuit board assembly line, most are designing their line to be 100% automated even though the technology, in some cases, is not ready. This includes planning for intelligent material handling systems, downloading information from a CAD/CAM system, and automatic inspection systems.

The one area of technology that benefits most from the trend toward inline manufacturing is robotics. Assembly robots that are accurate, networkable, easily programmable, and flexible are ideally suited to the competitive marketplace where new products must be introduced faster, product quantities are smaller, manufacturing life cycles are short, and quality requirements are high. One company, Control Automation, anticipated this trend in 1980 and configured a series of robotic and vision systems to be compatible with continuous inline flexible manufacturing. In addition, their market research indicated the greatest growth potential in electronic assembly, particularly in printed circuit board assembly, so they chose to concentrate in this area. The result is a family of robotic and vision workcells which are very reliable and user friendly that can be easily interfaced to both sophisticated and unsophisticated printed circuit board assembly lines. More recently other companies have recognized this trend and are designing similar systems.

In justifying a production PCB line using standard payback criteria, a typical manufacturing line manager today incorporates a mix of hard automation assembly equipment (for DIP's, SIP's, axials, radials, and other high volume parts), semi-automatic equipment (for inserting of power transistors, pots and other medium volume components), and manual labor (for capacitor packs, connectors, relays, heat sinks, pin grid arrays, and other low volume parts, as well as deposition of solder mask, cleaning, and other tasks). For high volume, low mix lines, high accuracy assembly robot workcells represent a cost effective method of automating semi-automatic and manual steps, particularly for insertion or placement of non-standard components. Low volume, high mix lines have trouble justifying any hard automation due to set-up and changeover times, and for these lines, flexible robotic workcells represent an ideal method for insertion of all types of components, both standard and non-standard.

In either case, packaged turnkey workcells, such as Control Automation's Flexcell progressive PCB manufacturing series are a necessary step toward cost reduction and high quality throughput. The cells can be configured to perform a wide variety of insertion or assembly tasks, and provide the user with fast, accurate, flexible solutions to problems which cannot be addressed by hard automation and less intelligent assembly robot alternatives.

Workcell Requirements

Rapidly changing needs and limited resources mandate a turnkey cell capable of quickly handling both current as well as anticipated needs. Features in a workcell optimized for today's PCB assembly requirements include:

- **Ease of Operation** - Turnkey cells must be capable of being rolled onto the shopfloor, integrated, turned on, and put into operation in a short time, with menu-driven set-up programs and push button/automatic control.
- **System Networking** - Each cell in the system should be capable of standing alone or communicating simply over standard serial lines with host controls of the user's choice. The workcell controller should be capable of synchronizing multiple cells simultaneously, networking over RS-232C or RS-422 to other peer workcell controllers, and networking to host computers using a wide variety of networking standards.
- **Integratable with CAD/CAM Databases** - The system should be able to accept absolute data from a CAD database and execute a new assembly program after coordinate transformation. Board types are identified with an in-line bar code reader, with corresponding, pre-programmed insertion programs downloaded from the workcell host CAD database via standard network protocols.
- **Integrated Material Handling/Parts Feeding Systems** - The system should be capable of simply

transferring PCB's in and out of the workspace from storage systems or previous workcells without operator intervention. For example, single boards or boards of various sizes and types mounted in flexible workholders can be transferred into the cell from a standard elevator type storage system using a simple slide transfer mechanism housed in a rigid superstructure to guarantee high accuracy performance. Such a design is easily expanded, exceedingly quick and reliable, and eliminates costly fixturing, belt conveyors, and delay times associated with general pick-and-place and other material handling systems. Intelligent feeders surrounding the workspace must feed in parts. Parts are sequenced into the cell via intelligent feeder systems, (bowl feed, stick, tape, vibratory track, or custom), with part preparation and lead forming steps performed on the system prior to pick-up. Feeders are equipped with sufficient buffering to minimize operator reloading, and are sequenced via preprogrammed insertion data downloaded from the user's database.

o **Calibration** - The system must have a simple-to-implement, multi-point calibration system to allow automatic adjustment to changes in the manufacturing environment. A calibration system provides the user with a simple, automatic, highly accurate technique for adapting quickly to changes in tooling, fixturing, feeding or other environmental changes, without having to reteach pick-up and insertion coordinates.

o **Intelligent Sensor Systems** - A complete system must have sensors throughout the system for inventory status, work-in-process monitoring, and a variety of other operational functions. System software status, continuous operation, station monitoring, and inventory reporting functions are gathered via sensors throughout the workcell, with data upload to the host controller when queried.

o **Programmable Tooling** - For multi-application tight tolerances insertion and placement capability in a single workcell, a high accuracy programmable end effector is the only reliable solution. Unattended automatic tool changing can have accuracy problems where repeatable tight tolerance assembly must be performed, and can unnecessarily extend assembly cycle time.

For example, Control Automation uses a Flexigrip™ programmable end effector that rotates automatically to the proper end tool. Programmable force sensing for each turret head provides the user with accurate, reliable insertion capability for anything from fragile pinheads to tough-to-seat multiple pin devices. Unsuccessful insertion due to component defects automatically activates a search program which will attempt to place the part before rejection.

o **Programmable Clinching** - Most users today partially or fully clinch inserted parts to insure part-in-place specifications throughout the line, especially for boards with unstable components such as SIP's. A programmable multi-position cut and clinch head activated by the robot's onboard controller and synchronized with the insertion head's movements, simultaneously cuts and clinches leads beneath the board during insertion. Four axis (X, Y, Z, O) movement allows for clinching of any component regardless of orientation.

o **Accurate Robot** - The heart of any workcell should be a fast, accurate, and reliable robot compatible with the characteristics of the workcell listed above. This means primarily Cartesian and SCARA type robots. The author's personal preference is the Cartesian type robot because of ease in programming, but regardless of the type $\pm$.001" reliability and $\pm$.002" accuracy within its work envelope is highly desirable, if not mandatory, for maximum efficiency .

Workcells should be designed to insert components at cycle times meeting or exceeding manual rates. Both standard components as well as non-standard components such as connectors, pin grid arrays, and relays should be inserted at rates of up to 3 - 6 seconds per insertion. Other cost reduction or cost avoidance opportunities are:

- o Reduced set-up and changeover time for batch production,
- o Increased quality farther up in the line than previously attainable, reducing inspection and test time in later production steps and significantly reducing rejects, rework and field failure,
- o Reduced raw material inventory and work-in-process,
- o Reduced floor space,
- o Reduced personnel training costs, OSHA costs, and Workmen's Compensation costs, and
- o Reduced production delays and subsequent costs associated with late or partial shipment.

Summary

If you are one of the manufacturers in the midst of this trend, designing a continuous inline manufacturing line, you should plan for the future and incorporate as much capability as possible. With workcells, as described in this article, the line can easily be configured to insert nonstandard components in the high volume line or to insert standard as well as nonstandard components for the low volume user. The additional modular workcells can be added as the manufacturing environment changes.

Integrated Manufacturing and Assembly

M. Weck, U. Dern, and D. Zuehlke

Werkzeugmaschinenlabor,
Technical University Aachen, West Germany

The question of the "Factory of the Future" can no longer be answered solely by the demand for a further increase in productivity. In almost all areas of production there is a growing demand for a fast and, above all, economical adaptation to the current market situation.
Therefore, modern production systems require a high degree of functional flexibility, which can be attained economically only by means of correspondingly intensive exploitation of all facilities. This consideration has led in the past few years to a constantly growing demand for automatized, flexible machine and system concepts.
This demand is not confined solely to the area of manufacturing. With the right selection and configuration of products, extensive automatization of assembly activities is also feasible.
The next logical step is the material and information-flow-specific linkage of both areas into integrated manufacturing and assembly systems. Such an integrated system has been developed at the Machine Tool Laboratory of the Technical University Aachen, West Germany.

1. Introduction

The automatization of manufacturing installations with the aid of computer-assisted systems makes possible substantial cost reductions, which help improve the competitiveness of the companies on an international scale.

Solutions based on the building-block principle consisting of largely standardized component systems and interfaces have contributed a great deal to the fact that such manufacturing systems no longer have the reputation of being prestigious but risky experimental installations; rather, they now constitute a practicable and, above all, an economical manufacturing concept also for small and middle-sized companies.

What is now required most are machine and system concepts - and this applies both to production and assembly - which combine an adaptive flexibility of the manufacturing equipment and a high level of automatization /1/. Quite generally, the demand is as follows: economic production even of small batches.

A production system which meets this requirement to a particularly high degree has been developed at the Machine Tool Laboratory of the Technical University Aachen and set up with the support of leading companies (Fig. 1).

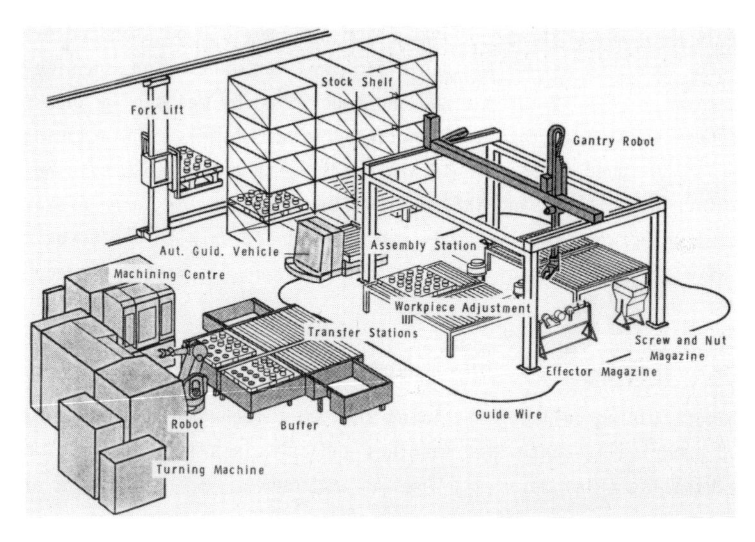

Fig. 1: Integrated Manufacturing and Assembly System

In this "integrated" system the manufacturing and assembly processes have been linked as regards both the material and the information flow, thus permitting a substantial reduction of turnaround time and the intermediate buffer inventory.

2. System Overview

As a product sample, a pipe valve is manufactured by means of this system in seven different sizes (Fig. 2).

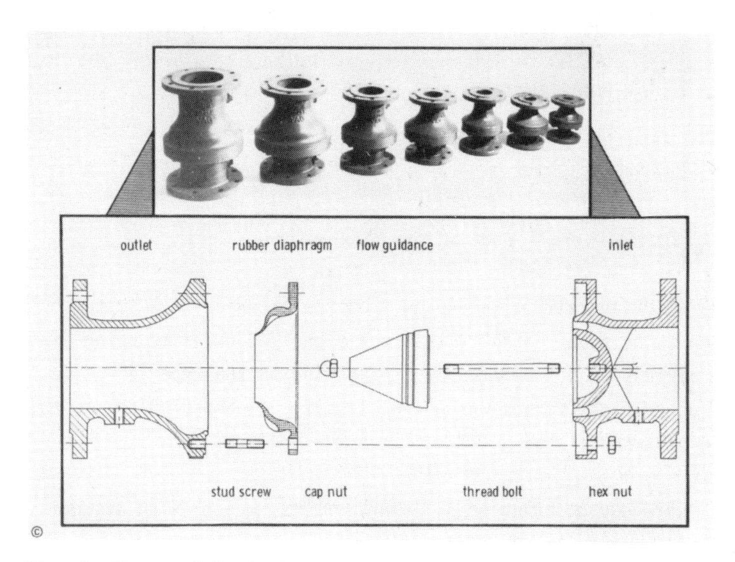

Fig. 2: Range of Products

For this purpose, cast iron blanks which are provided on standard pallets are first machined and then assembled into a finished product using outside bought parts. The entire process is automatized and supervised by means of a supervisory computer. The system is divided into two functional areas, namely manufacturing and assembly, which are designed as self-contained cells, with a view to providing defined interfaces for the material and information flow required for later expansions. The external link between the two cells and the connection with the associated storage as well as with other company areas is assured by an Automated Guided Vehicle System (AGVS).

2.1 Manufacturing

The manufacturing cell for machining the two components of a valve housing consists of two numerically controlled machines and one six-axis industrial robot for workpiece handling (Fig. 3).

Fig. 3: Overall View of the Manufacturing Cell

The manufacturing cell is uncoupled from the material flow system by means of a buffer which is of such capacity that while one job order is being executed the next job can already be put on standby. Depending on the type of valve, a job may consist of up to eight valves which are classified according to assembly groups and provided on two pallets.

A clearly defined position of all single parts, which is an essential prerequisite for automatic workpiece handling, is ensured by a uniform workpiece-specifically moulded plastic insert, which is held by a high-precision frame on the pallet. This insert accepts all variations of a family of parts and can be easily exchanged to hold another workpiece.

In the first operation, the flange coupling joints are turned. For this purpose, it is necessary for the chuck jaws to be first adjusted to the appropriate chucking diameter (with the aid of the industrial robot) before the workpiece is placed into the lathe chuck.

Upon termination of this machining operation the workpiece is turned over for a second chucking without being set down. This procedure is carried out by means of a special-type gripper within the working space of the machine.

The option of program-controlled exchange of the gripper of the industrial robot permits adaptation to diverse workpiece geometries and handling tasks. This, in conjunction with the possibility of chuck jaw adjustment, increases the flexibility of the system decisively /2/. Interchangeable jaws are also conceivable in this context.

For the subsequent operation - drilling/tapping of the holes for the bolting of the flanges and for the connecting of a bypass - it is necessary for the part to be transferred in the proper position to the second machine, which is a machining centre. For this task, a self-contained orientation station has been provided for, where the part is deposited by the robot. While scanning the outer contour the part is rotated into the defined posititIon and then picked up again by the robot.

On the machining centre all size variants of the part are chucked into a universal-type fixture. The position in which the part is clamped and the dimensional accuracy of the premachining can be checked with the aid of a switching probe and the results obtained be used immediately to correct the NC program.

Upon termination of the machining operation the part, while still in the working space of the machining centre, is cleaned of swarf and is again placed on the pallet.

2.2 Assembly

In the assembly cell a five-axis gantry-type robot performs all handling and assembly tasks (Fig. 4).

Fig. 4: Overall View of the Assembly Cell

Pallets with premachined workpieces can be provided at as many as four stations, which, like the buffer in the manufacturing cell, are designed as roller conveyors. Thus an uninterrupted job sequence is ensured in this area also.

The two principal assembly groups of the valve are assembled at two separate stations. For this purpose, the robot first picks up a valve chamber from one of the pallets provided and places it at an orientation station, similar to that in the manufacturing cell, in order to subsequently transfer it in the proper position into the chuck of a second station. At this station, the first assembly group is pre-assembled together with a flow guidance and the corresponding fastening bolt and nut.

The second assembly group is pre-assembled in the same manner as the first one at the orientation station. For the placing of the studs for fastening the two assembly groups together, the orientation station continues to be stepped around in accordance with the thread-holes.

For each one of these operations a special gripper or screwdriving tool is required which the robot can automatically pick up from a magazine. Variants-specific outside bought components are provided on the pallet, together with the premachined workpieces. Standard parts like studs and nuts are provided by separate magazining systems and are directly taken over by the handling device with the aid of the appropriate screwdriving tools.

When the pre-assembly is terminated, the robot puts both assembly groups together and bolts the two flanges. The hexagon nuts required for this purpose are taken one by one from a magazining system as in the case of the studs. Finally, the so assembled valve is again placed onto one of the pallets standing by.

2.3 Material Flow

Transportation of the pallets between the system storage, the manufacturing and assembly cells is carried out by an Automated Guided Vehicle System.

The vehicle carries, as do all the transfer stations on its travelling route, a roller conveyor, in order to be able to ensure pickup and transfer of pallets. The conveying capacity of this system is sufficient to also take care of additional cells or other shop areas.

The system storage, located in the vicinity of the production system, accomodates the pallets required for a given production period. The capacity of this storage can be adapted as required to cope with the number of stations in the system, the turnaround time of the various jobs and other workpiece-specific requirements. The storage is designed as a pallets shelf which is supplied by means of a charging device which is numerically controlled in the travelling and lifting axes.

3. Control System Structure

The material-flow-specific structure is reflected in the structure of the control system of the manufacturing and assembly system.

The individual components of the system are controlled by several microcomputer systems which are independent of each other. A supervisory computer coordinates and monitors all material flow, manufacturing and assembly-specific processes (Fig. 5).

Order files and NC programs are prespecified by a planning system via computer link or built up in operator dialogue. In this way it is possible to assign job priorities and have the system give preferential treatment to urgent jobs. All subsequent processes are controlled by the computer.

In the event of technical or organizational failures it is possible at any time for the operator to intervene and specifically manipulate the component functions.

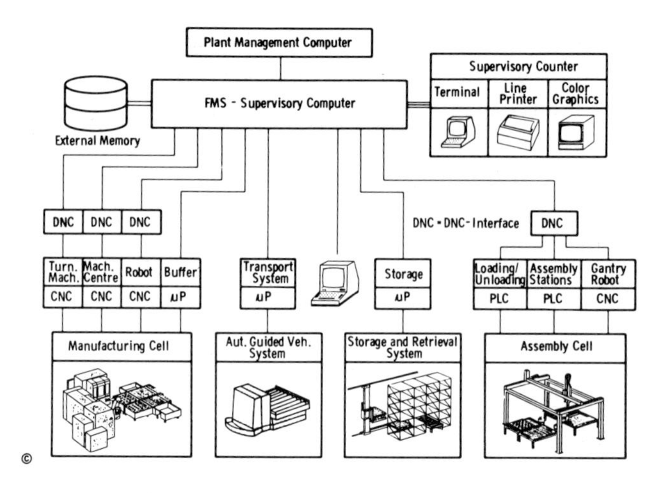

Fig. 5: Control System Structure

The software system thus realized includes the following functions:

- job management and follow-up of the job progress,
- NC-data management and distribution,
- organization of the material flow,
- coordination of the manufacturing and assembly processes,
- management of the storage and buffer inventories,
- communication with the operator staff,
- collection and preparation of the system data.

The controllers of the various component systems, if they are not provided with appropriate interfaces for the communication with the supervisory computer, are linked with the computer by means of a microprocessor-controlled DNC-interface developed in-house. Through additional memory capacity and programmable logic functions, this interface helps relieve the supervisory computer of coinciding tasks and facilitates independent coordination of the individual processes which take place in the cells.

This principle of hierarchical structurization into function modules and decentralization of intelligence, consistently adhered to, not only permits a partially automatized operation of the system in the event of failure or breakdown of the computer, but it also facilitates considerably the testing, putting into operation, adaptation and expansion of the system.

The data transmission between the supervisory computer and the controllers of the component systems is carried out via a fiber optic cable in telegram-mode within the framework of a transmission procedure. In this way a fast and above all reliable information exchange is guaranteed, independently of the distances between the systems and of the environmental conditions in the plant.

Both the software system of the supervisory computer and the programs of the microprocessor controllers have been prepared in the higher programming language PASCAL. The overall software is, to a great extent, of modular structure in order to facilitate adaptations and expansions.
All system files are set up by means of a configuration program with whose aid the user can parametrize the data in dialogue to suit company- and plant-specific requirements /3/.

The operator is supported by a comfortable information system. Of particular interest is a colour-graphic display in which all current data are represented during the entire process for the individual components or in the form of an overview.

4. Safety- and Monitoring Functions

In order to guarantee a high system safety and availability, all principal functions of the transportation, handling and machining are supervised with the aid of appropriate sensors and monitoring systems.

In addition to the functional reliability of the system, special importance must be attributed to the protection of personnel -- a factor that is frequently underestimated -- particularly in highly-automatized plants which, because of their motive power and speed features have a high potential of danger for the operator.

Habituation and non-observance of elementary precautions during the programming and test phases are frequently the causes of endangerment and accidents, because during these phases the operator's presence is required in the working space of the equipment, but his attention is concentrated on only a part of this.

Here, sample solutions are shown for the protection of personnel through suitable protection of the dangerous areas, supervision of the working spaces of the handling and conveying systems and reduction of the potential for danger when people are present within the protected areas /4/.

5. Conclusion

With the "Integrated Manufacturing and Assembly System" described in this paper a concept has been developed and realized which meets in a high degree today's objectives:

- maximum flexibility for a broad parts spectrum,

- extensive coordination of the areas of production and assembly,
- economically automatized production of single parts and small series,
- high system reliability and availability.

It is not our intention to offer a complete system solution, but rather a variety of specific solutions. The modules of the system described in this paper can be individually put together into other combinations, depending on the respective tasks and the production volume required.

This adaptive solution is made possible by a consistently modular structure of the hard- and software together with standardized material- and information-flow-specific interfaces. In particular, this offers the possibility of a step-by-step installation of the system with the option of expansion at a later date.

In this connection it has not been possible to give more than an overview of the multitude of details contained in the " Integrated Manufacturing and Assembly System" which has been developed at the Machine Tool Laboratory of the Technical University of Aachen.

References

/1/ Weck, M. et al.: Flexible Manufacturing Systems. Contribution to the Aachen Machine Tool Kolloquium '84, 7 - 9 June 1984, Aachen, W.-Germany

/2/ Toenshoff, H. K.; Weck, M.; Siemens, K.-J.; Engel, G.: Automatische Greifer- und Werkzeugwechselsysteme. Industrie-Anzeiger 101 (1979) No. 82, pp. 29/33

/3/ Weck, M.; Kohen, E.: Konfigurierbare Steuerungssoftware. Industrie-Anzeiger 106 (1984) No. 83, pp. 25/30

/4/ VDI 2853 (Draft Standard), July 1984. Beuth-Verlag GmbH, Berlin und Cologne

Controls II

Chairman: R. P. Sharma, Western Michigan University, Kalamazoo, Michigan
Vice Chairman: R. Sharan, National Institute of Foundry and Forge Technology, Ranchi, India

Remote Control of Robots with a Class or Codes

S. Ganesan and M. O. Ahmad

S. GANESAN

Department of Electrical Engineering
Western Michigan University
Kalamazoo, Michigan 49008

and

M.O. Ahmad
Department of Electrical Engineering
Concordia University
Montreal, P.Q. Canada

Summary

Methods of construction of a type of telecommand codes which are resistant to random intererence, for remote control of robots are presented. A command consists of time sequence of RF pulses. The time interval between the pulses carry the command information. Comparison of single stage, two stage and three stage codes, with maximum and reduced protection against interference is presented.

I. Introduction

Some types of robots are controlled by remote commands consisting of time sequence of RF pulses. The robot receiver is capable of receiving both the true command pulses and interfering RF pulses emanating from other robot's command transmitters or environment noise sources. In time interval coding scheme, the commands are defined by a set of pulse signals having identical or different carrier frequencies [1-8]. Each command code is identified by the time spacings between the pulses. By keeping the number of pulses in each command same and keeping the length of each command code as short as possible, complexity of the decoder/encoder hardware can be minimized. Eckler has described the procedure for the construction of missle guidance codes resistant to random interference [1]. Application of numbered graphs in the design of telecommand codes is described in [2]. Section 2 reviews briefly Eckler's single stage and two stage codes. Comparison of single stage and two stage codes, with full and reduced protection is given in section 3. Description of three stage and multi-stage codes is given in Section 4.

II. A Type of Telecommand Code

1. Single Stage Codes

A command is represented by a set of n pulses; the command information is contained in the (n-1) time spacings between pairs of successive pulses. There are K distinct commands. These K commands are encoded such that i false pulses cannot combine in any way with (n-1) pulses from any command to form either the

There is no efficient algorithm for choosing the optimal time spacings for various K and n values. A simple and systematic search technique for choosing the edge numbers, which requires far less search time than that required for a full search is described in [2,3].

II. 2. Two Stage Codes

Here security function of the code and the command function of the code are separated.
In the decoder the output from one (e.g. security) decoding operation becomes as input to the next (e.g. command) decoding operation. The symbol i/j denotes two stage code containing i. j pulses. The first stage decodes j clusters of i pulses each and the second stage decodes the cluster of j output pulses emitted by the first stage AND circuit. The two stages are designed according to single stage code rules, so that they are invulnerable to (i-2) and (j-2) false pulses. The upper bound to the number of false pulses that can be arranged in any pattern whatever with the i j true pulses without forming a false or repeated command is

$$M = \text{Min} \quad (i \quad (j-1) - 1, \; j \; (i-1) - 1)$$

Interleaved type of command and security pulses in two stage codes are designed by the following procedure. The j pulse command code spacings are chosen as multiples of i and also such that they are secure against (j-2) false pulses according to single stage code methods. The security code spacings $(t_1, t_2, \ldots \; t_{i-1})$ are chosen such that

a) They are secure against (i-2) false pulses and

b) None of the i (i-1) integers, as specified by equation 1, is a multiple of i.

The Table V gives the security code spacings satisfying the above two restrictions for various values of i.

Similarly i/j two stage interleaved code with reduced protection is formed such that:

1. The security codes are as short as possible (from Table 2)
2. The command code spacings are restricted to the multiples of i
3. The command code spacings are chosen from single stage code with reduced protection.

The i/j two stage codes described above, are secure against M false pulses from forming a different command.

$M - i \; (s + 1) - 1$ where s = security of the j pulse command codes (single stage codes) against false pulses.

Figure 2 shows two stage interleaved codes secure against 8 and 5 false pulses.

same command (shifted in time) or one of the other (K-1) commands. The maximum possible value of i is equal to (n-2), because (n-1) false pulses can always combine with one true pulse to form false commands in many different ways.

The various time spacings between successive pulses can be represented as

$$(t_1^j, t_2^j, \ldots t_{n-1}^j) \quad \text{where } j = 1, 2, \ldots K$$

The K commands are effective against (n-2) false pulses if and only if the following K n (n-1)/2 integers are all different [1].

$$t_i^j \quad \text{for } i = 1, 2, \ldots (n-1) \text{ and } j = 1, 2, \ldots K$$

$$t_i^j + t_{i+1}^j$$

$$\sum_{i=1}^{n-1} t_i^j \quad \text{for } j = 1, 2, \ldots K$$

Specifically for K = 2 and n = 3, the six integers t_1^1, t_2^1, $t_1^1 + t_2^1$, t_1^2, t_1^2, $t_2^2 + t_2^2$ all be different

Since it is desirable to keep all the commands as short as possible, the time spacings are selected such that they satisfy equation 1 and also such that the length of the longest command is minimum. For example, when K = 2, n = 3, the minimum length time spacings for the commands are (2, 3), (1 6).

In general, as the number of commands (K) increases, the length of the longest command, increases. Similarly, as the number of pulses (n) in the command code increases, the length of the longest command increases.

The length of the longest code can be reduced by reducing protection of n pulse code against false pulses. For example, for a 4 pulse K command code to be invulnerable to one false pulse, the following 4 K integer pairs must all be different [1].

$$(t_1^j, t_2^j) \quad (t_2^j, t_3^j) \quad (t_1^j, t_2^j, t_3^j) \quad (t_1^j, t_2^j + t_3^j) \text{ for}$$

$$j = 1, 2 \ldots K$$

Figure 1 shows 5 pulse single stage codes secure against 3, 2 and 1 false pulses when the numberof commands (K) is 3.

Table I to Iv show the time spacings for six pulse codes secure against 4, 3, 2, and 1 false pulses.

III. Comparison of Single and Two Stage Codes

Table VI compares single stage codes with maximum protection with single stage codes with reduced protection.

The value circled in the Table VI is the minimum length for the particular K value. It is found that for K number of commands and security against i number of interference pulses, by increasing the number of pulses in the code, and reducing the protection, the length of longest command reduces to a certain level. Afterwards with the increase in n, the length also increases for the particular K and n values.

Table VII summarizes the behaviors of two stage codes with maximum and reduced protection. It records the shortest known length of the longest command for different numbers of commands for various types of two stage codes all of which are secure against 5 false pulses.

From the Table VII, it can be seen that the total length of the code can be reduced by increasing the number of pulses in the code and by reducing the protection against false pulses.

It is also found that increasing the value of i and reducing the value of j, keeping the security against forming different commands by false pulses as constant, reduces the length of the codes in general, but increases the number of pulses in the code.

Table VIII compares a number of single stage codes and two stage codes with maximum or reduced protection, all of which are secure against 3 false pulses. It shows the shortest known length of longest code for different number of pulses in the code (n) and for various numbers of commands (K). The value circled is the minimum length for a particular value of K.

A n pulse two stage code (where $n = i.j$) can be considered as a n pulse single stage code with reduced protection against interference pulses, and also the restriction on the total length of the n pulse single stage code is relaxed to separate the security and command function of the code pulses for two stage codes. The n pulse two stage codes are longer than n pulse single stage codes The advantage of two stage code is that the time spacings can be chosen easily. All the K commands in two stage codes have the same security code spacings. Hence by changing the security spacings alone different systems (robots) can be guided and commar the command spacings from one system to the next need not be changed.

IV. Three Stage Code

A three stage code or multistage code can be formed by introducing one or more stages to the two stage code to perform either security function or command function. Let the symbol p/i/j represent a 3 stage code where p pulse code performs either security or command function in addition to the i pulse security codes and j pulse command codes. Interleaved p/i/j codes can be designed as described below. The j pulse command code spacings are chosen as multiples of $p \cdot i$ and also such that they are secure against (j-2) false pulses. The i pulse security code spacings are chosen as for i/j interleaved code spacings from Table V. The p pulse security code code spacings are taken as multiples of i with the restriction that

1. They are secure against (p-2) false pulses
2. None of the p (p-1) integers, as specified by equation 1, is a multiple of p.

The spacings with the above two restrictions are given in Table V. These values are multiplied by i to get the p security spacings.

The maximum length of p/i/j interleaved code is

$$L''' = p \times i \times L_j + L_i^I + i \times L_p^I$$

where

L_j = length of longest j pulse single stage code for K commands

L_i^I = length of i pulse security interleaved code

L_p^I = length of p pulse security interleaved code.

Instead of the above procedure the following modifications can be done. The i pulse spacings are chosen as multiples of p, with the other necessary conditions. The p pulse spacings are chosen from Table V. Then

$$L''' = p \times i \times L_j + L_p^I + i \times L_i^I$$

The process which gives the shortest length is chosen.

The p/i/j, p security/i security/j command codes, which are described above,are secure against p i (j-1) - 1 false pulses, i.e. p i (j-1) - 1 false pulses cannot convert one command to a different command.

The p security/i security/j command three stage codes have the same security as the p i security/j command two stage codes.

It is found that p/i/j code has shorter length than p i/j code.

Example: when K = 3; 3/3/3 codes have a max. length = 110

while 9/3 codes have a max. length = 140

Hence the length of p i/j two stage code can be reduced by dividing the security function into two and making them as p/i/j three stage code. In this process the security against forming the same command delayed in time is reduced.

Conclusion

In this paper a type of time interval coding scheme which is secure against random intererence pulses, for remote control of robots is described. This type of code can be treated as pulse interval modulated codes (PIM) in a channel that has randomly positioned interference pulses, but is otherwise noiseless. This channel model is often used with optical channel or RF channel.

References

1. ECKLER, A. R., 'The Construction of Missile Guidance Codes Resistant to Random Interference. Bell System Technical Journal, July 1960, pp. 973 - 994.

2. GANESAN, S. and AHMAD, M. O., 'Application of Numbered Graphs in the Design of Multi-Stage Telecommand Codes', Proceedings of 5th International Conference on Theory and Applications of Graphs, John Wiley and Sons, June 1984.

3. GANESAN, S., AHMAD, M. O. and SWAMY, M. N. S., 'Use of Numbered Directed andUndirected Graphs in the Design of Telecommand Codes' to be published.

4. GOLOMB, S. W. (Ed.), Digital Communications', Prentice Hall, 1964.

5. BARKER, R. H., 'Group Synchronization of Binary Digital Systems in Communication Theory', Jackson, W. (Ed.), Academic Press, N.Y., 1953.

6. GANESAN, S., 'Action of Interference in Control Radio Links Electro-Technology, India, June 1978, pp. 21 - 28.

7. MAKEY, R. C., 'A Synchronized Pulse Communication System with Pseudorandom Interpulse Period', IRE Transactions on Comm. Syst., CS-10, No.1, March 1962, pp. 109 - 113.

8. BLOOM, G.S. and GOLOMB, S. W., 'Applications of Numbered Undirected Graphs', Proc. IEE, Vol 65, No. 4, April 1977, pp. 562-570.

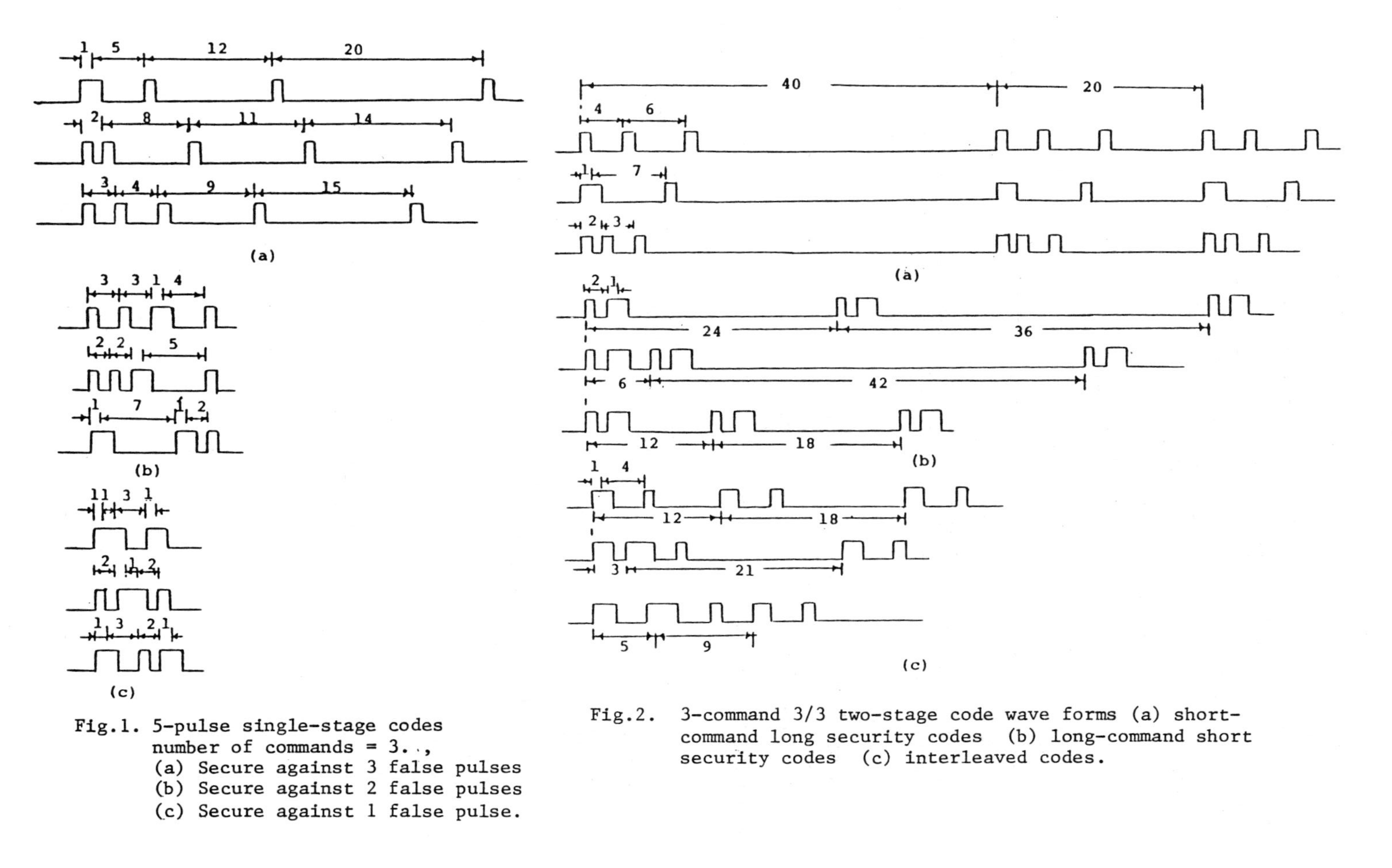

Fig.1. 5-pulse single-stage codes
number of commands = 3.,
(a) Secure against 3 false pulses
(b) Secure against 2 false pulses
(c) Secure against 1 false pulse.

Fig.2. 3-command 3/3 two-stage code wave forms (a) short-command long security codes (b) long-command short security codes (c) interleaved codes.

TABLE I

6 PULSE SINGLE STAGE CODE SECURE AGAINST 4 FALSE PULSES

No. of Commands K	Max. Length	Time Spacings
2	39	(14,7,2,13,3) (8,19,1,4,6)

TABLE II

6 PULSE SINGLE STAGE CODE SECURE AGAINST 3 FALSE PULSES

No. of Commands	Max. Length	Time Spacings
2	16	(1,2,6,1,3) (1,4,2,5,4)
3	20	(1,2,6,1,3) (1,4,2,5,4) (2,1,1,13,3)

TABLE III

6 PULSE SINGLE STAGE CODE SECURE AGAINST 2 FALSE PULSES

2	10	(1,2,1,1,3) (3,1,2,3,1)
3	12	(1,2,1,1,3) (3,1,2,3,1) (1,4,2,2,3)

TABLE IV

6 PULSE SINGLE STAGE CODE SECURE AGAINST 1 FALSE PULSE

2	8	(1,1,3,1,2) (1,3,2,1,1)
3	9	(1,1,3,1,2) (1,3,2,1,1) (2,1,1,3,2)

TABLE V

THE SECURITY CODE SPACINGS FOR 2 STAGE INTERLEAVED CODES

Number of Security Pulses i	Minimum Length	Time Spacings
2	1	1
3	5	(1, 4)
4	9	(2,1,6)
5	17	(6,7,1,3)
6	20	(4,5,8,2,1)

TABLE VI

LENGTH OF LONGEST COMMAND FOR SINGLE STAGE CODES SECURE AGAINST ONE FALSE PULSE

Number of Commands	Number of Pulses per Code (n)			
K	3	4	5	6
2	7	⑤	6	8
3	10	⑥	7	9
4	12	⑦	7	9
5	15	⑧	8	9
6	19	⑧	8	9
7	22	⑨	10	10
8	24	11	⑩	10
9	27	12	⑪	11
10	31	12	12	⑪

TABLE VII

LENGTH OF LONGEST COMMAND FOR VARIOUS TWO STAGE CODES ALL OF WHICH ARE SECURE AGAINST FIVE FALSE PULSES.
THE VALUE CIRCLED IS MINIMUM LENGTH FOR THE PARTICULAR K VALUE.

Number of pulses	Type of two stage code	Number of commands, K					Comments
		2	3	4	5	6	
8	2/4	27	39	55	63	79	4 pulse command codes have maximum protection.
9	3/3	26	35	41	50	62	3 pulse command codes have maximum protection.
10	2/5	(19)	(23)	29	33	37	5 pulse command codes have reduced protection; secure against 2 false pulses.
12	3/4	20	23	26	29	32	4 pulse command codes have reduced protection; secure against 1 false pulse.
12	2/6	21	25	27	31	35	6 pulse command codes have reduced protection; secure against 2 false pulses.
14	2/7	23	25	(25)	(27)	31	7 pulse command codes have reduced protection; secure against 2 false pulses.
15	3/5	23	26	26	29	(29)	5 pulse command codes have reduced protection; secure against 1 false pulse.
18	3/6	29	32	32	32	32	6 pulse command codes have reduced protection; secure against 1 false pulse.

TABLE VIII

MINIMUM LENGTH OF LONGEST COMMAND FOR VARIOUS CODES, WHICH ARE SECURE AGAINST THREE FALSE PULSES. THE VALUE CIRCLED IS MINIMUM LENGTH FOR THE PARTICULAR K VALUE.

Number of pulses	Type of code	Number of commands, K					Comments
		2	3	4	5	6	
5	5 pulse single stage code	22	38	48	56	68	Single stage code with maximum protection.
6	6 pulse single stage code	16	20	26	29	34	Single stage code with reduced protection.
7	7 pulse single stage code	15	19	21	22	23	Single stage code with reduced protection.
8	8 pulse single stage code	(11)	(12)	(15)	(17)	(17)	Single stage code with reduced protection.
6	2/3 Two stage code	15	21	25	31	39	Two stage code with maximum protection.
8	4/2 Two stage code	17	21	25	29	33	Two stage code with maximum protection.
10	2/5 Two stage code	13	15	(15)	(17)	(17)	Two stage code with reduced protection.
12	2/6 Two stage code	17	19	19	19	19	Two stage code with reduced protection.

Local Area Networks in the Factories of the Future and Its Impact on CAD/CAM and Robotics

B. Thacker

Universal Computer Applications
Southfield, Michigan

ABSTRACT

This paper briefly discusses the need for Local Area Networks (LAN) in the Factories of the Future. It outlines the activity at General Motors in this area known as Manufacturing Automation Protocol (MAP) which is based on the reference model of Open Systems Interconnection (OSI) developed by International Organization for Standardization (ISO). Next this paper discusses the potential impact of such factory floor local area networks on CAD/CAM and Robotics.

BACKGROUND AND NEED FOR FACTORY FLOOR LOCAL AREA NETWORKS

If American business is losing its competitive edge in the world marketplace, one remedy may be to improve productivity. Typically, productivity improvements have come by automating the control of a single process or of a single function but with very little, if any, communications with the outside world creating automation islands. Today General Motors has over 40,000 intelligent devices in the manufacturing area with a projected 400-500% increase in the next 5 years. Simply increasing the quantity of automation is not sufficient. Integrating these islands of automation is essential for the factories of the future to succeed in a world market demanding high quality, cost competitive products.

The factories of the future may have these processes and applications, islands of automation, integrated into one manufacturing system using a local area network. Some of the advantages of such a factory of the future are:

- o Lower networking costs of wiring, hardware interfaces, custom software etc.
- o More efficient use of resources.
- o Immediate and more accurate information for decision making.

This also makes it realistic for multiple vendors to attach and communicate with other vendors over the same network. In order to implement an integrated manufacturing system, it is necessary to have a viable protocol standard which would allow efficient communication between products supplied by many different vendors.

MANUFACTURING AUTOMATION PROTOCOL (MAP) AT GENERAL MOTORS

General Motors has taken a leadership role in trying to define a protocol standard for the factory floor local area network. In 1980 General Motors created a Manufacturing Automation Protocol (MAP) Taskforce.

"The purpose of MAP is to prepare a specification that will allow common communication among diverse intelligent devices in a cost effective and consistent manner. In order to meet this purpose, a set of objectives was established as follows:

1. Define a MAP Message standard which supports application-to-application communication.
2. Identify application functions to be supported by the message format standard.
3. Recommend protocol(s) that meet our functional requirements." [1]

Soon after its formation, the task force identified the International Standards Organization's (ISO) seven layer model for Open Systems Interconnection (OSI) as a basis for standardized networks. Even in 1981, the model was gaining wide support in the communications industry. However, since the model specifies function rather than protocols, compliance to the model doesn't assure multi-vendor communication. The strength and current support of the MAP specification is it's selection of existing or emerging Standard Protocols. All seven layers of the MAP specification will eventually be international standard protocols.

The MAP Task Force has concentrated primarily on evaluation of IEEE Project 802 and emerging ISO/NBS (National Bureau of Standards) specifications. Figure 1 illustrates the MAP protocol choices to date. GM specific upper layer protocols are necessary for a functioning interim MAP network. These interim specifications are jointly developed by GM and participating vendors with standards organizations' working papers as a base. Both GM and participating vendors are active in standards groups to influence future direction and encourage adoption. Standard upper layer protocols will be added to MAP as they mature.

At the 1984 National Computer Conference (NCC), a subset of MAP was demonstrated for the first time by several computer and programmable controller vendors. In 1985 few pilot plants will have MAP implemented. 1985 implementations may have several additional features like ISO/NBS Session kernel, draft ISO CASE (Common Application Service Elements), MMFS (Manufacturing Message Format Standard), (simple) Directry Services, (trivial) Network Management etc.

LAYERS	FUNCTION	MAP SPECIFICATION
USER PROGRAM	APPLICATION PROGRAMS (NOT PART OF THE OSI MODEL)	AS REQUIRED FOR EACH JOB
LAYER 7 APPLICATION	PROVIDES ALL SERVICES DIRECTLY COMPREHENSIBLE TO APPLICATION PROGRAMS	(EXISTING STANDARDS ARE UNDER CONSIDERATION)
LAYER 6 PRESENTATION	RESTRUCTURES DATA TO/FROM STANDARDIZED FORMAT USED WITHIN THE NETWORK	(EXISTING STANDARDS ARE UNDER CONSIDERATION)
LAYER 5 SESSION	NAME/ADDRESS TRANSLATION, ACCESS SECURITY, AND SYNCHRONIZE & MANAGE DATA	NBS SESSION IS UNDER CONSIDERATION
LAYER 4 TRANSPORT	PROVIDES TRANSPARENT, RELIABLE DATA TRANSFER FROM END NODE TO END NODE	ISO/NBS TRANSPORT CLASS 4
LAYER 3 NETWORK	PERFORMS MESSAGE ROUTING FOR DATA TRANSFER BETWEEN NON-ADJACENT NODES	ISO/NBS INTERNET IS UNDER CONSIDERATION
LAYER 2 DATA LINK	IMPROVES ERROR RATE FOR MESSAGES MOVED BETWEEN ADJACENT NODES	IEEE 802.2 LINK LEVEL CONTROL
LAYER 1 PHYSICAL	ENCODES AND PHYSICALLY TRANSFERS MESSAGES BETWEEN ADJACENT NODES	IEEE 802.4 TOKEN ACCESS ON BROADBAND MEDIA

PHYSICAL LINK

FIGURE 1

MAP SPECIFICATION SUMMARY BY LAYER

In long term, availability of VLSI chips from the semiconductor manufacturers for the lower level protocols will help in driving the prices down.

Many other large corporations have expressed their interest in MAP. Three meetings of MAP user's group has taken place in 1984 with more and more corporations and attendees participating. The last meeting had over 450 attendees from over 200 companies.

WHAT MIGHT BE CONNECTED TO MAP NETWORKS?

- ENGINEERING WORKSTATION: This engineering workstation provides product design information to be utilized by automated machine tools and other support systems such as scheduling and tool management.
- DISTRIBUTED PROCESS CONTROL: Realtime process control and monitoring will drive and gather information from a variety of plant floor devices such as robots, vision systems, and programmable controllers. These devices may be connected to the MAP directly or via a Gateway.
- MANUFACTURING OPERATIONS CONTROL ROOM: This will be the nerve center and brain of the factory of the future. Using production schedules as input, personnel will dynamically allocate resources, schedule maintenance and monitor production facilities.
- FLEXIBLE MACHINING CENTER: Combining several areas of high technology equipment such as robots, automated guided vehicles and computerized numerical control machines, flexible machining control will implement 'just-in-time' material control systems, rapid setup techniques and predictable material flow.
- DATA CENTER: Large plant data centers have historically been the location of the databases used in material requirements planning, personnel databases, process routings and other data required for plant floor operations.
- PLANT FLOOR WORKSTATION: It will provide manufacturing personnel with information to improve their decision making ability. This may include process data, engineering specifications, personnel data or realtime quality feedback.
- FLEXIBLE MANUFACTURING: The flexible welding and painting operations of today are just the beginning

of flexible assembly applications. Use of flexible systems will expand to meet existing and new needs for improved quality, reduced scrap and shortened downtime.

- o PBX's: They will provide the gateway to wide area networks. Voice/data integration will allow database access and information transfer to and from world-wide data centers using both public and private wide area networks.

IMPACT OF FACTORY FLOOR LANS ON CAD/CAM AND ROBOTICS

Factory floor local area networks like MAP will have very significant impact on the flow of information within the manufacturing organization.

MAP type local area network with appropriate application level protocols can make data and information transfer a reality even among various different vendor's CAD systems. Now it is possible to have true integration of CAD and CAM anywhere in the factory.

In the area of Robotics, the robots are no longer just standalone equipment. They are part of an integrated manufacturing system and decisions made elsewhere or new programs can be communicated to/from them. One can envision an application where several robots perform certain work on a subassembly. If this subassembly is defective because of any robot related problem then spare robot(s) can be programmed properly from another computer based on the known defect.

One can even go a step further and envision a situation where robots are programmed off-line using a graphics system in a graphic oriented language. Now one can simulate the performance of the robot on the graphic system to verify correct operation.

The next step can be simulating operation of several robots working on say a car body for any interaction related problems in case of malfunction of any one of the robots. If there are no problems then these programs can be stored in proper format on a file server machine. When the time comes for producing a different model car on a given assembly line, one can down-load the new programs from the file server machine and verify them for all robots on the production line. This can allow for a quick change over in production models on the same assembly line. Not too long ago, two very similar cars had to be produced on two different assembly lines.

CONCLUSION

Factory floor local area networks will help in providing cost efficient solution for the industrial automation needs by increasing the capabilities and decreasing the communication cost. Such local area networks (with its associated gateways) will provide us with an integrated manufacturing system by connecting various islands of automation.

REFERENCE

General Motors' Manufacturing Automation Protocol - A Communication Network Protocol for Open Systems Interconnection, April 1984

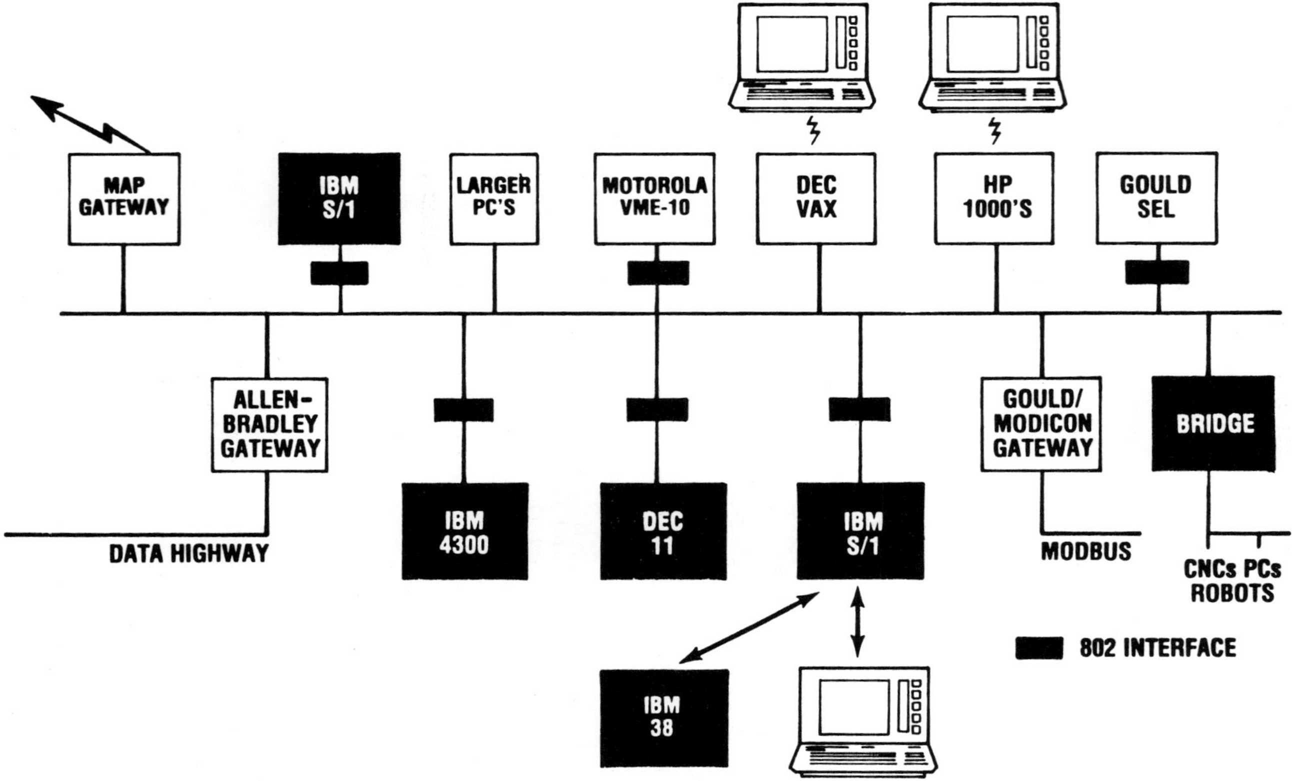

LOW COST HARDWARE
FIGURE 2

1ST QTR '86

Computer Aided Design of an Aircraft Control System

G. S. Alag

Gurbux Singh Alag*

NASA Ames Research Center
Dryden Flight Research Facility
Edwards, California

Abstract

Computer aided analysis and design tools for development of flight control systems for a variety of aircraft are widely used. The primary purpose of flight control systems is to improve the flying qualities of an aircraft. In case of a high performance aircraft maneuvering over a wide flight envelope, an adaptive control system has the potential for providing uniform handling qualities over the complete envelope and the computer-aided design tools for development of such a system are described.

I. Introduction:

The inability of simple mechanical linkages to cope with the many control problems associated with high performance aircraft has led to the present interest in digital fly-by-wire flight control systems.[1,2] Significant among the advantages of digital implementation are the weight and volume saving, ability to design complex controller structures, reliability of digital logic, and capability for time sharing multiple control loops.

An adaptive control system provides the capability of automatically compensating for parameter and environmental variations that may occur during operation. In case of an aircraft, such a system has the potential for providing uniform stability and handling qualities over the complete flight envelope and acceptable flying qualities despite external disturbances.[3,4]

In designing a parameter adaptive control system, consideration should be given to explicit adaptive systems in which online parameter identification is performed and implicit systems which do not

---

*National Research Council Senior Research Associate. The work presented here is supported in part by NRC Fellowship.

utilize direct parameter identification. In general, control methods which explicitly identify the aircraft have lower gain level requirements because the gains can be adjusted directly to their proper values. Recent studies have also indicated preference for explicit designs whenever the process to be controlled has non-minimum phase characteristics and/or high gain or bandwidth limitations.[5]

This paper discusses the computer-aided design and analysis tools used for the development and evaluation of an explicit adaptive control system for an aircraft.

II. Computer-Aided Deisgn Tools:

Many state of the art computer-aided design and analysis tools are currently in use.[6,7,8,9] These tools have been developed to utilize the interactive features of modern computers and help in reducing the effort involved in modeling and analyzing flight control systems. All the above referred programs are operational at NASA Dryden Flight Research Facility.

State space-oriented modeling and design procedures provide significant advantages over classical design methods. Most of the tools available use the state space approach for modeling, analysis and design of flight control systems and for designing linear feedback control laws and filters for linear time-invariant multi-variable differential or difference equation state vector models. Linear quadratic regulator synthesis procedure is the most general time-domain approach currently used for multivariable control formulation. The tools available use this procedure for design of continuous as well as discrete control laws for flight control systems.

Typically the tools discussed above will allow the engineer facility to perform following tasks for continuous and discrete systems.

- Root Locus, Bode and Nyquist Plots for Single-Input, Single-Output Systems
- Time Responses of State-Space System
- Linear-Quadratic Control Formulation
- State Estimation Using Kalman Filter

- Eigen-Value Placement
- Set of Supporting Programs of Linear Algebra for Matrix Manipulation

III. Adaptive Controller Designed Using Computer-Aided Tools:

Figure 1 shows the structure of an explicit adaptive control system for an aircraft. For development of an algorithm to be implemented, it is desired to estimate the varying parameters, estimate the states as measurements are contaminated with sensor noise, and compute the control based on the most recent gain computed. The tools allow the flexibility to estimate both the parameters and the states and based on most recent values of aircraft parameter, a gain is computed by using the linear-quadratic regulator theory. This gain enables the computation of control which would enable the desired performance from the aircraft. Some of the results obtained and details of the system are given in (10).

The tools can be used interactively with most computers and enable a significant reduction in time and effort. Comprehensive quantitative analysis of performance and the effect of significant parameters on the performance of an aircraft are facilitated by the use of computer-aided design methods discussed.

IV. Conclusions:

Some of the computer-aided design and analysis tools for flight control systems are discussed. These tools have a significant role to play in the future of control design practice. These tools run on most of the computers which allow interactive use of Fortran and most commonly used terminals are also supported for interactive graphics. These tools provide the capability to use methods for control design, system modeling and identification, and state estimation. A quantitative analysis of all parameters that affect a control system, e.g. sample rate, computational delays, noise, bandwidth, etc. is feasible by use of the above tools.

V. References:

1. Kass, P. J., "Fly by Wire Advantages Explored," Aviation Week and Space Technology, July 10, 1972, pp. 52-54.

2. Sutherland, Major J., "Fly by Wire Control Systems," AGARD Conference Proceedings, No. 52, Advanced Control System Concepts, Sept. 1968, pp. 51-72.

3. Smyth, R. and Ehlers, H. L., "Survey of Adaptive Control Applications to Aerospace Vehicles," ibid pp. 3-13.

4. Ostgaard, M. A., "Case for Adaptive Control," ibid pp. 15-27.

5. "Adaptive Control and Guidance for Tactical Missiles," TR-170-1, The Analytical Sciences Corporation, June 30, 1970.

6. Edwards, John W., "A Fortran Program for the Analysis of Linear Continuous and Sampled-Data Systems," NASA TM X-56038, Jan. 1976.

7. Armstrong, E. S., "ORACLS A Design System for Linear Multi-Variable Control," Marzel Dekker, Inc., 1980.

8. $MATRIX_X$ User's Guide, Integrated Systems, Inc., 151 University Ave., Palo Alto, California 94301, Nov. 1983.

9. DIGIKON IV User Reference Manual, Honeywell, Inc., Minneapolis, Minnesota 55440, NASA CR-11375, Jan. 1984.

10. Alag, G. and Kaufman, H., "An Implementable Digital Adaptive Flight Controller Designed Using Stabilized Single-Stage Algorithms," IEEE Transactions on Automatic Control, Vol AC-22, No. 5, Oct. 1977, pp. 780-788.

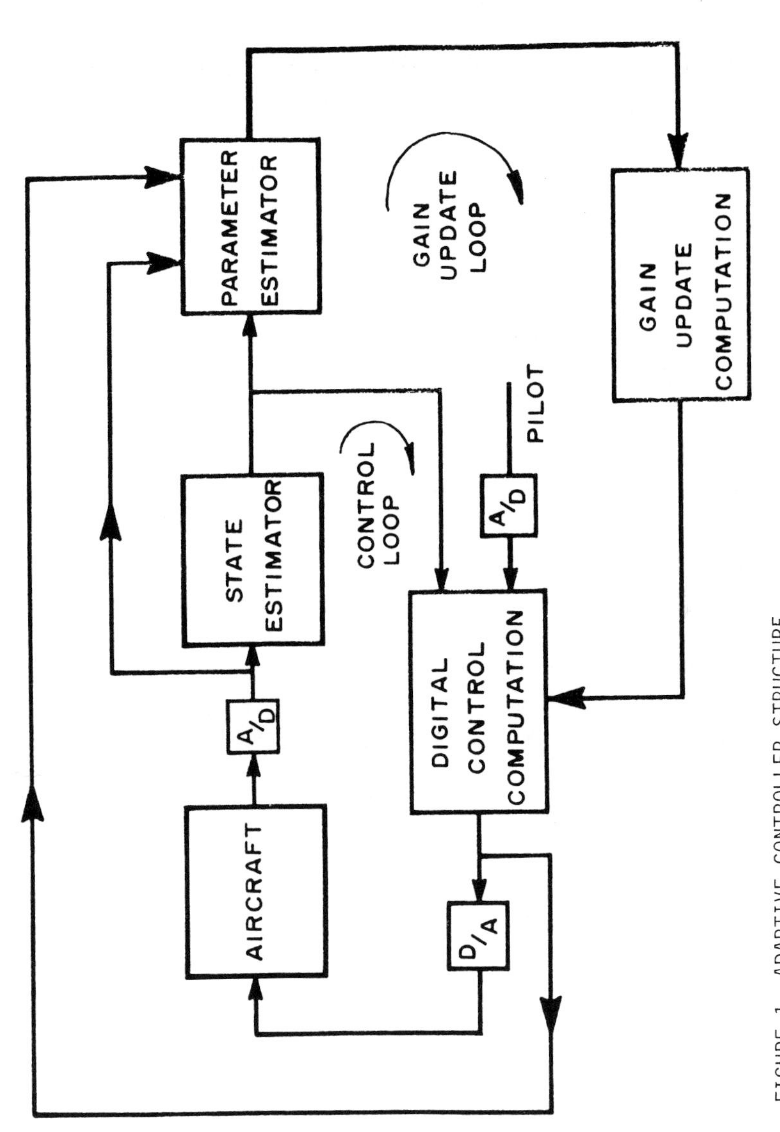

FIGURE 1 ADAPTIVE CONTROLLER STRUCTURE

A Multi-Microprocessor System with Distributed Common Memory for Computer Aided Manufacturing Applications

S. Ganesan, M. O. Ahmad, and R. P. Sharma

M.O. Ahmad
Department of Electrical Engineering
Concordia University
Montreal, H3G1M8, Canada

S. GANESAN

Department of Electrical Engineering
Western Michigan University
Kalamazoo, Michigan 49008

R.P. Sharma
Department of Mechanical Engineering
Western Michigan University
Kalamazoo, Michigan 49008, U.S.A.

Abstract

In this paper a tightly coupled multimicroprocessor architecture with distributed common memory and centralized common memory, for computer aided manufacture (CAM) application is described. Design of bus arbiter circuit and handshake sequence have been described. This multiple microprocessor system increases performance, reliability and matches the parallelism inherent in the CAM environment.

I. Introduction

Advances in factory computer systems, computer-aided design (CAD) and computer-aided manufacturing (CAM) will create a new industrial revolution. While automation was formerly used for large throughput systems with mass production, we can now automate plants, change models and batches, and still attain effective productivity through CAD/CAM. With the advent of the microcomputer, the computer intelligence can be brought to the individual machine. Intercommunication between these microcomputers and with an overall supervisory computer, and creation of CAD/CAM data base will bring in substantial productivity gain [1-4]. The flexible manufacturing systems of the future will combine the technologies of NC machine tools, computers, material handling systems, industrial robots, computer-aided design systems, and automatic warehouses.

A multimicroprocessor system suitable for CAM consists of two or more micros connected either through shared memory or via high-or low-speed data links [5]. The shared memory may be a multiported main memory, cache memory or a multiported disc [6]. The data paths may be either a bit serial or parallel bus connecting I/O ports of two computers or a shared bus to which two or more computers are interconnected in various ways.

Multimicro systems which employ the shared memory interconnect approach is called "tightly coupled." Here the common memory, I/O and other system peripherals, are shared by the micros. Interprocessor communication latency is low due to the potential access time being limited only by the actual memory access time. Loosely coupled systems do not have a shared memory. Tightly coupled systems require synchronization between co-operating processes. In loosely coupled systems concurrent processes can be performed asynchronously.

Depending on the application, data base handling in a multimicro system may be assigned to a special purpose back-end data base management processor. This data base can be accessed by all the micros in the system or the data bases may be distributed throughout the system in such a way that transfer of raw data is to the site of the data base. The data base updating is performed in a manner that minimizes load on the communications facility. It is possible to have either a centrally located data base directory that is frequently updated, to keep files or records locked or both, to avoid concurrent updates or interference. The critical data is usually stored in more than one location, to provide fault tolerance [7].

II. Multimicroprocessor System

A tightly coupled multimicroprocessor system is shown in figure 1. It has a number of microprocessor boards, a single common bus, a centralized common-memory (CMO) and necessary arbitration logic circuits. To each microprocessor a distributed common-memory, CMO, connected to the common-bus, is used to store data/information required by more than one microcomputer. For each microprocessor, the distributed common memory is seen as one single linearly addressable structure. The addressing scheme is designed such that any common-memory cell has the same address for all the microprocessors. For each microprocessor, the lower two block addresses select its private memory and the local distributed common-memory. The memory access control circuit connected to each microprocessor bus checks the memory address, and appropriately grants access to its private memory, local distributed common memory or passes the request to the common bus arbitration control circuit. The common-memory access control circuits [MAC] and common bus arbitration circuit [CBA] take into account priority of microprocessors and mutual exclusion, before granting access to the common memory (CMO, CM1 or CM2)[8]. The distributed common memory can be used to share information with a few microprocessors. To access the local distributed common memory, the local microprocessor does not require the use of common bus. So for some type of manufacturing applications, use of distributed common memory enhances performance throughput. A direct memory access facility is available to each processor to transfer data between private memory or local distributed common-memory and centralized common memory. The common bus arbiter circuit as shown in figure 2 has three parts-request queue, hardware semaphore and daisy-chain priority logic circuit. This interface circuit grants access to common memory taking into account the priority of the microprocessors and mutual exclusion. Master processor has the highest priority. The request queue flip-flops store the request for access to common memory from the processors. The requesting processor "busy waits" until grant signal is given by the bus arbiter. A multimicroprocessor architecture with hardware support for communication and scheduling is described in [9]. In the present system a hardware to support synchronization taking into account priority and mutual exclusion is used. The routine to access common memory is shown in figure 3. The system fault tolerance is achieved by memory management units, which relocate and protect programs/data against faults and programming mistakes.

Conclusion

Major effort of the designer in practical implementation of real time distributed microcomputer controllers for CAM application, is concentrated around the subjects of clever programming of the algorithm, proper scaling of all variables, selection of analog to digital and digital to analog converters. The multimicro system has received a widest acceptance in the process control industry and for CAM application.

References

1. RICHARD, C. Dorf, 'Robotics and Automated Manufacturing' Reston Publishing Co., 1983.

2. LERNER E.J., 'Computer-aided Manufacturing' IEEE Spectrum, November 1981, 34-39.

3. SUGARMAN, R. 'Blue Collar Robot' IEEE Spectrum, 1980, 52-57

4. KNO, M.H., 'Distributed Computing on an Experimental Robot Control System', IEEE 1981 Proceedings of Applications of Minicomputers, 330-335.

5. WEITZMAN, Cay 'Distributed Micro/Minicomputer Systems', Prentice Hall, 1980.

6. WEST,Trey 'Dual-ported RAM for the MC68000 Microprocessor', Motorola Application Note AN-881.

7. BOUCHET, P.and FEUVERE, A.J.M. and KURINCKX, G.A., 'PEPIN: An Experimental Multimicrocomputer Data Base Management System,' Int. Conf. Dist. Computing, April 1981, Paris, IEEE NO. 81 CH1591-7.

8. SUNDARARAJAN, D., AHMAD, M.O., and GANESAN, S., 'Interface Links 8 bit chips to provide multiprocessing', Electronics, October 20, 1983, pp 140-141.

9. AHUJA, S.R. and ASTHANA, A., 'A Multimicroprocessor Architecture with Hardware Support for Communication and Scheduling', ACM, 1982, NO. 0-89791-066-4, pp 205-209.

10. KIRRMANN, H.D., and KAUFMANN, F., 'Poolpo-A Pool of Processors for Process Control Applications' IEEE Trans. Computers, October 1984, pp 869-878.

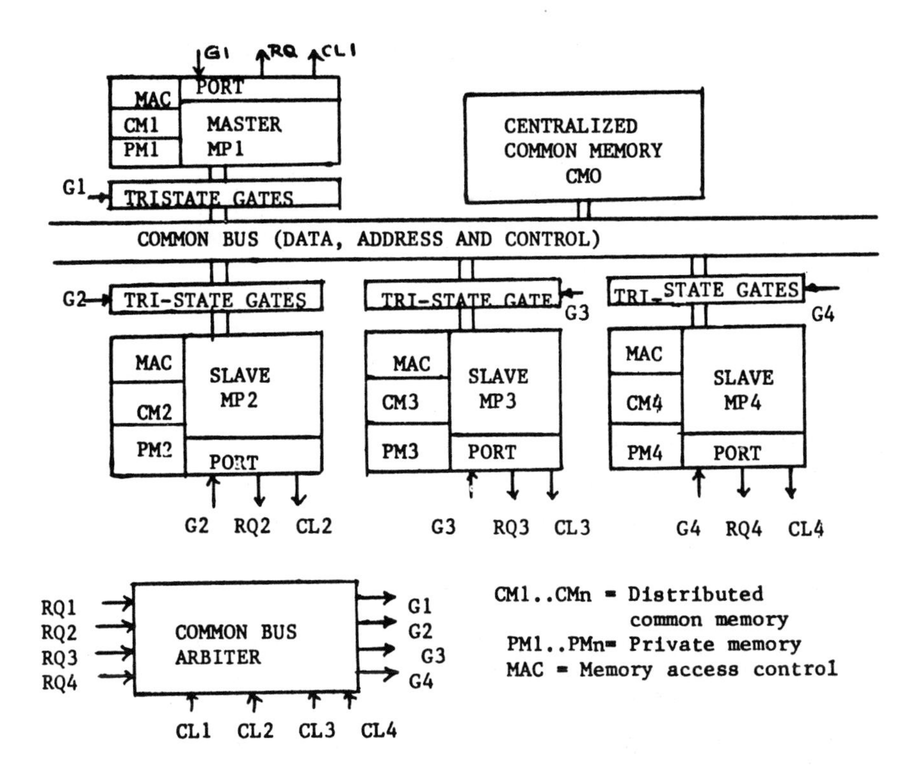

FIG. 1. MULTIMICROPROCESSOR SYSTEM

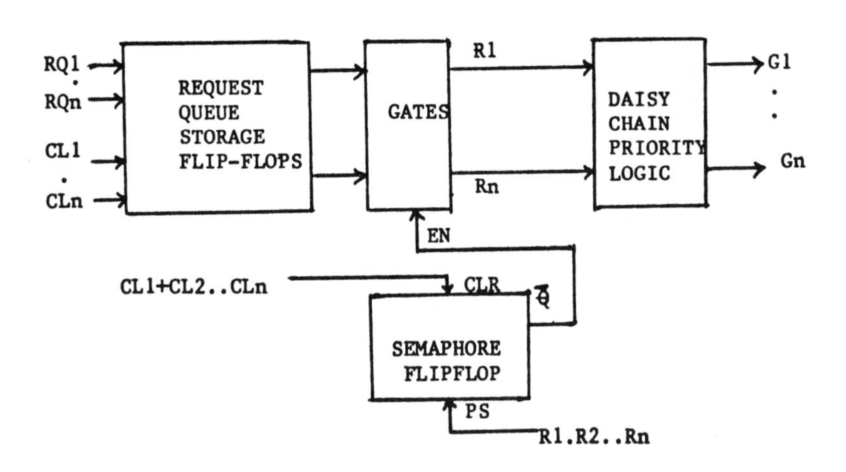

FIG . 2. COMMON BUS ARBITER CIRCUIT.

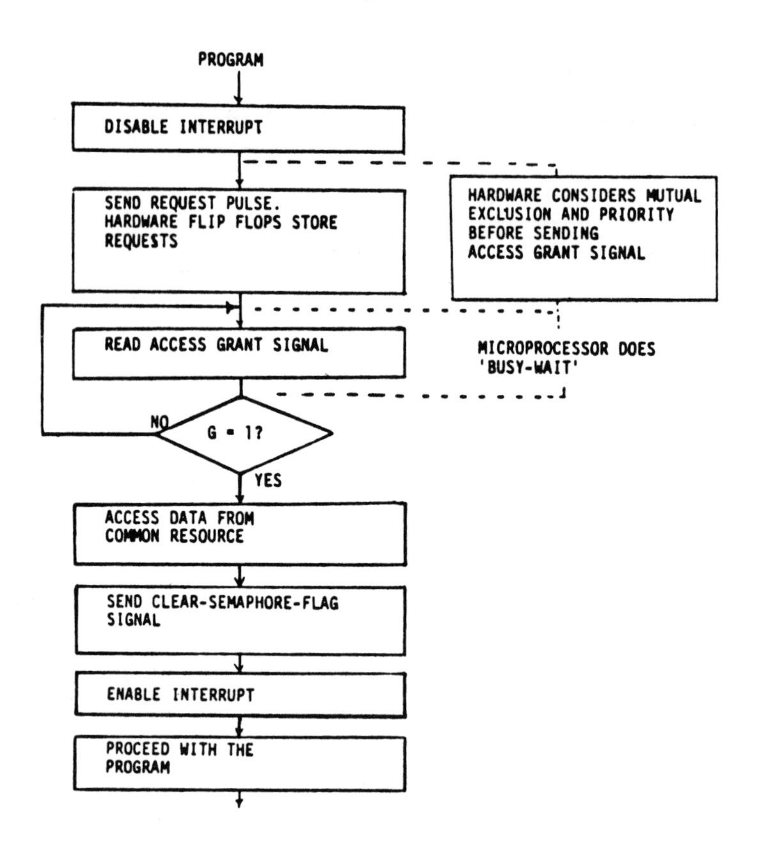

FIG. 3. ROUTINE TO ACCESS COMMON RESOURCE

The World of „Mechatronix"
Systematic Technology of Mechatronix Engineering and Education

K. Yamazaki and Hiroshi Suzuki

Production Systems Engineering Department
Toyohashi University of Technology, Aichi Japan

Abstract

Microprocessor application for mechanism control has recently been referred to as "Mechatronix". Much attention has been focussed to the addressed technology by the mechanical industry in general, especially in the manufacturing areas. A systematic methodology of "Mechatronix" engineering and related tools have been developed at Toyohashi University of Technology to be included as a part of mechanical engineering education. Continuing education program on "Mechatronix" for those working in the mid- to small-scale companies has also been created.

The developed methodology includes standardized software implementation and hardware fabrication procedures with the aid of standardized document management. The technology is particularly designed for use in developing special purpose mechatronix controllers according to customer's specification.

Hands on projects based on this methodology are undertaken by students as a part of educational requirement toward advanced degrees. Academic course works as well as continuing education programs consist of lectures, demonstrations and laboratory projects so that the participants can understand the details of microprocessor-based programmable digital servo and programmable sequence control functions.

1. Introduction

"Mechatronix" is a popular terminology originated in Japan representing the technology of a modern mechanical system tied up with electronics control. Since the Second World War, electric and electronic technologies and devices, especially computer system and its peripheral technology have been advancing rapidly and traditional manually operated machines have been replaced with the newly desingned machine system with intelligent control. When "Numerical Control" for a machine tool evolved, a digital control scheme was first introduced with the dedicated hard-wired electronic circuit.

Today's numerical control accommodates CNC(Computerized Numerical Control) which fully utilizes the advantages of a microprocessor. This technology has been useful to simplify the mechanical design. For example, a threading function of a lathe required a gear transmission mechanism by which the tool feed motion was synchronized with the main spindle rotation. However, a CNC lathe has the electronic synchronization mechanism thus eliminating the physical gear transmission. This has greatly simplified the tool feed drive mechanism as well as machine set up because a greater versatility has been provided to programmable threading operation. The microprocessor-based on-line calculation for synchronized motion allows cutting of specially designed thread such as tapered thread, consecutively variable(increasing or decreasing) lead thread which has been almost impossible in the conventional or hardwired NC lathe. This is a typical advantage of mechatronix technology obtainable by applying the advanced electronics to mechanism control.

The great impact of the mechatronix application can be found when the following criteria are economically satisfied.

(1) A conventional mechanism is simplified by utilizing microprocessor-based control system.
(2) A motion of the mechanism is programmablly controlled by the software.
(3) A new useful function or high performance, which can not be realized with the conventional mechanism, is implemented by software with no or few additional change in mechanism.

The essential technologies of mechatronix developed in machine tool control have been machine motion control and machine sequence control. As those technologies have been likewise considered useful to more generalized mechanism control, applications to the various machines have been proposed and demonstrated [1], [2], [3], [4]. The microprocessor in the mechatronix controller has also contributed to enhancement of control capability with its strong calculating power [5], [6] and total hierarchical control of the number of machines has been well developed by utilizing the large data handling capability (such as DNC systems) [7], [8].

The conventional way of developing commercially marketable mechatronix system has been a cooperative task whereby the mechancal part of the system has been designed and manufactured by mechanical industries and the mechatronix controller has been designed and fabricated mostly in electronics and control equipment industries. However, as the complicated and specially designed mechanism are increasingly replaced with the control software resulting in the reduced mechanism cost and increased functional versatility, the mechanical industries are rushing into the mechatronix technology to stay on competitive edge by developing the machine with custom-made contoller. This results in the great demands on the mechanical engineers who can freely handle the electronics, especially microprocessor application technology. This is causing a critical demand for human resource in mechanical industries today. In order to meet the demand, it is necessary to educate the mechanical engineers and students with mechatronix technology, having main emphasis on the microprocessor application to machine control. The mechatronix technology, however, is a new integrated area for which the current education system has not been well organized. The following steps should be taken to solve the problems:

(1) Establishment of "Mechatronix" as a new interdisciplinary academic area.

(2) Organizing the education of "Mechatronix" for the mechanical engineering students.

(3) Dissemination of "Mechatronix technology" for retraining or continuing education of in-service mechanical engineers.

This paper describes the methodology of systematized mechatronix technology and the educational programs currently developed at the Toyohashi University of Technology.

2. Mechatronix Controller Design

2.1 System Functions

The "Numerical Control Technology" can be considered as the origin of mechatronix technology. For Numerical Control with microprosessor system, the various architecture and configuration has been discussed [9], [10], [11], and the concept of the NC machine tool conroller has been applied to controllers of other machines such as robots, coordinate measuring machines and material handling devices, etc.

The functions required for generalized mechatronix control are:

(1) Programmable Automatic Control Function
(2) Manual Control Function
(3) Preparatory Function
(4) System Diagnostic Function

Each function consists of several subfunction modules as shown in Fig. 1, and it should become active when an appropriate operation is taken. Typical operations required in mechatronix control are:

(1) Automatic Operation
(2) Manual Operation
(3) Preparatory Operation
(4) Emergency and Initialization Operation

Each operation has one or more operation mode(s) based on the parameters which govern the corresponding function. For example, NC automatic function can be specified by input media so that one of the types of control continuity is selected (i.e. consecutive automatic operation or single step operation as shown in Table 1).

In order to adjust to the required change in operational condition such as an operator intervention and malfunctioning, an interacting mode transition rule should also be designed. This is called "Mode interlocking" and is very important for safe and reliable machine control. The interlocking is conveniently represented by "Mode Transition Chart" as shown in Fig. 2. The circles denote the mode state and the lines with arrow show the direction of state transition when either external or internal requests are generated by machine operator or system itself. Those requests are marked beside each line in an abbreviated form.

2.2 System Configuration

Required functions and operation modes should be implmented in the physical controller consisting of microprocessor-based hardware and software. Fig. 3 shows the generalized

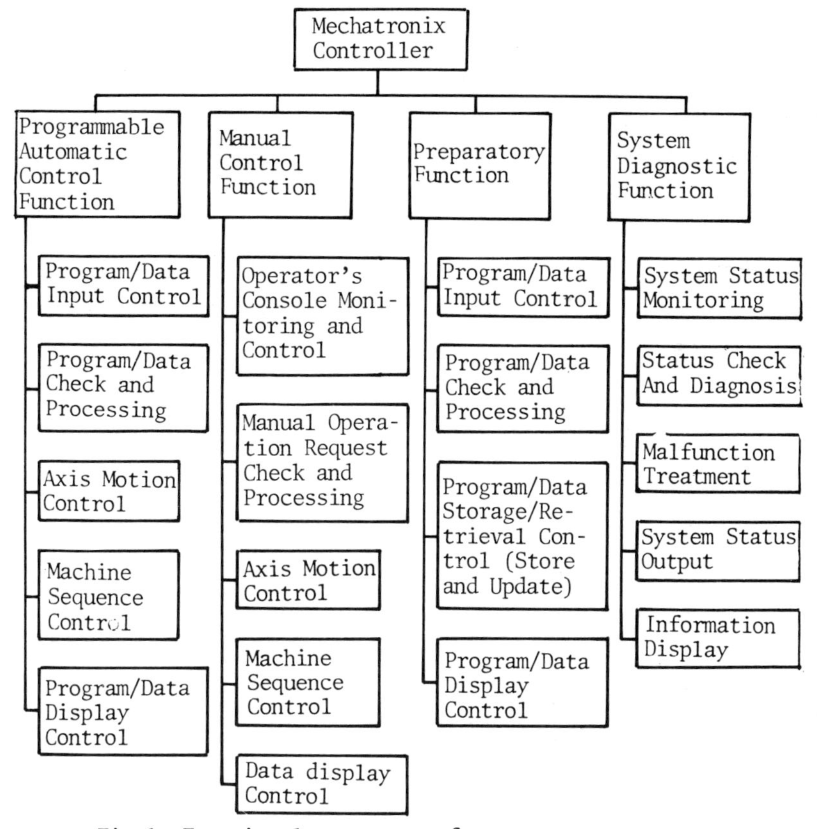

Fig.1 Functional structure of the generalized mechatronix controller

configuration of mechatronix hardware.

The upper part is input interfaces connected to various input devices, operator and data transmission from the related computer systems. The typical input devices located at machine side are position sensors, speed sensors, pressure sensors, force sensors, temperature sensors and keys and switches on the Operator's console. Appropriate interfaces should be provided for signal conditioning to take the input signals into the microprocessor based logic part of the controller. For examples, counters should be provided for counting relatively high frequency pulses input and an A/D converter is required when the input is an analog signal. All kinds of signals from different devices should be conditioned and converted into digital data after passing through the input interfaces. The input interfaces are connnected to system bus so that the CPU can freely access to the input data.

The memory part is generally classified into two parts with respect to its attribute. One is a system memory consisting of ROMs and RAMs. The ROMs part is dedicated for system control

Table 1 Example of the active modes of operation

1. Automatic Control Operation	3. Preparatory Operation
• Tape-Continuous • Tape-Single Block • Memory-Continuous • Memory-Single Block • MDI-Single Block • Feed Hold	• Machine Set up Data Input • System parameter Input • Tool Set up Data Input • NC Program Store • NC Program Edit • Manual Data Input • System Status Check
2. Manual Control Operation	**4. Emergency/Initialize Operation**
• Feed-(Rapid,Jog,Handle Step) • Spindle(ON/OFF,CW/CCW,Speed) • Coolant(ON/OFF) • Tool Select • Tool Clamp/Unclamp • Work Clapm/Unclamp	• System Reset • Emergency Stop

software which performs implemented control functions. RAMs in the system memory are dedicated for system constants and parameters setting. The other memory area is a user-RAMs for user program and data (e.g. robot motion program or NC part program, etc.). Some of the RAMs require battery back up for maintaining the data even if the power is off.

As a result of control, the various kinds of data and

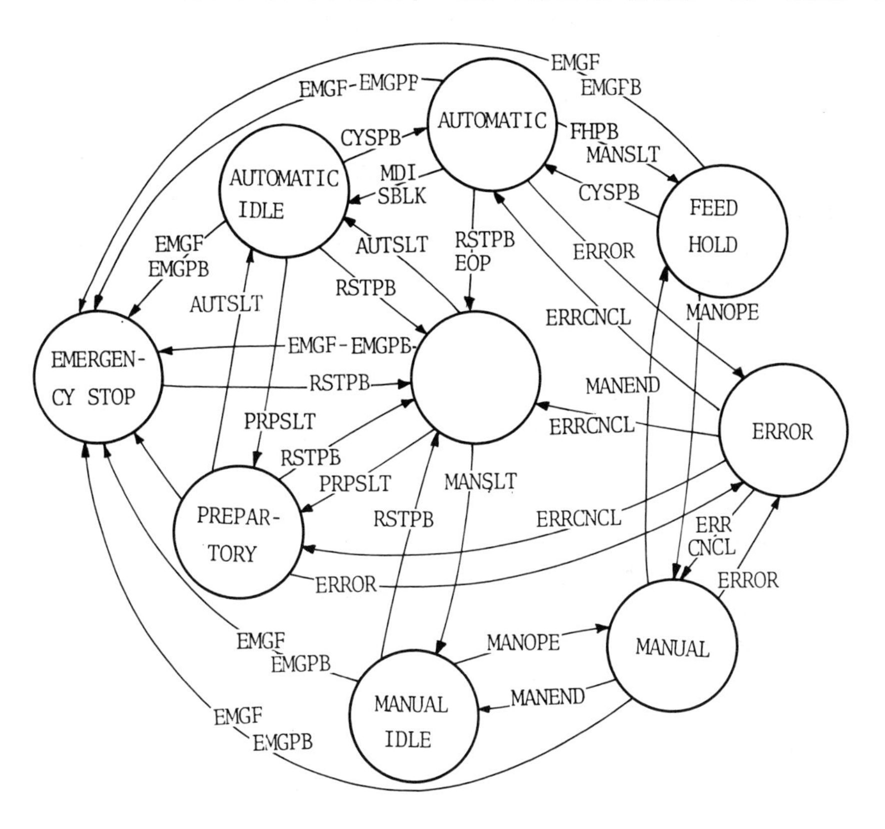

AUTSLT	:	Automatic Mode Select	CYSPB	:	Cycle Start Push button ON
EMGF	:	Emergency Failure Occurrence	EMGPB	:	Emergency Push button ON
EOP	:	End of Program	ERRCNCL	:	Error Cancel button ON
FHPB	:	Feed Hold Push button ON	MANEND	:	Manual Operation End
MANOPE	:	Manual Operation.	MANSLT	:	Manual Mode Select
MDI	:	Manual Data Input	PRPSLT	:	Preparatory Mode Select
RSTPB	:	Reset Push button ON	SBLK	:	Single Block Select

Fig. 2 An Example of Mode Transition Chart (CNC Control)

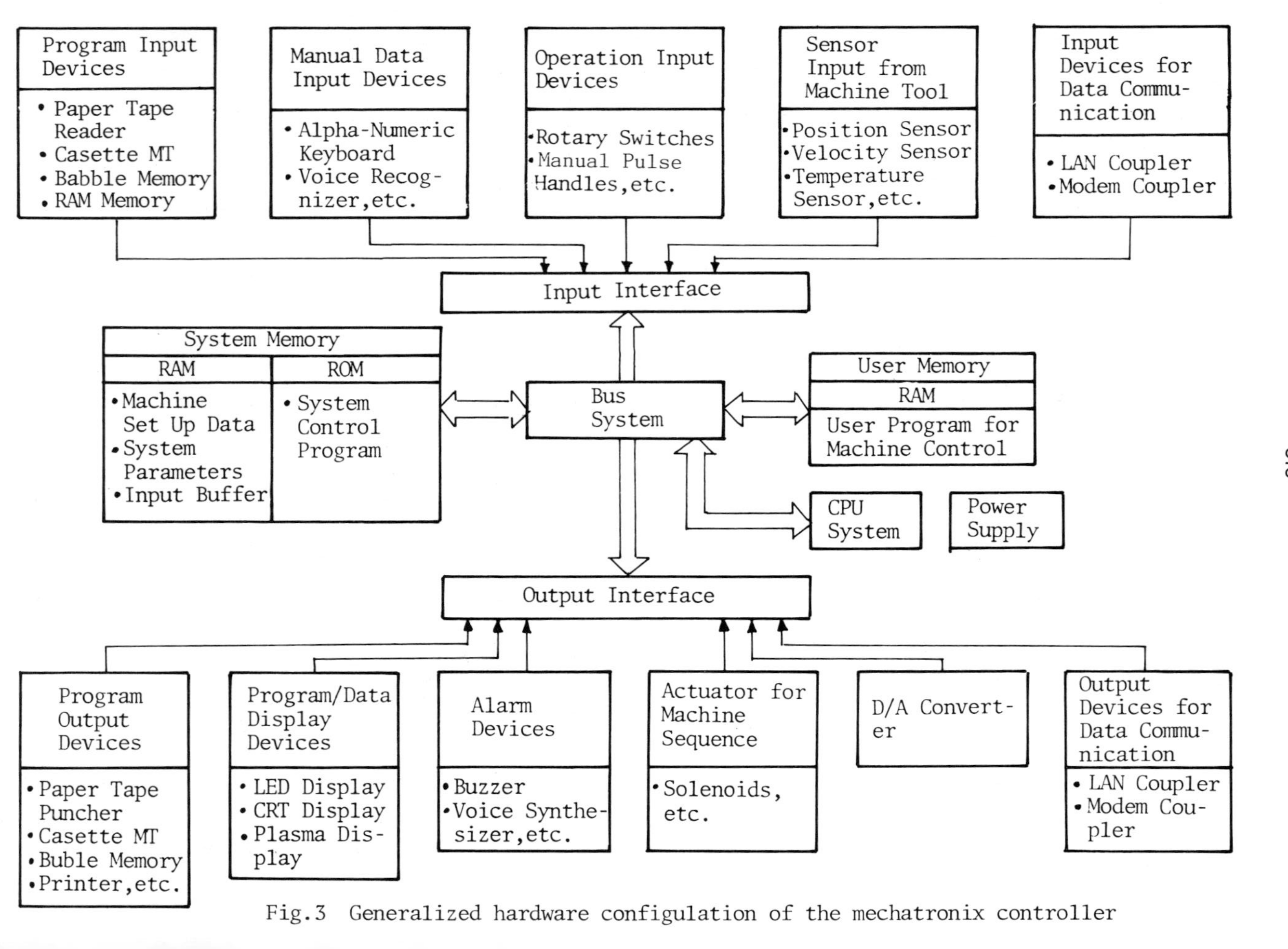

Fig.3 Generalized hardware configulation of the mechatronix controller

signals are output through output interfaces. The outputs to the machine are digital signals mainly for ON-OFF type(sequence) control and the analog signals mainly for continuous motion (digital servo) control. Other output peripherals are for man-machine interfaces such as display devices and for data communication to upstream computers.

To be cost effective, required functions should mostly be implemented in the form of software. Fig. 4 shows the general flowchart of the mechatronix system control program. After the power is turned on, the system is initialized and the system diagnosis is performed to check if malfunctioning exists. When the error is not detected, operator's console status and machine status are monitored to recognize the request for operation. Based on the monitor results, the operation mode is controlled by referring to the mode transition rule previously mentioned. Then the system goes to the processing of various control functions with respect to each operation mode required. After performing the control functions, data display is serviced for man-machine communication and the system goes back to the diagnosis processing. This repetitive loop of the system software is usually performed on the free time (without clock synchronization) basis. The control functions which require time-synchronized processing can be implemented in an interrupt-driven program. Typical interrupt-programs for mechatronix controller includes digital servo control for the continuous motion of mechanism, timer control for counting time, data communication service and system reset. The digital servo control program and timer program should be driven by clock interrupt and others are event driven type interrupt program.

3. Methodology for Mechatronix Controller Development

3.1 External System Specification

To achieve the efficient development, the procedure should be simplified and standardized [12]. It is also important to establish the straight forward method as much as possible. Fig. 5 shows the sequences of Mechatronix Control System Development.

System development starts when the requirement specifica-

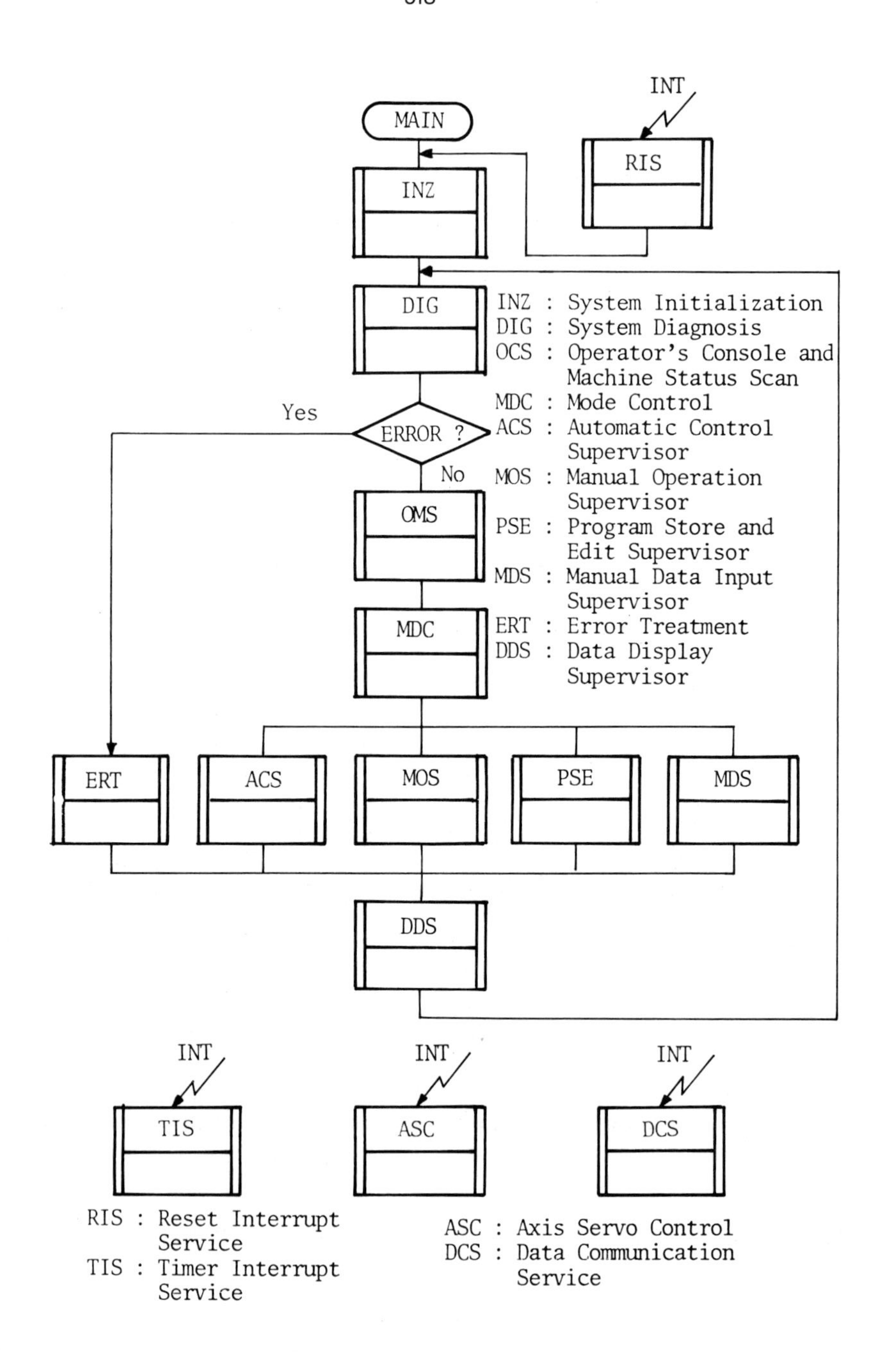

Fig.4 General flowchart of mechatronix system control program

tion is given by the end user of the system. In most cases, as the end user does not always have enough knowledge about the control system configuration, the requirement specification should be checked and rearranged into functional specifications and operational specifications by adding the items missing in the requirement specification. This process is usually worked out by negotiation between end users and system designers. Function specifications describe inputs and outputs to and from each function and what processing should be done to relate specified inputs and outputs. The performance of the function

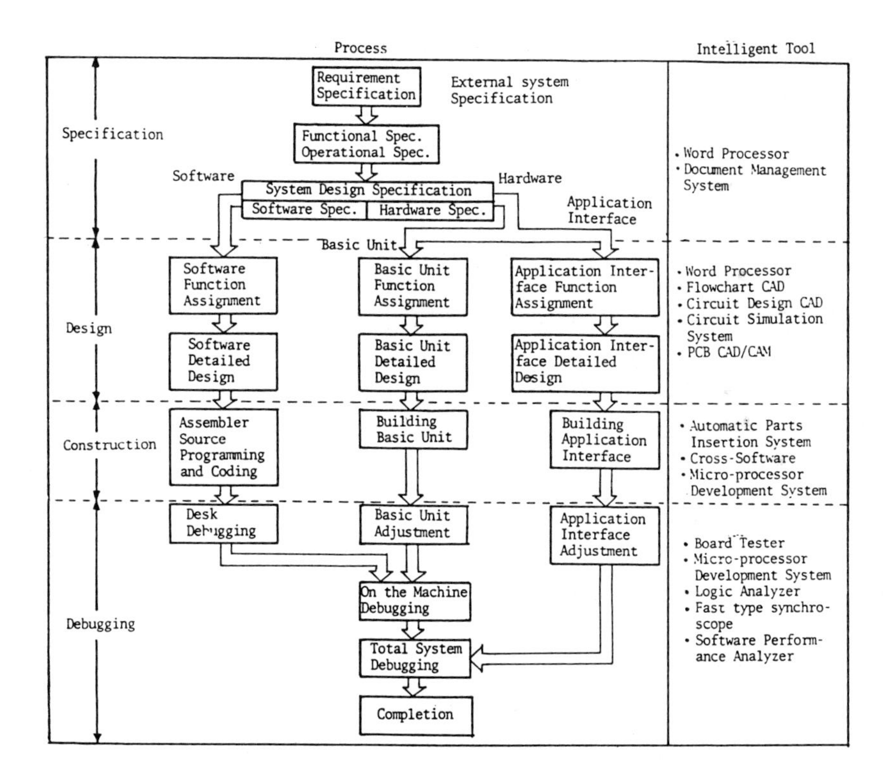

Fig. 5 The sequences of mechatronics control system development

(e.g. execution time and accuracy of calculation etc.) can also be specified if necessary.

Operational specification must give the explanation of how to activate a specific function by the machine operator. An operation mode transition (interlocking relationship) should also be specified in the operational specification.

If an end user has sufficient knowledge the functional and operational specifications may be written by the end user. The combination of functional and operational specification is sometimes called as the "external system specification". Once these specifications are obtained, system designer is ready to design the hardware and software configuration for the system required.

3.2 Internal System Specification

When designing the system configuration, it is important to define clearly what kind of method is taken for hardware and software fabrication.

The alternative methods currently available for fabrication are generally classified into the following three groups.

1. Chip Level Fabrication

Costom-designed hardware is fabricated from scratch. The software is also specially designed and implemented.

2. Module Level Fablication

Commercially available hard-ware modules (e.g. CPU boards, memory boards, and digital I/O boards, etc.) are selected and used. The software should be designed and implemented.

3. Unit Level Fablication

Commercially available control unit are selected and used. The typical hardware is programmable sequence control unit and axis positioning control unit. These unit are connected to personal or minicomputer which performs supervising the control functions. The machine motion and sequence control programs are written in easy language specially designed by unit manufacturer and the supervising program is written in the general language such as assembler or a high level language.

The results of the decision is represented in the system design specification consisting of hardware specification and software specification. This specification is sometimes called as "internal system specification".

3.3 Design and Construction

After obtaining the system design specification, the development process flow is divided into two streams. One is for the software development and the other is for the hardware. Passing through the software design stage the detailed software design is obtained usually in the form of logic diagram.

In the chip level fabrication, outputs from hardware design stages are circuit diagrams and mounting diagrams. Recently, the hardware design stage can be automated with a CAD system which allows the interactive circuit design, circuit simulation for verification, automatic mounting design, and PCB(Printed Circuit Board) design, etc.

The hardware consists of a basic unit and application interface. The basic unit is a fundamental microprocessor system including CPUs and memory circuits. A variety of interfacing circuits should be provided as an application interface. An application interface includes counters, digital I/O module, analog I/O module and data communication module, etc.

On the software construction stage, software source program is written in the assembler or a high level language. There are several ways to generate source program and to compile or assemble the program. The most popular one is to use the microprocessor development system. In recent years, the development system has been well developed and has a variety in terms of its cost and versatility. Some commercial type personal computer system equipped with the general operating system (CP/M, etc.) can also handle the software generation for a range of popular 8- and 16-bit microprocessors. The output from the software construction stage is the object program written in "microprocessor machine code". The object program list should be carefully checked by the designer.

3.4 Debugging

After the completion of hardware construction and adjustment, the generated software is transfered to basic unit hardware, and "on the machine debugging" is performed to check if the basic part of the hardware works properly with system software. Then finally the fabricated controller is hooked up with target machine system and total system is debugged. Debugging should be performed on both hardware and software with the appropriate debugging tools. Debugging procedure can be prepared with necessary test data by referring to the specification, design documentation and object module input/output list. If debugging reveals a hardware problem, the signals in the hardware circuits should be examined in order to find the pin point bug. The desired tool for the logic circuit analysis is a "logic analyzer" which can record and analyze simultaneously multi channel digital signals (usually 16 to 32 channels) at very fast sample rate (more than 50 MHz). The logic analizer offers a big help when analyzing the timing problem and noise problem.

When analyzing an analog signal problem, the high speed synchroscope is generally used. Recently the logic analyzer with analog analysis capability is marketed aiming at the microprocessor based system with digital and analog signal processing[13].

3.5 Development tools

The microprocessor development system mentioned in the above plays a great role in an efficient debugging. Fig. 6 shows the alternative ways of software generation and debugging using presently available tools. As shown in the figure, only the microprocessor development system can be consistently used throughout the process starting from software generation and ending with the real time debugging on the target system. The microprocessor development system, however, still has a weak point in that it is not a easy task to construct a microprocessor development system and requires a lot of time before the incircuit emulator (the most difficult part of the development system) for a new processor becomes available on market. Therefore, when a new microprocessor chip appears in the market, the development system for the chip is not always available from

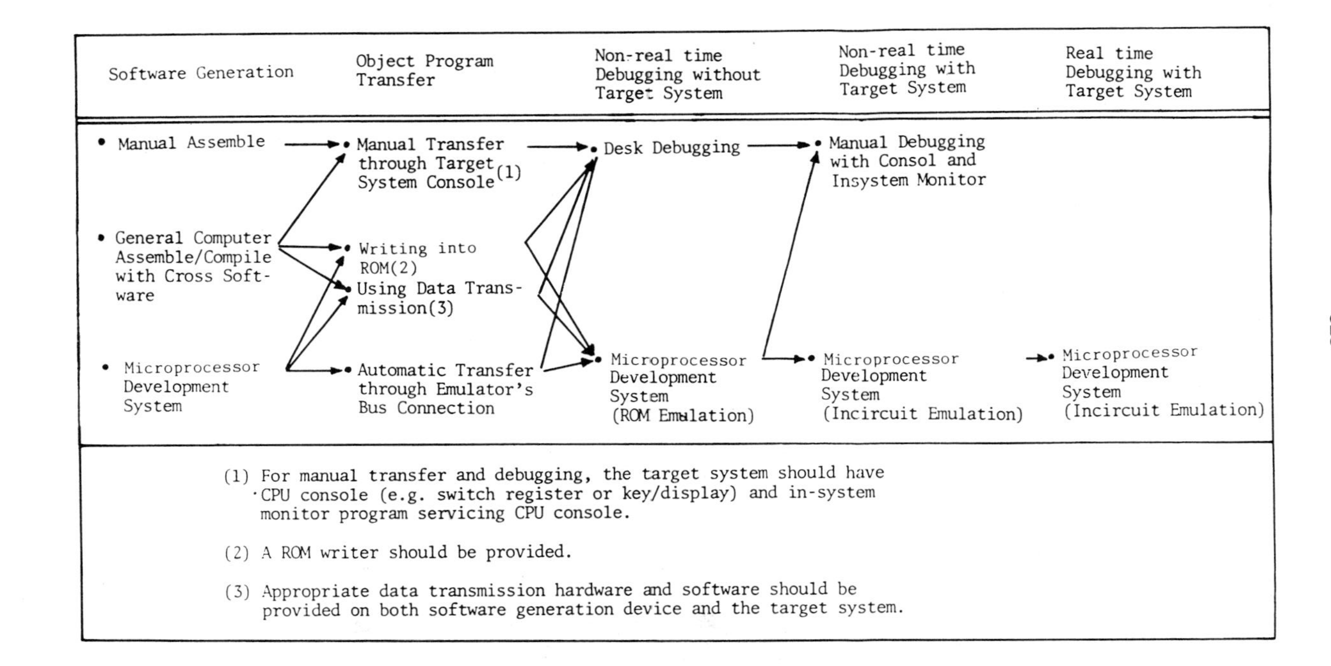

Fig. 6 Various methods for software generation and debugging

a development system vendor. The typical delay between chip release and the development system is some 6 months to one year. This causes the problem when the fast application of the new chip is required. This problem can be overcome by the following procedures:

(1) For software generation, the cross assembler for a new chip should be developed on the general purpose computer such as personal, mini and large computers.
Recently, the powerful method has been developed entitled "UDA (User Definable Assembler)"[14]. The UDA can automatically generate the assembler for the new chip when the specifications of the required assembler is given. These specifications should include operation code, operand information, macro instructions and grammatical rule of the assembler language. The development of the new assembler allows the software engineer to develop the source programs and to assemble them on the microprocessor development system. According to our experience, UDA method takes about 1 to 2 months to get the final version of assembler for typical 8 or 16 bit microprocessors.

(2) For hardware, some alpha-numeric keys and displays should be provided on the target system so that the primitive debugging operation can be done on the target system. A small size in-system debugging monitor should also reside in the target system memory for supervising the abovementioned hardware (keys and displays).
The monitor should have at least the following functions:

memory read/write
register read/write
execute
single step
reset

The generated object code can be transferred to the target system via ROMs or manual input. With the hard-scale industries. The contents of the seminar, as shown in Table 4, covers various aspects of the technologies necessary for both the design and fablication of the mechatronix system. The seminar has been a 12 days program consisting of 50 % lectures and 50 % hands on projects. The projects have been designed such that each participant can go through the major steps of the system development. The generalized mechatronix controller design and fabrication have been assigned to each group consisting of three participants.

The chip-level fabrication has been practiced by making the basic CPU unit (8 bit), digital I/O, keyboard display, counter, digital to analog converter and data communication interface. The completed controller has been hooked up to the AC servo motor and the position loop control has been demonstrated by implementing the digital servo control software as shown in Fig. 8, and Fig. 9.

5. Conclusion

As the mechatronix technology has drawn attentions in mechanical engineering, the technology for application of the

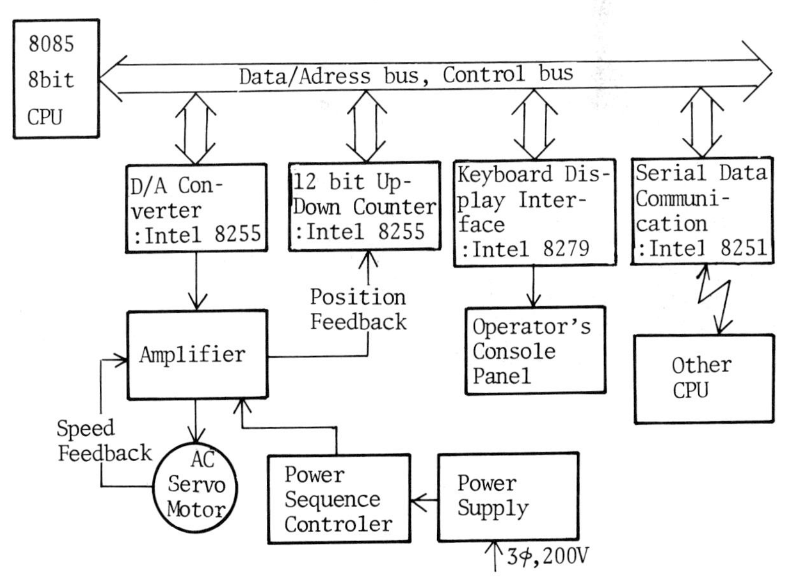

Fig.9 Hardware configuration of the mechatronix controller

microprocessor to the machine control is in great demand and comprehensive system configuration and standardized development methods have been highly desired by mechanical engineers. In order to meet those demands, the systematic mechatronix technology and the education system has been studied. The following are concluded from this paper:

(1) The general specification of the mechatronix control has been proposed as well as the unified system configuration.
(2) The standardized development procedures have been organized and the specific procedure in each development stage has been explained with respect to the levels of application and available tools.
(3) The mechatronix education comprising lectures and demonstrations has been established for mechanical post graduate students based on the systematized mechatronix technology.
(4) The dissemination program "Mechatronix Seminar" has been created in order to support the in-service mechanical engineers who want to get involved in the mechatronix technology.

REFERENCES

[1] French, D. and Ferreira, A., The Application of Numerical Control Techniques to Production of Gears, Annals of the CIRP, Vol.24/1/1975. pp.471-474.
[2] Weck, M., Bagh, P. and Holler, R., Measuring of Gears with a Numeric Controlled 3-Axis Measuring Machine, Annals of the CIRP, Vol.24/1/1975, pp.375-378.
[3] Weck, M. and D'Souza, C., Structure of Modular Control Systems for Handling Devices Using Micro-Processors --- Application in Manufacturing Systems, Annals of the CIRP Vol.28/1/1979, pp.385-390.
[4] Crossley, T.R. and Unsworth, R.E., Design of a Microprocessor-Based Draughting System, Annals of the CIRP, Vol. 29/1/1980, pp.363-367.
[5] Stute, G. and Hesselbach, J., Discrete-Time Position Control at NC Machines and System Optimization, Annals of the CIRP, Vol.24/1/1975,
[6] Stute, G. and Kapajiotidis, N., Integration of Adaptive Control Constraint (ACC) into a CNC, Annals of the CIRP, Vol.24/1/1975, pp.411-415.
[7] Crossley, T.R., and McCartney, D., A Decade of Direct Numerical Control, Annals of the CIRP, Vol.27/1/1978, pp.405-408.
[8] Crossley, T.R. and McCartney, D., Microprocessor-Based

lative to the immovable basis of the feeder. To obtain the required signal in the case of speed or displacement, two elements must be connected to the system, one to the bowl and the other to the base. The advantages of measuring acceleration are:

a) higher accuracy of the accelerometric feedback and a smaller possibility of introducing disturbances;

b) lower cost of fastening the sensor to the vibrofeeder.

In our experiments we used both speed and acceleration sensors. The former is based on the utilization of current induced in an immovable coil fixed to the base of the feeder by a vibrating magnet attached to the bowl.

We will continue our discussion with

(1) an analysis of the energy consumption of the adaptive vibrofeeder, as compared with that of conventional feeders.

(2) a presentation of the experimental results obtained with an adaptive vibrofeeder built at Ben-Gurion University of the Negev on the basis of an industrial AA model-5 vibrofeeder produced by Aylesbury Automation Ltd. (England).

The power N_o consumed by the system under consideration from the outer source can be defined as:

$$N_o = F\dot{x} \qquad (1)$$

The energy N used during half a period is:

$$N = \int_o^{T/2} F\dot{x}dt \qquad (2)$$

(We specify half a period because the electromagnet is only able to pull the armature during half a period, while during the other other half it does not influence the mechanical system.)

We will now describe analytically the model depicted in Fig. 2 and show the energy saving in comparison with a conventional vibrofeeder. According to the Figure:

X - the gap between the magnet and the armature when no electric current flows through the coil;

x_o- the initial gap between the magnet and the armature when the constant component of the current flows through the coil;

$|\Delta| = X-x_o$ - the deformation of the spring caused by the constant component of the electric current.

Taking into account the directions of these sections, we can determine that:

$$-\Delta = X - x_o \tag{3}$$

The following equations describe the work of the system.

$$m\ddot{x} + b\dot{x} + c(x + \Delta) = F_i + F_e$$

$$\frac{d(Li)}{dt} + Ri + \frac{1}{c}\int idt = U_i + U_e$$

where

b - lumped damping coefficient;

c - the lumped stiffness of the spring system;

C - lumped capacitance;

i - electric current;

L - lumped inductance;

m - the lumped moving mass (tray or bowl plus conveyed items);

R - lumped electric resistance;

t - time;

x - displacement of the bowl;

F_i - internal force developed by the magnet;

F_e - external component of control voltage;

U_i - internal component of control voltage;

U_c - external component of control voltage

The force F_i can be expressed as:

$$F_i = \partial W/\partial x \tag{5}$$

where W - magnetic flow equals $W = \frac{1}{2} Li^2$ (6)

while $L = \frac{L_o}{\alpha}\left(1 + \frac{x}{x_o}\right)^{-1}$ for $|x| \ll x_o$ (7)

where

L_o - initial inductance of the coil;

α - lumped coefficient of the dissipation of the magnetic field

Then

$$F_i = -\frac{L_o i^2}{2\alpha x_o}\left(1 + \frac{x}{x_o}\right)^{-2} \tag{8}$$

The voltage U_i, in our case, is generated by the circuit and is made proportional to the vibrating speed $\dot{x}$: $U_i = A_1\dot{x}$
Where A_1 is a contstant depending on the properties of the circuit. A constant magnetic field (generated by the constant current component) is used for the following reason. It is clear from equation (8) that the electromagnetic force, which is the

vibration exciter in the vibrofeeder under consideration, is proportional to the square of the electric current in the coil of the magnet. If the current is harmonic, i.e. if

$$i = i_o \cos\omega t \tag{9}$$

where i_o - the amplitude of the alternating current

ω - the frequency of the alternating

then the electromagnetic force will change with twice the frequency 2ω. To avoid this situation we have to ensure that the changing component of the current does not cross the zero axis (a negative value of the current is not allowed). By adding a constant current i* which is not less than the ampltiude i_o, we satisfy this condition.

On the basis of equation (4) and our own experience, we assume that the motion law of the bowl actuated by the current in accordance with equation (9) has the following harmonic form:

$$x = a\cos\omega t \tag{10}$$

where a - the amplitude of the bowl's oscillations.

Let us deonote:

q - electric charge;

q_o- amplitude of the charge;

ω_o^2 = c/m - natural frequency of the mechanical oscillator.

Now substituting equation (8) into equation (2) and remembering expression (10), we obtain:

$$N = \frac{L_o x_o q_o^2 \omega^3 a}{2a} \int_0^{T/2} \frac{\sin^3 \omega t dt}{(x_o + a\cos\omega t)^2} \tag{11}$$

To simplify the analysis of expression (11), we introduce a non-dimensional value $B = x_o/a$. It then follows from equation (11) that:

$$N = \frac{L_o \omega^2 q_o^2}{2} B \left[B \ln \frac{B+1}{B-1} - 2\right] \tag{12}$$

On the other hand, considering the mechanical part of the vibrating bowl as a one-mass oscillator, we can express the oscillation amplitudes as follows:

$$a = F_o/m\sqrt{(\omega_o^2 - \omega^2)^2 + 4n^2\omega^2} \tag{13}$$

where F_o - amplitude of the exciation force: and $n = b/2m$
From equation (8) we derive:

$$F = -\frac{L_o i_o^2}{2\alpha x_o}(1 + \frac{a}{x_o})^{-2} \tag{14}$$

Remembering that $i = \dot{q}$, we obtain:

$$i_o = q_o\omega \tag{15}$$

Substituting equations (15) and (14) into expression (13) we obtain:

$$q_o^2 = \frac{2\alpha a(x_o + a)^2 m\sqrt{(\omega_o^2 - \omega^2)^2 + 4n^2\omega^2}}{L_o\omega^2 x_o} \tag{16}$$

Substituting equation (16) into equation (12), we obtain an expression which describes the relationship between the power used and the excitation frequency in the following form:

$$N = 2[B\ln\frac{B+1}{B-1} - 2](x_o + a)^2 m\sqrt{(\omega^2 - \omega^2)^2 + 4n^2\omega^2} \tag{17}$$

Let us define $\lambda = \omega/\omega_o$ and $\eta = n/\omega_o$, thus simplifying equation (17), in the following way:

$$N \simeq 2[B\ln\frac{B+1}{B-1} - 2](x_o + a)^2 m\omega_o^2\sqrt{(1-\lambda^2)^2 + 4\eta^2\lambda^2} \tag{18}$$

It is now possible to show the dimensionless relationship of the required electric power for different λ while the value of B remains constant for the resonance regime $\lambda = 1$. Thus, the relationship of any power N_λ for any specific λ to the resonant power N_R (B is the same in both cases) is as follows:

$$N_g/NR = \sqrt{(1 - \lambda^2)^2 + 4\eta^2\lambda^2/2\eta} \tag{19}$$

Fig. 3 shows the dependence (19) for different damping conditions and presents the experimental results obtained with a vibrofeeder manufactured by Aylesbury Automation Ltd. (England). Comparing the calculated and measured curves, we see that the damping parameter in the experimental device is about 0.006. The natural frequency of the conventional vibrofeeder differs from the exitation frequency (50 Hz) by about 14%. In our experiments, the absolute power values for the Aylesbury Automation vibro feeder were:

I	N conventional	= 30	va	without load
	N adaptive	= 4.22	va	
II	N conventional	= 33.6	va	with load
	N adaptive	= 10.8	va	

Fig. 4 presents the dependence of the bowl's amplitude of vibration on the voltage change in the network. The Fig. clearly indicates that the adaptive vibrofeeder is considerably more stable than the conventional model.

Conclusions

(1) The energy consumed from the network by a vibrofeeder working permanently at its mechanical resonance frequency is considerably less than that of that required by a conventional feeder, the energy saving being 60-80%.

(2) The stability of vibrofeeding can be appreciably improved by the use of an adaptive self-tuning device, because the system constantly follows the "top" of the resonance curve even when the mass of the vibrating bowl is reduced to about 15-20% due to depletion of the feed material.

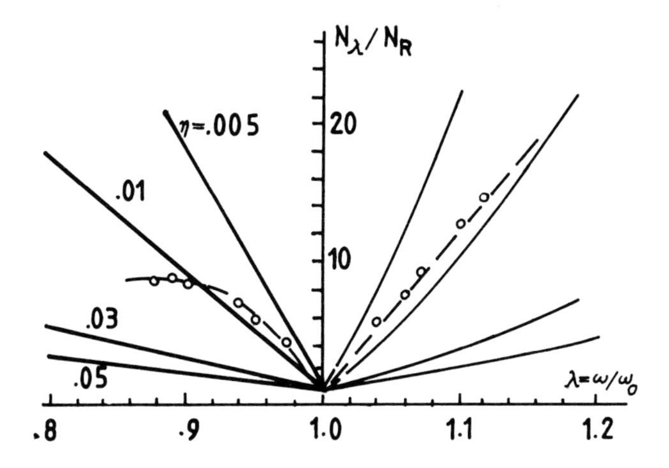

Fig. 3. Calculated dependence of the power consumption of the feeder as a function frequency of the voltage (for different friction condition). The dotted line represents the experimental energy consumption.

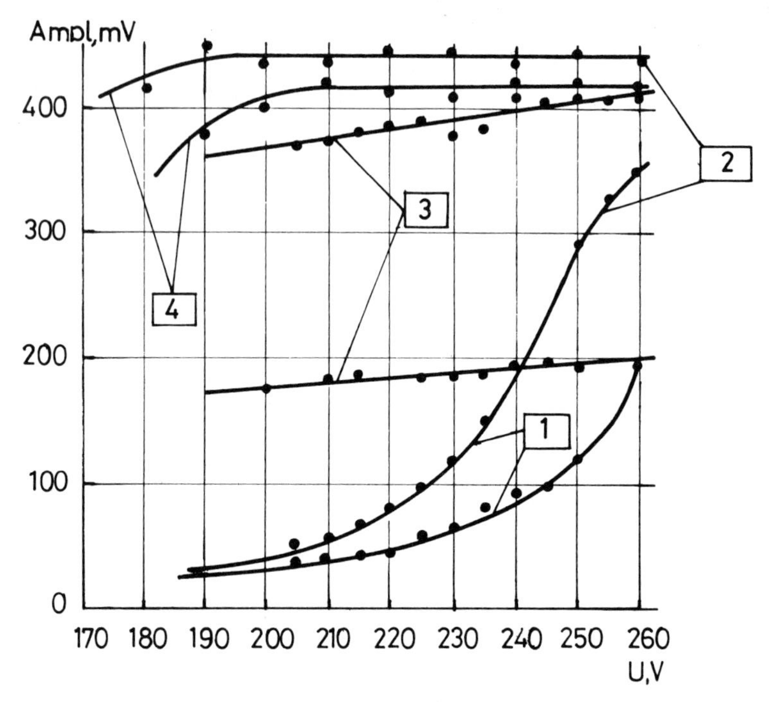

Fig. 4. Influence of changes in the network voltage on the bowl vibration amplitudes.
1 - conventional controller; 2 - unloaded feeder; 3 - adaptive controller (without stabilization of amplitudes); 4 - adaptive controller with amplitude stabilization.

References

B.Z. Sandler, Probabilistic approach to mechanisms, Elsevier, Amsterdam-Oxford-New York-Tokyo 1984.

Artificial Intelligence

Chairman: Kamal N. Karna, CC & GA Corporation, Cliffwood, New Jersey
Vice Chairman: David Bourne, Carnegie Mellon University, Pittsburg Pennsylvania

A Multi-Lingual Database Bridges Communication Gap in Manufacturing

David Bourne

Robotics Institute
Carnegie Mellon University
Pittsburgh, PA. 15213

Abstract

Flexible manufacturing must integrate and coordinate a diverse group of machines and people into one synergistic unit. This goal has never been reached to the satisfaction of the manufacturing community. We have developed a database system with language oriented primitives that help bridge the communications gap between machines and machines, and machines and people. These primitives facilitate automatic program construction and automatic language translation starting with an internal database representation. The multi-lingual database's first application controls ten different machine tools, ten different controllers of four types, and a range of human operators with different needs. This generic software system can be used in many ways to coordinate systems of machines, people and programs.

1. The Communications Gap

Machines and people working together must be able to communicate. This communication depends on a channel and an agreed upon code between the participants. Unfortunately, the machines in manufacturing share little or no agreement on these topics. The result is that there is no communications, except perhaps in the lowest common denominator. Most flexible manufacturing systems rely on a 1-bit handshake as the sole means of communication. One machine waits for a 1-bit signal while the other machine sends 1-bit after it has finished its part of the task. This restricts the communications channel to information only concerned with task synchronization.

Machines break and the details of the problem need to be sent back to a cell host. Machines go out of adjustment, and they often don't have enough information to detect the problem. The correction must come from the supervisory computer. Machines work with other machines, and they often need to know each others position (*e.g.* two robots). These examples are just a small sample of functions that cannot be performed without adequate communications.

It isn't good enough to be able to communicate between machines built by the same vendor. Corporations depend on special machines and controllers built by several vendors: this is what makes their product unique. There are large efforts underway to standardize the *how-to* of communications in multi-vendor systems lead by General Motors (MAP project [1]) and the National Bureau of Standards (AMRF project [4]). However, this is not the whole answer. A manufacturing system must also know *what-to* send, and this issue has been relegated to special purpose solutions.

The Cell Management Language (CML) is a multi-lingual database that automatically constructs and understands messages sent in a multi-vendor system. It includes software tools that are usually found in operating systems (scheduling, multi-tasking), database systems (relational operations, table data structures), production systems (rule management) and language primitives (parsing, semantic attachment, paraphrasing).

The synergistic effect of these tools make it easy to describe interpretive environments for each machine and person using the system. Once each machine (or person) has access to an interpreter tailored to its needs -- communications are accomplished by sending a program to CML and CML activates the tools that are needed to translate the program to an internal representation. For example, CML receives a message, parses it, attaches semantics to the parse output, generates internal structures, updates the internal model of the manufacturing cell, fires new rules as a result of the state change, and finally, automatically writes and sends new programs to whatever machines are effected by the change.

This paper shows how a typical machine tool interpreter can be programmed in CML and how a series of interpreters can be linked into a controllable system. The result of building such a system is a facility for communicating to each machine tool separately: for development and debugging, and finally an operating system for the manufacturing cell at large.

2. The Cell Management Language: Building an Interpreter

The software world in manufacturing is overflowing with programs known as *pre* and *postprocessors.* These are special purpose programs that convert data to the right representation before a program runs (*i.e.* preprocessor) and a program that converts the data to the right representation after a program runs (*i.e.* postprocessor). CML provides a programming environment with the tools to easily build and manage a set of I/O processors.

Figure 2-1 summarizes these programming tools and shows their logical connection. This figure also hints at some of the other tools that are available for language translation and language generation.

Suppose you wanted to build a preprocessor for English commands that would update an internal database. Here is what would be required in CML. Like any interpreter the lexicon must be defined first. What follows are a set of five tables that define the legal words and their syntactic categories.

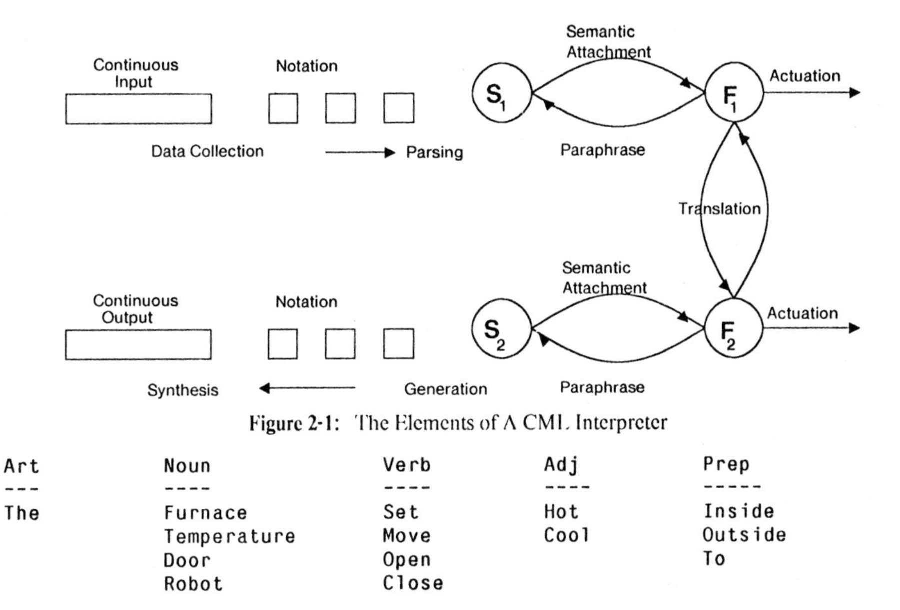

Figure 2-1: The Elements of A CML Interpreter

```
Art        Noun           Verb       Adj       Prep
---        ----           ----       ----      -----
The        Furnace        Set        Hot       Inside
           Temperature    Move       Cool      Outside
           Door           Open                 To
           Robot          Close
```

After the words are defined, the legal strings of words is defined using a syntax description language similar to BNF. This syntax description is contained in a table used by the parser to organize and label each word of the input. The following example shows how a set of declarative sentences can be described in CML. This description uses annotations after each "/" to denote labels for word position, optional or required words and phrases, and phrase definitions such as the prepositional phrase defined below[1].

```
Command F1   F2         F3         F4         F5         F6
-----------------------------------------------------------------------
Declare Verb Art/1      Adj/Opt    Noun/1/Opt Noun/2/Opt Command:PP/Phrase
PP      Prep Art/2/Opt  Noun/3/Opt Decimal/Opt
```

CML has a tool for *parsing* that takes as input the grammar description and the sentence to be parsed against that grammar. The output of the command is a parse table that labels each word. This is different than the output of most parsers, because only the leaves of the parse tree are kept, while the internal nodes of the tree are discarded. This economy in output is a great advantage in the semantic analysis.

CML> *parse command,Set the furnace temperature to 2250.*

---

[1]Note: Every legal sentence will probably not be a meaningful sentence, but this is a question of semantics and not syntax.

$Parse	Value
Verb	Set
Art-1	the
Noun-1	furnace
Noun-2	temperature
Prep	to
Decimal	2250

So far, the sentence has no meaning. The meaning is assigned to the sentence by matching the output of the parser with the arguments of a function set. CML provides *pattern directed function calls* that are roughly equivalent to rules the in a production system. The *if-part* of a typical production system rule are the arguments to a function (*e.g.* update) and the function is the *action-part* of the rule. In other words, "call the function if it has all of its arguments".

Rules	Call	Function	Arg1	V1	Arg2	V2	Arg3	V3	Arg4	V4
R1	eval	update	verb	=set	noun-1	x	noun-2	y	Decimal	z
R2	eval	move	verb	=move	adj-1	x	noun-1	y	Noun-3	z

The CML tool *vexpand* combines a data-set such as the *$parse* table with a set of rules and executes only those functions whose arguments are satisfied by the data. For example, the *update* function will only be executed if the *$parse* table has a row called *verb* whose value is *set*, and in addition has rows called: *noun-1*, *noun-2* and *decimal*. The values in the *$parse* table are then assigned to the appropriate types and a parameter table called *$sub* is built and used by the *update* function.

CML> vexpand $parse,rules

$Sub	Value
Verb	Set
Noun-1	furnace
Noun-2	temperature
Decimal	2250

Once one of the functions is matched, its parameter table built, the function is finally executed. The function called *update* causes a value in a table called *machines* to be changed to the value found in the parameter table, in this case, the value is *2250*.

Update	Function	Arg1	Arg2
E1	update	$sub:decimal:value	machines:$sub noun-1:$sub noun-2

Multiview Object Recognition System for Robotics Vision

Albert Bowers

The MITRE Corporation
Metrek Division
1820 Dolley Madison Boulevard
McLean, Virginia 22102

Abstract

This paper describes MITRE's robot vision system which uses simulated reference views generated from a wire frame object description data base. Multiple camera images are employed to recognize and refine the location determination of objects to be manipulated. Camera images, which often contain numerous extraneous features, are matched to this reference data base. A weighted best match criteria selects the most likely reference view corresponding to the object in the image. Rotation, translation, and scaling differences are corrected and a set of feature locations are generated which determine the location and orientation of the object. The system operates in a multiple object environment. However, there must be at least two standard viewing directions which present the desired object without any overlap with any other object in the scene. Some degree of artificial intelligence is employed by searching only the most likely regions of the image; and estimating the next best camera position based on the current degree of match achieved.

Introduction

The MITRE Corporation has established a robotics laboratory at its Washington Center office for the purpose of investigating various aspects of robotics related to object recognition and process planning, using a combination of deterministic and artificial intelligence solutions to solve these problems. One of the first problems studied was object recognition and location determination using multiple images from two or more camera locations. The objective of our research is to determine fast methods for object recognition which can be used for robot arm operation. This is achieved by finding the desired object in the work area and accurately determining its location and orientation using the least number of images. The location of the objects is not known in advance. We currently require that the object locations do not change during the period that the vision system is acquiring the images for object recognition and location. Our near-term goal is to develop a system that can recognize the objects in the work area and automatically generate a plan for assembling these objects into a desired structure. This plan will consider collision conflicts, structure stability, and the assembly of substructures. Substructures are treated as a new single object for the remainder of the plan implementation. Creation of the automatic structure assembly plan and its interface to the robot arm control system will begin in FY85.

This paper will not address the image processing or robot arm control modules of the system. Only those modules that directly support the object recognition and location functions will be discussed.

The workspace has been initially constrained to a flat black surface upon which objects having planar or circular surfaces are placed. A data base containing a wire-frame description of these objects was used to generate hidden surface images of 14 reference views of each object in isolation. These views are chosen to acquire a view approximately every 30° over the upper viewing hemisphere of each object.

The system does not have any restriction as to the number of objects that can be placed on the work surface. In theory, it does not even require that every object on the work surface be described in the data base; however, only those objects in the data base can be recognized. Other than the fact that a reference library of isolated views of each object is available, no other knowledge about the number of objects in the work area or their location is known to the system when it initiates its search for the desired object. Since only 14 views are currently stored in the reference library for each object, this does restrict the object to resting on a specific face of that object, but does not restrict its orientation otherwise. Overlap of the desired object with other objects in the workspace is greatly restricted. The system requires that there be at least two of its standard viewing positions that can obtain an image of the desired object to be recognized which does not have any surface of the object overlapped by any other object and the object itself does not overlap any surface of any other object.

To minimize processing time, only a table of feature parameter values is stored in the reference library; one such table for each of the 14 images mentioned above. No actual image data is retained in the data base. The number of features currently retained for each object is approximately 4 plus 6 for each corner on the reference image view. The first four features are used to quickly segment the image into areas that are of interest and areas of no interest. The features associated with each corner are used for detailed matching and recognition. To recognize the desired object, the system attempts to find significant corners on the object and associate these corners with their correct counterpart on the wire-frame description in the data base. Only the three best matched corners are used to determine the real-world position and orientation of the object. This means the system must be tolerant of extraneous information in the image and permit some mismatches and some poor matches, and above all, quickly decide what information in the image can be completely ignored.

Functional Description Of The Recognition And Location Modules

The image processing subsystem converts each image into a list of closed contours which represent the peripheral outlines of each object or group of objects in the image. A single object may be represented by more than one closed contour since irregularities in surface reflectivity can cause portions of a surface or all of a surface to be considered as part of the workspace background. This is caused by various factors associated with camera sensitivity, lighting directions, surface reflectance, and surface orientation relative to the camera. A single contour can also enclose several objects if one object is obscuring part of another object in the image.

To determine which contours to examine, the reference library has stored some simple features of the contours of the reference objects. These features are similar to the SRI recognition features such as the first and second order moments. These features are for quick segmentation of the image and are not used in the final recognition process. Only those image contours which have similar features are selected for examination. The image contours to be processed are ordered according to how well they have matched the desired quick reference features.

Since a single object can have more than one closed contour in the image, the reference library also contains a window template which is placed on the image in such a way as to align with the expected orientation of the desired object in the image and overlap all of its contours. The required orientation of the window is derived from the orientation of the principal axis of the image contour which was used as the basis for deciding that the desired object could be represented, at least in part, by this contour. Any image contour which intersects, or is wholly within, the window is considered in the subsequent feature analysis.

A corner finder is employed to locate and extract the most prominent corners from each contour. Curved surfaces are given arbitrary corners at intervals suggested by their rate of curvature. Such arbitrary corners would occur about every 20° of arc between adjacent corners. These corners are then subjected to feature extraction to obtain the feature values to be compared with those stored in the reference library. The feature matcher tolerates mismatches and can ignore extraneous features in the image. A similarity function is used to weight the significance of each feature in the final decision as to whether the reference image is the best match to the camera image. The similarity function can discriminate between overlapping N space regions based upon their means and standard deviations. A match is determined to exist only if the similarity function and the number of corners matched is above a specified threshold for the best matched reference image. This constitutes partial recognition.

The matched corners are then processed by the stereo vision processor. At this point pairs of images are considered. The projected rays from the camera to the matched corners are compared to determine their three dimensional miss distance; i.e., the shortest distance between two arbitrary rays in three dimensions. If the rays intersect or are closer than a specified distance threshold, then the miss distance is used as one component of that corner's position confidence sphere's radius. The center of the miss distance line is taken as one component of the corner's possible location. If the average of the confidence sphere radii of all corners yields a sufficiently small overall radius value then there is a good chance that the object has been properly recognized and the three best corner locations can be used to calculate the best overall location and orientation of the object. At least three corners must be used to determine position and orientation. More corners are not used because mismatches and poor corner detection generate significant deviations on some corners and these must be ignored.

If recognition is not satisfactory from two views, a third image is obtained and three sets of pairwise stereo computations are made. If recognition is not achieved again, a fourth image is taken. Each time an image is obtained after the third view, an analysis is made to determine if any of the reference images matched close enough that they might indicate a potential match and could, therefore, be used to suggest the next camera position and the most likely place in the resultant image where a match to the object might be found.

These techniques allow some degree of artificial intelligence to be applied to the recognition process. First, they allow the image contours to be ordered dynamically to permit the most likely regions of interest to be examined first. Second, they allow some capability for estimating where to move the camera to obtain the next view which can improve the chances of finding a match or eliminating possible overlap of objects in the image. Thus, the searches of the reference library are improved and the examination of the workspace is better than a simple preprogrammed sequence for taking the pictures.

Problem Areas

No recognition system can currently handle every problem, and our system is no exception. Any object which is primarily composed of non-linear edges and non-planar surfaces cannot be recognized. Objects that have a high degree of symmetry cause problems in determination of orientation, and sometimes even recognition, if the workspace contains several objects with similar characteristics. This is primarily due to our approach to recognition using features that have been normalized to remove the effects of rotation, translation, and scaling. A second source of variability is the stability of the principal axis of an object. It is not as stable as one would like it to be. For almost any object, there are many views where the object appears quite symmetrical. Such symmetry causes the principal axis to vary under rotation about the camera viewing axis. Therefore, the match criteria must be based on a similarity function which assigns a weight to each feature which is based upon its standard deviation from its expected mean value. This technique also reduces the sensitivity of the match to mismatches and extraneous corners found in the camera images. Total rejection of such corners is not desirable since corner count and corner proximity can be critical in distinguishing similar objects, or in quick rejection of dissimilar objects.

Conclusions

MTIRE's research is proceeding smoothly toward its goal of automated object recognition and process planning. Our recognition capabilities permit a reasonable object environment, but probably one that is only practical for a constrained production environment. We have not investigated alternative hardware configurations to increase the system speed, but have designed the system in modules that should easily lend themselves to a distributed processor architecture using various combinations of sequential, pipeline, and parallel processing techniques.

Advanced Applications of Robots

Chairman: Nicholas Shields, Jr., Essex Corporation, Huntsville, Alabama
Vice Chairman: Geary V. Soska, Cybotech Corporation, Indianapolis, Indiana

AUTOMATED BODY SYSTEMS FROM THE GROUND UP

FRANK A. DIPIETRO

Production Engineering, General Motors Corporation,

Warren, Michigan

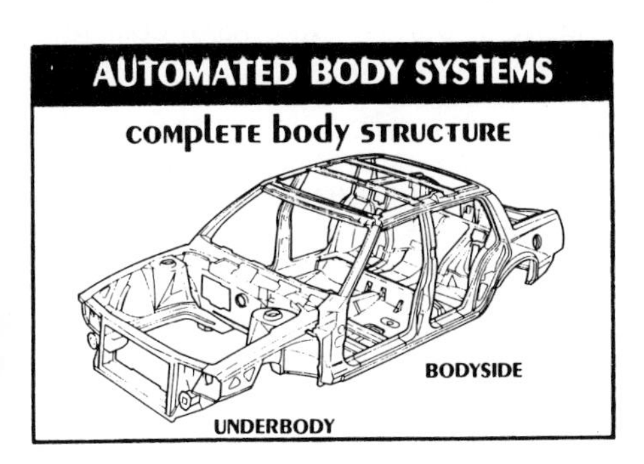

THE PURPOSE OF THE AUTOMATED BODY SYSTEMS IS TO BUILD THE COMPLETE BODY STRUCTURE OF THE INTEGRATED FRONT WHEEL DRIVE VEHICLE AS SHOWN HERE.

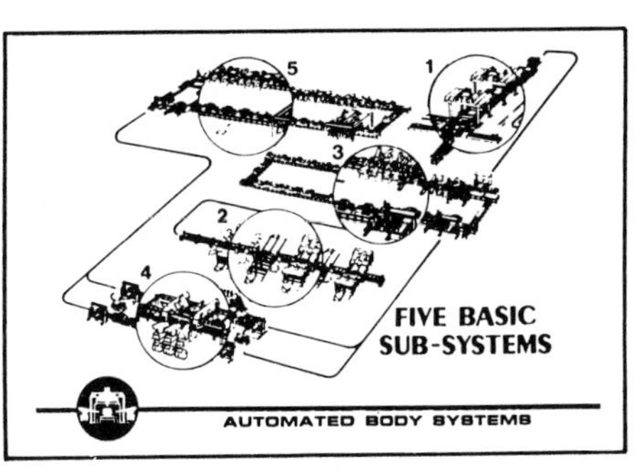

THE AUTOMATED BODY SHOP SYSTEM IS COMPRISED OF FIVE BASIC SUB-SYSTEMS:

1. AUTOMATED UNDERBODY SYSTEMS
2. ROBOTIC BODY SIDE ASSEMBLY SYSTEMS
3. CARTRAC I-UNDERBODY RESPOT SYSTEMS
4. ROBOGATE BODY FRAMING SYSTEMS
5. CARTRAC II-ROOF AND RESPOT SYSTEMS

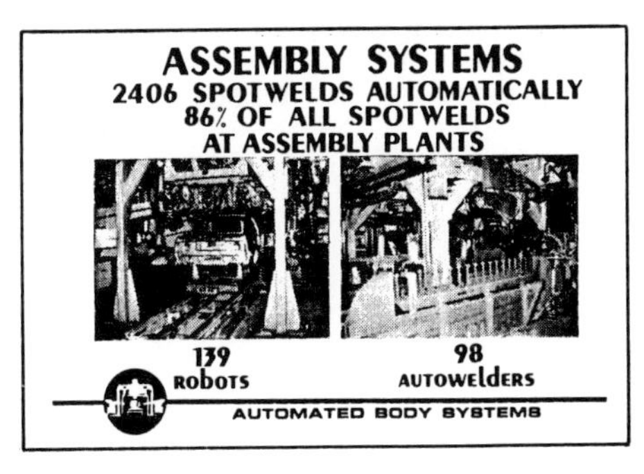

THE TOTAL SYSTEM INCLUDES 139 ROBOTS AND 98 AUTOMATIC WELDERS WHICH MAKE 2406 SPOT WELDS AUTOMATICALLY, REPRESENTING 86% OF ALL SPOT WELDS AT THE ASSEMBLY PLANT.

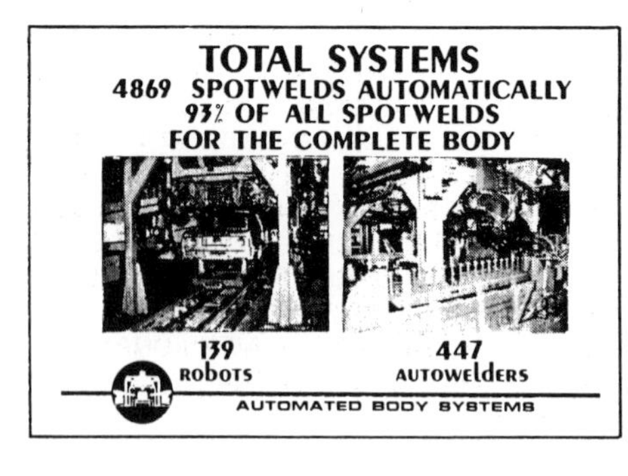

FOR THE COMPLETE BODY, INCLUDING METAL FABRICATING PLANT OPERATIONS, 93% OF ALL 4869 SPOT WELDS ARE MADE AUTOMATICALLY.

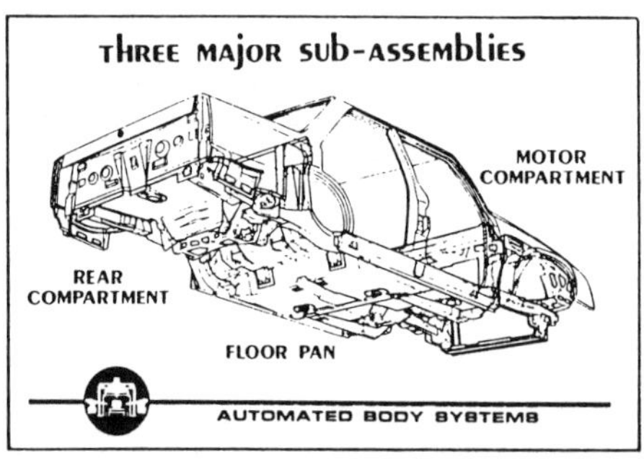

THE UNDERBODY CONSISTS OF THREE MAJOR SUB-ASSEMBLIES INCLUDING THE MOTOR COMPARTMENT, FLOOR PAN AND REAR COMPARTMENT PAN. TOGETHER, THEY COMPRISE THE UNDER-CARRIAGE OF THE BODY FOR THE FRONT WHEEL DRIVE VEHICLE. THE TRADITIONAL CHASSIS FRAME IS ELIMINATED, AND INCORPORATED AS AN INTEGRAL PART OF THE BODY STRUCTURE.

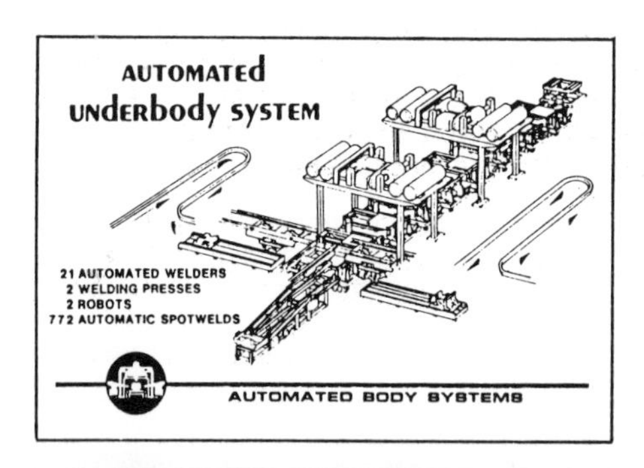

THE **AUTOMATED UNDERBODY SYSTEM** CONSISTS OF 21 AUTOMATIC WELDERS, 2 WELDING PRESSES AND 2 ROBOTS WHICH MAKE 772 SPOT WELDS AUTOMATICALLY.

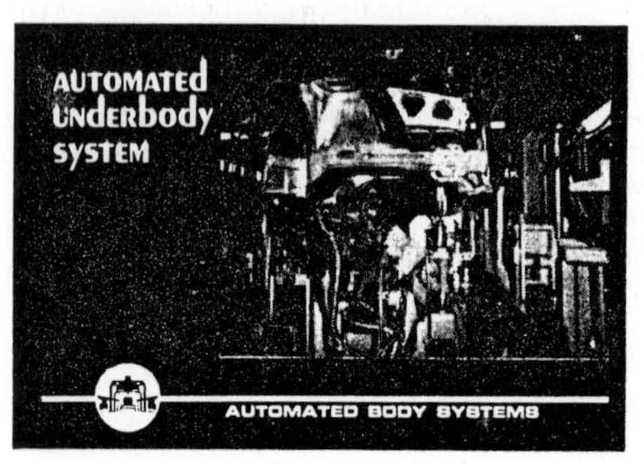

ALSO INCLUDED ARE AUTOMATIC PART HANDLING SYSTEMS WHICH LOAD PARTS AND TRANSFER THEM BETWEEN ELEMENTS OF AUTOMATION. THERE IS NO MANUAL SPOT WELDING PERFORMED IN THIS SUB-SYSTEM.

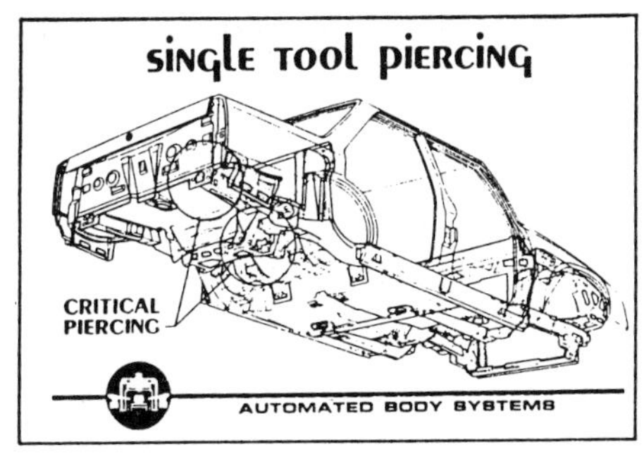

A UNIQUE FEATURE OF THE UNDERBODY SUBSYSTEM IS THE AUTOMATIC PIERCING STATION. THE REAR SUSPENSION AND BUMPER ATTACHING HOLES ARE PIERCED, UTILIZING A SINGLE TOOL AFTER THE BODY HAS BEEN WELDED. THIS RESULTS IN PRECISION LOCATION OF THESE CRITICAL HOLES. IT REPRESENTS AN EXCELLENT EXAMPLE OF QUALITY CONTROL THROUGH PROCESS CONTROL. WE ANTICIPATE DIMENSIONAL ACCURACY OF THE HOLE LOCATION WITHIN PLUS OR MINUS ½ MM.

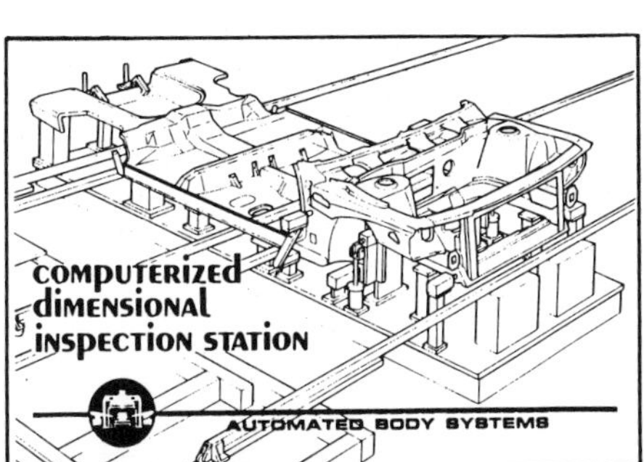

ANOTHER ELEMENT OF THE UNDERBODY SYSTEM IS A COMPUTERIZED DIMENSIONAL INSPECTION STATION TO VERIFY CRITICAL CHASSIS ATTACHING SURFACES AND HOLES.

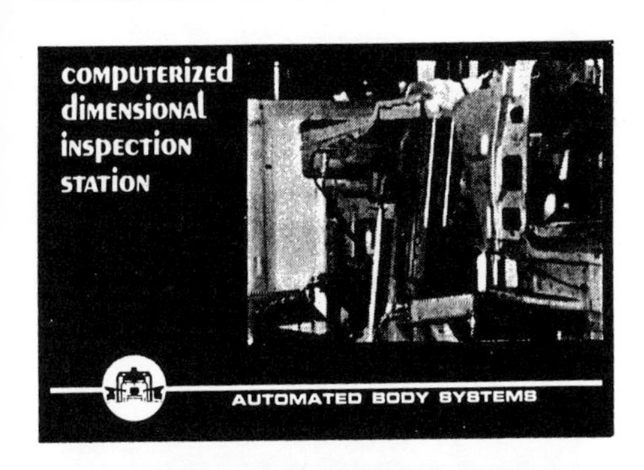

A FISHER BODY DEVELOPED CONTACT PROBE SYSTEM IS UTILIZED IN CONCERT WITH A MICROCOMPUTER TO PROVIDE REAL TIME DATA AND TREND ANALYSIS FOR STATISTICAL PROCESS CONTROL PROCEDURES.

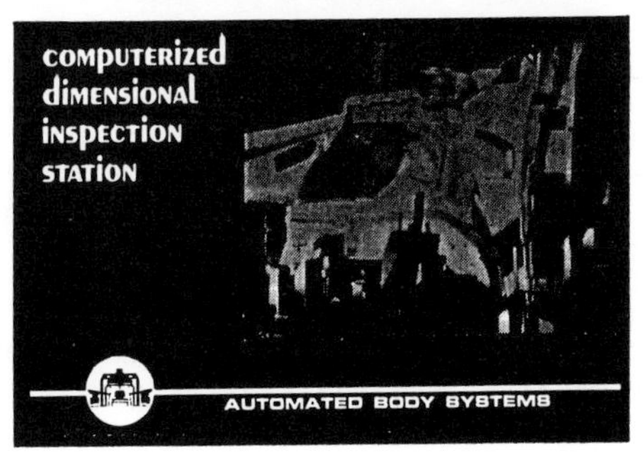

DIMENSIONAL MEASUREMENTS ARE SIMULTANEOUSLY TAKEN AT 30 PRE-DEFINED POINTS ON THE UNDERBODY AND STORED IN A MICROCOMPUTER BASED CONTROLLER. THE READINGS ARE PROCESSED AND REJECT/ACCEPT DECISIONS ARE MADE. THE RESULTS ARE THEN DISPLAYED PICTORIALLY AS WELL AS PRINTED ON A CONSOLE DEVICE FOR FUTURE REFERENCE.

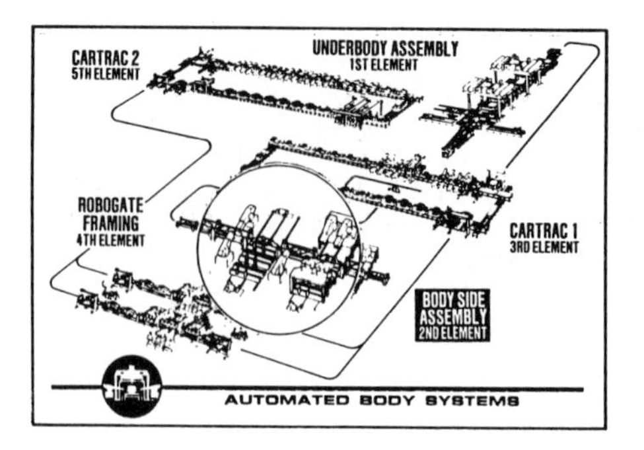

THE SECOND SUB-SYSTEM OF THE AUTOMATED BODY SHOP IS THE **"ROBOTIC BODY SIDE ASSEMBLY SYSTEM."**

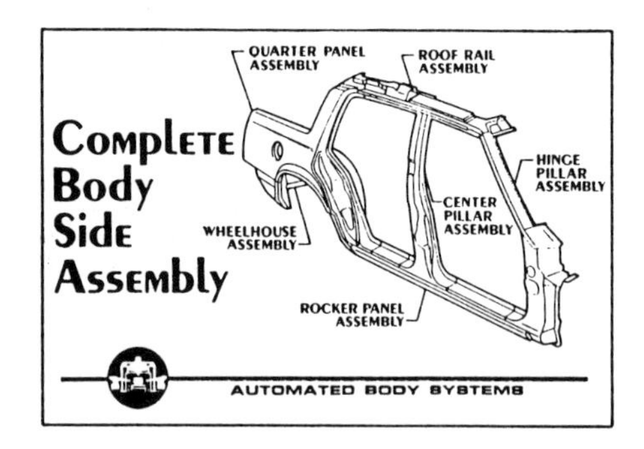

A COMPLETE BODY SIDE ASSEMBLY CONSISTS OF THE QUARTER PANEL, ROOF RAIL, CENTER PILLAR, ROCKER PANEL, FRONT BODY HINGE PILLAR AND WHEELHOUSE ASSEMBLY.

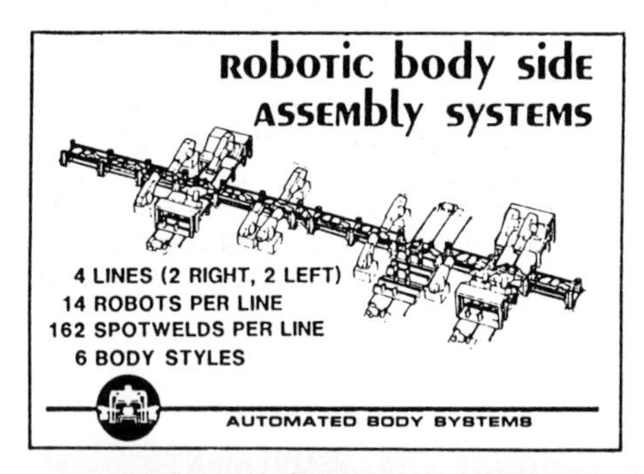

THE BODY SIDE AUTOMATION CONSISTS OF FOUR SPECIFIC LINES – TWO RIGHT AND TWO LEFT. THESE SYSTEMS CONTAIN 56 ROBOTS MAKING 162 SPOT WELDS. EACH LINE FEATURES 14 ROBOTS AND IS CAPABLE OF BUILDING SIX DIFFERENT BODY STYLES. THE CAPACITY IS 74 JOBS PER HOUR FOR EACH LINE.

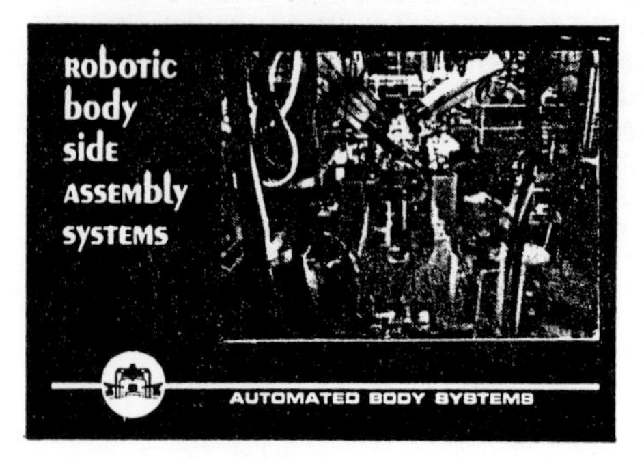

A KEY REQUIREMENT FOR THE ENTIRE AUTOMATED BODY SHOP IS TO PROVIDE FLEXIBILITY FOR MODEL CHANGEOVER TO MEET MARKET PLACE FLUCTUATIONS RAPIDLY. THIS IS ACHIEVED BY THE UTILIZATION OF THREE INDEXING TOOL TRAYS ON EACH LINE. THE INDEXING TOOL TRAYS EMPLOY THE SINGLE TOOL CONCEPT FOR ESTABLISHING THE CRITICAL DOOR OPENING DIMENSIONS IN THE BODY SIDE ASSEMBLY.

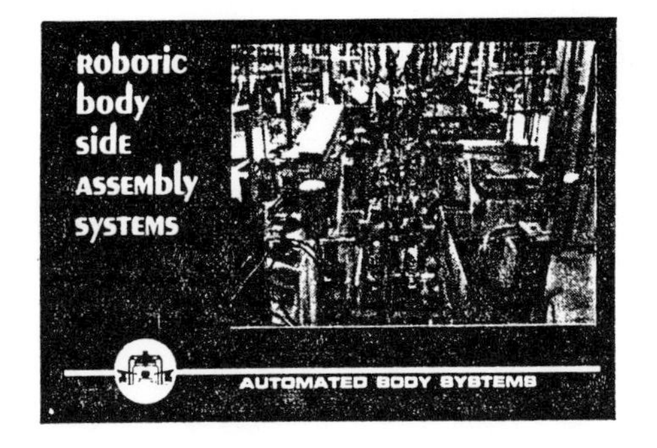

THIS REPRESENTS ANOTHER EXAMPLE OF QUALITY CONTROL THROUGH PROCESS CONTROL INSURING ALL DOOR OPENINGS FOR A GIVEN BODY STYLE ARE IDENTICAL. DESIGNED-IN FEATURES ALLOW FOR QUICK CHANGEOVER OF THE TOOLING PLATE ON THE INDEXING TOOL TRAY TO PERMIT EASY CONVERSION DURING MODEL CHANGEOVER PERIODS.

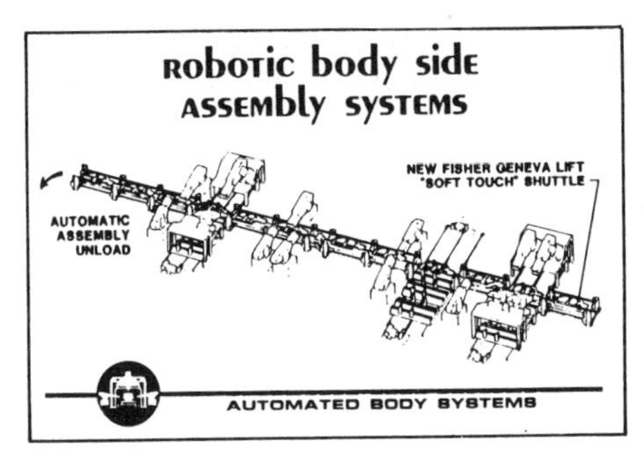

THE HANDLING OF THE BODY SIDE ASSEMBLY IN THE LIFT AND CARRY SHUTTLE SYSTEM IS CRITICAL TO INSURE NO SURFACE DAMAGE OCCURS DURING THE PROCESSING. A NEW FISHER BODY DEVELOPED MECHANICAL LIFT SYSTEM IS USED TO PROVIDE "SOFT TOUCH" HANDLING OF THE CRITICAL SHEET METAL COMPONENTS. FULLY AUTOMATIC UNLOADING OF THE BODY SIDE ASSEMBLY IS PROVIDED TO TRANSFER THE BODY SIDE ASSEMBLY TO AN OVERHEAD POWER AND FREE CONVEYOR.

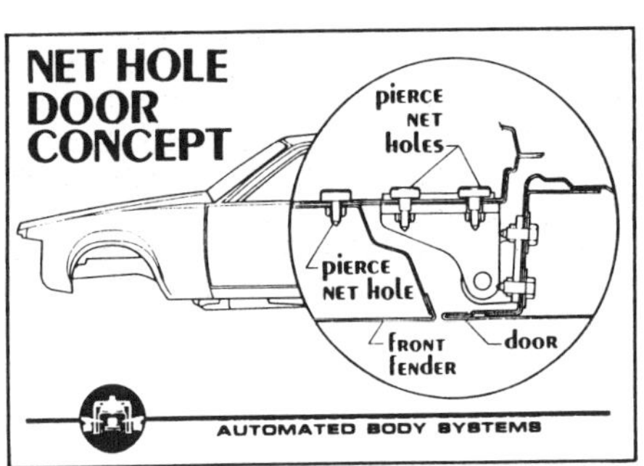

IN ORDER TO INSURE PRECISION DOOR AND FRONT FENDER FITS, WE ACCURATELY PIERCE NET HOLES IN THE BODY SIDE ASSEMBLY FOR THE DOOR HINGE AND FRONT FENDER ATTACHMENTS, PLUS THE DOOR LOCK STRIKER. PIERCING ACCURACY IS ACHIEVED WITHIN PLUS OR MINUS ½ MM. IMPLEMENTATION OF THE SINGLE TOOL CONCEPT INSURES CONSISTENT BODY DIMENSIONAL ACCURACY, PLUS UNIFORM DOOR TO FENDER FITS.

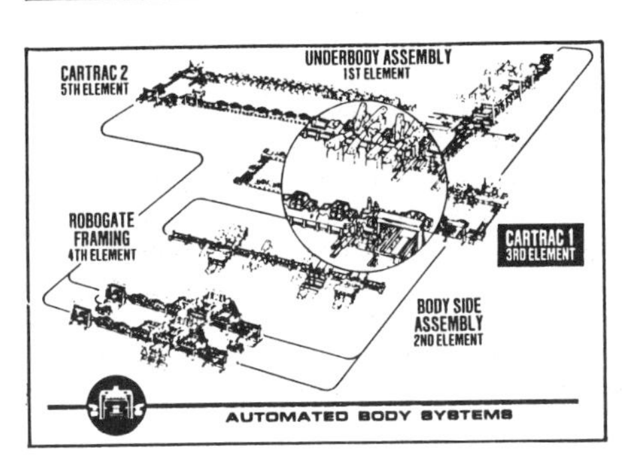

THE THIRD SUB-SYSTEM OF THE AUTOMATED BODY SHOP IS **"CARTRAC-I UNDERBODY RESPOT SYSTEM."** IT FEATURES 16 ROBOTS WHICH MAKE 220 SPOT WELDS ON EACH UNDERBODY.

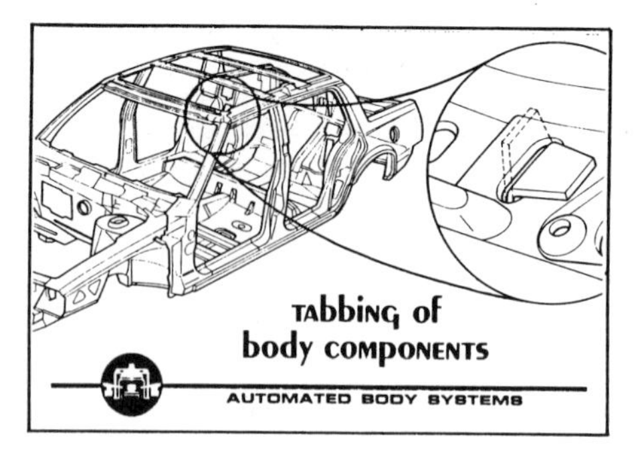

FOLLOWING COMPLETION OF THE UNDERBODY, THE MAJOR ELEMENTS OF THE BODY STRUCTURE ARE JOINED TO THE UNDERBODY WITH METAL TABS. THE TABS ARE A TEMPORARY METHOD TO INTER-RELATE THE MAJOR COMPONENTS OF THE BODY STRUCTURE PRIOR TO PRECISION GAGING IN ROBOGATE BODY FRAMING.

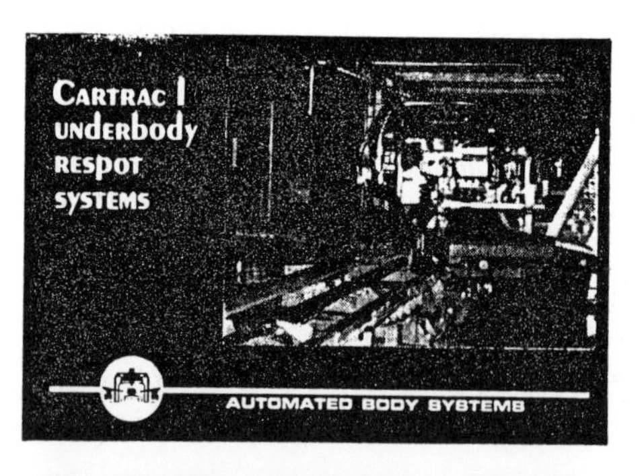

CARTRAC CONVEYOR SYSTEM IS A UNIQUE CONVEYOR. THE BODIES ARE ACCURATELY POSITIONED WITHIN PLUS OR MINUS 1 MM. FORE-AFT AND PLUS OR MINUS ½ MM. UP-DOWN. BEFORE EACH OF THE ROBOTS, THE BODY IS LOCATED ON A PRECISION TOOLING TRAY DESIGNED BY FISHER BODY AND IS MOUNTED ON THE CARTRAC PALLET.

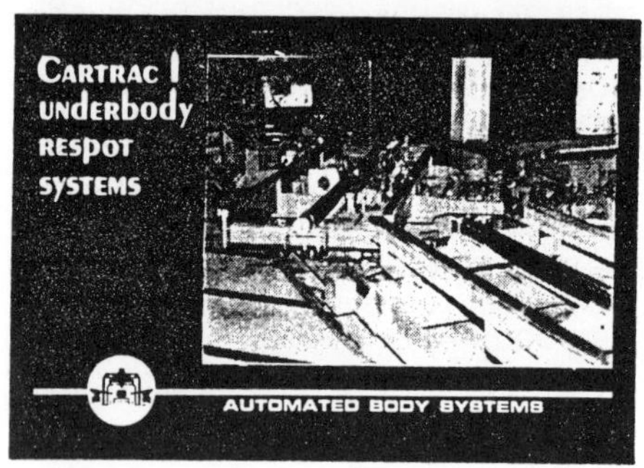

THE PALLET IS MOVED BY THE ANGULATION OF FIVE WHEELS DRIVEN BY A ROTATING CIRCULAR TUBE. THE MOTION OF THE PALLET IS A FUNCTION OF THE DEGREE OF ANGULATION OF THE WHEELS AGAINST THE ROTATING TUBE. THE SYSTEM HAS CAPABILITY FOR STOP AND GO AND/OR CONTINUOUS MOVEMENT AND ACCUMULATION AND PROVIDES FLEXIBILITY FOR BOTH AUTOMATED AND MANUAL ASSEMBLY OPERATIONS.

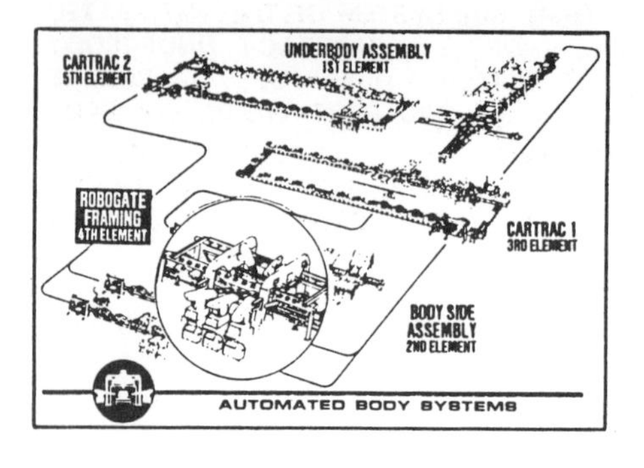

THE FOURTH SUB-SYSTEM IS THE "**ROBOGATE BODY FRAMING SYSTEM**" WHICH ESTABLISHES THE PRECISE DIMENSIONAL INTER-RELATIONSHIP BETWEEN THE MAJOR BODY SUB-ASSEMBLIES.

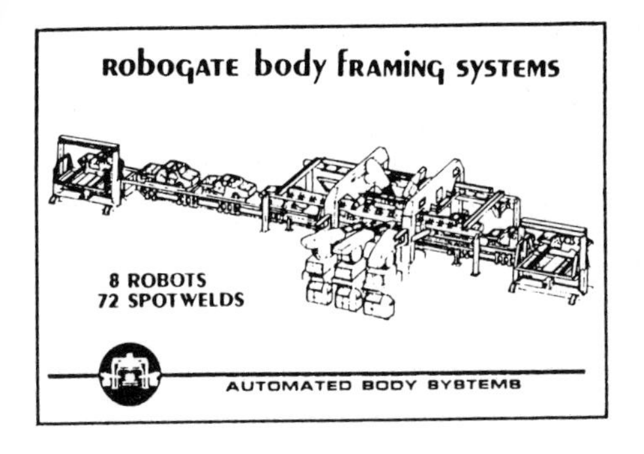

EACH AUTOMATIC BODY FRAMING SYSTEM IS COMPUTER CONTROLLED WITH EIGHT ROBOTS MAKING 72 SPOT WELDS PER BODY WITH NO MANUAL OPERATIONS.

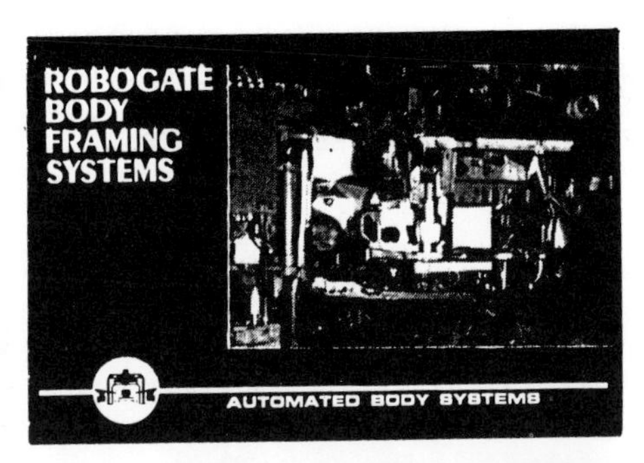

THE FRAMING STATION IS EQUIPPED WITH THREE SPECIFIC TOOLING GATES. THE SYSTEM INCLUDES A STOP AND GO, OVER AND UNDER SHUTTLE WITH (11) PRECISION PALLETS. THE BODY IS AUTOMATICALLY LOADED ONTO THE PALLET AND AUTOMATICALLY CLAMPED TO INSURE PRECISION LOCATION FOR THE FRAMING OPERATIONS. WE UTILIZE TWO ROBOGATE STATIONS TO ACCOMMODATE SIX SPECIFIC BODY STYLES, THEREBY MATCHING THE SIX BODY STYLE CAPABILITY IN THE BODY SIDE ASSEMBLY SUB-SYSTEM.

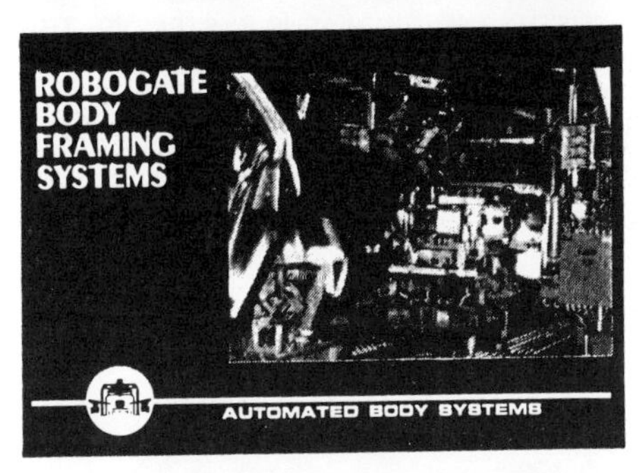

THE SINGLE TOOL CONCEPT IS UTILIZED BY HAVING A SPECIFIC TOOLING FOR EACH BODY STYLE. THE GATES ARE AUTOMATICALLY INDEXED FOR EACH BODY STYLE. AUTOMATIC PNEUMATIC CLAMPING IN THE SPECIFIED SEQUENCE INSURES PRECISION DIMENSIONAL PART LOCATION. PANELS ARE PRECISELY LOCATED AND AUTOMATICALLY CLAMPED, TO INSURE THAT EACH PART IS IN CORRECT LOCATION EVERY TIME. PRECISION CROSS CAR BODY DIMENSIONAL ACCURACY IS ACHIEVED WITHIN PLUS OR MINUS 1 MM.

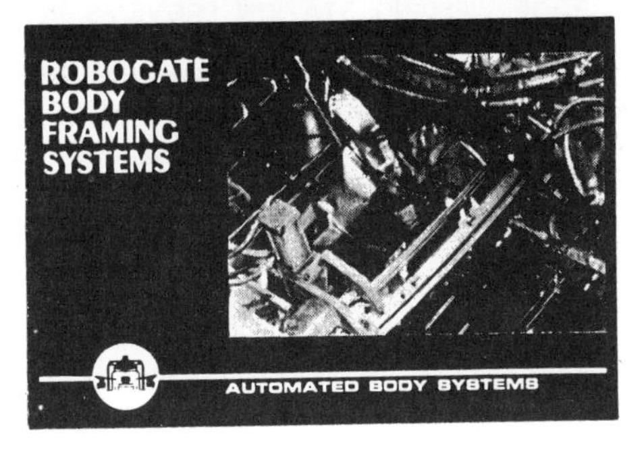

OUR PRODUCTION LINE FLEXIBILITY IS ACCOMPLISHED BY PROVISIONS WHICH ALLOW FOR CHANGING GATES WITHIN 6 SECONDS. IN ORDER TO MEET MARKET PLACE FLUCTUATIONS, WE HAVE DEVELOPED MODEL CHANGEOVER FLEXIBILITY TO IMPLEMENT ROLLING MODEL CHANGEOVER WITH MINIMUM PLANT DOWNTIME.

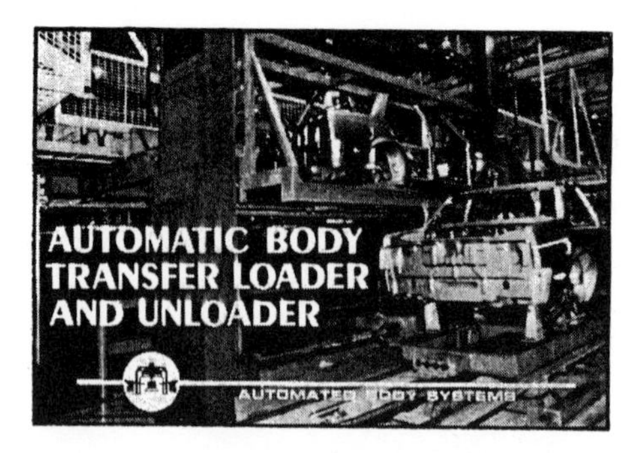

AUTOMATIC BODY TRANSFER LOADERS AND UNLOADERS ARE PROVIDED AT THE BEGINNING AND THE END OF THE ROBOGATE FRAMING LINE.

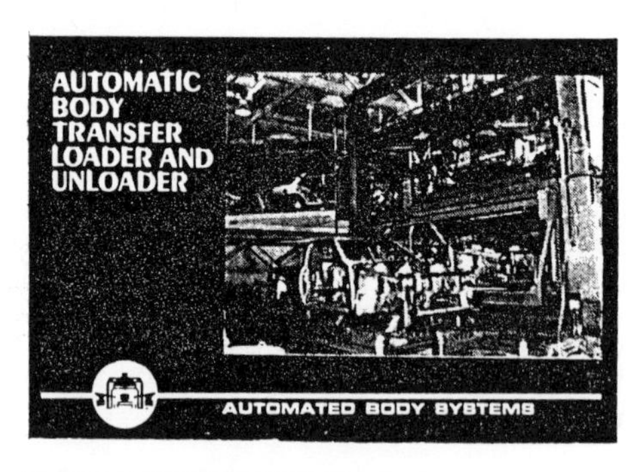

PRODUCTION CAPACITY OF THE ROBOGATE FRAMING STATION VARIES BETWEEN 70 TO 100 JOBS PER HOUR DEPENDING ON BODY COMPLEXITY AND WELDING REQUIREMENTS.

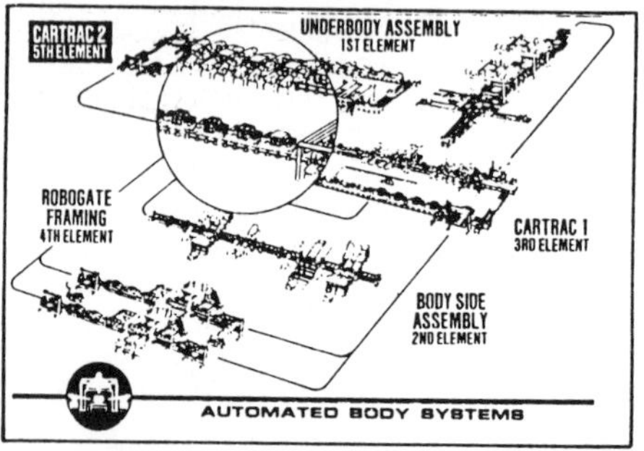

THE FIFTH AND FINAL SUB-SYSTEM IS THE **"CARTRAC II ROOF AND RESPOT SYSTEM."** IT AUTOMATICALLY LOCATES AND WELDS THE ROOF BOWS AND ROOF PANEL TO COMPLETE THE BODY STRUCTURE.

THE ROOF WELDING STATION PROVIDES INDEXING TOOLING FOR EACH BODY STYLE TO LOCATE THE ROOF IN A PRECISION MANNER. WE PLAN TO HAVE WINDSHIELD AND BACK LIGHT OPENINGS WITHIN PLUS OR MINUS 1 MM. DIMENSIONAL ACCURACY.

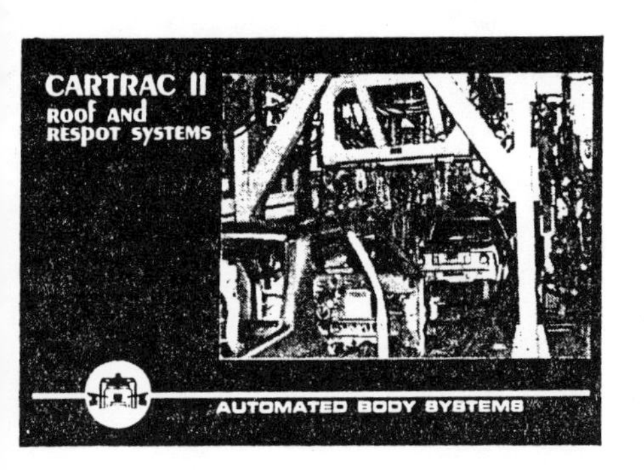

THE FINAL SECTION OF THIS SUB-SYSTEM INVOLVES THE USE OF 46 ROBOTS WHICH MAKE 633 SPOT WELDS PER BODY. IN ORDER TO INCREASE PRODUCTIVITY, WE USE 20 FISHER BODY DEVELOPED DUAL-TIP WELDING GUNS WHICH MAKE TWO SPOT WELDS SIMULTANEOUSLY. FINALLY, THE BODY IS AUTOMATICALLY TRANSFERRED FROM CARTRAC II TO AN OVERHEAD POWER AND FREE CONVEYOR FOR SUBSEQUENT BODY ASSEMBLY. THIS COMPLETES THE OPERATIONS PROCESSED IN THE AUTOMATED BODY SHOP.

Robotic Nondestructive Inspection of Aerospace Structures

Gary L. Workman and William Teoh

University of Alabama in Huntsville
Huntsville, Alagama 35899

Summary

The industrial robot has been demonstrated to be a useful tool for the nondestructive inspection of aerospace structures. Real time x-ray imaging systems, in combination with robotic manipulation, is an extremely versatile inspection tool which will become more prevalent in the future.

Introduction

Nondestructive inspection of critical components and systems such as used in power generation and aerospace systems requires a consistent scanning capability in order to determine structural integrity in a cost effective manner. The scanning system may be single-ended such as in pulse echo ultrasonics or eddy currents or double-ended as in through-transmission ultrasonics or x-ray imaging applications. Many of the scanning systems developed in the past for automated nondestructive inspection of large structures such as aircraft wings or boilers usually are designed specifically for that application and can be very expensive. An added deficiency in such systems is that there is no built in flexibility for scanning other types of structures. Hence there is a need for general purpose scanning systems such as the industrial robot.

The nondestructive inspection of large structures has traditionally been performed manually and represents a labor intensive activity dependent upon the level of inspection required. For example, an inspection procedure for weldments in a critical structure may call for 100% coverage using x-ray techniques, while a less critical structure may require only 80% coverage with dye penetrant inspection. The dye penetrant inspection would only find cracks which have propagated to the surface while x-rays or ultrasonics can find cracks which lie below the surface of the structure. X-ray inspection provides the most thorough inspection procedure, if it is applicable for the materials used in the structure. Unfortunately, it is also the most expensive when the cost of film, processing and the time to acquire exposures of the proper contrast and orientation is included. The expense of such an operation provides a definite incentive to achieve the same level of capability without film and processing costs.

Real-time x-ray imaging has come into its own as a more cost effective and versatile inspection method, particularly when the image processing capability of today's systems is considered. The primary problem is that the resolution of today's image intensifier and camera systems does not match industrial grade x-ray film. When inspection procedures require very high resolution, only film can satisfy the inspection specifications. Even state-of-the-art image processing systems cannot make up for the lack of resolution in the original image acquisition. Also considering the initial costs of including the real-time capability with the x-ray equipment, radiation safety facilities for the inspection, etc., the total costs can be quite high. In very critical inspection procedures, such as solid rocket engines where a particular part geometry is assured; hard automation combined with real time x-ray imaging has been implemented successfully. Since the systems were fabricated for specific part geometries, the system will not work with other part geometries. The NDT industry is used to designing manipulators for a particular inspection and not how to utilize general purpose scanners for the same purposes.

The industrial robot presents a unique compromise in scanning accuracy, speed, and dexterity which is a little different for each robot. Consequently, between the large number of choices available today, many inspection tasks are well suited for choosing an industrial robot as the scanner. One manufacturer (General Electric) has designed a robotic x-ray film system which optimizes the manual labor problems but does not allow for the flexibility of the robot and real-time imaging combination. The large structures such as are being used in aerospace; such as the aft skirt of the space shuttle solid rocket booster, provides as excellent example of a structure which can be cost effectively inspected through the use of real-time x-ray imaging and robotics.

The solid rocket booster sub-assembly such as the aft skirt, forward skirt and frustrum can be inspected by rotating the structure on a turntable, and performing vertical scanning motions using manipulators for the x-ray source and image intensifier. When suspicious areas are observed via the imaging subsystem, then the versatile robotic arms allow angular manipulation for another angle of view in order to perform stereoscopic analysis for a three dimensional perspective of the detected flow.

In order to determine the feasibility of the proposed system, we have used a Cincinnati-Milacron T^3 robot in conjunction with a x-ray real-time imaging system from Ridge Inc. (Atlanta) to perform programmed and manual scans on aerospace structures. These tests demonstrated that the programmability, availability, dexterity of control, and ease of use of today's computerized robots does provide an extremely flexibility scanning tool for nondestructive inspection. Experiments were performed with varying image processing capabilities, including the use of personal computers.

Pre-programmed scanning modes or manual scans in the teach mode of the robot allows well controlled scans for aerospace structures. Pre-programmed modes of varying angular orientations allow a stereoscopic reconstruction of the three dimensional perspectives of the suspicious regions. Preliminary calculations show that we can determine the depth of a defect to within 50 thousandths of an inch with minimal image processing.

The remainder of this paper describes the experiments conducted in which we attempted to show that:

a) A combination of robotics, x-ray radiography and image processing provide a viable, cost effective method for nondestructive inspection of large structures.

b) The dexterity and flexibility of the robots allow for detailed defect characterization.

Based on the results obtained, a nondestructive inspection cell is designed. This cell would be suitable to inspect large aerospace structures.

Experimental Setup

Because of its large size, it is not possible to place a section of the solid rocket booster into the Robotics Laboratory here at UAH. As a compromise, a nose cone is used instead. The nose cone, measuring about 68 inches across at the base and 70 inches high is made of the same aluminum alloy as the rest of the solid rocket booster. Figure 1 shows a picture of such a cone. Grids are placed on a small section of the outer surface;

these grids are used for tracking purpose only, and served no other useful function otherwise.

A 250 kilovolt Phillips x-ray machine was provided by Ridge Inc. This is a self-contained unit. The x-ray source is connected to the rest of the equipment by a 20 foot cable. Hoses of the same length serve to supply water to cool the x-ray tube.

Of all the available robots at the UAH Robotics Laboratory, only the T^3 has the lifting capability and reach required in the present work. Thus, the T^3 is used throughout this work. Since two such robots are not available, it was decided that the x-ray source would be mounted at the end of the robot arm, while a stationary tripod is used to support the image intensifier inside the nose cone. The physical arrangement is shown in Figure 2. As shown in this figure, the nose cone is placed on a table and cannot be rotated. All the above constraints mentioned means that only a small portion of the cone surface can be scanned. For feasibility studies, this was considered acceptable. The video signals are directed to an IBM PC microcomputer equipped with a TECMAR video board for processing. A second IBM PC is used to track the beam as well as to control the robot.

In the present arrangement, the x-ray source is placed 18 inches from the outer surface of the cone, while the image intensifier is placed 6 inches from the inner surface of the cone. Using this geometry, the digitized image displayed on the CRT is approximately twice its actual size. By varying the geometry, various effective magnification can be obtained.

Preliminary Results

The T^3 robot is preprogrammed to execute an up/down scan. A routine is provided which, when invoked, would cause the robot to scan a grid to one side of the present grid. It must be mentioned that the nose cone is a property of NASA, and we are not at liberty to arbitrarily introduce defects to it. In order that we can carry out the experiment, an artificial defect consisting of a small clamp measuring 3/8 x 1/4 x 1/8 inches is taped to the inside surface of the cone.

During the experiment, whenever the 'defect' is detected, the robot enters the routine with the result that an adjacent grid is scanned without changing the orientation of the robot. The second image is then digitized and subtracted from the previous one, thereby providing a pair of stereoscopic images as shown in Figure 3. A rather simple model is used to analyze the images. Referring to Figure 4 it can be shown that the depth x of the defect (measured from the outer surface in inches) is given by:

$$x = (Yh - yH + Yt) / (y + Y)$$

and its error Δx is given by:

$$\Delta x = [H / (Y + y) + \frac{(Yh + Yt - Hy)}{(Y + y)^2}]s = 0.05''$$

where s is a scale factor in inches per pixel. Note that this simple model is based on the assumption that a point (micro-focusing) x-ray source is used. In the present work, the x-ray source used is certainly not a point source, thus the accuracy is expected to be somewhat worse.

The next question that must be answered is, can such a system detect a real crack. To approach this problem, the nose cone is removed and replaced by a small section of another solid rocket booster skirt in which a crack is known to exist. In fact, the crack is sufficiently extensive that it is visible to an unaided eye. The robot is maneuvered to a suitable location and orientation. The x-ray image is digitized and examined. The presence of the crack is unmistakable.

Encouraged by the success of this geometry, attempts are made to examine other smaller structures made of various materials. First, a weld sample made by NASA, Marshall Space Flight Center, Huntsville, Alabama, was used. In this sample, dislocation is deliberately introduced. Intentional weld defects are made that provide a region of high porosity and a region of excess materials, as shown in Figure 5a). The corresponding digitized image is shown in Figure 5b). Note that even without further computer enhancement, the regions corresponding to the above mentioned defects can immediately be identified.

Other materials we have examined include rocket engine nozzles made of carbon epoxy composite materials, as well as ceramic honeycomb structures used in automobile catalytic converters.

All of the above mentioned experiments point to the fact that a combination of robotics, x-ray radiography and image processing is a powerful tool for nondestructive inspection. Among its many advantages are speed, accuracy, and economy. Using a rather primitive image processing system, meaningful results can be obtained in less than 15 seconds per frame. Employing an image intensifier means that the x-ray pictures (digitized or otherwise) may be archived on magnetic media rather than photographic films and can be readily recalled and re-analyzed at will. Further, the entire process can be automated, thereby eliminating the problem of personnel safety associated with the use of x-rays.

Cell Design

Focusing on the problem at hand, we need to design a robotized work cell for nondestructive inspection of solid rocket boosters. Such a cell will eventually be implemented in Kennedy Space Flight Center, Florida. Figure 7 is a diagrammatic representation of this cell. In this diagram, two robots are used. One robot is placed inside the structure to be tested, and is used to hold the image intensifier. The structure sits on a turntable so that in programming the robots, only up/down scanning needs to be used. The x-ray source is attached to a second robot outside the structure and this robot may be moved along the floor on air bearings. This arrangement is necessary as different sections of the solid rocket booster has slightly different diameters. According to our estimation, choosing a commercially available image processing system would allow this cell to completely inspect a section of the booster in about 15 hours, with minimal human intervention.

One very important consideration in the choice of robots is the fact that they must have off-line programming capability. In this figure, two Cincinnati Milacron T^3 robots are shown; they are chosen for the weight lifting ability, reach and dexterity. At present, no off-line programming capability is available for this robot, although it is expected in the near future.

The need for off-line programming capability is immediately obvious. At all times, the two robots must not only be in perfect synchronization with each other, they must also maintain a fixed geometry with respect to each other. In particular, the x-ray source must be normal to the image intensifier plane, and the distance between the x-ray source and the intensifier must be fixed. Traditional methods of robot programming, such as using a teach pendant, is time consuming, inaccurate and error prone. It is much more expedient to create a mathematical model that describes the cell. It is a relatively simple matter to constrain the two robots in the model to maintain the required geometries. The positions of the two robots can be easily extracted from the model and used to generate the 'points' of the robot programs. These programs can then be downloaded to the respective robots for execution.

This approach has additional advantages. By not requiring the x-ray beam to be perpendicular to the structure's surface, one can examine portions of the surface that may be partially obscured by the presence of underlying structures, such as stress bearing beams, pipes and so forth. Further, as long as the distance between the source and image intensifier remains fixed, the distance between the x-ray source and the surface may be varied. This provides a real-time zoom-in/zoom-out capability which may be useful at times. All these features cannot be achieved until accurate off-line programming capability can be implemented on the robots.

Figure 1. Solid Rocket Booster nose cone.

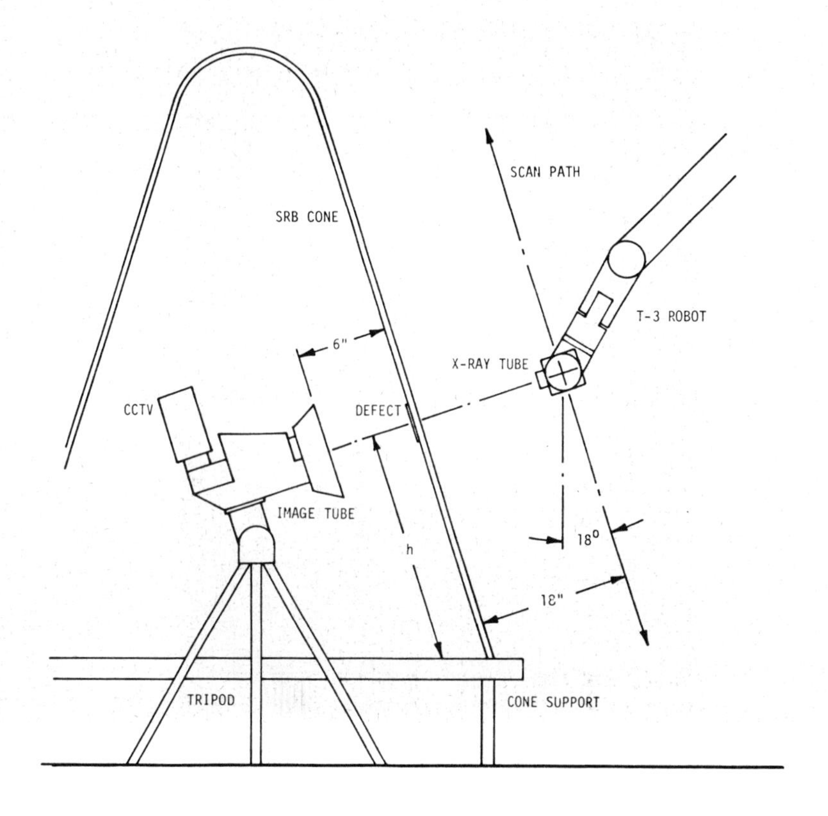

Figure 2. Experimental setup.

Figure 3. Digitized image of clamp taken at two different positions.

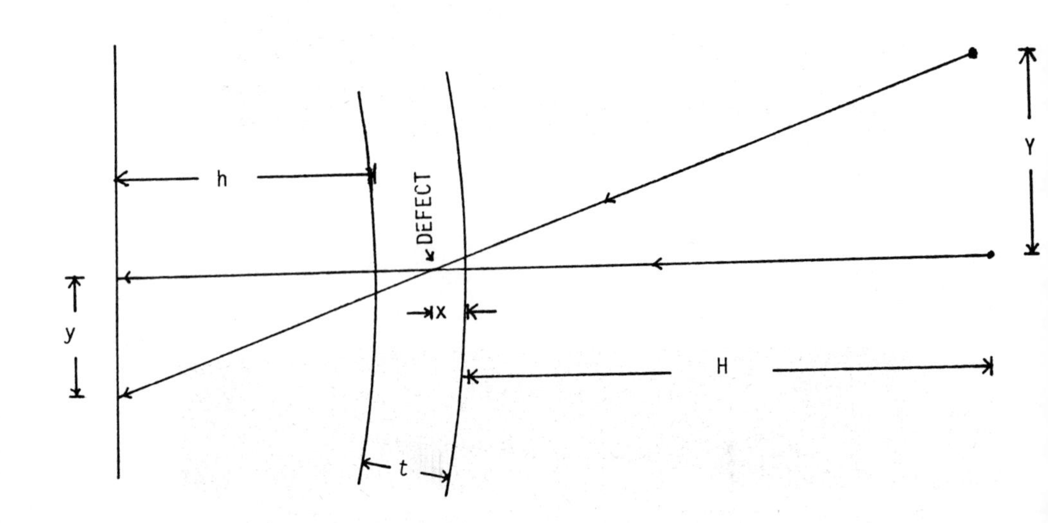

Figure 4. Geometry used for stereoscopic calculations.

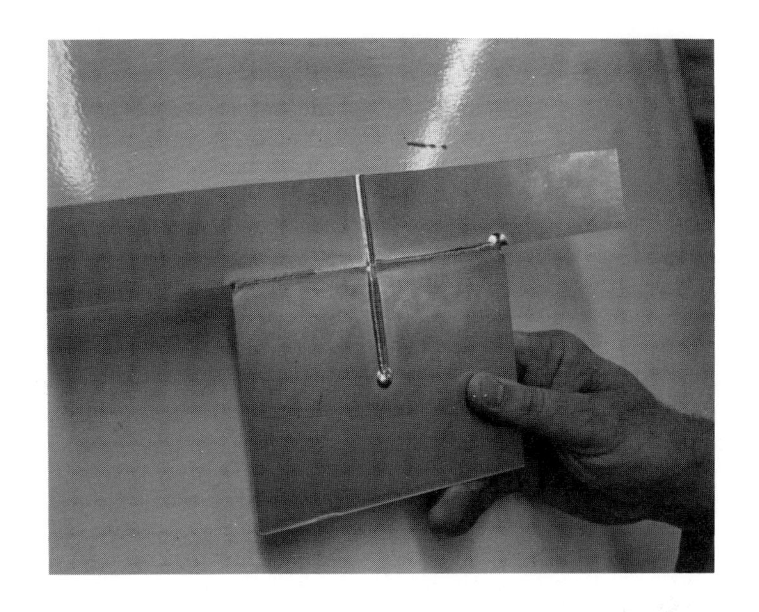

Figure 5a A weld sample.

Figure 5b Digitized image of the weld sample.

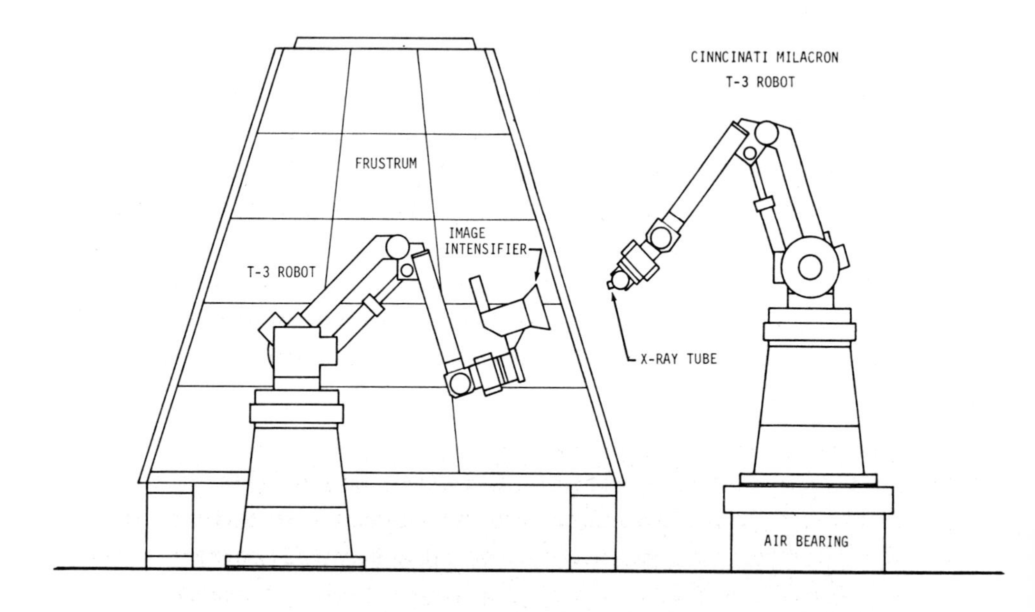

Figure 6. Nondestructive Inspection Cell.

An Overview of Human Factors and Robotics in Military Applications

Anna Mavor and H. M. Parsons

Essex Corporation
333 North Fairfax Street
Alexandria, Virginia 22314

Introduction

Most of the work in the field of robotics has been concentrated on the engineering technology of building robots to perform selected functions. Specific areas of concern have included locomotion, manipulation, sensing (vision; tactile), and intelligence/decision-making. The introduction of robots into the workplace raises a number of issues that are not addressed by the technology. These issues focus on the interaction between humans and robots in terms of such factors as supervision, monitoring, control, maintenance, and safety.

Until now, the military has shown a greater interest than has industry in studying the human factors aspects of human-robot interactions. Both the Army and the Air Force have funded specific projects to explore human factors considerations in robotics. Although the contexts and operational functions are different, many of the experiences in these projects should be relevant to interface design and safety considerations for the factory of the future.

In this paper we will discuss several robotic systems under development or being proposed by the Army, the Navy and the Air Force. Each type of system will be described and the human factors issues identified. Many of the illustrations will be drawn from our work with the Army's Human Engineering Laboratory on the human factors implications for Army battlefield systems. The categories of system types covered in this paper are:

- Autonomous land vehicles such as the Engineer Topographic Laboratory (ETL) Robotic Vehicle, the U.S. Marine Corps Ground Surveillance Robot (GSR) and Robot Defense Systems' PROWLER.

- o Explosive Ordnance Disposal Systems used in both civilian and military operations such as Hadrian (used by the British Army and the Los Angeles Police Department), Hunter, RMI(3), etc.
- o Autonomous Weapons systems such as the Tank Effectiveness Augmentation by Remote Subsystem (TEARS) developed by Whelan at Rand.
- o Logistic support systems such as BRASS which proposes the use of robotics in the automation of Army Amunition Supply Points that serve units in the battlefield.
- o The Integrated Computer Assisted Manufacturing (ICAM) project conducted by the Air Force and construction/manufacturing systems for welding, sanding, painting, pipe fitting, and structural shape processing used in Navy ship building.

Autonomous Land Vehicle

At the present time there are no autonomous land vehicles, however, the Army, the Navy and DARPA are working on programs leading in that direction. Autonomous vehicles or vehicles under supervisory control are extremely important in combat applications because of their ability to exist in hostile environments. One test bed for developing autonomous vehicles is the ETL Robotic Reconnaissance Vehicle facility. The ETL vehicle is composed of a robotic platform with a recon sensor system that is tele-operated from a remotely located control van by means of RF communication links. The terrain navigation subsystem for this vehicle is expected to evolve, through research, from teleoperation to autonomous operation. During this evolutionary process there are several human engineering design issues relating to the control functions, control consoles, and the control van layout. Current plans call for two operators: a navigator who plans the vehicle route and monitors performance and a driver who keeps the vehicle from encountering obstacles. The navigator has two graphic CRTs which can display terrain maps, computer-recommended routing, vehicle status and pictures of what the vehicle's cameras are seeing. The driver sees what the vehicle's cameras see plus status information on vehicle speed and attitude. Human factors considerations include the specific design of displays to facilitate task performance, the design of controls for ease of operation, the information processing workload on the operator, the interaction procedures needed to insure effective and efficient coordination between the navigator and the driver, and the level, type and spacing of training. Additionally, as the system evolves, human factors studies should be conducted on the allocation of decision-making/intellectual tasks to the human operator and to the robot.

Future Contributions

Human factors can make significant contributions to the design of robotic systems for industrial/manufacturing applications. Much of what has been learned through work on military robotic systems can be transferred. Even though the contexts are different there are similarities such as the need for human factors engineering in the design of control and monitoring facilities, the requirement to determine which functions should be automated and which should be left to the human, the question of skill levels and training needs for operators and maintainers, and the issue of safety when humans and robots are working in the same environment. These needs can all be addressed by human factors methods and tools. A general list of tools is shown in table 1. Application of these methods can lead to the following types of benefits:

- o Increased productivity
- o Increased reliability
- o Effective/efficient training and manpower distribution
- o Improved safety and environmental factors
- o Operator/employee acceptance of automation

It is hoped that industry will take a leaf from this Department of Defense book and directly incorporate human factors investigations and applications into robotics in manufacturing.

TABLE 1.
HUMAN FACTORS TOOLS

- o Task analysis
- o Functional job analysis
- o Communications analysis
- o Interviews (formal and informal)
- o Critical incident technique
- o Operation sequence diagram (OSD)
- o Time-line analysis
- o Error analysis
- o Systematic observation of work operations (video tapes)
- o Use of records
- o Simulation
- o Experimentation
- o Human factors information resources (literature, handbooks, standards, experts, etc.)
- o Workload analysis
- o Accident/injury analysis
- o Biomechanical/anthropometric measurements
- o Environmental/industrial hygiene measurements
- o Safety analysis (failure mode effects analysis (FMEA), fault tree analysis (FTA), hazards)
- o Human factors checklists
- o Cost/benefit analysis
- o Mock-ups

From Macek, A. et al Human Factors Effecting ICAM Implementation, Tech report AFWAL-TR-81-4095. Minneapolis, MN: Honeywell, 1981.

References

1. Howard, J.M. ICAM decision support system case study. In A. Macek, L. Heeringa, B. Somberg, D. Saur, R. Robbins, and J. Howard (Eds.) Human factors affecting ICAM implementation. Wright-Patterson Air Force Base, OH: Materials Laboratory, Air Force Wright Aeronautical Laboratories, Air Force Systems Command, Report No. AFWAL-TR-81-4095,1981.

2. Howard, J.M. Human factors issues in the factory integration of robots. Paper presented at the Robots VI Conference and Exposition, 1982. Dearborn, MI: Society of Manufacturing Engineers, Report No. MS82-127, 1982

Robotics Applications in Advanced Space Systems

Nicholas Shields, Jr.

Essex Corporation
3322 South Memorial Parkway
Suite 8
Huntsville, Alabama 35801

Summary

While the power, speed and accuracy of Earth-based robots are being proved daily, applied research in space-based robotics is proceeding toward that period in which machines can take over orbital assembly, servicing and maintenance tasks from humans. The same rationale for installing robots in modern factory environments can be used for replacing conventional extra-vehicular activity by astronauts with autonomous and semi-autonomous machine systems in space. The current arguments for robotics applications are safety, cost, reliability, operating cycles between rest and refurbishment, repeatability and accuracy.

In an environment where operating time can be substituted for operating power, where there are extremes of light and dark, vacuum, temperature, radiation, where weight is not a factor and there is no up nor down, we are confronted with some significant variables which influence the design and operating protocols of robots and the human's ability to effectively work with automated and remote machine systems.

Research at the Integrated Teleoperation and Robotics Evaluation Facility at the Marshall Space Flight Center is identifying those tasks and engineering operations which are most amenable to space-based automation and robotic applications. This paper provides a general description of the robotics laboratory facilities located at the Marshall Space Flight Center and a discussion of the research and application issues in space robotics.

Introduction and Background

During the past two decades, managers of terrestrial factories have been installing robot systems on the factory floor to reduce human risks, increase system output and accuracy, and decrease production costs. At the same time the aerospace industry has been investigating robotic applications for space. The interest in autonomous and semi-autonomous machine systems for orbital operations arises from the same concerns that we have on Earth, namely safety, economy and increased production.

Since the beginning of 1972, the National Aeronautics and Space Administration has sponsored a technology development program for teleoperators, robots and other autonomous and remotely managed systems, to determine the feasibility of performing orbital tasks without the conventional reliance on extra-vehicular activity (EVA). Candidate tasks are those which require long duration exposure, are outside the normal operating parameters of EVA crew, call for multiple-repetitive operations or are situations in which the tasks are very well defined. Activities such as construction beam fabrication, materials transfer, large space system assembly, payload deployment and retrieval, nominal servicing and repair have all been investigated with an eye toward robotic and automated functions.

With the advent of the orbital Space Station there will be an increased interest in space materials processing, space fabrication and manufacturing and a parallel increase in the demands to employ robots in this most elegant "factory of the future." Indeed, one of the initiatives directed by the U.S. Congress was for NASA to develop a Space Station Robotics Application Program.

What this will mean for the research, development and application of robotic systems for future space missions is limited only by the imagination.

Approach

So much has already been accomplished and demonstrated with EVA as a primary and contingency operating mode for space operations that it will serve for some time as the mode of choice and then as the baseline model for other approaches. This active intervention of crew members using EVA was instrumental in the recovery of a damaged Skylab mission. The 20 other unplanned, contingency EVA tasks performed by the crews of Skylab kept the laboratory and its instruments in operating order and provided additional science capability. More recently, the EVA repair of the Solar Maximum Observatory actually returned a failed spacecraft to operating order. The versatility of human performance during contingency operations practically begs the question, "Why then, robotics?".

During the planned missions and programs through the end of this century the scope of the work to be conducted in space will expand greatly. Larger payloads, greater mass, and longer duration activities are driving mission planners and systems engineers to find ways to support EVA tasks. The Manned Maneuvering Unit recently used during the Solar Maximum Repair Mission is one such support system, as is the shuttle-attached Remote Manipulator System which captured "Solar Max."

Now, NASA has requested the aerospace industry to study preliminary design requirements for an Orbital Maneuvering Vehicle (OMV) which will be capable of leaving the shuttle, rendezvousing with a distant satellite and docking with it--all under remote control. Once docked, the OMV will have the capability to perform remote servicing, maintenance and repair with one or more dexterous manipulators. Initially, the degree of OMV autonomy will be limited by technology and cost, but following demonstrations in space more autonomy will be required as humans and machines begin to share a sense of the space environment. Expert systems and artificially intelligent programs will independently check out and test payload components for proper operations. Sensitive end effectors will perform extremely dexterous manipulations as they feel for latch fasteners, extract fasteners, grasp and remove modular components of a payload, properly orient replacement components and insert them for payload refurbishment.

Visual systems will sense the conditions of the payload and use this information internally based on object recognition or relay it to a remote operator for a more subjective interpretation of the condition of the spacecraft or its components.

Mobility systems will compute masses, orientation and attitude and will reposition a spacecraft in its orbital environment or bring it back to a servicing platform for repair, resupply or upgrading.

With the appropriate combination of machine intelligence and flexibility, human control or supervision from a ground based remote work station, and payload design features which accommodate robotic interaction, we should be able to conduct around-the-clock orbital operations in support of our satellite fleet and the Space Station.

The approach taken to date has been to involve several government and industry research laboratories in a comprehensive program to identify key research issues for robotics applications in space. The National Bureau of Standards has contributed technology for end effector proximity sensing and positioning. The Jet Propulsion Laboratory has contributed technology toward developing a highly sensitive end effector for dexterous manipulation. The Johnson Space Center and SPAR Aerospace have developed the large scale Shuttle Remote Manipulator System. The Marshall Space Flight Center has acted as the lead NASA center for teleoperator and robotics research for the past dozen years. The Langley Research Center has focused on issues of large space structures assembly with automated systems and on the computer models for robot control. Every major aerospace company is pursuing robotic programs for space applications, notably Grumman Aerospace, LTV Aerospace, Martin-Marietta, Boeing, McDonnell Douglas and Westinghouse. What each of these organizations is pressing is the frontier of active, autonomous robotic systems which will be required to operate in an unstructured, and in some cases an unpredictable, environment.

This last point--the degree of prediction in the environment--might be the chief difference which concerns designers of terrestrial robotic systems and those who design orbital systems. In the ground based environment, we can consider the physical space in which we work as part of the system, providing horizontal and vertical references, stable work platforms and fixtures, and we can draw considerable resources, such as power, from our surroundings. In the oribtal environment, initially at least, we will have no fixtures on which to attach all of our robotic systems, there is no up nor down as we experience it on Earth, payload and robot mass interactions and contact dynamics are critical and the consumable resources are limited.

Challenge and Opportunity

The augmentation and supplementation of conventional human activities in space offers the inventive a chance to study human performance in order to understand what actually is being done in a physical, sensory and cognitive sense and to translate those findings to machine systems in both hardware and software models. Building factories of the future in space offers the creative an opportunity to develop a robot "breeding program" wherein the necessary robot technologies which have been demonstrated on Earth are combined into a robot system capable of movement, sensing time and position, vision, decision making, communications, self diagnosis, self repair, manipulation, tactile and force sensing, determining range and rate, using specialized tools, and in most circumstances integrating the inputs and outputs of these capabilities. The challenge is to select those attributes of several robot systems to be incorporated in a single machine system which can then operate and survive in the space environment.

Those of us involved in robot vision systems will have to contend with blinding illumination and deep shadows in the same scene. The video bandwidth, frame rate and communications link to remote human supervisors will be extremely limited. For recognition of distant targets, the vision system will have to be capable of selecting satellites from among the background of stars, or selecting a particular satellite from among the thousands in orbit.

Manipulator technology development will have to accommodate both the generic tasks of orienting, positioning, grappling, rotating, pushing and pulling and the specialized tasks involved in cutting, stripping, unscrewing, opening and closing, inserting and extracting, bolting and threading. These two classes of manipulator tasks will require general purpose end effectors and specialized end effectors, either of which can be exchanged for the other. Within the class of specialized end effectors, interchangeable tool effectors will have to be designed in a kit which can be carried by the robot for dealing with complex manipulative tasks. Beyond effector kits, the manipulator arm itself will have to achieve complex geometries in order to reach serviceable items and avoid delicate spacecraft appendages such as antennas, rocket thruster pods, and solar arrays. Contact avoidance detection will be required in both the manipulator hardware and software. The required precision and speed of movement of manipulator arms will have to be advanced with the requirement to move masses of hundreds of kilograms.

The software network which will be required to integrate the several robotic subsystems represents the greatest challenge and opportunity for designers of autonomous space systems. Internal sensors, external sensors, tactile feedback, navigation, docking mechanisms, communications, expert systems, alternate path decision logic, learning programs, and collision avoidance will have to be integrated with one another and with a central processor in order for orbital robots to successfully carry out their missions. The work being accomplished at SRI, Carnegie-Mellon, MIT, IBM, Langley Research Center, and Marshall Space Flight Center is designed to address the issue of robots and artificial intelligence, but there are still enormous opportunities for participation in specific space oriented robotic programs.

Research Issues

In 1983, the Marshall Space Flight Center completed construction of the Integrated Teleoperation and Robotics Evaluation Facility. Built upon a dozen years of research, the facility is designed to permit the resolution of research issues in a simulation facility which permits full scale simulations of orbital robotic applications. The centerpiece of this facility is a 4000 square foot epoxy flat floor. This floor, constructed to a level of .08 cm across its 30 meter diagonal provides a flat test bed (±.0013 cm/.1 sq. meter) over which full scale spacecraft mockups can be floated on air bearing pads. Powered by cold gas thrusters and drive motors, these spacecraft mockups can be flown by remote operators through an operational scenario involving visual inspection, rendezvous and docking, remote servicing, satellite capture and retrieval, payload change-out or similar orbital activities. The facility can accommodate several mockups on the flat floor permitting the study of vehicle interactions during docking or remote manipulation. The facility also has a twenty foot robot arm which overhangs the flat floor and provides computer resolved motion for target mockups mounted at the end of the arm. This 6-DOF manipulator provides the capability to perform "fly-around" maneuvers in conjunction with another vehicle floating on the floor. This is an operation which has great importance in space robotics but is generally excluded from Earth based robot requirements.

During the next several years, data from this particular facility and others in the network of space-oriented facilities will have a significant impact upon our understanding of the appropriate applications of autonomous and semi-autonomous systems in orbital operations. Additionally, data on the appropriate roles of humans and machine systems will help to identify new research areas to which we can all contribute.

Recommendations

As we work toward successful and appropriate automation in our Earth-based factories, we should maintain an awareness of the research activities directed toward bringing automation to future space facilities. The technology developed from this research can also find applications in terrestrial factories; and as we develop ever more sophisticated Earth-oriented robotic technologies, we should not ignore the opportunity to translate these to orbital applications. There should develop the same reciprocal relationship between Earth and space-based robotic development and research as there has been between the Earth and space system life sciences.

A System for Shortcreting Underground with Sensor-Controlled Travelable Industrial Robot

Hong Cheng and Charles H. Kahng

MECHANICAL ENGINEER, RESEARCH INSTITUTE OF JIANGSU COAL MINE CONSTRUCTION CO., XUZHOU, CHINA, AND, PROFESSOR, UNIVERSITY OF COLORADO, BOULDER, COLORADO, USA

Abstract

The art of shotcreting has been known to engineers since the turn of the century but only in recent decades has its full potential been realized. Shotcrete offers advantages over conventional concrete in a variety of new construction and repair work, especially in the field of tunnelling and mining.

In the workplace of shotcrete the working conditions are very bad; the job of shotcreting is repetitive and uncomfortable, the work part to be moved is heavy and there are potential dangers to a human operator because of the dust and air pollution.

Dealing with this problem, a system has been designed giving a survey on further developing the shotcreting technique and introducing the application of industrial robots that will be distinguished from those of special mechanisms and manual processes.

The robot can be programmed to carry out a sequence of continuous path motions to complete the shotcreting operations. Sensors and adaptive controls have been installed that give considerably greater capacity, safer and cleaner working conditions and a superior finished product.

The full mechanized underground wet-shotcreting system is shown in Fig. 1 and the operational set-up block diagram of a robot for a shotcreting system is shown in Fig. 2.

The robot can also be used in a shaft. The application of shotcreting by means of a robot in a shaft is shown in Fig. 3.

1. Introduction

Shotcrete is mortor or concrete conveyed through a hose and pneumatically projected at high velocity onto a surface, the force of the jet impacting on the surface compacts the material. The quality of shotcrete depends largely on the skill of the application crew much the same as the quality of a weld depends on the welder. The nozzleman, the most important member of the crew should hold the nozzle at the proper distance (usually between 0.8m and 1.2m, i.e., around 1m) (as shown in Fig. 4), and nearly parallel to the surface, or at an angle of 80°-90° from the direction of advancing (see Fig. 5), as the type of work will permit; to secure maximum compaction with miminum rebound and maximum strength of products by following a sequence routine of continuous path motions (refer to Fig. 6) that will fill all the surface with sound shotcrete using the maximum practicable layer thickness (around 7mm) without sagging.

Now this job will be performed with full control of the process at all times, giving higher productivity and superior products, better and cleaner working conditions with maximum protecttion during operation.

2. Data and Description of the Industrial Robot System

2.1. Robot Body and Technical Data

Permitted handling weight 30 kg

Performance:

Element	Degrees of Freedom	Speeds
Arm rotation	220°	90°/s
Arm reciprocating	650 mm	1 m/s
Arm up-down	400 mm	0.5 m/s
Arm bending	±30°	90°/s
Elbow bending	±90°	120°/s
Wrist turning	±120°	120°/s
Wrist bending	±60°	120°/s
Wheel driving	Forward and Reverse	0.5 m/s

Robot body description and dimensions are given in Fig. 7.

Repeat accuracy at wrist ±2 mm

(Higher accuracy for the purpose of shotcrete is not necessary or worthwhile.)

The working range of the robot is shown in Fig. 8.

Environment:

The robot and controls mounted on it are designed to withstand ambient temperatures from +5°C to +70°C. It is also designed to be well protected from danger of dust by a dust-proof cover and water screen.

2.2. Component Parts and Operation Description

The industrial robot system is made up of five main parts (refer again to Fig. 7).

. . Control system

. . Measuring and servo system

. . Mechanical system

. . Sensor and vision system

. . Hydraulic system

The control system, which includes the robot controller and the shotcreting conditions controller units, is shown in Fig. 9.

The microcomputer issues position instructions to the servo circuit of each robot axis. It can also freely read axis positions.

The functions necessary for manual operation and teaching can be achieved by the pendant or joystick. Using these, an operator can instruct the robot from positions near the work area, or remote control the robot on the ground with the help of a vision system.

A wet-process shotcrete machine is used with this robot. The shotcrete machine controller regulates compressed air pressure and volume and controls the switch to guarantee that the power connection will be cut off before the air supply in an emergency mostly in the case of a jammed hose.

To shorten the calculation time of the software, a multiplier, divider and sine table were added to the hardware.

The measuring and servo systems include servo amplifiers and stepping motors with a feedback unit.

Position regulation is controlled by means of a following system which is illustrated in Fig. 10. The transmission principle of the following system is shown in Fig. 11.

A stepping motor drives the following valve, changes the oil direction and quantity according to input signals. Since the rigid feedback between the input and the following valve exists, the deflection which exists between input and output of the system can be automatically eliminated, i.e., the performers would act following the input signal until the elimination of the deflection, stop, and then be located properly.

After turning an angle, the stepping motor will transmit the core 3 of following valve via special gear 2 to move a distance, and an opening will be formed within the valve. Compressed oil will be able to enter the cylinder (either reciprocating or turning) to move rack 6 (or to turn the gear for the turning cylinder), by means of gear series 7. This movement then becomes the same movement of the special gear 2 and turns the core 3 in the opposite direction until the opening of the valve becomes zero, then the cylinder will stop. Therefore, the cylinder and the stepping motor synchronize perfectly.

The mechanical system includes the robot and the transmission which converts the rotation of the motors into the required motion as mentioned above. (The cylinder for elbow bending is placed in the rear part of the arm, the motion is transmitted by a gear-sprocket-chain transmission unit.)

For keeping the nozzle at the proper distance, a sensor system is designed as shown in Fig. 12. The pressure P2 is controlled by the position of the flapper (i.e., steel ball 4). When the flapper is closed, pressure P2 is equal to the supplied air pressure P1; when the flapper is open, the pressure P2 approaches the ambient pressure P0. The distance between A and B is designed to equal the proper distance of the nozzle as desired (shown in Fig. 5). Under P1 action, piston 8 moves until it contacts a rock surface. Simultaneously, the hollow piston of the upper cylinder moves in the same direction until the flapper touch sensor rod 2 opens the flapper. The result is P2=P1=P0, both piston 6 and piston 8 move back under the action of spring 7. The flapper then closes by means of spring 5, P2=P1>P0, the cycle of the sensor system will continuously cycle to keep the proper distance between the nozzle and the rock surface.

The vision system provides both the size and contour of the environment and nozzle orientation information to the robot control system. The principal components of the vision system are a solid state camera, a structured light source, an incandescent lamp which is located off-center from the camera axis, and a camera interface system (refer to Fig. 7). Figures, taken by the camera, convert into signals via the interface system and are displaced on a TV screen (shown in Fig. 13). The operator can monitor the performance of the robot underground constantly via the three input parts: key board, joy stick and pedal or disk unit. Commands are passed and checked by the surface compter and received by the radio, underground, then they are translated into a 16 bit word and passed to the computer on the robot. The robot will then carry out all commands immediately.

The hydraulic system is simple (see Fig. 14). The A.C. motor activates the pump, the oil pressure can be controlled by a spill-way valve, all cylinders, except 2, 6 and 8, are controlled by following valves which are activated by stepping motors.

3. Program Flow For Spraying Execution

Figure 15 is a rough flow chart of the automatic spraying process for this robot using the system described above.

Initially, the user inputs the dimensions of the cross-section of the tunnel and standard spraying conditions, that is the conditions of the normal surface of the rock, to the computer from the operating console. Next, the user teaches the robot the location of original points on the cross-section to allow it to determine the position of the nozzle. When this operation has been completed, the robot begins calculating the nozzle angle, trajectory, spraying velocity, etc., then the controller automatically drives the robot to the point where spraying will begin and executes a spraying movement. During this operation the robot can accept manual modification as described above. When one spraying pass is completed the controller changes the starting position for the next pass and reverses the spraying.

4. Summary

We have presented the design of a system for shotcreting underground which will withstand the rigors of the underground environment as well as surface and general tunnel usage. Problems still exist and efforts are continuing to solve them and to complete a detail design of a perfect system through continued experimentation.

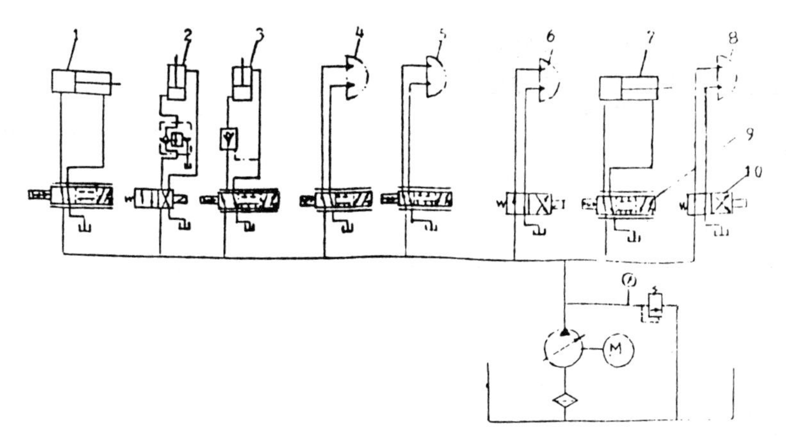

Fig. 14. Hydraulic System

1. Cylinder for arm reciprocating
2. Cylinder for arm up-down
3. Cylinder for arm bending
4. Cylinder for arm turning
5. Cylinder for elbow bending
6. Cylinder for wrist turning
7. Cylinder for wrist bending
8. Cylinder for wheel driving
9. Directional value
10. Following valve

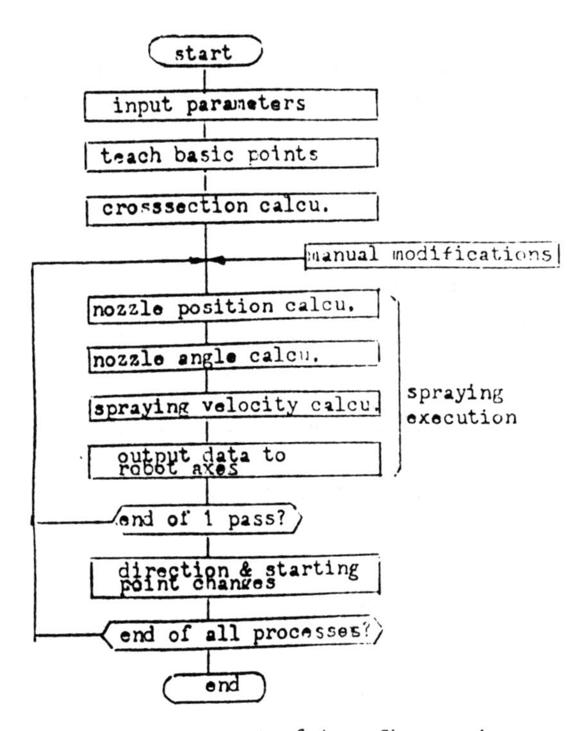

Fig. 15. Flow Chart of Auto-Shotcreting

Automation in the Textile Industry – Prospects and Impacts

D. R. Buchanan

Department of Textile Materials and Management
School of Textiles
North Carolina State University
Raleigh, NC 27695-8301

The history of the textile industry is one of continuing automation and technological change. Beginning with the Industrial Revolution, which occurred in the mid-18th century, the textile industry was transformed from a cottage industry, in which nearly every task was performed manually, to a factory-based industry that utilized substantial machine replacement of human labor. Ever increasing automation of textile processes has occurred to this day. The application of robots, which already had made a substantial impact on the automobile industries of Japan and this country to certain rather simple textile tasks has raised the question of whether robots represent a continuation of the 230 year automation trend in textiles or whether they represent some drastically new technological change. Any discussion of this issue benefits from a distinction between robots and hard automation.

Robots vs. Hard Automation

A rational definition of the term "hard automation" might be the following:

> Custom-engineered automated manufacturing machinery, built to accomplish a specific set of tasks and incapable of doing other tasks without disassembly and rebuilding.

This definition seems to correlate well with what we observe in the textile industry of today. Although there is no question that many processes are highly automated, at least relative to the days of the spinning wheel and the hand loom, it also is clear that these machines are very specialized and could not, in fact, perform any other function than the one for which they are designed.

A formal definition of a robot has a been agreed upon by the Robot Institute of America, who define a "robot" in the following way:

> A reprogrammable multi-functional manipulator designed to move material, parts, tools, or specialized devices through variable programmed motions for the performance of a variety of tasks.

Notice that the emphasis here is on the concept of variability, denoted by the words "reprogrammable", "multi-functional", "variable programmed", and "variety of tasks". Clearly, a robot is something quite different from a piece of machinery that might be classified as hard automation. The difference, however, will not be in the job it can do, but rather will be in the possibility of its being suitable for a variety of tasks accomplished through presumably simple reprogramming.

Revolutionary vs. Evolutionary Concepts

In the textile industry, we have a number of examples of hard automation. This category is conveniently divided into two subclasses. Those devices that represent a major departure from the previously existing technology and also accomplish tasks that were difficult or impossible for humans, we will call *revolutionary concepts*. On the other hand, there are as many examples of *evolutionary concepts*, which can be thought of as the simple replacement of a manual operation by a machine operation.

Examples of revolutionary concepts in yarn manufacture automation might include: the spinning mule, which represented a radical departure from the spinning wheel both in process and in product; the spindle and flyer mechanism which led to ring spinning as we know it today and also represented a radical departure from previous technology; and, finally, open end spinning, which also represents a technological break in both the spinning process and the yarn product that it produces. On the other hand, revolutionary concepts in hard automation would include such inventions as automatic knot tieing devices, which do what humans had done previously, although somewhat more reproducibly (modern techniques of yarn splicing, however, probably should be considered revolutionary concepts since they represent a process which humans cannot duplicate), faulty end detection using computer-driven detection systems, and production monitoring using computer-controlled systems. Also in this category are some of the rather inventive materials transfer schemes seen at the last ITMA Show, an example of which is the automatic transfer of ring spinning yarn packages to a winding frame.

Fabric manufacture also has its share of hard automation inventions. In the revolutionary category, the fly shuttle loom should be considered. Although its mechanism was similar to the hand loom, the possibility of much greater fabric widths as well as its great effect on productivity probably justify its categorization as revolutionary. Similarly, Jacquard pattern mechanisms certainly represent a revolutionary application of technology, as probably do fluid jet looms. Evolutionary concepts in fabric manufacture might include automatic quill changing, as well as simpler techniques for weaving and knitting pattern control.

In the field of robotics, there seem currently to be no revolutionary concepts operating, that is, no application of robots that has been disclosed does anything that was not previously done by humans. There are, however, a number of evolutionary concepts which center about the question of material handling. In particular, these include doffing of winders, doffing of open end spinning frames, and doffing of extruders. They also include the loading and unloading of autoclaves, particularly in yarn dyeing processes, warehousing of palletized yarn, and perhaps the transport of yarns, beams and fabric rolls by computer-guided vehicles.

Effects on Productivity

The effect of this continuing flood of inventions, mostly in the hard automation class, since the Industrial Revolution has been a dramatic effect on the productivity of the textile industry generally. Since 1760, essentially at the beginning of the Industrial Revolution, productivity in these two textile categories has increased tenfold in each 70 year period, on the average. That this has occurred on a fairly steady basis over a period of 230 years is not only poorly understood by the general public, but it also suggests that the introduction of robots into this scenario will serve to continue this trend into the near future. On this view, robots should be considered simply as another major technological innovation in the industry.

Reasons to Consider Robots in Textile Manufacturing

The few current uses of robots are based largely on their perceived reliability, versatility and reprogrammability, and their ability to use at least primitive feedback and control mechanisms. From management's point of view, however, it would seem that there are three important reasons to consider spending the considerable sums of money that introduction of robotics involves. These include: reduction of labor costs, improvement of quality, and improvement of marketplace response through enhanced flexibility.

Labor costs are affected because the labor costs associated with a robot tend to be less influenced by inflation than the labor costs associated with human beings. Based on an estimate from the automobile industry that current robot hourly costs are (or can be) roughly $6.00 an hour, these are competitive now with textile manufacturing labor costs. However, in the future, there should be a greater edge in favor of the robot because of the effects of inflation on salaries and wages. Generally speaking, one robot replaces two people in applications where such a simple correspondence can be made, as estimated by the North Carolina Department of Labor. In certain instances, other labor-related considerations also are important. For certain jobs, for example, the labor pool for those jobs is continually decreasing, either as necessary skills are lost in the general population, or as a job is perceived as being less desirable relative to the other kinds of work available in the community. Also, for jobs that have high turnover rates for some reason, it is quite possible that the training costs of new employees are substantial and should be added to the effective human labor cost for those jobs.

The improvement of quality in processes that use sophisticated automation and particularly robot applications is largely a result of the enhanced repeatability and reliability that is possible. Depending on the job, it is possible not only to do it more efficiently, but to effect savings in parts or supplies because of the greater efficiency of the robotic application. This is a particularly possible result when the system can be interfaced with feedback and control loops for rapid response to external events. Finally, there is a view that human intervention in the manufacturing process should be restricted to the greatest extent possible, in that many human actions (for instance, handling) are actually detrimental to quality.

The enhancement of flexibility occurs particularly with robotic applications because of the possibility of reprogramming in a rather simple fashion when circumstances dictate. Since there is a definitely identified trend in the textile industry (as well as in other consumer oriented industries) towards increased levels of manufacturing flexibility, flexibility of automation can be very valuable depending on a firm's closeness to the marketing and fashion end of the business. While humans represent the ultimate in flexibility and adaptability, it is quite conceivable that robots may be able to replace some human functions in manufacturing processes requiring very high levels of flexibility.

Note that it is not likely that humans will be totally replaced, but the skill levels required of those working in conjunction with robotic or other highly automated processes may be quite different than is the case today.

Development Trends in Robotics

Currently available robots can have most, if not all, of the following characteristics:

1. Spatial flexibility with up to 6 degrees of freedom.
2. Teaching and playback capability.
3. Memory of any reasonable size.
4. Program selection by external events.

5. Position repeatability to 0.3 mm.

6. Weight handling capability to 150 kg.

7. Point to point or continuous path control.

8. Synchronization with object movement.

9. Interfacible with external computers.

10. High reliability (typically 400-500 hours between failures, on the average).

On the other hand, textile uses for robots will probably not become revolutionary, instead of evolutionary, until most of the items in the following list have been developed to the point where they are reasonably available at a reasonable cost:

1. Vision - for recognition, parts orientation, and flaw detection.

2. Tactile sensing - for recognition, orientation, and physical interaction.

3. Real-time computer-based interpretation of visual and tactile data.

4. General purpose versatile end effectors ("hands").

5. Mobility - the ability to move from work station to work station.

6. Self-diagnostic error tracing.

7. Inherent safety.

Robots and the Automated Factory: Robot Systems

If these characteristics can be routine components of robots of the future, then entire robot systems will probably be developed. This is nothing more than than the automated factory concept, in which integrated and interrelated technologies operate with a minimum of human supervision and intervention. In such an automated factory, there will be four general components. Hard automation will be very much a part of such an automated factory, as there seems no way (nor any reason) to substitute some of the successful highly automated textile machines with robots. These will be accompanied, however, by robots, particularly in material transfer, assembly, and special operations applications. At the front end of this factory, it could be expected that computer-aided design and tranfer of instructions and information directly to the manufacturing machines will be a feature. Finally, at the output of such a system, management information from a sophisticated computer-based system will be sent to management for use in its decision-making.

Future Textile Industry Uses for Robot Systems

It probably is somewhat dangerous to attempt to guess what the next uses for robots in the textile industry will be, but common sense might indicate that they should show up in several general areas. The first of these would be material transfer. We can expect to see more sophisticated versions of those systems available now, particularly with respect to mobility so that, for instance, a number of spinning frames can be served by one system and the output can be directed to a number of places. These systems also will have much more sophisticated sensing systems and may, in fact, even be able to monitor quality as they are performing their material transfer tasks. We also should include more sophisticated versions of the driverless vehicles now available for material movement. These may have even more sophisticated computer control than currently used, and probably would not be dependent on floor tracks for their guidance system. If the latter were to be true, this would represent the ultimate flexibility for these devices.

Another major area will be that of inspection. Effective inspection will take place automatically at many more positions in the textile process than it now does. It will utilize rapid response vision and/or tactile systems in some cases; in other cases it will utilize transducers for the measurement of physical properties. In addition the overall inspection system will be programmable to allow decision-making based both on the plant specifications and also on customer specifications, and these may be changed nearly instantaneously as the product being manufactured changes. Finally, such a system has to provide a complete management information package on demand and this certainly will be a built-in feature of such systems.

A third area would be that of process control. In some cases, the monitoring of quality with feedback to a control point can be accomplished with hard automation, in other cases it probably ought to use some form of robotics. Real-time, efficient sensing systems will be needed in either case and they will be able to operate at nearly every phase of the textile product manufacturing process. In addition, they will have the ability to maintain complete records of the product history and its complete component and process identification. This will be extremely useful when production difficulties do occur.

Conclusions

Robots represent a logical extension of the hard automation developments in textile technology that started with the Industrial Revolution. The importance of future developments will depend on the degree to which robots and hard automation can be combined with computer control of design and information systems to form flexible automated factory systems capable of producing high quality, low cost products that can be sold profitably.

CAD/CAM and Flexible Manufacturing

Chairman: Ezat T. Sanii, N. C. State University, Raleigh, North Carolina
Vice Chairman: Joseph Elgomayel, Purdue Universtiy, Lafayette, Indiana

Computer Aided Process Planning

Han Bao

Department of Industrial Engineering
North Carolina State University
Raleigh, N.C. 27695-7906

Summary

This paper reviews the common bases for process planning and discusses the characteristics of the manufacturing logic necessary for the creation of a generative process planning system.

Introduction

Process planning is that function within a manufacturing facility that establishes which manufacturing processes and parameters are to be used to convert a discrete part from its initial form to a final form. A good survey of process planning systems is given in [4]. In general, these systems can be separated into two groups known respectively as Variant and Generative. The former group is based on the similarity of parts and standard process plans for the major categories of parts. The latter group aims at developing a plan according to some pre-programmed strategies.

The question of which system to use depends on the type of parts being made. If the parts are similar enough in their physical look and process requirement to form a family, then a variant system is usually desirable. On the other hand, if the part design changes often so that it is more practical to consider it as a unique product, then a generative system appears to be more desirable. Whichever process planning system is considered appropriate, the technology is still considered crucial in providing a bridge between the design and manufacturing needs for producing the part.

The objective of this paper is to explore the manufacturing logic leading to the formation of a process plan. It will also be shown how this logic was applied in a number of typical computer-aided process planning systems currently being used.

Requirements of Process Planning

The process plan, also known as routing sheet, contains all information required for the realization of a part design into an actual product. Typical engineering details available from the process plans are: stock material, process (es), sequence of operations, departments involved, operation time data, machine tools, inspection requirements, and types of jigs and fixtures needed. The information mentioned above should be contained in the process plan, irrespective of whether the plan is realized manually, with the aid of a computer, or automatically by a computer.

The success of a process plan depends on the following requirements of the manufacturing logic [2]: consistency, realism, and practicability. The manufacturing logic should be consistant so that even a younger and relatively inexperienced planner can interact with the system and still obtain a viable process plan, as if it were made by a more experienced planner. It should be realistic in the sense that it is based on existing machines, tools, and processes in the shop, and not on some hypothetical ones even though the latter may offer a more effective means of production. It should be practical in consideration of process capability and current machining practices. Also various aspects of the material flow through the plant and through individual work stations must be considered.

Basis for Process Planning

For the purpose of this paper, there are four general approaches in computer generated process planning:

A - Operations

B - Group Technology

C - Part Attributes

D - Part description

A. Operations Approach

Operations are described in a hierarchical manner as indicated in Figure 1. The principal processes are forming, cutting, casting, and assembly. Forming and casting are usually followed by cutting thus making the latter the most common process in the manufacturing industry. Within each principal process there are several major operations that may be classified either by the manner in which they are performed, or their importance in the sequence. Eary and Johnson (8) recognized the following major operations: 1) Qualifying Operations; 2) Critical Operations; 3) Secondary Operations and 4) Requalifying Operations. The Qualifying Operations are those which may be required to get the workpiece "out of the rough" and to provide an acceptable plan on the workpiece for location and support for following machining operations. The Critical Operations are those needed to accomplish some unique characteristics on or from some surface of the workpiece. Critical Operations are classified as Product Critical Operations and Process Critical Operations. The former operations are necessary to the functioning of the product, and typically, are dictated by specifications of surface finish, flatness, concentricity, size tolerances and so on. The latter operations are involved when the surfaces they create serve as registering surfaces for the location system. Secondary Operations also have a functional purpose on the workpiece, but are generally performed to standard part print tolerances. Finally, Requalifying Operations may be needed to return the workpiece to its original geometry. Figure 1 has the essential relationships between process requirements and orderly arrangement of operations. However, as of today, this operations approach has no yet been applied in any known system of generative process planning.

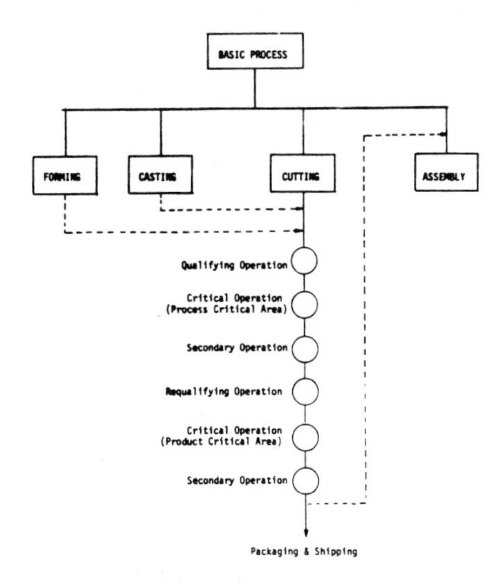

Figure 1 - Operation Classification

B. Group Technology Approach

Group Technology is a method of identifying and bringing together related parts, so that design and manufacturing can take advantage of their similarities. Group technology results in a rearrangement of the manufacturing plant into cells or clusters of machine tools. The main function of which is to process a unique family of parts. With such an arrangement, the material flow through the plan would become smoother, easier to trace hence easier to control and to improve. Group technology, through a systematic method of classification and coding, leads to standardization and reduction of component varieties, makes retrieval easier and eliminates unnecessary duplication of design (12).

On the manufacturing side, one of the major benefits of group technology is the realization of automated process planning. Without exceptions all of the commercially available process planning systems in the United States are based on some kind of classifications and codings derived from group technology (6,7,10,16). The CAPP system (6) is perhaps the best known variant process planning system in the U.S. Its basic approach is to compare the new part with standard parts the process plans of which were already worked out and stored away in a file. If there is some degree of similarity, then the process plan of the standard part can be recalled and all necessary editing can be made to fit the new part. The structure of CAPP is given in Figure 2. Before CAPP can be put in operation, five data files must be prepared first. These files are respectively the part family matrix file, the operation code file, the standard plans file, the process plans file, and the cross reference file linking past number and classification code. A typical process planning session is shown in Figure 3.

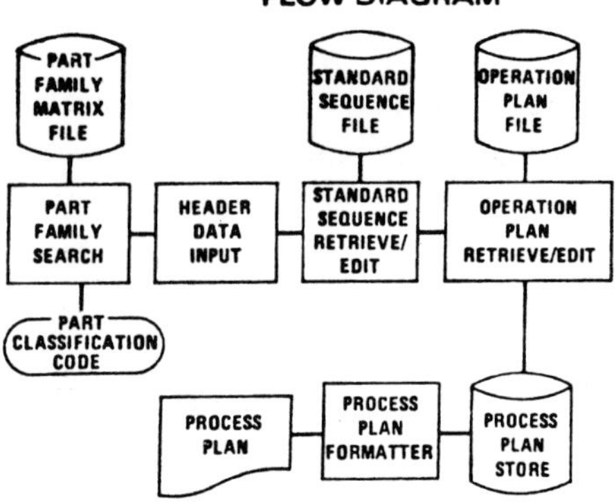

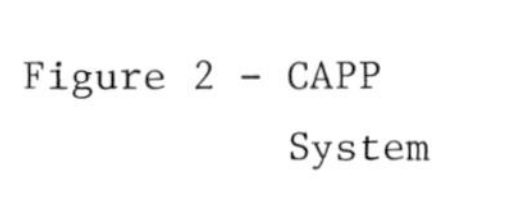
Figure 2 - CAPP System

TYPICAL PROCESS PLANNING JOB

AIM: TO DETERMINE THE SEQUENCE OF OPERATIONS FOR AN ITEM THAT HAS BEEN CODED AS 12361000

ASSUMPTIONS

1 - THERE ARE 4 PART FAMILIES IN DATA FILES
2 - THE CODED ITEM MATCHES WITH PF#4

USER COMMAND	SYSTEM RESPONSE
CALL UP CAPP	'MAIN MENU' IS DISPLAYED
MS/12361000	PART FAMILY # 4
SP/4	HEADER SECTION OF PF#4 IS DISPLAYED
FI	CURSOR IS SHIFTED TO TOP LEFT CORNER OF SCREEN
OS	OPCODE SEQUENCE IS DISPLAYED
EX/10	OPERATION PLAN FOR THIS OPCODE IS DISPLAYED
FI	CURSOR IS SHIFTED TO TOP LEFT FORNER OF SCREEN
RV/10	COMPLETE OPPLAN FOR OPCODE 10 IS DISPLAYED
SC	PROCESS PLAN IS STORED PERMANENTLY AS A COMPLETE PLAN

Figure 3 - Typical Process Planning Session with CAPP

Autoplan (16) qualifies as a generative process planning system because it has the ability to generate a process plan for a specific part without retrieving a standard plan and invoking some editing function. AUTOPLAN organizes its parts in a classification tree (Figure 4) which involves major part groups first followed by sub-groups then by features associated with each sub-group. The tree and its ramification is by necessity built up after an extensive review of historical process plans and part design information. Beside the Part Classification tree, the other two essential components of the AUTOPLAN system are the Data Base File and Manufacturing Logic Block as illustrated in Figure 5. The Data Base File contains relevant information on each type of operation such as operation number, work station number, planned time, actual time, operation code, etc. The Manufacturing Logic Block is the crucial link between the Classification Tree and the Data Base File. This will be discussed in more detail later.

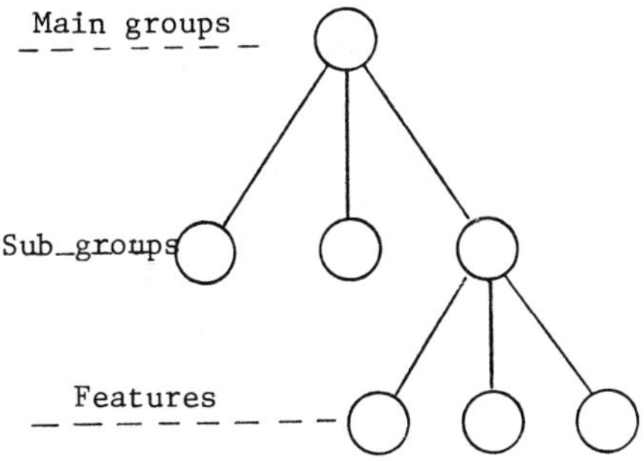

Figure 4 - Part Family Tree

C. Part Attributes Approach

In this approach the attributes of the part constitute the driving force for the CGPP system. Process Planning is performed in piecemeal manner, concentrating on a particular major machining process, such as drilling, turning, milling, etc. In the case of the APPAS system, only the hole drilling operation is addressed (17). The kind of attributes that must be specified are mainly the geometric tolerances, such as straightness, roundness, concentricity, etc. The particular part classification system to capute these attributes is known as COFORM (13). Besides the description of these attributes, a data base file containing process availability, accuracy and limitations is needed since process selection is based on a comparison of the coded requirement with information in this data base file.

D. Part Description Approach

The approach is the result of a body of theory and techniques called "Solid Modeling" (11, 14). Developed in the early 1970's, Solid Modeling has received enough exposure and understanding to be regarded as the "missing link" between a CAD system and a CAM system. The fundamental difference between a Geometric Modeling System (CMS) and other CAD system is that GMS considers any complicated solids as various ordered "additions" and "subtractions" of simpler solids, such as cylinder, sphere, block, cone, etc...by means of modified Boolean set operators - union, difference and intersection. In other words, unambiguous solid models are the building blocks of the new system and, once they are internally represented in the computer through the input processor, will evoke special procedures to calculate such things as mass properties (volume, etc.) or generate a mesh for finite element analysis. The initial and current impact of Geometric Modeling System is understandably in the area of part design and analysis, however, it is only a matter of time before GMS will also be used as a vehicle to provide other manufacturing functions, such as creating process plans and generating NC tapes. Among the currently available technologies for generative process planning based on the Part Description Approach, there are only two systems that we know of which appear to be successful in linking the solid representation of the part with its processing requirements. They both come from Germany and are known as the CAPSY system and the AUTAP system. (9, 15) CAPSY is a product of the GMS system called COMPAC (15). Using an extended version of EXAPT2, which is an APT-like language for Numerical

(11) A.R. Ness, "Using Integrated Manufacturing Schema for Solid Geometry Part Description and Analysis", Proceedings of the CAM-I International Spring Seminar, Denver, CO. , April 15-17, 1980.

(12) R.H. Phillips and J. Elgomayel, "Group Technology Applied to Product Design", Educational Module, Manufacturing Productivity Educational Committee (MAPEC), Purdue University, 1977.

(13) D.W. Rose, "COFORM - A Code for Machining", unpublished Master's Thesis, School of Industrial Engineering, Purdue University, 1977.

(14) A.A.G. Requicha and H.B. Voelcker, "Solid Modeling: A Historical Summary and Comtemporary Assessment", IEEE Computer Graphics and Applications, Vol. 2, Number 2, 1982.

(15) G. Spur, F. L. Krause et al., "A Module for Processing Part Geometry", The tomputer and Automated Systems Association of SME, MS-79-194, Dearborn, 1979.

(16) S.A. Vogel and E.J. Adlard, "The AUTOPLAN Process Planning System", 18th Numerical Control Society Annual Meeting and Technical Conference, Dallad, Texas, May 17-21, 1981.

(17) R.A. Wysk, "Automatic Process Planning and Selection - APPAS", Ph.D. dissertation, Purdue University, May 1977.

The One Step Process

Michael P. Stanislawski

Milwaukee Area Technical College
1015 North 6th Street
Milwaukee, WI 53203

Summary

This paper discusses the use of a CAD/CAM system to generate an APT geometry data base. Once the geometry is defined it can then be transformed into CNC data by using the basic language on another computer. This method can be used for wire EDM, plasma cutters, milling pockets and profiles, and turning applications. A specific example is cited using Computervision CAD/CAM, Numeridex Graphic Numerical Control system and a Bridgeport Vertical Mill. A specific example will detail the milling of a profile using the one-step method.

Introduction

The present process of constructing CNC data on a CAD system is a costly, time-consuming process that consists of four steps. They are: part geometry, tool path creation, cutter location and finally, post processing to CNC data. With the use of a Computer-Aided Design (CAD) system and a mini-computer system, it is possible to condense this process to one step. The steps are part geometry and the one-step processor.

By eliminating the tool path creation, cutter location file creation and post processing, it is easy to see that productivity can be increased dramatically. Also, this process can be done on a much smaller computer, again saving valuable computer time for more complex tasks.

Operator Information

In working with this procedure it should be noted that the manufacturing process which takes place at the machine tool must be controlled. This starts out with some basic information for the operator contained in a job folder. The job folder that will be discussed here is for a Bridgeport Boss Eight Controller on a Vertical Milling Machine.

The job folder used here contains:

1. Engineering Drawing
2. Setup Plan
3. Coordinate Drawing
4. Tool List

Engineering Drawing

The engineering drawing is the part print that the operator needs to check the part after it has been machined.

Setup Plan

The setup plan depicts to the operator the holding devices needed to machine the part along with the approximate location of the devices on the table. Along with the setup plan, the programmer conveys to the operator the tool change position, clearance plane and retract plane.

Tool Change Position

This is the position on the machine where the tool will be changed, either manually or automatically. It can be further defined as the first line of information for every tool path. A tool path will always start and stop at the tool change position.

Clearance Plane

In this example the clearance plane resides in the Z axis. It is defined as a value in which the tool can move without obstruction, from the tool change position to the first point of the part geometry in the rapid mode.

Retract Plane

The retract plane is a value above the part to which the tool will move within a tool path. The move from clearance plane to retract plane is a rapid move. This value will usually be .250.

Coordinate Drawing

The coordinate drawing can be defined as an engineering drawing reduced to just the tool path geometry needed to machine the part All values on this drawing are expressed as coordinate values from the part zero which will be used for machining.

Tool List

The tool list depicts the tools to be used to machine the part in the order they will be used. Speeds, feeds, length, diameters are also listed. If any special tooling instructions should be noted they should be mentioned here.

At this point we have completed the job folder containing the information necessary for the operator to machine the part.

Machine Tool Format

At the machine tool, there will be two types of moves the cutter will make, rapid and feed moves. Rapid moves generally will be out of the metal, while feed moves will be with the cutter moving into the metal or in metal.

The format for a rapid move is:

N___GOO X__Y__Z__

Rapid Move Format

The N___ stands for a line number which must be present for each line of information that is entered into the controller. The controller keeps track of information stored in it by using this N number. Usually a CNC program is numbered by fives or tens.

The GOO tells the machine to move in rapid to the coordinate values specified in the line.

The coordinate values must have the axis designation along with a decimal point. Example:

N10 GOO X5.

This command tells the cutter to move 5 inches in the X positive direction.

It is also possible to move in a negative direction. Example:

N10 G00 X-5.

Feed moves will use this format:

N__G01 X__Y__Z__F__

The basic format discussed earlier will prevail, along with the new characters in the command string, the G01 and the F. The G01 indicates to the controller to move in the cutting mode. The F expresses the velocity in Inches Per Minute (IPM). An example of a line would be:

N10 G01 X-5. F10.

Another command that will be utilized is circular interpolation. Circular interpolation will drive the cutter around a circular geometry. The commands are G02, G03. G02 drives the cutter in a clockwise direction. G03 drives the cutter in a counter clockwise direction. The line format is:

N__G02 X__Y__I__J__F__

N__G03 X__Y__I__J__F__

The elements that are needed for this command are:

Line Number

Circular Command

X and Y - define the end point of the geometry

I and J - define the center of the circle

F - defines the feed rate

A sample line of information would be:

N5 G02 X1. Y.1 I.5 J.5 F10.

Another series of commands which will be used are G41, G42 and G40, along with a tool number. The commands are:

G41 - Cutter Left

G42 - Cutter Right

G40 - Cancels G41, G42

The purpose of these commands is to position the cutter right or left of the part geometry. These commands enable the programmer to use geometry on the part print to machine a part as opposed to calculating the cutter center line. The G40 command will cancel the G41 and G42 mode of operation.

In using this command the cutter must also be defined in the memory of the machine tool. The values the controller is looking for are the tool number, the diameter and the length. This information will be found on the tool list. A sample line of information would be:

N__ G41 T__

The N__ gives the block number. G41 tells the controller to plac the cutter left of the part geometry. The T__ references the cutter in the controller memory to the part program.

The final series of commands which will be used are the M03, M06 and the M02 command. The M03 command turns the spindle on. The M06 command tells the controller to do a manual tool change. The M02 tells the controller to end the program.

Tool Path Defined

A tool path contains all the necessary information for the machin tool to cut the desired geometry, starting and stopping at the tool change position. The tool path at this point will take on a definite format having three parts. The first part of the program contains the necessary information to start the tool, defining the tool and positioning it at the first point of the part geometry. Part two is the geometry for the tool part. Part three is the geometry for the tool part. Part three removes the cutter from metal and returns to the tool change position.

The information contained in part one is as follows:

N5 G00 X4. Y4. Z1. M06

The first line of the program defines the tool change. The machine tool will move to the point specified, waiting for the operator to complete the manual tool change.

N10 G41 T1

The second line defines the cutter left or right of the part geometry and the tool to be used.

N15 G00 X__ Y__

The tool is at the clearance plane level which is Z1. This information is contained in line 5. The X Y values expressed here are the first point of the tool in metal. The machine tool will move in rapid, from the tool change position at the clearance plane level to the first point of the cutter in metal.

N20 G00 Z.250

The tool moves from the clearance plane to the retract level in rapid.

N25 G01 Z-__ F__

This line will position the cutter to depth for cutting our part geometry.

Part two of the CNC program contains all the feed moves to cut the desired geometry, which the programmer defines. This can include any legal moves that the machine tool will allow.

Part three of the CNC program contains the information necessary to leave the part geometry and return to the tool change position. This data will occupy the last two lines of our part program.

The second from the last line will move the tool in rapid from Z depth to clearance plane.

N__ G00 Z1.

The last line of the program takes us from clearance plane above the part to the tool change position, ending the program with a M02.

N__ G00 X4. Y4. Z1. M02

In this section of our presentation it has been shown that a tool path can be divided into three parts containing definite information in each section.

Input File

In summary then, first, at this point we have set up a procedure for interpreting information at the shop floor. Secondly, the machine tool programming format has been defined. Thirdly, we have set up a definite format for our tool part configuration.

This leads us to the next step, which is the creation of an APT geometry file. It will be used as the input file in the one-step processor. This file is created on the Computervision CAD/CAM system.

The part is called up and positioned on the screen normal to the spindle. The part geometry will be placed in a file in the order that the part is to be machined, starting with a point.

These are the only rules that apply:

1. Part normal to the spindle.
2. Must start at a point.
3. Must select geometry in order of machining.

One-Step Processor

The one-step processor is constructed so that it will take in an APT geometry file and output a CNC data file, in the format described when the tool path was defined. This processor is written in CPM Basic and resides on the Numeridex NICAM system. Also a series of questions must be answered to set up the first five lines and the last two lines of the tool path.

Therefore, parts one and three of the CNC data come from answering questions, while part two will come from the APT geometry file created on Computervision.

The questions the one-step processor asks are:

Tool Change Position X Y Z
Enter Tool Number
Cutter Left or Cutter Right
Retract Value
Cutter Depth
Spindle Speed
Feed Rate

Decimal point answers must be given for coordinate values.

After this is done the processor will handle the tool path geometry translation.

The translation is based on the concept of taking an APT geometry definition and converting it into a line of CNC data. This transfer can take into account the basic elements of point, line and circle; one geometry element will be translated into one line of CNC data.

By using a one-to-one relationship between APT geometry and CNC data it is possible to string together any geometry configuration.

The only problem encountered in working out this logic was the circular moves. Basically, the end points of the circle had to be calculated along with outputting G02 and G03 for circular interpolation.

The math calculation for the end points is done in the following manner. In looking for the ending X value, the equation is: X coordinate center of the circle plus the radius multiplied by the cosine of the ending angle.

The Y coordinate is calculated in a similar manner. The Y coordinate for the center of the circle is added to the radius multiplied by the sine of the ending angle.

Once this is done the circular interpolation block is complete. The ending points have been calculated. The remaining element in the line that must be completed is the logic for the G02 and G03 commands. This is based upon the angle values generated by the CAD system in the APT geometry file.

Condition one is where the ending angle is greater than 360 degrees. This logic will give us a G02 move, which is clockwise.

The second condition is when the ending angle is less than 360 degrees. This will indicate a G03 move, which is counter clockwise.

A third condition exists when the starting and ending angles are both greater than 360 degrees. If this condition prevails the output will be a G02 move.

In summary then, the one-step processor receives information from two sources. Manual input completes the first five lines of the part program and the ending two lines. The processor then receives the APT geometry file and converts APT geometry into CNC data, transferring elements of geometry into one line of CNC data.

Operation

The operation of the one-step process starts at the shop floor with a detailed set of instructions for the operator. This information is nothing new for CNC programmers and should be supplied no matter how the CNC data is arrived at.

Once the job folder is developed the programmer selects the geometry for the tool path (starting at a point and selecting the geometry in the order of machining). An APT geometry has been created and now resides on the Computervision system.

The Numeridex NICAM system, using CPM Basic and the Numeridex communications package, is used.

The communications package on the NICAM system logs into Computervision and the file will be transferred and will be resident on the NICAM system. I Program, called I COPY, is used to transfer files back and forth between CPM and NICAM format.

Once the file is transferred, the one-step processor is called up. Seven questions are answered and the APT file generated on Computervision is processed, giving us the following CNC data file.

The CNC data file is then transferred back to NICAM format to be sent to the Bridgeport machine tool. Before it goes to the machine a plot of the tool path is run on the NICAM system to verify the path before machining.

Once the data is verified the information is sent to the Bridgeport machine tool.

The operator sets up the machine tool, loading a .500 diameter mill that will be used to cut the entire part. The first tool that will be loaded into the machine's tool table is the 1.375 diameter mill. This will give us a cut around the part at a distance of .6875 from the part geometry.

The .5 diameter mill remains in the spindle while the tool table is changed. Tool 1 is now defined as having a diameter of 1.000. The second roughing cut is taken.

The .5 diameter mill remains in the spindle while tool 1 in the tool table is again changed. The finish cut is now taken, with the tool being defined as .500 diameter.

This same process can be used for cutting a pocket or an open profile if desired. Also, wire EDM machines, flame cutters, lathe work and drilling operations can be performed using the one-step process. Naturally, the CPM Basic program will have to be changed to conform to the particular machine tool and geometry configuration.

In conclusion then, this paper has presented an effective way to do a CNC program for any geometry configuration in a two and one-half dimension world. This method can save time and money for the conscientious, progressive company interested in combining CAD/CAM and Operational Control.

Generalized Concepts and Features Towards the Development of a Processor for Application to Fluid-Thermal-Structural Problems

Kumar Tamma

Department of Mechanical and Aerospace Engineering
College of Engineering
West Virginia University
Morgantown, West Virginia

Summary

The paper demonstrates the feasibility of implementing systematic generalized concepts to display pre and post computer graphic models for general applications, namely, fluid flow models, heat transfer models, and structural models. Concepts for generating neutral files to feed information pertaining to the generic models as required for the associated display processor are also discussed. Finally, the capabilities and features of a graphic processor developed is described and demonstrated through applications to general problems.

Introduction

With the widespread attention of the finite element method as the 'principal tool' in the emerging computer-aided design (CAD) world, the technology for creating geometric models within a CAD framework is becoming ever increasingly popular. During recent years the finite element technique has spurred tremendous interest amongst practicing engineers as well as developing researchers. Currently, there are available to users several general purpose programs capable of being implemented on a variety of computers. The continued development and applications of finite element technology encompass a broad spectrum of engineering applications including aerospace, automotive, naval, mechanical and civil engineering design. Although the method originally started with applications to structural/solid mechanics areas, at present, research in thermal sciences and fluid mechanics is proceeding at a rapid pace. In this regard, some positive attributes of the finite element tool, for example, are ease in modeling complex geometries, consistent treatment of natural (differential type) boundary conditions and the capability of being pro-

FMS and Statement of the Problem

A simulation study is carried out on the ENSAM (Ecole Nationale Supérieure d'Arts et Métiers- Paris) flexible manufacturing cell, Fig.1.

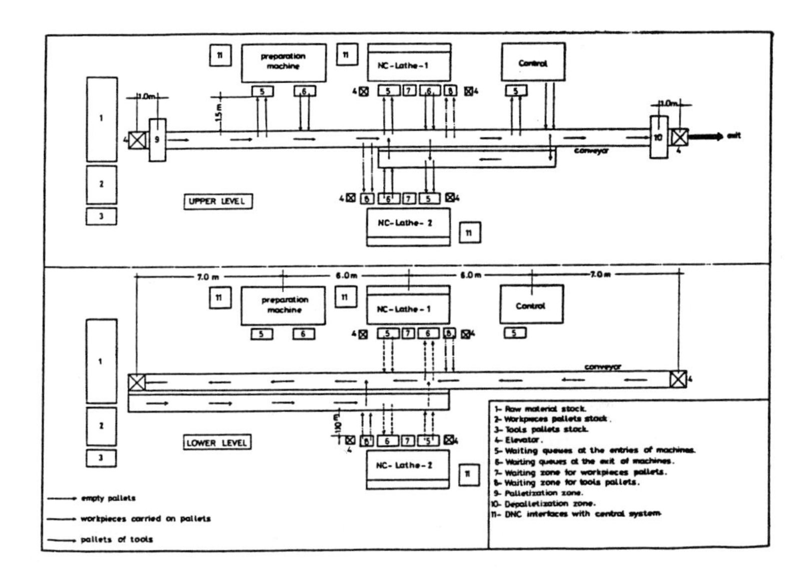

Fig.1. Implantation of the ENSAM FMCell.

The flexible cell groups two numerically controlled lathes, a control station, loading/unloading robots, loading/unloading stations, as well as a preparation machine. Work stations are linked by a double level multi-switching automatic conveyor. The two levels are linked by means of elevators. Parts are transported on pallets with preset fixtures. Between operations, parts have to be fixtured to different pallets. DNC terminals pilot the machines. Queueing areas at the entry and exit of machines are conceived to assure a production autonomy and area for in-process inventory. Raw pieces and pallets are available at the entry of the flexible cell. All the above units are under the control of a central computer. Six types of workpieces belonging to the same family are machined in the system.

The major task is to determine the schedules of the different parts simultaneously so as to assure the flexibility and productivity of the flexible cell. The hierarchical structure of information processing in the problem is illustrated in Fig.2. The structure comprises three levels, i.e. determination of loading sequences of parts, selection of workpieces pallets and selection of cutting tools.

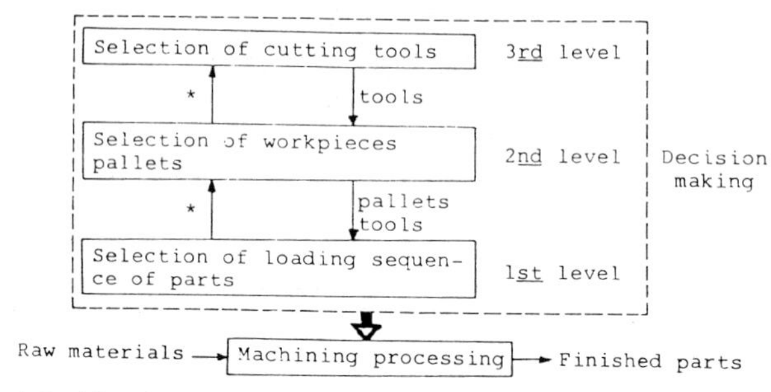

Fig.2. Structure of information processing

Simulation Model of the FMS

Q-GERT network modelling technique is used to study the scheduling problem of the ENSAM FMCell [4] . The model is analyzed by the Q-GERT analysis program [5] . The complete Q-GERT network of the flexible cell is shown in Fig.3. The network depicts the flow of transactions and all the potential processing steps associated with them.

Scheduling Procedure

Due to the complexity of the problem, it is difficult to find an optimum solution. The work provides a decision rule to obtain a good solution for a large-scale problem in a practical manner.

Control of the system proceeds as follows: when a machine indicates queue vacancy, the central system selects; by means of the decision rule given as (UF,16) ; within the various pieces to be fabricated or in-progress, those whose operating sequence necessitates their passage on this machine, giving priority between these candidate pieces permitting moving the most appropriate one to the machine. The choice of workpieces is to be made maintaining the representation of all types in the system, and allowing the decision algorithm to control their production according to the present situation. The flow chart of Fig.4. shows the

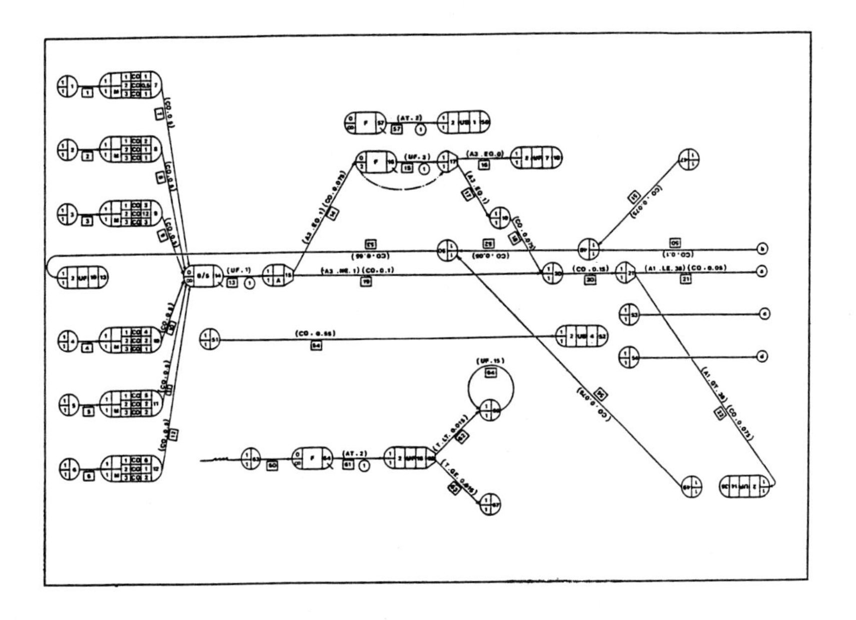

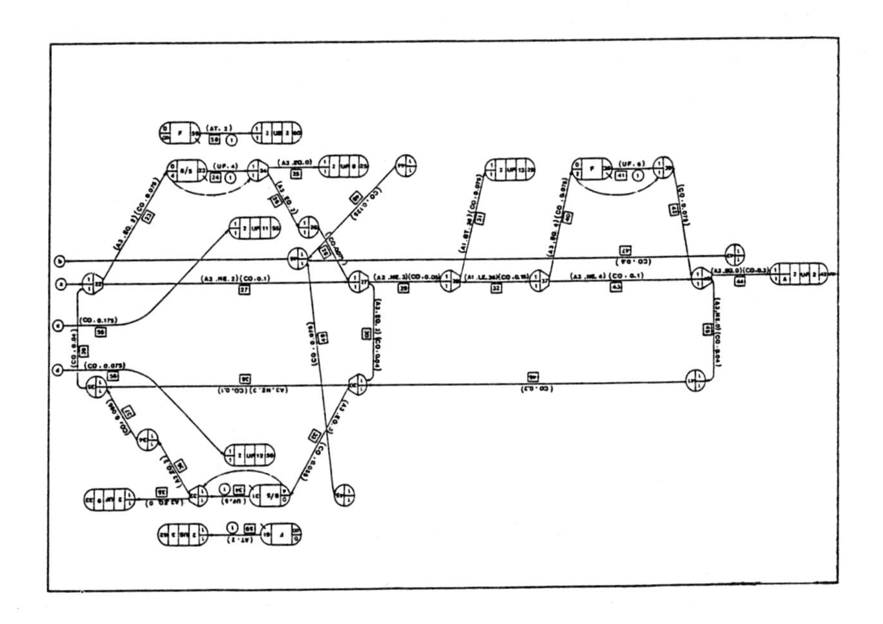

Fig.3. Q-GERT network of the ENSAM FMCell.

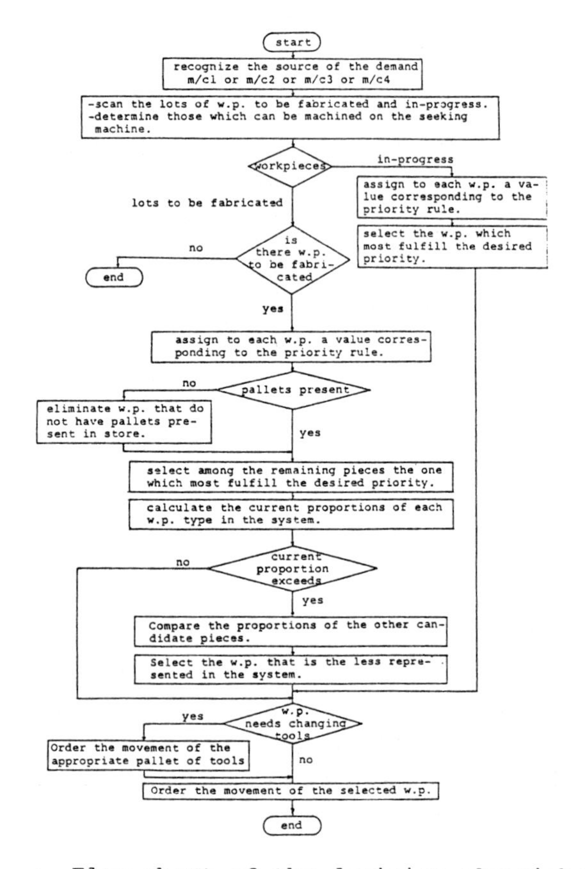

Fig.4. Flow chart of the decision algorithm.

logic associated to the decision rule (algorithm).

Priority rules inserted in the decision rule govern the assignment of jobs to machines. Twelve rules are tested as shown in Table 1 .

The steps mentioned in the flow chart complete one cycle in the scheduling process. The following cycle begins with the selection of a workpiece when a place is once more vacant in a queue. Similar cycles repeatedly proceed using the decision rule until working time is terminated, at which a solution to the scheduling problem is obtained.

The simulation is carried out for 480 min. , the CPU time on UNIVAC is approximately 60 s. for each run.

very concisely. Furtheremore, the classification and coding technique renders itself very handsomely to computer applications.

The Universal Tool Classifcation Code (UTCC) is a classifcation and coding system that provides the means for identification and specification purposes of cutting tools. The UTCC system consists of twenty two positions that describe the type, geometry, dimensions, material, manufacturer, and the other factors related to the specification or application of a cutting tool. Information content of each position of UTCC was decided upon after careful consideration and inspection of all the attributes related to tools. The information content, the coded presentation, and the sequence of different positions of the UTCC codes are compatible, to the extent possible, with the ANSI standards and other widely accepted practices that were available for tool identification purposes.

The UTCC is structured in such a way that it can be combined with the classification codes of other items in manufacturing. This would make the establishment of an aggregate classification and coding system possible and will facilitate the incorporation of UTCC into the aggregate system. The first position of UTCC, therefore, divides the broad spectrum of manufacturing item into subgroups with each subgroup consisting of one major category such as raw materials, purchased components, manufactured piece parts, machine tools, assemblies and tooling. The major category of tooling is further divided into some major divisions with a numeric code assigned to each. At the present time, the first two major divisions of the tooling category are established in details. The first major division consists of inserts, tool bits, tool blades, thread chasers, and other similar cutting tools. The solid cutting tools, insert holders, and tool shanks form the second major division of tooling category. The third position of the UTCC code further divides each major division into some subdivisions. The subdivisions for the tooling category consist of the different types of tools as related to their application in various operations. The UTCC positions four through twenty two have been established to identify the tool and specify all of the important features and dimensions for each major type in the form of a polycode structure.

A tool classification and coding system has potential applications in almost all activities of a manufacturing environment such as tool inventory, purchasing, design, and process planning. By using the UTCC, a firm can establish an all inclusive file of tools. This tool data base can be sufficiently detailed so that it could provide the input for systematic selection of cutting tools. The utilization of UTCC will lead to standardization of cutting tools and will make the optimum or preferred tool selection possible.

Robotics and CAD/CAM Education I

Chairman: Jack D. Lane, GMI Engineering and Management Institute, Flint, Michigan
Vice Chairman: Richard A. Zang, RCA Technical Excellence Center, Princeton, New Jersey

A Manufacturing Program for High Technology

Joseph E. Kopf

New Jersey Institute of Technology
Department of Industrial and Management Engineering
Newark, New Jersey 07102

A look at "High Technology" in manufacturing reveals a new approach and a new way of thinking. Computers are here to stay in processing data, controlling machinery and assisting in decision making. They may get smaller, more powerful and more sophisticated but will probably get more "user friendly" so that they can be useful to those with little programming knowlege. Robots, while not new are offering a new challenge in the applications area. The future robots will be more highly sophisticated and use voice control, have multiple sensing devices, information feedback, be capable of some decision making, have vision, have multiple arms and more accurate motion repeatability. Computer numerical control (CNC) is now an old science and many different types of machines are computer controlled. The challenge of the future is to take all these individual high technology developments and tie them together into a working system as a manufacturing cell or into a computer controlled automated system.

Looking at the Factory of the Future we see several possibilities. We see a completely automated factory with no production workers, but a swarm of highly trained electro-mechanical maintenance mechanics. Next, we see a long assembly line with a large number of computers used to control the parts flow, feed back data and make the necessary adjustments to maintain continuous flow, with robots doing assembly work and doing material handling chores between CNC machine and a number of production people here and there working between and along with the robots and the CNC machines on the same production line. We also see a small shop with CNC machines and robots working with more production workers around groups of machines in manufacturing cells. Complete automation will be used in those few industries with the production quantities necessary to make the system cost effective. Also possible will be industries with less quantity production and where the cost of complete automation may be prohibitive. We will also see the "job shop" were automated machines can be justified but the small production quantities and the need for fast changeover is necessary but again the cost of automated systems is prohibitive.

How do our institutions train our manufacturing people for this new world? First, we developed an upper two year program in manufacturing that will meet the future needs of industry (see Fig. 1).

JUNIOR YEAR

FIRST SEMESTER				SECOND SEMESTER			
CIS	202	Computer Programming & Business Problems	2-2-3	Mt Sc	311	Prop. of Mat'l	3-0-3
ENG	342	Technical Report Writing	3-0-3	OS	371	Supervision & Employee Rel.	3-0-3
MATH	108	Math. Analysis I	3-0-3	MATH	209	Math. Analysis II	3-0-3
IET	315	Industrial Statistics	3-0-3			Elective (Hum or SS)	3-0-3
IET	414	Industrial Cost Analysis	3-0-3	IET	420	Quality Control	3-0-3
IET	317	Manufacturing Operations Analysis (Using Computers)	2-2-3	IET	318	Mfg. Proc. Design	2-2-3

SENIOR YEAR

IET	416	Production Scheduling	3-0-3	OS	472	Mtg. & Org. Behavior	3-0-3
IET	405	NC for Machine Tools	2-2-3			Elective (Hum or SS)	3-0-3
IET	423	Motion & Time Study	2-2-3			Elective (OS or Tech El.)	3-0-3
		Elective (OS or Tech Elective)	3-0-3	IET	422	Tool Design	2-2-3
		Elective (Humanities)	3-0-3	IET	424	Fac. Planning	2-0-2
				IET	426	Fac. Planning Lab	1-3-2

Several years back we purchased a 2 H.P. CNC milling machine and taught the students manual parts programming and began to teach computer programming. Next, we purchased a small robot that would interface with an Apple computer and let the student program the robot to do standard type moves. Recently, we purchased a small computer controlled lathe. We still, however, require our students to do manual programming so that they understand the basic principles involved in CNC.

Our next problem centered around the controversy of training on small equipment versus using actual production machines and the cost of such equipment. We felt that exposure to the large production machines was important, however, an understanding of the basic principle can be taught on small machines. With the mastery of the basics the transition to large machines is relatively simple.

A newly formed "Industrial Consortium in CAD/CAM-Robotics" at our Institute has resulted in money being available for additional equipment. Several industries have given money to the consortium and in return faculty and students work on projects for the companies. This has allowed us to obtain the following equipment: a U.S. Robots- Maker 1 Robot; a Schrader-Bellows Motion Mate Robot; a Copperweld Auto Robot (donated by ATT Technologies); a Fanuc Model 1 Robot (on loan from Westinghouse Electric Corp.), and a PRAB-FA (donated by Westinghouse Electric Corp.). Two small robots equipped with teach pendants will be available soon as well as several material handling conveyors. These will be used by students to simulate manufacturing cells tying together various small machines and work areas to simulate computer integrated manufacturing.

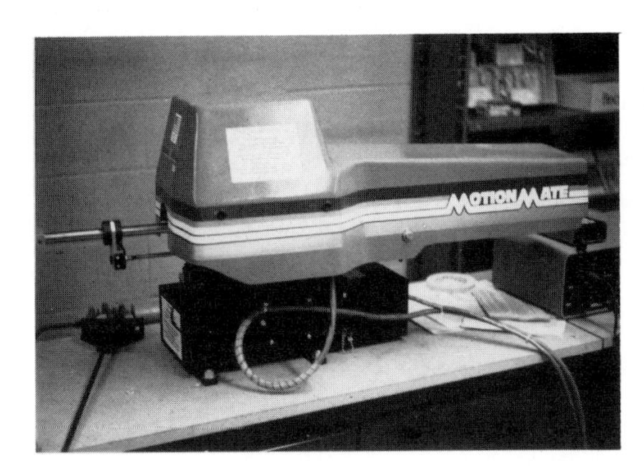

Photograph No. 1

A smaller industrial robot used by students. Five degrees of freedom. Electric drive-pneumatic arm.

Photograph No. 2

A small industrial robot. Five degrees of freedom. Electric drive and electric arm.

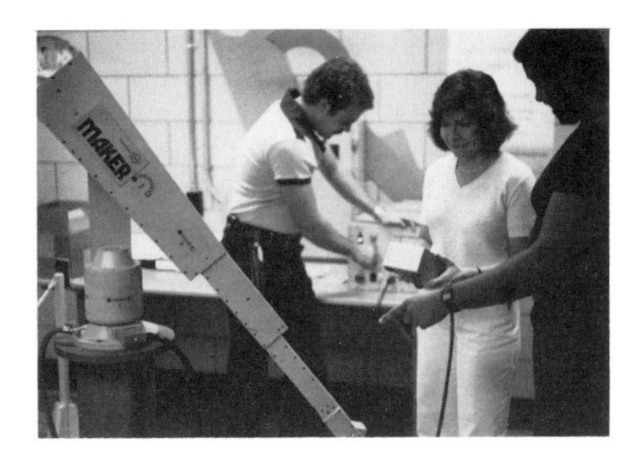

How does this manufacturing program interface with the Factory of the Future?

1- The program teaches the basics needed in high technology with a strong emphasis in the use of the computer for problem solving.

2- The use of High Technology in Manufacturing emphasizes productivity and cost reduction to gain a competitive edge. The technical courses in the program are designed to this end. Technical courses such as Tool Design, Motion and Time Study, Production Control and Facilities Planning emphasize productivity and cost control.

3- Industrial Statistics, Quality Control and Industrial Cost Anaylsis are important supportive courses in manufacturing.

4- A good course in Properties of Materials is of general use and necessary in Tool Design.

5- Mathematics, technical writing and management skills are necessary in working in a high technology environment.

6- Any well rounded program will contain humanities and social science courses.

7- The technical courses contain laboratory work where the students get hands-on training on the latest equipment. The equipment is available to the student at any time and extra training is gained when working on industrial projects with faculty.

8- The close relationship with industry, through the consortium, continually exposes students to the latest developments in manufacturing.

Conclusion

1. Educational programs in manufacturing must be seen as a business. Educators can no longer go to industry and beg for equipment because the equipment they get will most likely be the machines that are to be replaced or otherwise obsolete. Educators must go to industries and sell their expertise and use the income to buy equipment for training purposes and equipment for doing manufacturing development work. This work should be done with student participation. Industrial people and educators must work together.
2. The course work stresses the basics and educates the students for life-long learning.
3. The basics can be taught on small training equipment and then applied to industrial size equipment.
4. In general, laboratory equipment is under utilized and expenditures of large equipment is not cost effective.
5. If funding is limited, institutions should form a partnership with industry so that students will get some exposure to production size machines.

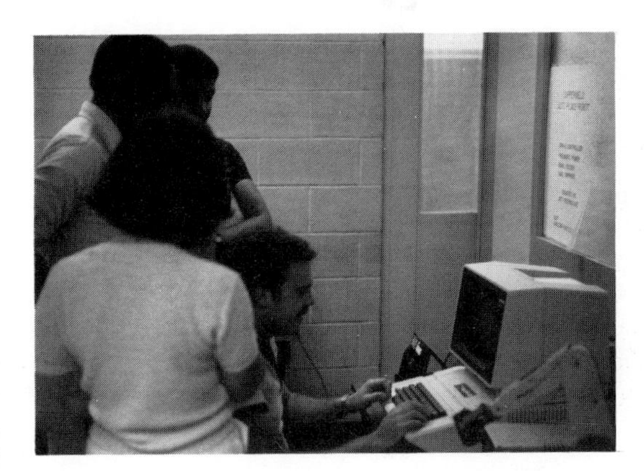

Photograph No. 3

Students programming a small robot. Sutdent gets training on this set-up prior to working with larger robots.

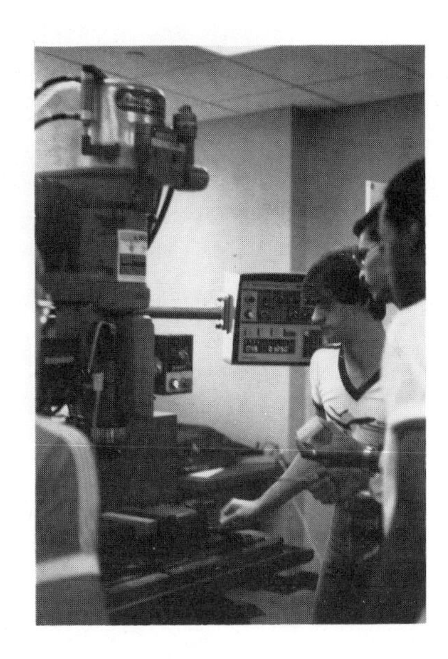

Photograph No. 4

Students working on a CNC machining problem using a production 2 H.P. CNC Milling Machine.

Photograph No. 5

Student programming a small CNC lathe. Programs can be written for any size part and the small lathe will turn a scaled model.

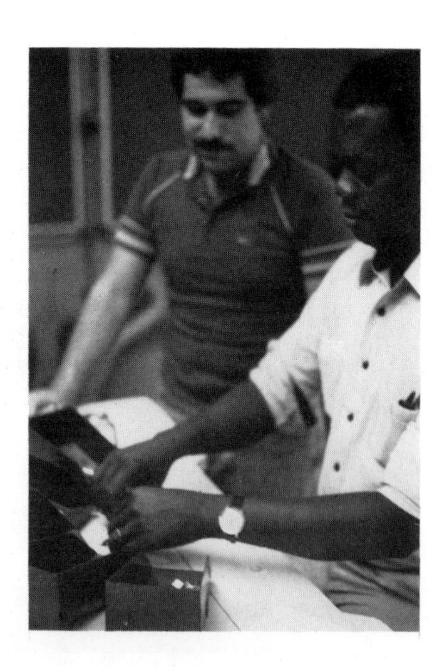

Photograph No. 6

Students working on a work place layout to determine the best arrangement for efficient production.

References

1. Johnson, Ph.D., Clay G., So You Want to be a Robotics Technician. Journal of Tau Alpha Pi, 1984.

2. Schreiber, Rita, R., The New Manufacturing Engineers, Manufacturing Engineers, June 1984.

Applied Robotics: A Videotaped Course

Richard A. Zang

RICHARD A. ZANG

Corporate Engineering Education
RCA Technical Excellence Center
Princeton, New Jersey

RCA found plenty of information on robot coordinate systems and other technology details, but precious little on coordinating and orchestrating the use of robots in an actual manufacturing enviornment. This one-of-a-kind course bridged the gap.

Abstract: This paper describes the development of a thirteen session videotape course on Applied Robotics in Manufacturing. The primary objective of this course is to provide RCA engineering and technical personnel with the information and understanding necessary to make informed decisions leading to successful robotic applications. The course starts with an introduction to applied robotics and proceeds all the way through how to organize a robot project. The course covers such topics as: Coordinate Systems, Performance Characteristics, End Effectors, Power and Servo Systems, Robot Control, Robot Programming Languages, Vision, Flexible Manufacturing Systems, Applications, and Future Trends. This paper provides insight into each session by looking at: objectives, topics for discussion, homework assignments, and summaries. Further, insight is provided into the overall development of the course; instructor recruitment, textbook selection, general approach, and philosophy.

Economic needs and the development of supporting technologies have produced a rapid growth in robotics technology. By 1983, more than 30 robots were installed at various RCA manufacturing sites. Unlike the engineers at companies who manufacture robots (e.g. Unimation, PRAB, Cincinnati Milacron, and Bendix), engineers at RCA are not interested in designing robots, but rather, in how to apply them. Thus, the need arose for continuing education on applying robots in manufacturing.

Sending hundreds of RCA engineers to workshops and seminars didn't appear to be cost justified. Having a course on videotape seemed like a good idea. The search was on; somewhere out there, somebody must have a videotaped course on applied robotics! But no; we came up empty handed. Thus, we took the first steps to develop our own internally produced videotaped course. Eighteen months later we were at journey's end with a complete video course package (tapes, study guide, and text) in hand.

The video course package consists of 13 internally produced videotapes (i.e., sessions), an internally produced study guide, and a textbook. A brief synopsis of each of the 13 sessions follows.

Session 1, Introduction to Applied Robotics, gives the participant a bit of history leading to robotics technology, and destroys some common myths surrounding the subject of robots.

The next four sessions give the basic facts and terminology about robots and their configuration. Session 2 covers Coordinate Systems, gives a survey of the basic coordinate systems, and by the use of examples, gives the participant a good feel for their relative advantages and disadvantages. Session 3 gives Performance Characteristics, and covers "figures-of-merit" such as load capacity, speed, resolution, accuracy, and repeatability. Session 4 illustrates the variety of end effectors that can be used with robots, gives an understanding of various wrist designs, and shows the capabilities gained by utilizing force and optical sensors. Session 5, on Power and Servo Systems, surveys the various power systems used in robots and looks at the advantages and disadvantages of each.

At this point, the course participant has a solid foundation in robot technology. But, as they learn in this course, robots should not be seen as islands in a manufacturing environment. The necessary tools and methods for integrating the robot into the industrial environment is the subject of the next five sessions. Session 6 on Robot Control, shows several ways in which the complexities of an industrial process can be controlled, including the programmable controller and computer. Session 7 identifies what RPLs (Robot Programming Language) are, what they can do, and perhaps of most importance, what they cannot do. Session 8, Vision and Other Sensors, explores the various types of presence detectors, outlines the nature of a machine vision system, and looks at the fundamental difficulties which limit the performance of vision and other types of sensors. And session 9, devoted to Flexible Manufacturing Systems, indicates the changes in product design and organization necessary to create flexible manufacturing systems, in which robotics technology plays a catalyzing role. Finally, session 10, Robot Automobility/Odex, introduces the concept of walking machines and demonstrates the progress made in the field, highlighting Odex I.

Nearing the end of the course, participants are ready to put their new-found knowledge to work. The last three sessions are devoted to applications and project management. Session 11, Organizing a Robotics Project, gives guidance in the choice-of-scope of projects, selection of personnel and vendors, and warns of pitfalls. Session 12, Applications Within RCA, takes the participant on a walking tour of robotic applications at various RCA manufacturing sites and RCA's Princeton Laboratories. And session 13, Applications Outside RCA/Future Trends, takes a look at various robotic applications outside RCA and goes on to indicate some of the trends in robotics development.

FORMAT/PHILOSOPHY

Corporate Engineering Education (CEE) at RCA, has as its principal charter, providing primary corporate leadership for the continuing technical education of RCA engineering and technical personnel. There are several services offered to RCA technical personnel by CEE, but the primary service, and the one of particular interest to this paper is, CEE Video Courses.

A typical video course consists of between ten to fifteen two-hour weekly sessions. Although each individual RCA facility is at liberty to schedule courses as they see fit, most locations offer CEE courses on split-time (i.e., half on company time, half on employee time).

A typical two-hour session breaks down as follows:

- 40-60 minutes of video presentation, and

- 60-80 minutes of supplemental lecture, class exercises, questions and answers, and class discussion led by an on-site Associate Instructor (AI); usually an employee of the location offering the course.

CEE's video courses are purposely designed to serve as a core to be added to and customized for a given location's individual needs. The AI will know his/her location's needs and will taylor the course accordingly.

THE BEGINNING

After the need for a course on applied robotics was established, we at CEE set out to find an already existing videotaped course. Since none was found, the decision was made to produce our own home-grown videotaped course on the subject.

In the search for an already existing videotaped course, it was discovered that The Center for Professional Advancement (East Brunswick, NJ) offered a live course entitled "Applied Robotics for Industry". That in itself didn't help us much; because it was live as opposed to videotape; but it did give us a good lead to an instructor for our course.

It turned out that R. Michael Carrell (the live instructor) was an RCA-Princeton Laboratories employee. And wouldn't you know that CEE is located in Princeton as well. Making a long story short, Mike agreed to be our video instructor. It should be noted that going from, speaking to a live audience to, speaking to a camera, is not always an easy transition. However, we lucked-out because Mike adapted well and did an admirable job.

One of the most common complaints of any course seems to be the text. Why? It's this author's opinion that textbooks are too often adopted after-the-fact. To eliminate this potential "problem", it was decided to adopt a text up-front, so to speak.

Finding a text was not an easy task. To appreciate this, let us keep things in perspective:

- The search for a text was early '83.
- Robotics technology was only starting to emerge from the infancy stage.

Thus, *designing* robots was on its way to becoming a mature discipline, while *applying* robots was lagging a few years behind. The net result: it was fairly easy to find textbooks on designing robots, but next to impossible to find one on applying robots.

The following four books are what we ended up considering:

1) Ayres, R. and Miller S., *Robotics: Applications & Social Implications*, Ballinger Publishing Company, MA (1983)
2) Dorf, R. C., *Robotics and Automated Manufacturing*, Reston Publishing Company, Inc., VA (1983)

3) Tanner, W. R., *Industrial Robots: Volume 1/Fundamentals*, Second Edition, Robotics International of SME, MI (1981)
4) Tanner, W. R., *Industrial Robots: Volume 2/Applications*, Second Edition, Robotics International of SME, MI (1981)

The choice was quickly narrowed to the first two books. The last two books by Tanner are compilations of various papers, and are quickly becoming dated. The first two books have a strong applications flavor. We decided on Dorf's book because it devoted more attention to the industrial environment.

At this point in the course development (we were about 5 months into it), we had come up with what we felt was an excellent course outline. All that remained was producing thirteen videotaped sessions and a study guide (i.e., almost everything).

THE COURSE:

Introduction to Applied Robotics (Session 1)

After completing session 1, Introduction to Applied Robotics, the course participant is able to:

1. Describe the place of robots in the history of technology.
2. Apply the formal definition of a robot to various machines and decide which are robots and which are not.
3. Show similarities and differences between industrial robots and the androids of fiction and amusement parks.
4. Describe some differences in US and Japanese use of robots and other manufacturing technologies.
5. Describe some of the robotics projects in RCA.

A summary of the first session (as it appears in the study guide) is as follows:

> While robotics technology is getting a lot of attention, its technical roots go back hundred of years. It is only in the past decade that the economic needs and the capacity of the supporting technologies have combined to produce a rapid growth in the industry. These trends are expected to intensify. Japan leads the world in the systematic use of robotics. There are many startup companies in the US and major corporations are entering the market. RCA has over 30 robots installed.

Basic Facts & Terminology (Sessions 2-5)

Without a solid foundation in the fundamentals, topics such as Robot Control (session 6) make little sense. Thus, the next four sessions are devoted to building this solid foundation needed for later sessions.

After completing session 2, Coordinate Systems, the course participant is able to visualize the shape of the robot workspace and the coordinate system of various robot configurations; determine the optimum coordinate system for a particular job, and judge the importance of the choice of coordinate systems against other factors; and determine the number of degrees of freedom necessary for a robot application.

Space does not permit giving the Homework Assignment for each of the thirteen sessions. Instead, to give a sample, the assignment from this session (as it appears in the study guide) is given below:

> How many degrees of freedom are necessary to place a tool at any point in the robot workspace, and in any orientation at that point? Is this count affected by the basic robot configuration? Write a short essay justifying your answer.
>
> In the application you spotted for the last session (or another of your choice), what configuration is best? How many degrees of freedom are needed? Write a short essay justifying your choice.
>
> How many degrees of freedom are in your hand? (Marvelous machine, isn't it?).

To summarize this session: Describing robots is much like classifying animals in a zoo; while most have heads, tails, and limbs, there are vast functional differences in the overall 'design'. One way of classifying a robot is by the way the basic skeleton is formed, giving it a natural coordinate system. Basic coordinate systems are surveyed with examples of each and some comments on their relative advantages and disadvantages.

After completing session 3, Performance Characteristics, the course participant is able to distinguish among resolution, accuracy and repeatability as they describe a robot's performance, and to properly use these figures-of-merit in an applications study; and understand the relationship among load carrying capacity, speed and acceleration and the tradeoffs affecting cycle time.

To summarize this session: Users want figures-of-merit to compare the performance of robots. Like horsepower and fuel economy ratings of cars, such numbers can be deceptive unless their basis is understood. Thus load capacity, speed, resolution, accuracy, and repeatability may appear in robot specifications. It is important to understand what these mean and how they relate to the basic design of robots.

After completing session 4, End Effectors, the course participant is able to illustrate the variety of end effectors which can be used with robots; to understand the variety of wrist designs and their influence on a robot's utility for specific jobs; and to show the capabilities gained by grippers equipped with force and optical sensors.

To summarize this session: Many wrist configurations have been used to orient the end effectors. Many of these have problems when three degrees of freedom are needed in the wrist, for the endmost motor must produce enough torque to support the load; a motor of such capacity may be heavy and decrease the net load capacity of the robot. An exception is the three roll wrist, which is strong and light and may be driven by remotely located motors.

End effectors are the interface between the general purpose robot and the specialized job at hand. Robot vendors would prefer users to supply the end effector, and users often want the vendors to do it. The result is negotiation, with the vendor or a third party providing consulting services and demonstrations as part of the customer service, usually at added cost.

Where end effectors are equipped with sensors, the robot can monitor the process environment, calibrating itself to a changing workplace. In some cases a more sophisticated robot can reduce the cost of installation and ancillary tooling.

And finally, after completing session 5, Power and Servo Systems, the course participant is able to identify the advantages and disadvantages of various power systems used in robots; and is also able to show the advantages and disadvantages of different servo implementations used in robots.

The summary of this session (as it appears in the study guide) is as follows:

Pneumatic, hydraulic and electric drive systems are all used in robots, with hydraulic systems providing the highest performance in terms of acceleration and load capacity.

Electric drive systems primarily use variants on the permanent magnet DC motor. Stepping, brushless, and coreless motor designs each remedy one of the disadvantages of the simple DC motor.

DC motors are usually used with gear reducers. The Harmonic Drive is a special reducer widely used with robots which provides high reduction ratios without backlash in a compact package.

Servo systems use motors inside negative feedback loops so that the robot members can be precisely moved with a high degree of stability. The feedback loop may use analogue or digital sensors. Incremental shaft encoders and digital feedback loops with microprocessor control are widely used.

Integration into the Industrial Environment (Sessions 6-10)

At this point in the development of the course (we were about 12 months into it), talking of applications makes little sense unless the course participant understands that "no robot is an island". The next five sessions develop this understanding and provide the necessary tools and methodology for integrating the robot into a manufacturing setting.

At the completion of session 6, Robot Control, the course participant: knows that tasks which seem simple actually contain complexities which must be anticipated in automation design; is able to show several ways in which the complexities of an industrial process can be controlled, including the programmable controller and computer; and understands the need for special languages as interfaces between humans and computers.

To summarize this session: Simple tasks, like making a sandwich, contain hidden complexities when all the exception conditions are accounted for. In designing an automation system, these must be anticipated and provisions made in the controller. If the controller uses traditional relay logic, it quickly becomes costly, unreliable and difficult to change. Programmable controllers, which are microprocessors in 'battle dress', provide a very effective way of managing industrial control problems. They express decisions in a ladder diagram 'language' used for relay wiring.

Further, ladder diagrams are clumsy for tasks involving counting, comparison, timing, or arithmetic; a more general approach is needed. Binary digits (1 & 0) provide a way of doing this. 1's and 0's can be used for control, logic, and computation. If they are represented by different voltage levels

Initiating University Robotics Education Through Industrial Students Project

Wayne W. Walter

WAYNE W. WALTER, PH.D.,P.E.
Mechanical Engineering Department
Rochester Institute of Technology
One Lomb Memorial Drive
Rochester, N. Y. 14623

The virtual explosion of recent industrial interest in robotics and computer-aided-manufacturing has created a serious challenge to universities to not only stay abreast of this high technology, but to develop new courses in this area for their students. As U. S. industry makes the factory-of-the-future concept more of a reality, the demand for graduates with quality manufacturing engineering degrees is certain to increase dramatically. This poses some very serious and immediate problems for U. S. universities. Universities must quickly develop new programs, provide the opportunity for their faculty to acquire the necessary expertise and experience in this area, and acquire the required hardware and software to support these programs. These three areas of concern are discussed below.

Many universities have already developed new manufacturing engineering programs or have created manufacturing engineering options within their existing mechanical or industrial engineering programs. In most cases, such options within existing programs appear to be built on a core of conventional fundamental courses with additional new courses at the upper end. Upper division courses are normally of a design flavor if the option is offered in a mechanical engineering department, or they will most likely have a systems emphasis, if the option is within an industrial engineering department. Such a program offered as an option in an established department appears to be the easiest alternative to implement, and has the added benefit of not requiring a new ABET accreditation as a new program will.

Developing the necessary faculty expertise in the manufacturing area is a difficult problem, especially for faculty members experienced in the conventional subdisciplines in mechanical engineering such as stress analysis, dynamics, and design. Much of this expertise is not available from texts, and can only be acquired from experience. For most of us, what is required is a three-to-six month leave of absence at a major manufacturer in order to work with the hardware and software currently being used there. I have personally found this very helpful having just returned to RIT after spending six months in the manufacturing development department of a major U. S. automaker. As a member of a group responsible for robotics and flexible manufacturing systems, I worked with machine vision-guided robots, and robotic simulation software. As a result of the experience gained, I have been able to offer a new robotics course in our program. My experience there has also provided many of those little stories that are so important in the classroom. It's an experience not without peril, however. As an academic, it took some time for my industrial colleagues to consider me as a significant member of their group. My initial weeks were spent on a learning curve, and it takes time to begin to make a contribution in such a situation. Maintaining a good attitude appears to be the key to success here. A willingness to accept my group colleaques on an equal basis helped them to do the same with me. As difficult and as humbling an experience as this is in some ways, it is a rewarding one and a viable means to update and refresh our faculty with state-of-the-art technology.

The problem that is most difficult to deal with is the substantial expense involved in purchasing software and hardware which can be effectively used for teaching. In robotics, in particular, our experience at RIT shows that table-top educational robots are just not adequate to prepare our students for the industrial types they will encounter in industry. Many of these educational robots require assembler language (a language not particularly user-friendly). They are often quite slow, so that simulations of machine loading or assembly tasks are difficult to do meaningfully. It is also difficult to design fixtures and tooling for these types, and to utilize external sensor input via an IO port. Lastly, and probably worst of all, they often foster a "Star-Wars" type impression of robots in the student.

Prices of industrial robots are usually prohibitive for a limited-use university environment. With typical prices between $30-100K, such purchases on a typical university budget are out of the question. Although many universities look to industry for donations, the present economy often prevents this. Some manufacturers are reluctant to donate an item that they did not manufacture themselves. In addition, many manufacturers with an unused robot must first look for potential users within their own organizations. Only after exhausting all potential internal applications is it possible for them to donate the robot to a university.

The Mechanical Engineering Department at RIT has pursued one solution to this problem for the last year. Through the means of student robotics projects at a local U. S. automaker's facility, students have been exposed to state-of-the-art industrial robots and associated equipment without any expense being incurred by RIT. In addition to designing robotic work cells, tooling, grippers, and fixtures, students have had a chance to work on projects involving machine vision inspection systems, robotic sensors, voice digitizing systems, and robotically-loaded CNC machining centers. A very brief description of each project is given in Table I.

Most of our student projects are conducted on an independent-study basis. In each case, a development engineer at the automaker's facility is designated as the project leader. Students meet with their project leader on a need basis, although it is expected that this will occur at least once a week. The development engineer, together with the student's faculty advisor, provides the students with guidance and constructive criticism regarding their projects. The project leader also helps the students to establish the project priorities and design objectives. As a result, students get an appreciation for the concepts of design for manufacture and assembly, and design for production and factory-floor environments.

Each project team submits a set of design drawings to their project leader at the end of the design phase. Although some students do their drawings on conventional drafting stations, many design teams construct their drawings on the automaker's CAD terminals. If appropriate, after their drawings are approved, students monitor the build-phase of their project. They consult with toolmakers at the automaker's facility who construct prototypes from the students' drawings. Students then have the opportunity to set-up and debug their devices in the automaker's development lab by mounting them on an appropriate robot. During their interaction with the skilled tradesmen, students get exposed to many of the attitudes and concerns of union members regarding such topics as the impact of robotics and automation on the workplace, worker productivity, and remaining competitive in the present world market.

The question of patent rights is often an issue of concern to students at the beginning of a project. Patent policy in this type of situation is still evolving. Since students do their projects almost exclusively at the automaker's facility, this situation is somewhat different from industry-sponsored research done on a university campus. The automaker has taken the position that students should not be prevented from pursuing a patent for their work if the automaker is granted a no-fee license to the patent. The automaker has no intention of marketing their ideas, but does want to use them at no cost. To date, one patent has been sought for a quick-change end-effector developed by a graduate student. In this case, the student declined the opportunity to apply for exclusive patent rights due to the substantial expense involved. Instead, the automaker applied for the patent and named the student as a co-contributor. It is important that the patent issue be discussed at the beginning of a project, and that the students involved understand the policy under which they will be working.

Although all of those involved to date have been mechanical

or industrial engineering students, it is our intent to expand this program to include electrical engineering students. It is hoped that both groups will collaborate on joint projects, allowing a technical interaction which does not often occur on the campus.

The program has so far worked out well for both the students and the automaker. The students work with high-tech equipment which the university cannot afford on real-world problems of great interest to the automaker. In addition, while working in small design teams, students have a chance to interact with development engineers and toolmakers at the automaker's facility. Students learn how to effectively work with both groups, as well as how things get done in a large organization. The automaker gets the student's time and ideas for free, and probably of more importance to him, gets a chance to look these students over for possible permanent employment without any obligation on his part. The automaker also gets a chance to make its high-tech activities known among the student body - a positive factor during recruiting activities.

To date, approximately twenty undergraduate and two graduate students have participated in this collaborative project effort. All projects so far have been well received by our students. Interaction of this type between industry and universities can be so beneficial to both parties.

TABLE I
BRIEF DESCRIPTION OF STUDENT ROBOTICS PROJECT TO DATE

PROJECT NO.

#1 Design of grippers and fixtures to machine load small disc-shaped valves using an IBM 7535 robot; system set-up and de-bug, including the necessary programming of the robot in AML language.

#2 Conceptual design of a robotic work cell to insert fuel injectors in a fuel system; design of a material-handling system required.

#3 Design of the grippers and tooling required for Project #2.

#4 Design of the inserter mechanism for Project #2.

#5 Design, build, and de-bug a compliance device for a robotic gripper.

#6 Design of a clamping mechanism for a robotically-loaded large CNC machining center tombstone.

#7 Fixture design for a vision sensor for robot guidance.

#8 Fixture build, set-up and de-bug for Project #7.

#9 Design a light-weight, quick-change gripper, including connections for all pneumatics and all electrical connections.

#10 Inspecting castings for defects using a machine vision system with gray-scale processing.

#11 Design, build, and test a six-handed robot wrist for high speed assembly applications.

#12 Determine the wear properties of a foam plastic to be used for the material handling system of a robotic work cell.

#13 Design, build, and test fixtures and tooling for a PUMA work cell.

Industry, Government and Education Collaboration in Robotics Education

Lloyd R. Carrico

Lloyd R. (Dick) Carrico, CMfgE, Director of Technical Services

INFAC - The International Flexible Automation Center
3914 Prospect Street
Indianapolis, Indiana 46203 U.S.A.

OVERVIEW

Due to the advancements in manufacturing technologies and equipment within the past several years, there exists a tremendous need for knowledgeable personnel to fill positions in industry as qualified engineers, technologists, maintenance technicians and technical trainers.

In no area of industry has this probably been more true than the area of Industrial Robotics. This has been brought about by the nature and characteristics of these industrial machines.

As most industrial personnel know, the robots of today are "...reprogrammable, multifunctional manipulators...." which sets them apart from what we have come to think of as standard industrial equipment of the past. Industrial robots are, in most cases, "general purpose" machines that have been designed to be applied in a wide variety of production environments to perform various tasks and applications.

In addition to standard "part handling" tasks, more and more industrial robot applications are becoming "process" oriented, creating the requirement of involved personnel to have a more indepth knowledge of production concepts.

PROBLEMS OF THE PAST

The realization that more knowledgeable personnel were needed started about 1977-1978, when the Robotics industry was still basically in it's infancy but expanding rapidly. To meet the demand for knowledgeable personnel at that time, there were two basic types of training being performed:

1. Vendor training.
2. Community College/Technical School Training.

The vendor training being performed at that time consisted mainly of familiarizing the users personnel with the features of the vendors products. Courses were taught in Operations, Teaching and Programming, and Maintenance.

At that time, this type of training appeared to provide the immediate needs of the users and vendors. But, several problems began to be recognized:

1. Basic skills of maintenance and engineering personnel fell far short, thus causing the vendors training to appear complicated and confusing.

2. Vendor training related mainly to the product and not necessarily the application being performed. As a result, some personnel were familiar with the product as a "stand-alone" machine, but could not relate to the robot with its peripherials or the process involved.

The results of these training efforts, although admirable, provided industry with a work force that was only partially knowledgeable, much less qualified, to work with industrial robots.

Recognizing this problem, post secondary schools (mainly community colleges and technical schools), attempted to remedy the situation. In a very short period of time these institutions started providing programs to "train people in robotics by offering their curricula courses with titles such as:

1. Electronics Technician with Robotics Option.
2. Electro Mechanical Technician with Robotics Option.
3. Robotics Technician.

These "robotics courses" were found to be inadequate also. Utilizing established programs, robotics courses were "tacked on" to meet industry's demand and to eliminate the time consuming process of developing new curriculum.

The established programs used, dealt with electronics theory or mechanical technology. Both of these types of programs provided information needed in reference to robotics, but in very few instances were both technologies presented together in relation to robotics.

In addition, the robotics information being presented was very lacking. At that time, there were basically no robotics texts available for reference.

The information presented was garnered from technical papers and/or from articles found in trade magazines and journals. Although informative, these documents made it very difficult in establishing a good, complete, definitive course.

Another topic missing from these courses -- probably the most important topic -- was the information of how robots were implemented and applied in reference to production processes.

In most institutions, there were very few, if any, instructors with "real-world" experience. They were not only lacking in production process knowledge, but had never had opportunities to work with or around industrial robots. But yet, it was their responsibility to train people in these areas.

Compounding these problems was a lack of knowledge on the part of local, state and federal governments. This lack of understanding made it difficult for institutions to successfully reap the benefits of grants and other monies required to establish effective programs. Most grants, etc. being awarded at that time, were benefiting institutions dealing in research and development projects. This was providing industry with a large amount of theoreticians, but very, very few manufacturing or production type individuals capable of working in the real world of industry.

Another problem compounding this situation was a lack of communication between industry and academia. Academia was attempting to inject this new technology into their existing programs without asking for - or receiving - input from industry.

Industry, at the same time, was expressing its concern in reference to the lack of practical knowledge of recent graduates. This was causing industry to implement basic, generic, conceptual training programs which proved to be very expensive, due to the time and money it took. It also proved to be costly due to the fact that the personnel being trained would leave for other positions shortly after their training. These people were in some instances "over-trained" for the entry level positions they were filling, and were unchallenged in their jobs. Upon completion of apprenticeships and their companies' real world training, they would opt for the more challenging positions offered elsewhere.

This brings us to the second solution - or partial solution - to equipment needs. Knowing the benefits of learning on "life-size" equipment most robotic manufacturers have implemented several types of programs to ease the major expense of equipment purchases. One program is providing equipment at a discount. Thus, some major institutions can better afford the expenditure.

For institutions not so fortunate as to have funds to expend on discounted machines, some manufacturers are providing equipment on a "consignment" basis. The equipment would be owned by the manufacturer, but would be installed at the institution. Thus, both could use it - the institution during its courses of instruction, and the manufacturers on a scheduled basis, so as to benefit their personnel and customers in that immediate area.

Users are also providing equipment for training purposes. As robots become more "high tech", older machines are being replaced with the newer, up-to-date machines. In some cases, these older machines are being made available to the learning institutions.

Another solution to the lack of funds of learning institutions has been the establishment and utilization of grants from professional societies and organizations, such as the Society of Manufacturing Engineers. SME and similar organizations have been working with academia and industry in establishing "educational grants" to be used for the purpose of equipment acquisition.

One last key area where funds, grants and contributions from government and industry are being used is in the area of acquiring knowledgeable instructors. In the past, those possibly qualified were lured to industrial-type positions because of the higher wages and benefits. But, in the recent past these funds have created the opportunity for some learning institutions to offer more attractive incentives for the much-in-demand technical educator.

LOOKING TO THE FUTURE

As just mentioned, within the last several years, there has been a major effort between institutions, industry and government to provide adequately trained, up-to-date personnel in the area of robotics.

But, what about the future? Everyday the state-of-the-art technology is modified or made better; equipment is being improved and redesigned almost as fast as it is being produced. And what about the processes involved?

Remember that earlier in this discussion I mentioned the fact that people are having difficulty in relating to robotics in relation to processes. It has been realized that robots are just part of the production process in a facility; they are a part of what is becoming known as the "automated system". This is causing the same type of problems and situations previously discussed.

Fortunately, the efforts just described are already addressing these problems. One specific example is that of a government grant awarded for the development of a "Robotics/Automated Systems Technician Training" curriculum model. This grant, provided through the U. S. Department of Education to the Center for Occupational Research and Development will provide learning institutions not only with a program guide, but with course outlines in a modular form.

Also addressing this area of "automated systems" are "Productivity" and "Technical" centers that are being established around the United States so that technology transfer can take place. Most of these are joint efforts between industry, government and academia. But, there are a few being established through the private sector. One such center is INFAC - the International Flexible Automation Center, whose objective it is to provide a continuing flow of information in reference to state-of-the-art technology - not only in robotics, but in all aspects of "flexible automation" and processes.

SUMMARY

As the result of technology's inpact on today's and tomorrow's - industrial needs in reference to knowledgeable manpower; industry, government and education are dedicating themselves to the joint effort of establishing the most up-to-date training programs available.

Through the continual use of advisory committees composed of academicians and industry representatives, programs will be maintained in an up-to-date status.

Through the availability of grants, donations, and contributions, the latest state-of-the-art equipment will become more readily available to learning institutions.

And lastly, but most importantly, through the combined representation of all concerned people on committees such as Robotics International's Education and Training Division, and Human Factors Division, information will be continually made available to academia, industry, and individuals, providing them with assurance that the aforementioned problems will be continually addressed.

produced at lower cost. However, this important advantage has a major drawback that will be examined next.

The Problem

Suppose the manager of an auto parts manufacturing plant is offered, for, say $50,000, a robot which could do the work of three semi-skilled workers. Using the robot would not only assure uniform quality and increased production, but would also eliminate the costs of three salaries, of three sets of fringe benefits, and of training and supervising three workers. Is it likely that the manager would hesitate to replace those workers with a robot?

What will then happen to these workers? What will they do when faced with the prospect of permanent unemployment?

Nor does the problem stop with these few workers: these few will be multiplied many times over as the number of robots employed in industry swells. What will these men and women do when a prolonged period of unemployment renders them unable to buy food for themselves and their families? When they are unable to meet the house payments and to pay medical bills? How long will it be until the pressures of these concerns begin to exact their toll, and alcoholism and depression and violence become widespread throughout society?

How long will government be able to sustain these workers on welfare rolls before they resort - as many did during the Industrial Revolution - to rioting and pillaging in order to survive?

Some industrialists and economists reject this concern. They dismiss it as groundless, and point to both the prosperity which followed the Industrial Revolution and Japan's current economic boom. It is therefore sensible to examine these two events in more detail.

Historical and Logical Perspective

Oddly enough, those who downplay the social impacts of total automation point to the opportunities and prosperity that were brought about by the Industrial Revolution. It is true, of course, that many benefits derived from this socioeconomic upheaval. Increased factory production demanded increased supplies of raw materials: the market for cotton and other natural fibers expanded dramatically. Increased factory production demanded increased supplies of fuel: coal production soared. Increased factory production demanded better means of shipping and transportation:

highways, waterways, and rail systems were improved and expanded, as were the vehicles that were used to traverse them.

All of these demands created new service jobs. Drivers, ship and train workers, salesmen, import-export specialists and merchants, bankers, secretaries, communication workers, convenience stores, operators, and restaurateurs - these and many, many other kinds of workers were now needed in great numbers.

Thus, the Industrial Revolution provided the impetus by which industrialized nations entered a period of unprecedented growth and prosperity. Growth and prosperity, however, were not without their costs: this new era was punctuated by, among other things, unrest, sabatoge and riots, as well as by the abuse of child labor.

During the early years of the Industrial Revolution children worked side by side with adults, often working as much as thirteen hours a day in "sweat shops" that were poorly lighted and poorly ventilated. Virtually no effort was made to protect the health or safety of the workers. Wages were shamefully low.

Nevertheless, in spite of the fact that the Industrial Revolution in its early stages brought great wealth to management at the expense of laborers' increased hardships, the latter group had, by mid-nineteenth century, begun to realize improvements in their working and living conditions.

In the United States, the Civil War has brought government contracts to manufacturing, spurring industrialization, and manufacturing expanded in this country at an amazing rate. By the time the Industrial Revolution peaked in the early years of the present century, America was producing more than thirty-five percent of the world's total manufactured goods. A revolution indeed!

But was it? Scholars argue as to whether the Industrial Revolution was a true revolution. In its strictest application, **revolution** refers to a sudden, rapid, and pervasive change in the existing order, a change usually marked by violence. The Industrial Revolution, then, was more evolutionary in that it occurred over a long period of time. Roughly two centuries, from 1750 to 1950 - though these dates are somewhat arbitrary - are the time frame most often used to describe this movement.

As has been noted, however, violence did indeed accompany the Industrial

Revolution, nor is there any question as to its pervasiveness: economics, government, religion, education, the very social structure itself - all were if not actually transformed, at least touched by the Industrial Revolution. **Revolution** then does seem to be an apt name for a movement of such magnitude.

Believing that total industrial automation, when accomplished, will effect a similar - i.e., pervasive - social change, the authors have chosen to refer to this change as the Robotic Revolution.

Like the Industrial Revolution, the Robotic Revolution will not merely touch virtually all areas of society, it will radically change, or transform, them. Perhaps nowhere will this effect be more clearly seen than in the world of work. Robots will increasingly replace laborers in industry, imposing heretofore undreamed of amounts of free time upon the laborers. In short, the Robotic Revolution will render manual labor as we now think of it largely obsolete.

The ramifications of this in relation to society are staggering: workers will have to rethink the so-called work ethic. In America - where worth of the individual is often equated with one's answer to the question "What do you do for a living?" - men and women will be forced to find self evaluators other than work as we know it today.

Further, what people choose to do with their new free time will both shape and be shaped by responses of the political, economic, and social, structures to this change in work.

Unlike the Industrial Revolution, though, which, as has been noted earlier, gave rise to new and increased service jobs. The Robotic Revolution will provide no such impetus. With existing service facilities regularly operating below capacity, the Robotic Revolution will not stimulate the construction of new highways and bridges and canals or the laying of new railroads or the developing of new airlines. While experts disagree as to the kinds and numbers of jobs that might be generated by the Robotic Revolution, one of the present authors, himself actively engaged in robotics research and development, believes that any "new" jobs that might be created as a result of roboticization will be highly technical in nature. Requiring advanced levels of training and ability on the part of technicians, they will by their nature be limited in number.

One job category that can be envisioned, is temporary increase in robot assemblers, robot operators, and computer programmers. It should be pointed out, however, that one robot programmer/operator can manage a large number of robots. He/she needs to teach only one robot, then the program can be recorded and used by thousands of other robots. Thus, every robot need not have its own operator. The demand for this type of job, although new, will be very limited. Robots will have sophisticated self-diagnostic features that will not require highly trained repairmen. Thus an operator can easily be a repairman as well. A repair job will chiefly consist of changing well packaged modules and will not require sophisticated tools.

A very positive distinction between the Industrial and the Robotic Revolutions is found in the circumstances that each imposes upon workers. The Industrial Revolution took men and women off the farms where they had grown their own food and enjoyed the benefits of open space and clean air, and put them into crowded cities where food had to be bought, and where they spent their days performing repetitious, monotonous tasks in an environment often marked by polluted air and hazardous working conditions.

The Robotic Revolution, on the other hand, will remove workers from jobs where their lives are endangered by the working conditions, such as working with or near dangerous machinery or chemicals or in gas-filled coal mines.

Another significant difference between the two revolutions lies in the time spans involved. Unlike the Industrial Revolution which occurred over an extended period of time, the Robotic Revolution will occur quickly. To expect roboticized industrialization to become a reality within as few as twenty years is not unrealistic, given the present rate of robotic growth.

Some observers will no doubt feel that fears of a Robotic Revolution are groundless for a different reason. They will point to Japan's current prosperity as an example of the benefits to be gained - with impunity - by total automation.

Actually, the success of Japan's roboticization and the prosperity that it has brought have resulted almost entirely from that country's American market and have not been an intrinsic factor of Japan's own industrial/economic system. If Japanese imports were banned in America and Western Europe, a spectacular tailspin of that nation's economy would very likely follow. This claim can be substantiated by comparing Japan's

How to Increase Productivity

Binh Ninh, J. P. Sharma, and Suren N. Dwivedi

Department of Mechanical Engineering
University of North Carolina At Charlotte
USA

Abstract

Productivity is one of the most serious economic considerations which have to be dealt with carefully. Without efficient production, a company or a profit - oriented organization cannot survive in the existing economy.

Increased productivity is very vital for all manufacturing organizations. The search is on and many methods are being tried for booting productivity; but, with technology advancement, new effort, and new strategies are required to increase productivity with emphasize on quality and efficiency.

The purpose of this paper is to examine the problems faced by the manufacturing, and to plan strategies for increasing productivity.

Introduction

What is productivity? Productivity is defined as the ratio of the output to the input. The output might be merchandises and services; the input can be raw material, capital and labor.

The objective of the productivity is to use wisely all the available scientific and technological resources in order to obtain the most competitive and highest quality results.

Increasing productivity gives indication of a higher quality product. For instance, in an assembly line, when an employee produces ten quality units in one hour compared to that of six units produced earlier in one hour, then it could be said that productivity is increasing. In case he still produces six units, yet with a better quality, one then classifies productivity as improved productivity.

When productivity is improving or increasing, the output value is always greater than the cost of the input. Naturally, this results in good profit.

Today, almost all industries in every nation seem to be working hard to push products for quick sale in the highly competitive market and to overcome industrial problems linked with productivity because of cost consciousness. All these are linked with several factors which have to be carefully examined according to the existing environment and opportunity of that nation.

A systematic study of productivity and its various controlling factors are shown in figure 1

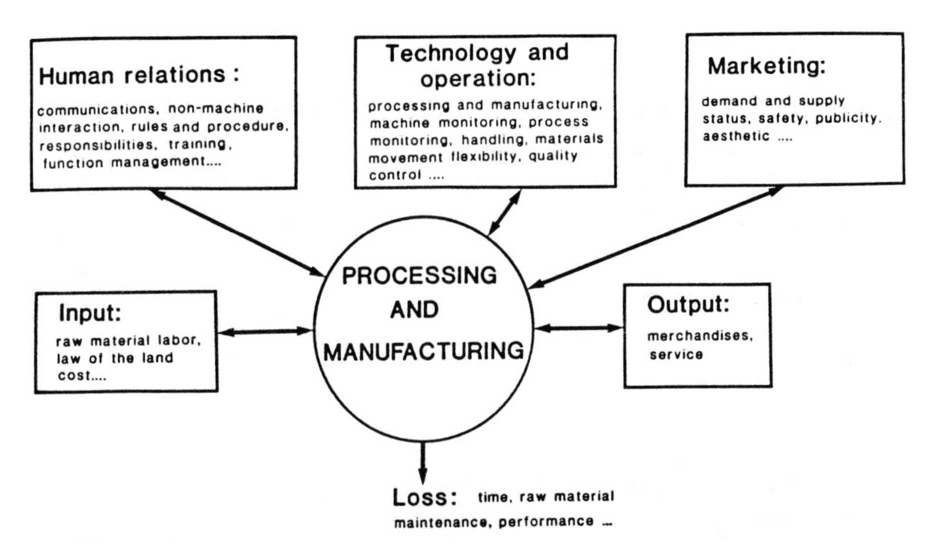

FIGURE 1: SYSTEMS SHOWING VARIOUS CONTROLLING FACTORS FOR PRODUCTIVITY

These mentioned factors can be put into four groups:

Group I: Human relations and management.

Group 2: Technological group

Group 3: Marketing

Group 4: Investment

Group I: Human relations and management.

In this group, important factors to be considered are:

1.1/ Communications

1.2/ Man-machine interaction

1.3/ Rules and procedures governing the organizations

1.4/ Responsibilities to complete the job

1.5/ Incentives for leadership

1.6/ Function management

1.7/ Training

Group 2: Technological group

Factors belonging to this group are highly technical and are as follows:

2.1 Processing and manufacturing

2.2 Machine performance monitoring

2.3 Process monitoring

2.4 Handling

2.5 Materials movement

2.6 Research and development

2.7 Flexibility

2.8 Quality control

2.9 Maintenance

2.10 Failure auditing and design review

2.11 Automation

Group 3: In the marketing group, various factors responsible for productivity are:

3.1 Public reaction to product

3.2 Demand and supply status

3.3 Safety

3.4 Publicity

3.5 Aesthetic

Group 4: The fourth group in the form of investment is very vital for productivity All the above three groups will require investment.

A proper investment plan is to be worked out depending upon the size of the industry for improved and increased productivity.

The links between the above groups is shown in figure 2

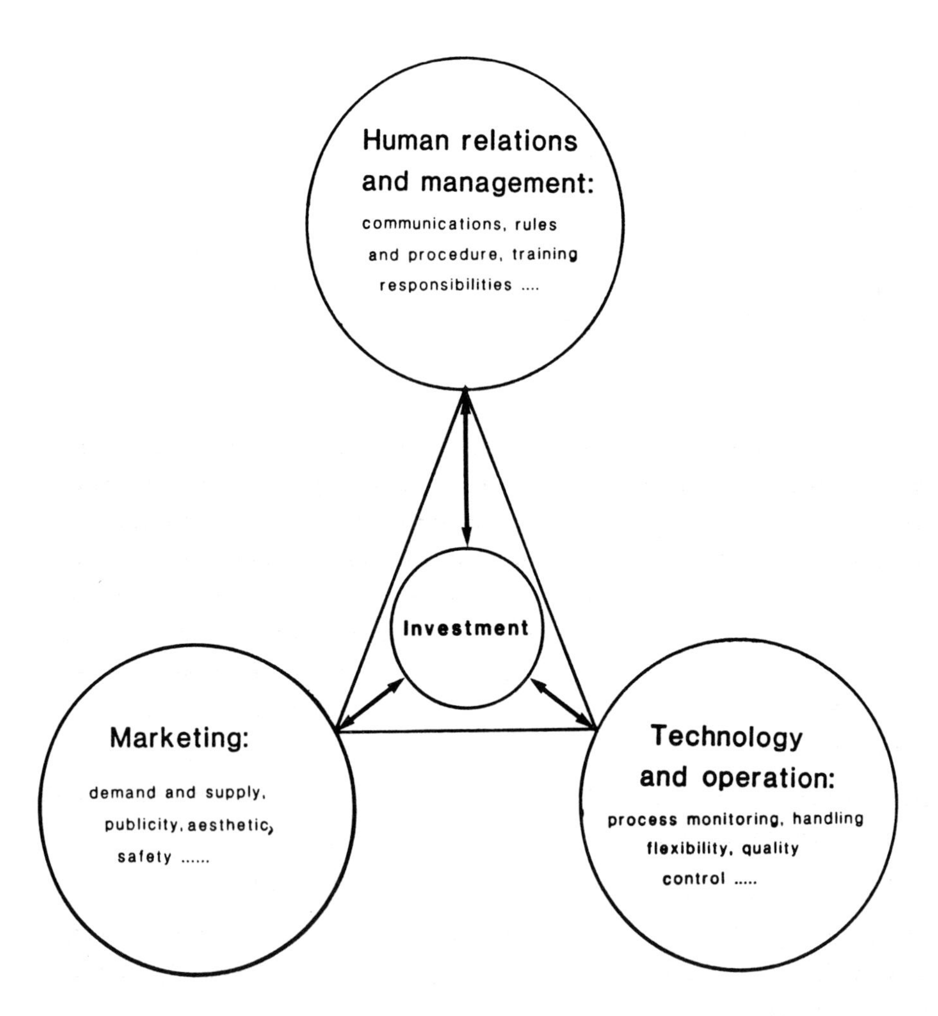

FIGURE 2: INVESTMENT ON GROUP PLANS FOR PRODUCTIVITY

a day to satisfy the needs of the customers, and supplies very good customer service.

The application of automation in industry brings about more opportunities for employment and advancement. Automated systems gain a lot of profit for the company. As the company has grown, new jobs have been created and more employees are hired to perform the new projects. This means more income for the families so automation indirectly increases the standard of living. On the other hand, automated systems are the ways to the shorter weekend. If the workers just have to work forty hours per week, they have more leisure time and a higher quality of life. People need to work smarter but not harder, and the automation is the answer for all the above problems.

B/ Training for increasing productivity

Productivity can increase easily when the quality of the products increases. This purpose needs a good training program. Eventhough they have the robots, trained workers are still a must. A helpful program provides for the workers the skills that will enable them to work more efficiently. Training must have a clear objective; it must be understood by the individuals who are to be trained.

The work force is one of the most important resources of the economy. When the producer has a good training program for them, they will be a skilled labor force for the industrial base. People must know how to use them by giving everybody a skill and giving everybody a chance to do his best. If people can do so, they get closer to improved productivity.

A training program is not only for the skilled workers, but the owners must also expect to have a good management team and excellent leader from the training. For instance, if the top executive manager sets up a wrong plan for the productivity, can the robots or the skilled workers increase the productivity? Naturally, they cannot. When people have excellent business leaders, they get closer and closer to their aim.

C. Investment for increasing productivity

There is much modern equipment on the market, and many new techniques are invented; but people don't want to buy them. They don't want to spend their money. They don't want to invest in their business. Can the productivity increase? No, it cannot.

Financial investment is an important factor. People must have enough financial investment to assure that they have what they need to improve productivity.

In the year 2000, people cannot keep a machine which was made in 1984 and say that they have enough features for increasing the productivity with this machine.

Conclusion:

An effort has been made in this paper to supply an overview for improving productivity. People need too many vital factors to achieve that goal. Training programs, financial investment, and new techniques.

The concentration on the application of new techniques like robotization or automation is becoming one of the most effective ways to reduce input to increase output. Automation has improved process and so people no longer have a big problem with high cost of labor, lack of skilled workers or lack of workers who are willing to work in uncomfortable jobs such as forging, heat treatment, die casting, and paint spraying. In increasing output, the automation provides a better quality, a faster service, and more products than manual operations.

No wonder automation is now an accepted term in Japan, United States, Germany, France ....The advantages of this new technology are so attracting the attention of socialist countries such as USSR, China, Poland....

It goes without saying that robots and automation do a great job in improving the productivity, but they cannot do it alone; cooperation from the other elements such as technology, equipment and people is needed.

The managers, the engineers, the workers, and the robots if work together as a winning team will increase productivity.

References:

1. BELYANIN, P.N. "Automation of production process based on the use of Industrial Robots", Polish scientific publishers Warszavwa 1977, pp 449 - 451

2. BOSKIN, M.J. "The economy in the 1980", Transaction books, 1980, pp. 43 - 47

3. DORF, R.C. "Robotics and Automated Manufacturing", Reston Publishing Company, Inc. 1983, pp. 26 - 28, pp. 118 - 136

4. FITCH, J.C. "Developing Automation Applications: How to Identify and Implement high yield project", IE Magazine, NOVEMBER 1981, pp. 47 - 48

5. FREEMAN, R.G. "Automation and Telecommunications The key to improved productivity", Prentice Hall Inc, September 1981, pp. 78 - 81

6. GLENNEY, N. "Modular Integrated material handling System facilitates automation process" IE Magazine NOVEMBER 1981, pp. 123 - 124

7. GROOVER, M. P. "Automation, production systems and Computer - Aided - Manufacturing", Prentice Hall Inc., 1980, pp. 3 - 7

8. GROOVER, M.P. "Job shop automation strategy can add efficiency to small operation flexibility", IE Magazine, NOVEMBER 1981, pp. 74 - 75

9. LEBOBUF, M. "The productivity challenge", Mc Graw Hill Book Company, 1982, pp. 8 - 10

10. MACAROV, D. "Worker productivity", Sage Publications 1982, pp. 43 - 47

11. MC BEATH, G. "Productivity Through People", Halsted Press Book, 1974, pp. 71 - 77

12. QUEMENT, J. L. "National Robotics Research and Development Program in France", Scholium International Inc. 1981, pp. 9 - 11

13. SUTER MEISTER, R. "People and Productivity" McGraw Hill Book Company, 1976, pp. 3 - 9

14. Sharma, J.P., Dwivedi, S.N. "Computer Applications for the factories of the future", Proc. ASME, Las Vegas 1984

Status and Strategies of Automation Through Robotization and R & D Integration in Developing Countries

D. Pershad, J. P. Sharma, and Suren N. Dwivedi

D. Pershad
Head, Industrial R&D,
I.I.T, New Delhi 110016
INDIA.

Jagdish P. Sharma
Visiting Professor

Suren N. Dwivedi
Associate Professor

Mechanical Engineering & Engg.Sc. Department
University of North Carolina at Charlotte,
Charlotte, NC 28223, (U.S.A.)

Abstract

In the past two decades many developing countries have built-up the base for development of their economy through industrial growth. Automation has also made an impact in increasing the productivity of these industries. Automation through robotization require input of high technology through R & D efforts. This has not made sufficient progress because of different patterns of socio-economic structure as compared to that of industrialized nations. In this paper the existing status in developing countries has been discussed. A strategy has been evolved for introduction of "Friendly Robots" for creation of new jobs through integration and commercialization of R & D with automation.

Introduction

"Robot"-a term derived from Czech word "Robotnic" is a subject which is getting special attention by scientists and engineers for research and development in automation. This is because robot is a reprogramable machine which can perform repitative and untiring work over a long period of time in manufacturing and work space located in difficult environments. This has brought fundamental change in manufacturing and productivity concepts in Japan, USA, Germany and Sweden where robots are used to perform automotive work. Automation in industry is now linked with robotization. But, careful planning and improvements are required for automation through robotization in areas linked with capabilities in mechanical systems, sensory feed back controls, artificial intelligence, vision and communications. The advances made in interfacing and utilization of electronics and microprocessors with electromechanical sensors, controls and mechanical manipulators are exhibiting excellent capabilities and potential of robots to perform effective human skills This led to a general feeling,all over the world, that automation through robotization in industry may eliminate human jobs in both the skilled and the semiskilled categories. Though the demand of robots for automation is exponentially growing in more than twenty developed countries and it is expected to go up from 27000 robots in 1982 to 200,000 robots by 1990.

This growth pattern and trend shows that developing countries like India, China and other nations who are busy in updating their policies towards industrialization cannot afford to do away with the deployment, manufacturing and development of robots. But, the question arises that how and where it can effectively be utilized? These countries do not think of replacing human skills by use of robots, but will certainly make an effort to use robots to perform automatic work in difficult environment. These areas are off-shore drilling for oil and gas exploration, space studies, flood controls and development of remote and desert areas. These areas exhibit extremely difficult and harsh environment from safety considerations for human beings to perform effective work. Thus the deployment of these robots in these areas can play a special role in accelerating the growth of the economy and will open several ventures for creation of habitat in remote places in developing countries. This would thus promote creation of new jobs and help to expedite difficult resources untapped so far.

Deployment of these robots with special nomenclature-"friendly robot" in developing countries for the purpose mentioned above will be a welcome addition to the society as shown in figure 1.

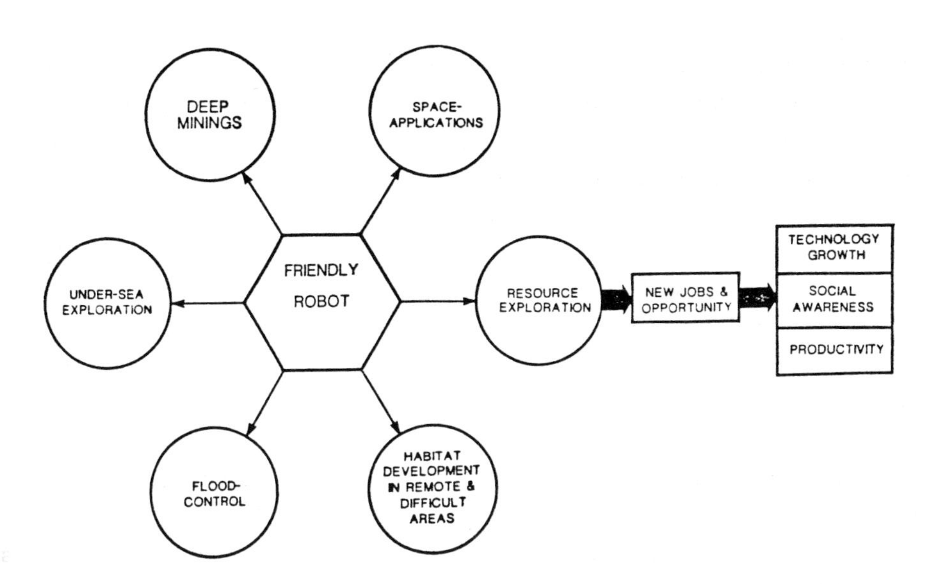

FIGURE 1 : DEVELOPMENT OF FRIENDLY ROBOTS FOR DEVELOPING COUNTRIES

There are many special socio-economic problems in developing countries all over the world which need special attention for deployment of computers and robots for automation. Factors affecting automation through robotization are shown in Fig. 2.

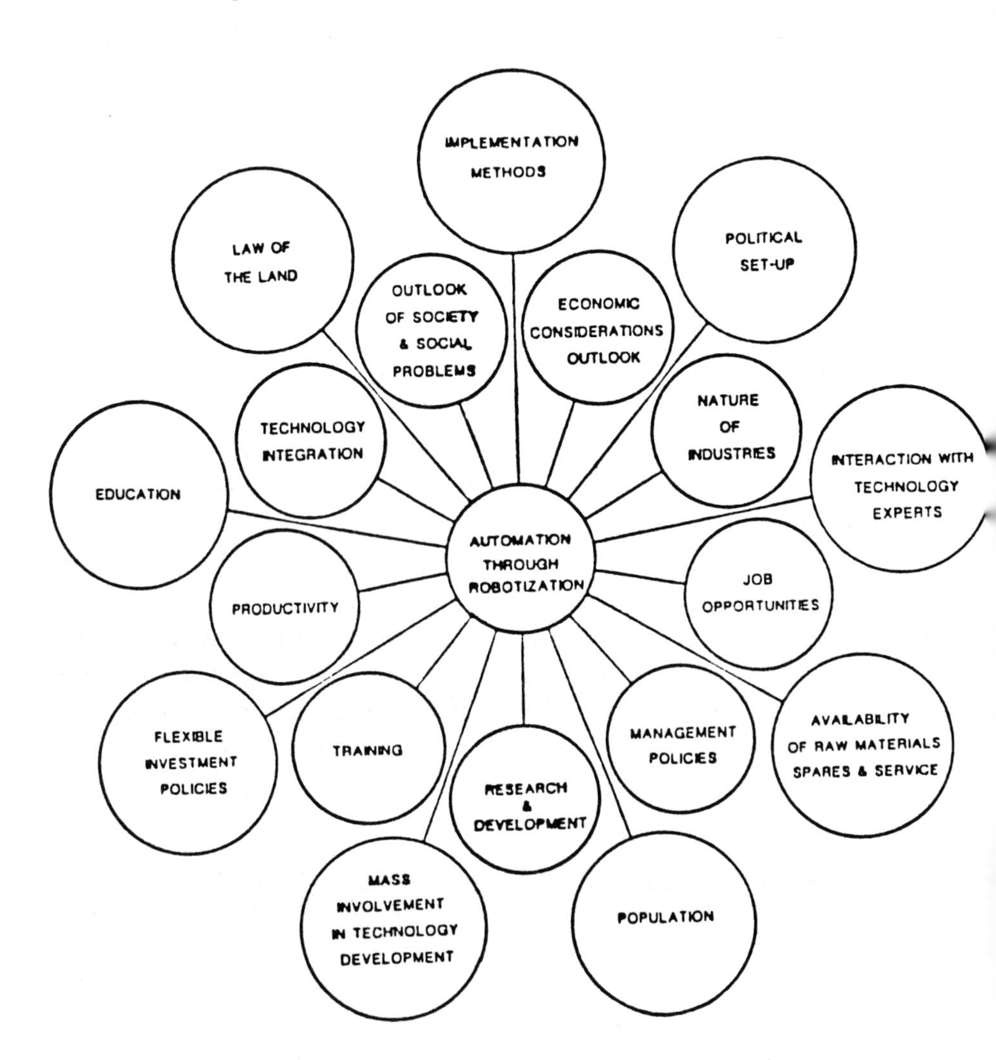

FIGURE 2 :: FACTORS AFFECTING AUTOMIZATION THROUGH ROBOTIZATION

These socio-economic problems linked with special need of more job opportunities amongst masses in countries where population density per square mile is high. Some of these problems need careful analysis, synthensis and planning with respect to the following points:

1) Surplus Scientific and Technical Manpower: Most of the developing countries have surplus technical and scientific manpower without sufficient job opportunities to absorb their skills. An effort is required to use their talents in development of computers, robots and robotization for development of resources in areas of difficult environments. This would require incentive and organization on the part of community to develop commercialization of Research and Development business through use of national laboratory facilities. This would provide self-employment and job opportunities to scientists and engineers who would then use the expensive instruments on time sharing basis available in the nation's laboratories for developing their scientific talents and growth in technology transfer. New innovative ideas and its application by self employed scientists will help in economic growth, productivity and technology transfer. This would require special legislation to allow the use of such time sharing facilities during the time when equipment is not in use and its utilization factor is low. Some of the countries have already started thinking in this direction to update their capabilities with Hitechnology. This would require a long term liberal loan and investment policy on the part of financial institutions and government for promotion of R & D industry. A group of industrialist is already thinking for starting an R & D based industry for manufacture of Robots in the private sector. This is definitely a good sign towards keeping abreast with the development of modern technology.

2) Availability of limited scientific tools for development: Though, there are considerable enthusiasm in manufacturing organizations in developing countries for development of automation through robotization, but the expected growth may not be as fast as that in developed countries. This is because, the availability and accessibility of scientific tools for research and development efforts are very low. This poses biggest problems of focusing R & D efforts attuned to industrial need. In some developing countries these are really socio-economic problems related to under utilization of equipment and human fear that the expensive tool if used excessively for want of spares may make it inoperative. A proper legislation of contractual system for scientific personal for commerical utilization of expensive equipment and facilities in national laboratories will certainly

help in growth and transfer of technology. Automation concept through development and utilization of modern technology by mass envolvement of scientists and engineers in commercialization of Research and Development business will effectively result in creation of new jobs, new knowledge and skill.

3) Shortage of jobs

More men and less jobs always create a sense of insecurity. Very often this idea travels fast in the masses to resist any R & D effort which could promote automation and robotization. As such there is a need on the part of the government to frame legislation to promote R & D business effort in institutes to develop robots for deployment in hostile environments to explore untapped resources for creation of more jobs, habitat development in far remote desert areas and mountaineous terrains and flood control. Friendly robots as companions should be developed as tools for deep minings where human life face danger of hostile environment and safety hazards.

4) Industry integration with R & D organization

Because of socio-economic reasons there is very little rapport between industry and R & D effort for integrated development of Hi-tech products. This gap can be reduced by liberal investment policies on the part of financial institutions towards R & D and technology transfer goals. Industry should also participate in liberal investment on R & D and allow private scientists and engineers to get self-employed as a support to industry on predetermined time bound programs to be widely circulated by them. This will inculcate a sense of responsibilities among the scientists and technical communities to deliver quality work and help the country in increasing the productivity. This will generate more jobs among educated scientists and Engineers. Some developing countries are formulating policies which will create technological environment, additional jobs and thus increase productivity through automation and robotization.

5) Law of the land

The law of the land is formulated by the government. Its method of implementation will have definite influence on R & D integration with industry in development of Hi-tech, automation and robotization. These laws are essentially formulated on the basis of the immediate human need, local socio-economic problems, growth of industrialization and goals set for the growth of national economy and prosperity of the nation. In democratic and developing countries technological environment have to be created within the frame work of the law of the land.

aspects will be discussed below.

Technical and Economic Considerations

The primary motivation for the use of industrial robots in undoubtedly economic because the cost of operating a robot is significantly lower than the cost of manual labor at the present time. The increase of industrial productivity during the past decade had been rather slow and greater use of robots and other computer-aided methods in the manufacturing industries should be a definite help. The use of robots also results in improvement of product quality. The products become more dimensionally accurate and more uniform in quality.

Improving Industrial Safety by the Use of Robots

Apart from the above techno-economic considerations, another major motivating factor is the desire to replace human workers by robots in dangerous jobs and hostile environments. Robotic devices are being used in the nuclear industry since its very inception. The first robotic installation in the manufacturing industry was designed by Unimation in 1961 [2] and the robot was used to remove hot castings from a die-casting machine. This is one of the applications where robots are very popular because the working conditions are very poor. Some other examples where robots are used in conditions which are uncomfortable or hazardous for human workers are as follows:

- operations in foundry (fettling, moulding by lost-wax process, etc.)
- spot welding
- arc welding
- handling of the workpiece during heat-treatment
- loading and unloading of forging machines and presses.

Work is going on to robotize other jobs of this type [3] and in the future, we will see robots being used in dangerous work, such as firefighting, mining, repair of high-voltage electric lines, cleaning of windows of tall buildings, etc.

By relieving the human worker from monotonous or dangerous work or in unhealthy environments, robots have already saved lives. Robots should be rocognized as an important tool in promoting occupational health and safety.

HAZARDS INTRODUCED BY ROBOTS INTO THE INDUSTRIAL WORKPLACE

When robots were introduced into the industry a few years back, it was never realized that this kind of equipment could cause new types of occupational safety problems. But after a few years of experience with industrial robots, we realize that though on one hand, robots enable us to remove workers from unsafe workplaces and thereby promote safety, they themselves, on the other hand, introduce new hazards, heretofore unknown in the industry. Preliminary examination of accident statistics indicates that the accident rate of robots is higher than that of other automated machines such as numerically controlled machine tools. This problem needs to be examined thoroughly.

Analysis of Accident Data

Reliable date on accidents involving robots are difficult to find since this information is not yet classified seperately by the governmental authorities. But there had been many reported cases of accidents and near-accidents in robotic installations. There had even been five fatal accidents involving robots since 1978 [4]. Four workers have lost their lives in Japan and one fatality has been recorded in the United States. The problem of robot safety has suddenly become more critical after these tragic events occured.

We have been able to obtain detailed information regarding many of the accidents involving robots [5]. After analysis of the facts pertaining to each of the fatal accidents that have occured in robotic installations, it was found that the following chain of events happened in almost all cases:

a) A minor problem occured in the robotic installation (in the interfacing equipment and not in the robot itself)
b) A worker entered the danger area to correct the problem
c) The person was experienced and, through proper training, should have been made aware of the danger present
d) The person either was not properly trained or else wilfully violated the safety measures or he inadvertently returned power to the robot while still in the danger area
e) The robot struck the person from behind. That is, the person was unaware that the robot was moving until it was too late.
f) The robot alone is usually not responsible for the death of the person. Rather, the robot pushed him into another machine or crushed him against something else.

Based on surveys conducted in factories using robots, some interesting data on accidents or near-accidents in robotic installations in Sweden and Japan are now available [6,7]. These accidents were not as serious as in the above cases and none of them were fatal. These studies, conducted between 1978 and 1982, provide us with some interesting information about the reasons why accidents occur in robotic installations. Improper operating method was identified as the principal factor. These could be programming errors, errors in operating the switches on the control panel or getting into the working zone of the robot while doing adjustments and test running and being hit by the moving parts of the robot. Equipment deficiency was not an important factor in these cases, though in general, unplanned motion of the robot due to equipment malfunction could be an important hazard.

Some Particularities of Hazards Caused by Robots

The distinction between robots and conventional machines is significant from an accident prevention point of view. In robots, a digital computer is integrated with an electro-mechanical device creating a new type of operating situation. Conventional machines usually have the following characteristic features:

- Through some of them move very fast, they perform repetitively and predictably
- Their danger zones are obvious and fixed in one place
- They operate in a cyclic manner and their motion paths do not usually change from one cycle to another
- They seldom dwell for varying lengths of time
- They usually will not operate without the action of an operator or an obvious input signal.

Robots, on the other hand, possess a large work envelope which is sometimes poorly delineated in space as in the case of anthropomorphic robots. Because of their flexibility, their motion paths may change from one cycle to another and they can reach any point within their working volume. If not moving, the robot may be awaiting a signal from another machine in the workcell. It may unexpectedly leap back into motion and prior motion paths may abruptly change. Robots, therefore, even when operating correctly as programmed, may surprise (and injure) anyone near them. That is why it is important that new safety regulations be developed specifically for robotic installations.

Types of Hazards Introduced by Robots

Because of their particular construction, robots introduce some specific hazards into the workplace, which are briefly discussed below:

a) Mechanical hazards

- collision of the robot arm with a person
- ejection from the EOAT (for example, a tool or molten metal). In case of a power failure, the object held by the EOAT should not be allowed to drop
- pinch points between the robot arm and a fixed object, such as another machine in the workcell
- injuries caused by tools (such as a grinding wheel) held on the robot arm

b) Non-mechanical hazards

In this category are included hazards of electrocution, of burns, exposure to electric arc, to lasers, X-rays and other ionizing and non-ionizing radiations.

In this context, all hazards introduced by the robot during its normal and during aberrant behavior must be considered.

Hazards during Different Phases of the Use of a Robot

From the standpoint of hazard analysis, all types of man-machine interactions must be examined in detail and four distinct phases can be identified in this context:

a) Test run in the manufacturer's plant

b) Programming

c) Operation in the automatic mode

d) Maintenance activities

During most of the lifetime of the robot, it operates in the automatic mode and it is not necessary for humans to come in close proximity of the robots. Robots should hence be placed in areas protected by barriers to prevent entry of all unauthorized persons.

The point-to-point robots are usually programmed by the leadthrough method using a teach pendant whereas the continuous path robots are programmed by the walkthrough method. Programming and maintenance are two types of activities where there is greatest amount of interaction between the humans and the robot and it is during these operations that accidents occur most frequently. Efforts should be directed particularly at improving the safety situation during these two phases.

METHODS OF MAKING ROBOTIC INSTALLATIONS SAFER

Robotic installations are getting increasingly more sophisticated and robots are being integrated with other equipment to form an automated system controlled by computers. Such systems are being used in the automobile industry now and will be used more widely in the future. In order to assure the safety of the human workers in these systems, actions must be taken in two areas. First, the robots should be designed to be intrinsically safe machines. Secondly, safety should be a major consideration in doing the layout of the installations and in establishing standard working procedures.

Intrinsic Robot Safety

Here we will discuss the robotic safety features that are inherent to the robot design. We will examine the following robot subsystems [8]:

a) Mechanical hardware

b) Electrical/electronic hardware

c) Control and operational system software

a) Mechanical hardware - Sound engineering practices should be used in robot design. For a robot of a particular size, the designer should strive to reduce the kinetic energy of its moving parts to a minimum, so that the control is easier and more reliable. In order to optimize the arm design to reduce its inertia, heavy components such as motors and gears could be placed close to pivots and the arm could be constructed of lighter materials such as composites. Suitable damping and braking systems are necessary and mechanical stops should be used to put positive limits to the motion

Robotics and CAD/CAM Education II

Chairman: Fred Sitkins, Western Michigan University, Kalamazoo, Michigan
Vice Chairman: Louis Galbiati, SUNY College of Technology, Utica, New York

Robotic Education Growth in the U. S.

Louis Galbiati and Atlas Hsie

State University of New York College of Technology
Utica, New York

SUMMARY

The progress that has been taking place in robotics education in the U. S. during the past four years and the geographical growing pattern of robotics educational activity will be covered in this paper.

Robotics is having a profound effect on the educational programs in all the industrial countries of the world. It is proving to be the leading-edge technology catalyst and the focal point for advances in the state-of-the art in all the traditional technical disciplines such as electrical, industrial/manufacturing and mechanical.

The Robotics educational activities at a college grow very rapidly as the institution, the students and the industrial community gain experience and insight on the impact of Robotics on their future.

It is clear that robotics education covers the complete educational spectrum; to associate degrees, baccalaureate degrees and graduate studies. The subject is too broad and deep to cover entirely at the undergraduate level. The Robotics Master's program will have to have the capability to accommodate the student who may desire to go in either of two directions. One direction would be a narrow speciality with great depth dealing with a specific technical area such as vision, adaptive systems, dynamics, intelligence, or manipulators. The other direction would be for the student who wants to study robotics as a system where the students would be concerned about the industrial environment, the economics, and the utilization to improve productivity and reduce costs.

ROBOTIC SURVEY AREAS

In 1978 it was decided to start accumulation data on robotic education in the United States and Canada. The study was to include two aspects of the movement; one was the changes that were occurring with time and the other was the geographical distribution of Robotic educational activity. Periodic surveys were used to obtain data on the changes with time. The data was processed by geographical region; the country was divided into seven regions, as shown in Figure 1. Generally speaking, Area 1 is New York and Ontario, Area 2 is New England, Area 3 is the Northern Central region from Minnesota to New Jersey. Area 4 is the Eastern Southern region from Virginia to New Mexico; Area 5 is the Central U. S. from Arizona to Montana and North Dakota. The West Coast states and Hawaii are in Region 6 and Western Canada is in Region 7. The regions were selected to provide insight on different industrial loses.

EDUCATIONAL GROWTH WITH TIME

The first survey in 1979 indicated that there was a total of 24 colleges in the U. S. with Robotic activity. Three were at the Associate level and 21 at the Bachelor and Graduate level. The Robotics International of SME survey taken in 1982 (i. e. two years later) indicated that the total had increased by 400% with a 900% increase at the Associate level and 340% increase at the Bachelor level.

The 1983 SUNY College of Technology survey taken in March, 1983, indicates that the dramatic increase in robotics education activity is continuing. The total activity has increased by 180% with a 207% increase at the Associate level and a 170% increase at the Bachelor's level. At this point in time, the SUNY survey indicates that there are 56 colleges with Associate degree programs, 119 with Bachelor and Graduate programs, or a total of 175 colleges have Robotics courses.

In 1983, there were 17 colleges offering degrees in the Robotics field. Ninety-one colleges and universities have active research activities in Robotics. In Canada, there was a total of ten colleges with Robotics programs, four are at the Diploma level and six at the Bachelor or Graduate level. A new survey is currently underway and is expected to provide insight on changes which are taking place in the past two years.

The numerical data presented in this paper can be considered as a lower limit of the true condition since the survey material was distributed in March and some replies were received after the data was processed. My estimate is that replies had been received from at least 90% of the principal colleges.

The rapid expansion of Robotics courses throughout the country is surprising in that considerable time is required to add a course or courses at most Universities. I feel that this is very position evidence as to the importance of Robotics in the technical academic program. This conclusion is substantiated by the replies to Question 9.

ACTIVITY BY GEOGRAPHICAL LOCATION

Insight on the changes that have occurred by region is given in Table 1. It can be seen that there has been a doubling of the Robotics educational activity in Regions 1, 2, 5, and 6 in the 1982 to 1983 period. This is the result of the central and southern U. S. industries being more aggressive relative to implementing Robotics in their industrial base in the early period while the other parts of the U. S. and Canada are catching up. In Canada there was a large increase during the 1982-83 year, that is from one institution to nine institutions. In general, approximately 50% of the institutions in all areas responding to the 1983 survey currently have Robotics courses.

Tables 2, 3 & 4 provide data on questions 3, 4 & 5 by geographical area. The numbers in Areas 3 and 4 may be higher due to the fact that there was no way to normalize the regions. However, it is certainly clear that the Robotics educational activity is not restricted to any part of the country and that there are programs at all three educational levels, i. e., Associate degree, Bachelor's degree, and Graduate degree.

The survey data indicated that a very large number of institutions have or are acquiring robots. Many of these were in the very low priced category and could have a long term adverse educational impact in that the transition to advanced intelligent robotics or the implementation of Robotics application courses may prove to be more difficult. My feeling at this point, is that it may be prudent to give higher priority to robots having compatibility to IBM PC language if you wish to have greater capability to a total integrated CAD/CAM system in the future.

There was a high degree of consistency in the replies to Question #9. Approximately 75% of the replies indicate that robotics was very important in academic programs whereas only some two percent of the replies indicated that Robotics was not important in their academic program. It should be noted that some of the replies in this category were from colleges that specialized in non-technical programs. The percentage response of the feeling was relatively constant in all the areas.

CONCLUSION

The data indicate that the development of Robotics educational activity is a nationwide movement and that it is being implemented at all levels of the educational system. The movement originally started as a fragmented part of the established disciplines, but the emergence of degree programs indicates that it is being recognized as a complete discipline which crosses traditional lines and utilizes the technologies of all the disciplines.

It may very well be unrealistic for an educational institution simply to try to keep it as part of one of the established disciplines and courses since it would tend not to get the emphasis that the Robotics movement requires. Our experience at the SUNY College of Technology in developing the Robotics Program indicates that progress is greatest when separate courses and laboratories are established.

The data indicating that approximately 75% of the institutions replying to the survey were planning to purchase robots is consistent with the response to Question 9 on the importance of

Robotics in the educational programs. I feel that this is a leading indicator on the future growth of Robotics educational activity. The next survey will be expanded to include vision and application of robotic vision to industrial/manufacturing processes.

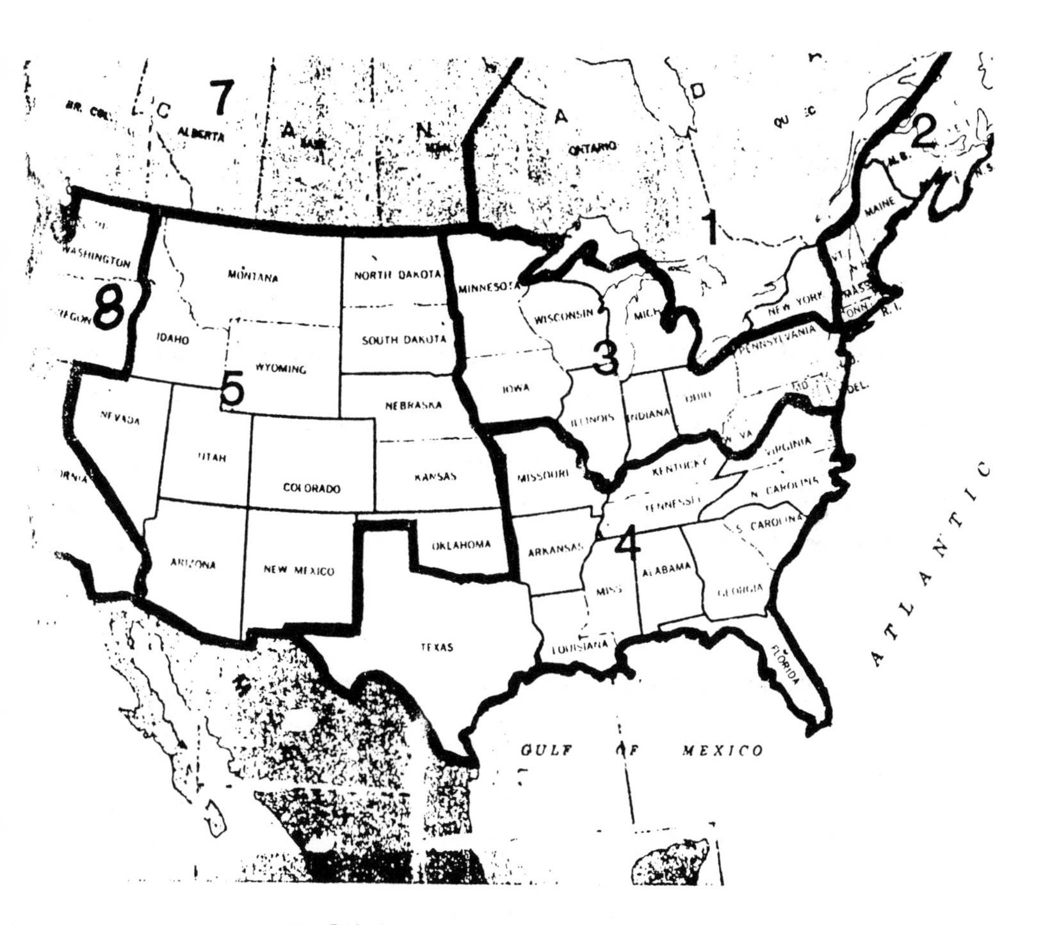

1. **GEOGRAPHIC LOCATION OF THE SURVEY AREA**

ASSOCIATE PROGRAM

AREA	1	2	3	4	5	6	TOTAL
82	1	1	17	7	1	0	27
83	5	3	28	14	4	2	56
%	500	300	160	200	400	200	207
	BACHELOR AND GRADUATE PROGRAM						
82	9	4	31	15	5	6	70
83	15	9	39	28	12	16	119
%	166	225	125	187	240	266	170
	TOTALS						
82	10	5	48	22	6	6	97
83	20	12	17	42	16	18	175
%	200	240	140	191	266	300	180

Table 1

ROBOTICS INCLUDED AS PART OF OTHER COURSES

Area	1	2	3	4	5	6	Total
Colleges	37	16	65	50	20	21	209
%	62	59	65	56	44	58	59

Table 2

INSTITUTIONS HAVING ROBOTS

Area	1	2	3	4	5	6	Total
#	27	16	63	48	22	19	195
%	45	59	64	54	49	53	56

Table 3

INSTITUTIONS PLANNING TO PURCHASE ROBOTS

Area	1	2	3	4	5	6	Total
#	45	24	80	63	23	22	257
%	75	89	82	71	51	61	73

Table 4

1983 SUNY ROBOTICS SURVEY

Name of Institution: ______________________________

Address: ______________________________

Degrees Awarded: ___Associate; ___Bachelor; ___Graduate

NOTE: Use back of sheet for additional comments.

	YES	NO
1. Do you offer a degree in the Robotics field?	___	___
2. Do you offer any courses in Robotics?	___	___
If yes, how many_____.		
Titles ______________________________		
______________________________		
3. Do you include Robotics as part of other courses?	___	___
What courses______________ ______________		
4. Does your institution have any Robots?	___	___
How many_____		
What makes ______________________________		
______________________________		
5. Do you plan to purchase any robots in the next year? If yes, how many__________	___	___
6. Are robots a high priority item if laboratory equipment funds become available?	___	___
7. Do you have any faculty with knowledge in robotics?	___	___
8. Do you have any Robotics research?	___	___
Annual Amount___; Manyears $____		
Type of Research Basic____; Applied____		

9. Your feeling as to the importance of Robotics in academic programs:

____Very Important; ____Slightly Important; ____Not Important

INDIVIDUAL'S NAME AND TELEPHONE NO.: ______________________________

Proposed Plan for the Development of Curriculum and Laboratory Facilities in Automated Manufactoring Engineering at Old Dominion University

Alok Verma

Mechanical Engineering Technology Department
Old Dominion University
Norfolk, Virginia 23508

Abstract

The dilemma faced by the educational institutions in view of the demands of the current job market has been discussed. Different schools have adopted different approaches in facing the challenge.

Old Dominion University is responding to this challenge by modifying its curriculum and opening an automation laboratory. The paper discusses the proposed plan for development of automated manufacturing engineering at the school with its effect on existing programs.

I. INTRODUCTION

A. The Challenge:

The last few years have seen a tremendous explosion in the area of automated manufacturing engineering catalyzed by the developments in computers. Industrial activity in the area of computer-aided manufacturing has increased with such a fast pace in the last decade that educational institutions find themselves lagging behind in providing technically competent persons for utilizing the emerging technologies.

B. Strategies Adopted by Other Universities:

This rather sudden change in the job market place has createtd a challenge for the educational institutions who have attempted to respond to this issue by mounting major efforts to build needed expertise and laboratories to provide the relevent education and training to their students. This resulted in a major effort to recruit faculty with the desired expertise. It soon became apparent that the available pool of personnel was not only small, but insufficient to meet this demand. Thus, most of the institutions are directing their energies to develop required expertise using existing faculties.

C. The University:

Old Dominion University is an urban university with the primary mission of meeting the educational and professional needs of its students and the region through excellence in teaching, scholarly research, and leadership in community service.

The school is organized into six schools. The School of Engineering comprises the Departments of Civil Engineering, Civil Engineering Technology, Electrical Engineering, Electrical Engineering Technology, Mechanical Engineering and Mechanics, and Mechanical Engineering Technology. The Bachelor of Science Program in Mechanical Engineering Technology (MET) currently has two options, Design and Systems.

D. Proposed Strategy:

The four-year degree program in MET is an ABET-accredited program which has been in existence for some time without any major modifications. In view of the rapidly changing environment in industry it was recognized by the author that if Old Dominion University has to maintain leadership in providing education and training in high technology areas, it must:

a. Evaluate its existing programs and their relevancy to the needs of current and future job markets.
b. Take definitive steps to develop the expertise of its faculty in the area of automated manufacturing engineering.
c. Plan methodically to develop the curriculum and laboratory facilities.

d. Interact with industries in the area and offer them assistance in the implementation of new technology.

II. CURRICULUM DEVELOPMENT

A. Present Curriculum in Mechanical Engineering Technology:

At present the Mechanical Engineering Technology program offers two options to students: Design and Systems. Both options require 129 semester credit hours. The Design Option has courses in machine design, tool design, mechanism, inspection theory, numerical control, engineering specifications, etc. The Systems Option has courses in the area of thermal power systems, electrical power and machinery, automatic control system, air conditioning and refrigeration, etc.

B. Modifications in Present Curriculum:

Automated manufacturing engineering is a broad term which encompasses several disciplines. It not only includes conventional areas like production planning and cost estimation, but also overlaps such modern areas as system integration, microcomputers and computer-controlled machines.

The range of topics covered by automated manufacturing engineering is so large that it would be impossible to include every subject into the curriculum; however, there are certain areas which cannot be ignored:

a. Numerical Control
b. Robotics
c. Automated Material Handling
d. Plant Layout and Cost Estimation
e. Modern Manufacturing Processes

It was realized that new courses in the above areas must be included for the manufacturing option. Present MET curriculum already includes a course in numerical control; however, more emphasis would be required in the area of engineering materials, manufacturing processes, digital electronics, and electrical power and machinery. Less emphasis will be required in the areas of mechanism, machine and tool design.

New courses will be required in the areas of robotics, production planning, plant layout and material handling, computer-aided manufacturing, and engineering material.

C. Proposed Schedule of Course Offering:

The proposed curriculum for option in manufacturing contains the following courses which are not currently offered:

a. Engineering Materials
b. Modern Manufacturing Processes
c. Introduction to Robotics
d. Robotics Application
e. Plant Layout and Material Handling
f. Computer-Aided Manufacturing

III. LABORATORY DEVELOPMENT:

For a university to remain at the leading forefront of high technology it must have up-to-date laboratory facilities to conduct research and development work in this area.

Industries at present are cautious in embracing full-fledged automation. Management in the industries seems to be more comfortable with islands of automation. However, as the reliability of the integrated automation systems increase and cost decline, it is predicted that Flexible Manufacturing Systems will become the key to productivity within the next two decades.

A state-of-the-art laboratory facility will serve two purposes:

a. Students can be trained hands-on in the high technology areas and will prove more productive in industry.
b. Faculty will have the opportunity to do research in automated manufacturing and offer the local and regional industries the expertise to implement these new technologies.

Research possibilities in this area are enormous and diversified. Machine vision, artificial intelligence, adaptive control, tactile sensing and integration of advanced machining processes like laser welding and water jet cutting with robotics, are just a few to name.

Section B presents a proposal for laboratory facility followed by Section C which deals with integration of this laboratory into the curriculum. Sections D and E deal with the budget and space requirements. In Section F it is proposed to start a new interaction program between the University and regional industries in the high technology areas.

A. Present Laboratory Facilities:

The present laboratory facility in Automated Manufacturing Engineering at ODU is inadequate. Three teachmovers from Microbot are currently used in the laboratory for instructional purposes in the course titled Introduction to Robotics. A CNC lathe and a vertical milling machine have been ordered and will soon become part of the laboratory. Some programmable controllers, servo motors, limit switches and micro processors from Electrical Engineering and technology departments can be used in the proposed lab.

B. Proposed Laboratory Facility:

The proposed laboratory will be utilized for two primary purposes.

a. Education, Training, and
b. Research

Until a few years ago educational equipment was not available in automated manufacturing and most of the universities who offered programs in this area had to rely on expensive industrial equipment. Fortunately, some of the companies have started offering miniature educational equipment which are much less expensive than their industrial counterpart, although they may not perform as well.

Research however, requires state-of-the-art equipment and therefore educational equipment cannot be substituted for industrial equipment. Thus, the laboratory must have a balance of equipment in both the categories.

It is envisioned that at least two industrial robots will be required to satisfy the need for research and training. Utilization of the equipment has been discussed in the next section.

Selection of the equipment has been based upon (a) the equipment's proven record in industry and other universities, and (b) positive comments from some colleagues at other universities with whom the author has been in touch. Following is a list of equipment which will be required over a period of three years.

a. CNC milling machine, Dynamyte 2000
b. Computer-aided Part Programming System
c. Unimation Puma 500 Series Industrial Robot
d. ASEA IRB6 Industrial Robot
e. CNC Programming Work Stations

f. Rhino Educational Robot
g. SI Material Handling System, Starter Kit
h. Spectralight CNC Lathe
i. Unimation Machine Vision System
j. GE Programmable Controllers

Figure 1 shows the proposed layout of the laboratory.

The proposed laboratory will be located in a new building adjoining the two present buildings which house the School of Engineering Departments. The architectural plan of the new building has been completed and the construction should be complete in approximately two years.

C. Integration of Laboratory Into Curriculum:

Development of laboratory and curriculum should be simultaneous and complementary to each other. Of the six new courses proposed for curriculum development, at least four will utilize the laboratory extensively along with the course on numerical control which is already being offered.

The educational robots (Microbot and Rhino) will be used extensively for the introductory course in robotics. The two industrial robots will be used only for demonstration purposes in this course. Laboratory work will involve programming the robot, to simulate simple industrial operations, study of the elements of a robot, and evaluation of their performance.

The second course in robotics entitled "Applications in Robotics" will involve both educational and industrial robots. Small group of students will simulate industrial processes in welding, machining, grinding, palletizing, assembly and laser cutting using these robots.

The course entitled "Computer-aided Manufacturing will utilize most of the equipment in the laboratory with demonstration using industrial robots and hands-on experience with CNC milling machine, CNC lathe, material handling systems, programmable controllers, and educational robots.

D. Budget Requirement:

The cost of equipment discussed in Section B varies from a $1000 to $65,000. If the equipment acquisition is scheduled on the basis of an equitable cost distribution over a period of three years, the budget requirement would be approximately $75,000/year. Cost of installation and peripheral structure would be about 15% of the cost of the equipment, which will make the total requirement about $86,000/year.

E. Space Requirement:

It is anticipated that 1,317 sq. ft. of area will be required over a period of three years of development. This calculation takes into account the space occupied by the equipment, working space for students and instructors. The space gap required between equipment from safety considerations has also been accounted for. This is especially true in case of robots, where safety screens must be provided to prevent any accidents.

Space requirements for individual equipment are based upon the laboratory layout as shown in Figure 1.

F. Involving Local and Regional Industries:

It is proposed that while efforts are channelled into the development of curriculum and acquisition of equipment for the lab, a simultaneous move be made in the direction of pursuing local and regional industries to donate equipment to the University. It is also proposed that a program of faculty involvement in industry be initiated which will give the faculty hands-on experience in the factory environment. This will prove beneficial in the long run in developing expertise of existing faculty enabling them in turn to help industries implement new technologies. Steps have already been taken for negotiations in this matter with the Naval Air Rework Facility with the help of Dean Ernest Cross and Chairman, Mr. Gary Crossman.

IV. Conclusion:

James S. Albus, in a recent book on the effect of automation, wrote "The human race is now poised on the brink of a new industrial revolution which will at least equal, if not far exceed, the first industrial revolution in its impact on mankind. The first industrial revolution was based on the substitution of mechanical energy for muscle power. The next industrial revolution will be based on the substitution of electronic computers for the human brain in the control of machine and industrial process."

Educational institutions already find themselves lagging behind in providing education and training in the rapidly expanding field of automated manufacturing. The wisdom of university leadership lies in foreseeing the changes in future and making timely modifications in response to the need of time.

Proposed Layout of the Laboratory

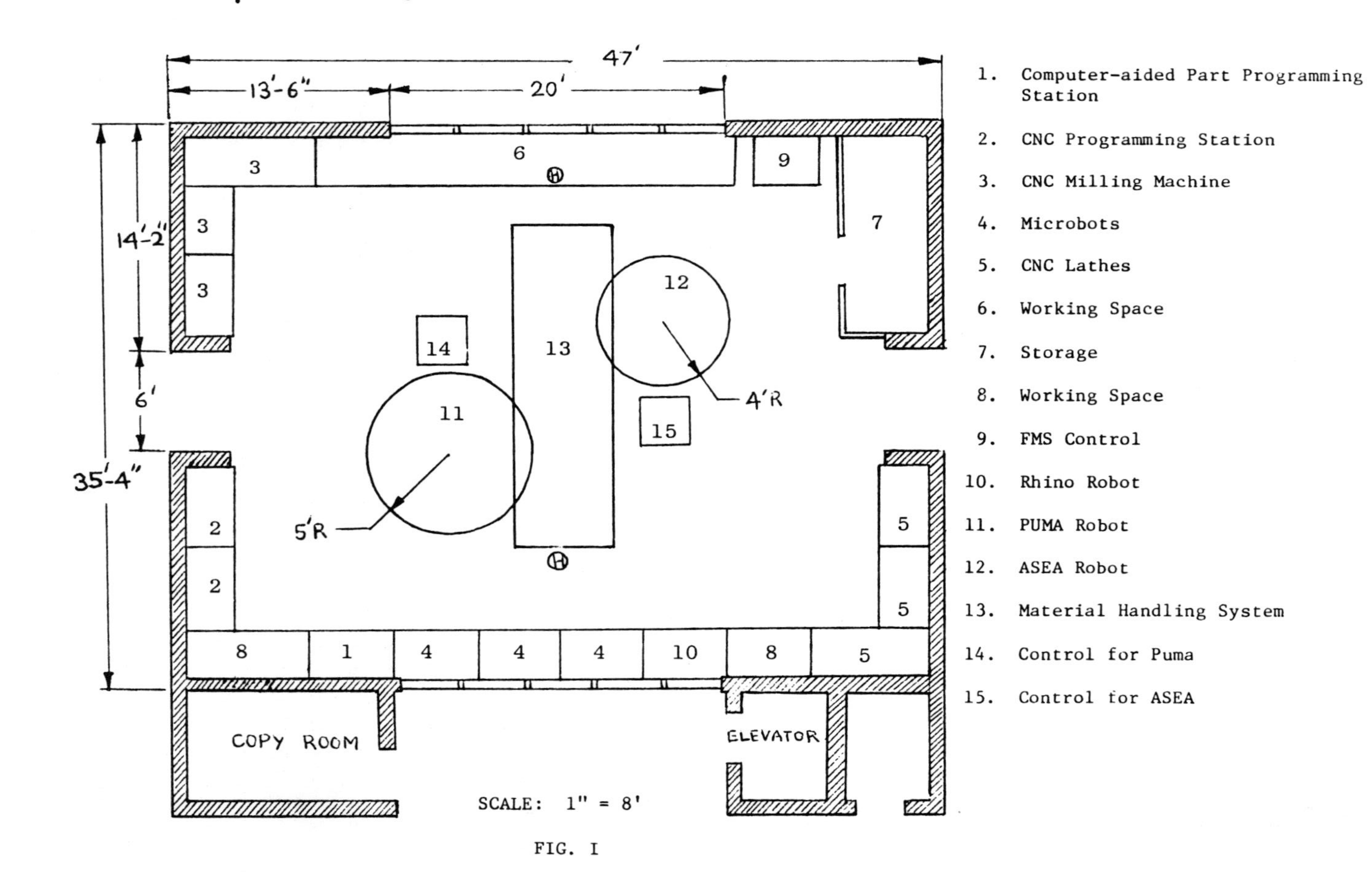

SCALE: 1" = 8'

FIG. I

Design of an Experimental Robot for Classroom Use

Y. Hari and Suren N. Dwivedi

Associate Professor
Department of Mechanical Engineering &
Engineering Science
Univeristy of North Carolina at Charlotte
Charlotte, North Carolina.

Abstract

The design of an experimental robot for class room use, built at the University of North Carolina at Charlotte is described. The mechanical robot arm provides a five degrees of freedom motion very similar to that of the human arm except that the YAW motion, which is the rotation of the hand about a vertical axis at the wrist, is not provided for. A microcomputer controls the generation of the pulses which drives the stepper motors. The interface circuit for the stepping motors and the microcomputer was designed and software developed to move the robot arm in any direction given any number of steps allowing development of specific tasks. A summary of the mechanical design, interface circuits and software description of future work is given.

Mechanical Design Procedure

The first step in the design process was to choose the robot arm configuration. The revolute configuration was chosen because of its basic structural features. The revolute configuration is the one most used in industries where parts handling and assembly are involved. Once the configuration of the arm was chosen, the physical size of the mechanism was decided. The objective was to design a robot model just large enough to be of value to industry yet small enough for class room demonstration. The design was such that a larger model could be made by "scaling up" the dimensions of the original to meet demands of a heavier load and longer reach. The end result was an arm with maximum reach (without hand) of 27 inches, and a static load capacity of five pounds.

The major parts of the robot are the base, shoulder, upper arm , elbow, forearm, wrist and mounting plate. A schematic diagram of the robot arm is shown in figure 1. The dimensions of each part were designed by considering the critical case of loading and then using a safety factor of three. The robot arm is designed to be driven by four separate motors. Each shaft is driven by a separate motor and a fourth motor rotates the base. The three shaft motors are mounted on the base in-line or staggered according to the size of the motors used.

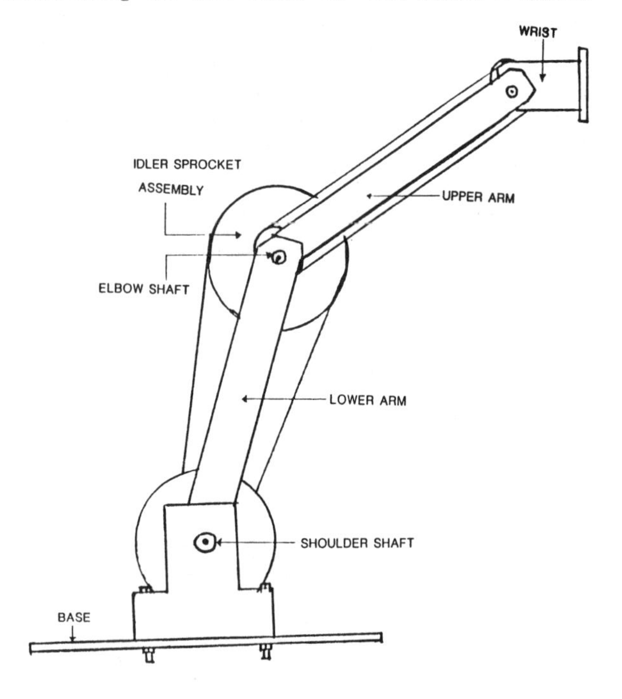

FIG 1: ORIGINAL ARM DESIGN

The motors, driving the elbow and wrist, have a sprocket mounted on their shafts. Chains connect these motors to the shoulder sprockets. The third motor, driving the shoulder has a sprocket on its output shaft and is connected by chain to the sprocket fixed to the shoulder. On the shoulder there are five sprockets. The larger fixed sprocket drives the shoulder while the other sprockets are idlers in the sprocket train used to drive the elbow and wrist. Moving up the elbow there are three sprockets. Once again, the larger fixed sprocket drives the elbow while the other two are idler sprockets in the sprocket train used to drive the wrist. The wrist has only one sprocket, which is fixed and drives it. The sprockets are chosen such that the driving sprockets are smaller and the driven sprockets are larger. This arrangement provides the maximum reduction in angular velocity of the shafts while providing the maximum increase in torque.

The various movements of the robot arm are accomplished by driving each individual motor either clockwise or counterclockwise to attain desired motion. The motors are placed back far enough on the base plate so that they will provide a counter-balance when the robot arm is fully extended with a load.

After a preliminary designed robot arm the following mechanical modifications were done. A mean to remove the slack from the drive chains was devised. Secondly, the overall weight of the arm was reduced to reduce the size of the driving motors. A turntable on which the arm and its controlling motors would rotate was also designed. Finally, some type of hand mechanism had to be added so the arm could manipulate its load. A typical Hand Assembly is shown in figure 2.

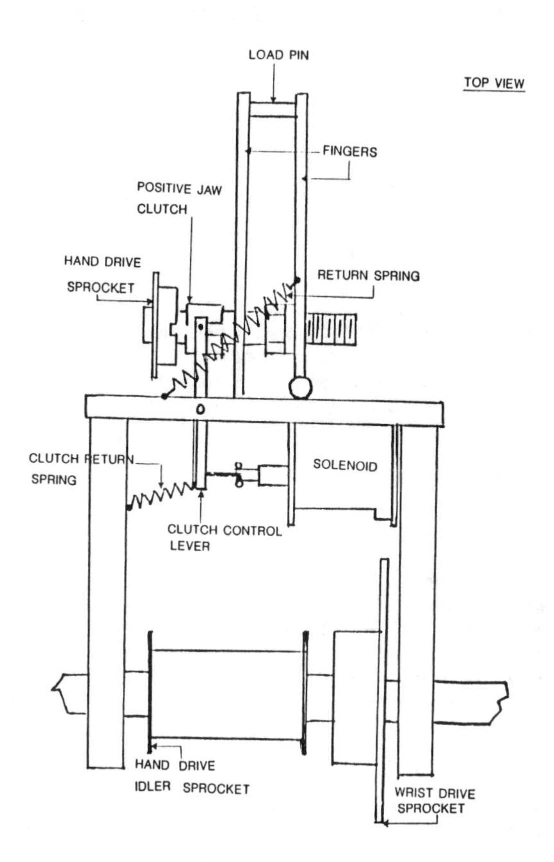

FIG 2: HAND ASSEMBLY

The kinematic analysis of the robot is included in the engineer's notebook which relates the effective working point of the hand to the individual joint coordinates or, in this case the joint axes angles.

Design calculations to determine the required output torque of the driving motors and also other design calculations for design of various parts of the robot arm mechanism were done.

The maintenance of this robot is very minimal. All that is required is the lubrication of the sprocket chain and of the bronze bearings. This can be done with mineral oil of SAE -20.

Computer System

A Commodore VIC-20 Microcomputer which uses the 6502 microprocessor is used to control the motion of the robot arm. The robot arm is driven in the following manner. First stage is the microcomputer controller which is the 'Brain' of the mechanical robot arm. The controller sends digital information to a translator which converts this digital information into usable electrical voltages and currents that operate the stepper motors which drive the arm. The complexity of motion of the robot arm is determined by the complexity of the software and is restricted only by the mechanical limitations.

Electrical Design Procedure

The objective was to design and implement an interface which could be controlled by a microcomputer and deliver phase switched DC power to stepper motors. It was not certain what features would be adequate for an electrically stable interface and still be cost effective. It was decided to translate the TLT output signals of a microcomputer to high power signals capable of driving stepper motors. The interface would consist of a digital translator and a high power drive circuit.

Translator Design

Three alternatives for translator design were considered. The first and most inexpensive alternative was to implement two flip-flops wired to the user port on the VIC-20 feeding the drive circuit. This was not a good solution because four lines would be needed to energize the drive circuit. An even more important consideration was the complexity of the software needed to provide the proper step sequences. This alternative was ruled out due to software consideration.

The second alternative was to implement a Cybernetics CY512 intelligent stepper motor controller. This alternative was the most appealing of the three in that it had a built-in stepper motor control language. Once a specific task is developed, the software can be burned into ROM and the CY512 will run the program with its built in microprocessor.

The third alternative was to use a Sprague UCN-4202 or UCN -4204. The main things that make the Sprague chips appealing was that they require only two lines from the port to step one motor, are NMOS and CMOS compatible and have power-on set. A

Research in CAD/CAM and Robotics in Ohio University

Jay S. Gunasekara, Kenneth R. Halliday, Roy L. Lawrence, Eugene Adams,
John C. Collier, and Billie Collier

Ohio University, Athens, Ohio

Summary

This paper highlights some of the current research and development activities in the areas of CAD/CAM and robotics at Ohio University. The main thrust in CAD/CAM research has been in the areas of extrusion of 'difficult to extrude' materials --both metals and polymers-die design and manufacture, and the finite element modeling of the extrusion processes. Extensive software has been developed for the design and manufacture of dies. The flow of material has been modeled using a package developed by Battelle in collaboration with the U.S. Air Force Materials Laboratories and a package developed in-house. Results are readily displayed on an Intergraph System with color enhancements. An expert system is being developed as part of this on-going research program. The main thrust in robot research has been in the area of incorporating vision and tactile sensing systems in assembly operations.

1. Introduction

Two years ago the College of Engineering and Technology at Ohio University developed a commitment to expanding its emphasis on CAD/CAM and manufacturing research and education. An eight faculty membered informal group was structured to cooperate on research efforts; external funding coupled with endowment income and other internal funds were used to purchase an Intergraph CAD/CAM system with a later set of almost entirely external grants funding expansion of the system. Holguin drafting system was added to a HP 1000 computer, shared Manufacturing Laboratory facilities for the new Stocker Engineering Center (to be occupied in Summer 85) were agreed upon, the nucleus of graduate and undergraduate manufacturing options in most departments was identified, and experimental graduate and undergraduate courses in CAD/CAM applications were offered.

Current interdisciplinary CAD/CAM and robotics research efforts include melt transformation extrusion and coextrusion; flow modeling and die design by finite difference and finite elment techniques; computer control and data treatment; CAD/CAM design, analysis, tool path generation and modeling of streamline dies and inserts; and use of vision and tactile systems in robotic operation.

2. Extrusion of 'Difficult to Extrude' Alloys

The recent development and application of metal-matrix-composite materials such as Al 2024 with 20 vol% SiC whiskers in the design and manufacture of aircraft structural components have increased the demand for superior die design and more effective control of extrusion parameters. The metal-matrix-composite material can reduce the weight of a typical aircraft component by as much as 40%. However, it has been found extremely difficult to manufacture complex shapes from this material using conventional manufacturing technology because of the breakage of the SiC whiskers and the distinct loss of critical mechanical properties.

2.1 Shear Die Vs. Streamlined Die

In conventional extrusion of high-strength aluminum alloys (2000 and 7000 series) shear (or flat-faced) dies are used. This results in internal shearing of the material inside the die and, in turn, a considerable temperature increase at local points within the deforming region. Hence, the extrusion must be carried out at a sufficiently low ram speed to avoid adiabatic heating and hot shortness in the product.
Lubricated extrusion through streamlined dies has numerous advantages. The streamlined die permits smooth entry and exit of the material without abrupt velocity change and, hence, can increase the quality of the products for non-conventional aluminum alloys and metal-matrix composites.

Hoshino [1] conducted a series of experiments on streamlined dies of various shapes, using lead as a model material. Observations made on grid distortion and extrusion load confirm that the streamlined die with optimized geometry (die length, etc.) and proper lubrication can reduce the extrusion load and, more importantly, produce a more homogeneously deformed product. The theoretical investigation conducted by Gunasekera, et al. [2], using the rigid visco-plasticity analysis of extrusion confirms Hoshino's [1] experimental observations. Although this analysis was confined to axisymmetric and plane-strain situation, the streamlined die produced more homogeneously deformed products than other dies.

2.2 CAD/CAM of Dies for Metal Extrusion

A new method of die design has been developed by Gunasekera [2] to design simple aircraft structural sections such as angles. Using this method, Gunasekera [2] developed a fully interactive die-design package called "STREAM" which is capable of generating conical, convex parabolic, and concave parabolic die shapes in addition to streamlined die shapes. Present (and future) versions of STREAM and SHEAR (a package for designing shear dies) are capable of generating the surface of the dies used to produce the following products: round to round, round to common structural shapes, round to any structural shape, multi-hole dies, and dies for hollow parts. Packages for the first three items have been developed and are presently undergoing testing; for the last two, the principles are still under development.

The system is fully interactive and asks the user a number of questions such as the product type, length of the die, material, etc. The three-dimensional geometry of the die for more complex components is designed and modeled using the computer. The program can be used to fit any type of spline. However, it requires the length of the die and dimensions of the orifice. This information is obtained using the results of process simulation.

Perspective projections of EDM electrodes used for die sinking are shown in Figure 1. STREAM also generates a die-coordinate file which is compatible with the Applicon CAD/CAM

System at Wright-Patterson Air Force Base and the Intergraph system at Ohio University for manufacturing dies.

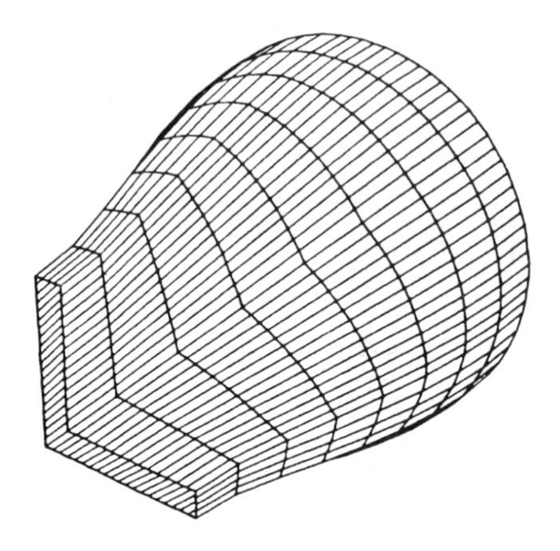

Fig. 1. Perspective projection of EDM electrode for streamlined extrusion die sinking

3. Polymer Melt Transformation Extrusion

Melt Transformation is a continuous process, developed at and patented by Ohio University, that impacts high levels of orientation in one and two directions in polymeric extrudates (3-9). Strengths of polyethylene and polypropylene, comparable to aluminum have been demonstrated. In the Melt Transformation Extrusion Process, a plasticating extruder supplies molten polymer to specially designed dies, the melt is conditioned, experiences elongational flow to impact orientation in the desired direction(s) and, at least, the outer portion of the extrudate is crystallized in the land of the die.

Since the strength developed in Melt Transformation Extrusion is dependent on the development of macroscopic orientation during converging flow, streamline dies are preferred. Prediction of the streamlines is complicated by the fact that polymer melts are typically non-newtonian and viscoelastic. Furthermore, in Melt Transformation Extrusion the melt is being transformed to an asymmetric fluid and crystallized inside the die, giving rise to directionally dependent properties and slip at the wall.

Higher line speeds were achieved when an uncoated retrofitted die was used to coextrude a three layer ribbon extrudate consisting of a polypropylene core and polyethylene skin layers on either side (9). By proper control of temperature gradients in the land of the die, the polypropylene core was crystallized after orientation but before leaving the land of the die, whereas the linear polyethylene skin layers had not yet crystallized when exiting from the die. Since the skin layers remained molten inside the die, they functioned as lubricating layers, and, after extrusion and crystallization, as crack inhibiting layers.

3.1 Flow Visualization

A side viewing die has been designed on the Intergraph, analyzed for leakage using MSC/NASTRAN, and is being built. This die will enable flow streamlines to be determined using birefringence and tracer techniques and uses interchangeable inserts for the shaping and land sections. Because of the designed use flexibility, this die permits viewing the flow streamlines in single layer extrusion and in both feedblock and multiple manifold coextrusion.

3.2 CAD/CAM and Finite Element Modeling

Once the initial streamline studies mentioned above using the side viewing die have been accomplished the observed flow will be modeled using the finite element technique now being developed and displayed using the Intergraph CAD/CAM system. Various constitutive equations will be used to model the flow. Once a good description of the flow patterns in coextruded materials has been achieved, the CAD/CAM system will be used to optimize the geometry of the shaping section for maximum orientation development. Subsequently, shaping section inserts will be machined to the specification developed in the optimization study and the predicted flow behavior and properties compared to the actual flow behavior and properties. The machining of the optimum shaping section inserts will be accomplished by generating numeric code on the Intergraph system that will then drive the machining tools necessary to make these sections.

3.4 Apparel Production

Current advanced apparel production techniques apply numeric control to drive cutting tools on multiple layers of fabric. However, the current practice does not integrate the actual garment design, visualization and pattern layout with the numeric coding of the laser cutting devices. With the Integraph CAD/CAM system, including the numerical control packages, the flat patterning package, the nesting package, and the sculptured surfaces package (all currently resident on the O.U. system), it is possible to integrate the entire apparel production operation. Two students are currently working on supplementing and adapting the available flat patterning and nesting software to allow: the design of a garment in a three dimensional color perspective with seven other views (currently possible); then after specifying the seam lines, calculate the compression and expansion of the fabric during flat patterning (an adaptation of current metal bending capability), lay out the pattern (possible now for simple shapes); nest the pieces to minimize waste while maintaining fabric pattern and grain lines (an expansion of the nesting capability); visualize the laser or knife tool paths (could use the currently available pocketing visualiza- tion); transfer the numeric code to the controller (as will be done in the milling research mentioned above); and laser or knife cut the fabric layers. The laser system in a pulsed mode will also be used to form a bonded seam in the material (if it contains some synthetic fibers) by melting and merging selected portions of intentionally overlapped layers of fabric.

4. Robotics Research

The robotics research program at Ohio University centers on the adaptive use of standard robots to assemble and perform other "difficult" operations. The primary equipment is comprised of a Unimation Puma 560 robot, solid-state one and two- dimensional cameras and frame grabbers, a 68000 series computer running UNIX and C for computational and generalized supervision purposes, a second 68000 series computer for image processing and data acquisition, grippers with position control and force sensing and a variety of force sensing and positioning instruments. Various computer controlled fixtures and movable platforms complement the set up. Much of the present effort is devoted to the development of techniques for assembling small mechanisms such as tape recorder drives.

In this area, a project is also underway to explore the technology and applications of touch sensors for industrial robots. To date, numerous possible types of touch sensors have been examined. These include piezo-electric, variable conductance and deformation devices. The most promising of these have been modelled using dynamic simulations. A final design for a model referenced compliant surface touch sensor has been synthesized, and construction of the sensor, a gripper and the control system are underway.

When completed this gripper will be used as both a stand alone sensor device for the PUMA and as an integrated device working in cooperation with the vision system currently under development. It is expected that the sense of touch will be most useful in tasks such as acquisition of a workpiece, assembly of components and the pass-off of workpieces in multi-aim cooperating architectures.

5. Manufacturing Education

In the area of manufacturing education, Ohio University has made significant strides to plan and implement a multi-disciplinary manufacturing option for graduate students of the mechanical, chemical and industrial engineering departments. The option, which allows students to study intensively in the manufacturing area, is designed to satisfy the degree requirements in each specific department. A shared manufacturing and CAM laboratory, to be administered by the college, is planned for the new Stocker Engineering Center due to open in the Fall of 1985.

6. Conclusions

Ohio University has developed a strong background for and commitment to an interdisciplinary approach to CAD/CAM and manufacturing research and education. Central to this effort is the continuing research on extrusion of difficult materials, i.e. SiC reinforced and unreinforced aluminum alloys extrusion and polymer Melt Transformation Extrusion. An Intergraph CAD/CAM system with a dedicated milling machine is being used to design, analyze and manufacture streamline dies for these extrusion processes involving complex fluids. Furthermore, the CAD/CAM and robotics equipment and expertise is also being extended and applied to other selected manufacturing areas such as apparel production. This concerted research effort is consistent with the manufacturing options and enrollments developing within the associated departments.

References

1. S. Hoshino. Extrusion of Non-Axisymmetric Sections Through Converging Dies, Ph.D. Thesis, Monash University, Australia, 1981.

2. J. S. Gunasekera, H. L. Gegel, J. C. Malas, S. M. Doraivelu, and J. Morgan. Computer-Aided Process Modeling of Hot Forging and Extrusion of Aluminum Alloys, Annals Int. Inst. Prod. Eng. Res. (CIRP), 1976.

3. J. R. Collier, T. Y. T. Tam, J. Newcome and N. Dinos, Poly. Eng. and Sci. 16, 204, 1976.

4. J. R. Collier, S. L. Chang, and S. K. Upadhyayla, Flow-Induced Crystallization in Polymer Systems, Monographs #6, Gordon and Breach, 1979.

5. J. R. Collier, K. Lakshmanan, L. Ankron and S. K. Upadhyayla, Structure-Property Relationships of Polymeric Solids, Ed. A. Hiltner, ACS Monograph, 1982.

6. J. R. Collier, M. Barger, B. Pandya, and S. Oh, "Melt Extrusion and Flow Induced Orientation and Crystallization," Houston, A.I.Ch.E., Spring National Meeting, March 1983.

Integration of Robotics and CAD/CAM in Education

Glen Boston

Department of Manufacturing Engineering
Miami University
Oxford, Ohio 45056

Integration of Robotics and CAD/CAM in Education

Competition Provides Incentive

"Competition" from external sources has drastically reduced sales and caused unemployment in many of our basic manufacturing industries (Steel, Autos, Applicances, Clothing, Machine Tools, etc.). This situation results in current economic and social problems with potential escalation in the future. Even though competition to this degree is considered economically sound from a world or macro viewpoint, we, as a nation, must participate fully in this economic betterment so that our manufacturing base has substance and entity. Participation is obviously contingent on improvement.

Required Techniques

Analysis of causal factors contributing to our problems reveals the following:

1. Labor costs
2. Productivity
3. Product Quality
4. Management Insight and Planning
5. Cost of Money
6. Government Support--Taxation--Regulation

Managers and engineers can affect the first four (4) of the above items with improved methods in modernizing and building the "Automatic Factories of the Future". Extended techniques in automation involving "flexibility" with "Design and Process Improvement" in product manufacture must be employed. These techniques involve utilization of computers, robotics, sensor defection, simulation and systems development. The engineering educator has the opportunity and the responsiblity to prepare the student for this industrial environment.

Inclusion in Curriculum

Currently a problem exists in all engineering curricula--how to include all required engineering courses in a four-year period. These courses consisting of mathematics, the basic sciences, engineering sciences, engineering design and biological and social sciences all vie for their curriculum inclusion at a sufficient quantity and quality to be beneficial to the student's ability to cope, contribute, and earn a living in our society. Prerequisites and orderly learning (via stepping stone principles) tend to minimize the ability to add new courses based on developing techniques.

At Miami University, in our Manufacturing Engineering Program, we are approaching the curriculum quantity dilemma with the inclusion of basic programming knowledge and a course directed at computer overall understanding coupled with "Integration" of the CAD/CAM, Robotics, Simulation and methods improvement techniques into our existing courses. Typical but not inclusive courses are "Quality Planning and Control," "Manufacturing Processes," "Graphical Analysis," and "Electric Circuit Analysis." This approach not only helps solve the curricula program but requires total faculty development in these new techniques rather than selected faculty development. Representation of the details for the integration approach will be shown for the Manufacturing Processes course.

Integration in Manufacturing Processes

Robotics and computers are introduced into process manufacturing as additional automated equipment that provides improved productivity and quality, with reduced manufacturing costs. Application then becomes paramount so that this concept can be demonstrated in the following areas:

Computer/Robot Applications

- Machine Loading and Unloading
- Welding
- Stacking
- Inspection
- Testing

Computer Planning Applications

- Production Planning
- Alternate Process Evaluation
- Numerical Control
- Factory Information Systems

Simulation of above applications is made using a microcomputer and/or robot as shown in Figure 1.

Figure #1 - Computer and Robot

The robot is capable of five (5) axis motion, cable driven arms and base coupled with stepping motors, and internal computer circuitry of 4K with external interface capability input and output ports. The micro-computer has 128 K capacity with CRT and printer capability, with double disc storage.

The robot has a limited number of program motions within itself, but unlimited motion movements when interfaced with the micro-computer.

Configuration requires compatible transmission rates, data format, and settings for standard interface signals between the micro-computer and the robot's microprocessor. Programming requires basic computer language coupled with a series of commands and values to the stepping motors that the robot's microprocessor deciphers and transmits movement through the stepping motors. The configuration utilized is shown in Figure 2. This equipment setup permits program documentation through print-out, and allows storage for future use.

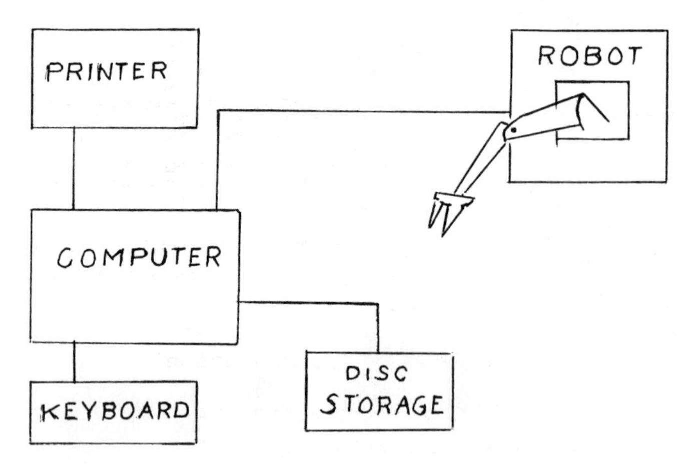

Figure #2 Configuration of Computer/Robot

The students are given a lecture-laboratory demonstration of the computer/robot language and programming. Also, a lecture is made of typical robot applications in industry. Students in groups of five or less are assigned a project that requires them to formulate a typical industrial application and simulate this application on the computer/robot configuration. Documentation and a technical laboratory report explaining the application are required. Figure 3 is a typical student documentation of the computer/robot making six (6) welds on an assembly.

```
3 OPEN "COM1:1200,N,8,1,RS,CS0,DS0,CD0" AS #1
5 INPUT "HOW MANY RUNS DO YOU WANT";Z
6  PRINT #1,"@RESET":INPUT #1,I
7 FOR I=1 TO Z
11 PRINT #1,"@READ"
12 INPUT #1,I
13 INPUT #1,A,B,C,D,E,F,G
14 PRINT A,B,C,D,E,F,G
15  PRINT #1,"@STEP 221",0,-378:INPUT #1,I:REM  STEP TOWARD TOOL
16  PRINT #1,"@STEP 221",0,0,-700,-381,-381,700:INPUT #1,I
18  PRINT #1,"@STEP 221",0,250,220:INPUT #1,I
19  PRINT #1,"@CLOSE 240":INPUT #1,I:REM  CLOSE ON TOOL
60  PRINT #1,"@STEP 221",-462,-528,0,0,0,-50:INPUT #1,I:REM  MOVE TOWARD
61  PRINT #1,"@STEP 221",0,0,0,0,0,-700:INPUT #1,I:REM  THE FIRST WELD SPOT
62  PRINT #1,"@STEP 221",0,700,-495,0,0,-250:INPUT #1,I
63  PRINT #1,"@STEP 221",0,0,0,430,430,-150:INPUT #1,I:REM  OVER 1ST SPOT
65  PRINT #1,"@STEP 221",0,85,0,0,0,-50:INPUT #1,I:REM HIT 1ST SPOT
70  PRINT #1,"@STEP 221",0,-181,0,0,0,-50:INPUT #1,I:REM RETRACT TOOL
75  PRINT #1,"@STEP 221",-44,-57,115,0,0,-50:INPUT #1,I:REM  MOVE OVER 2ND SPOT
80  PRINT #1,"@STEP 221",0,75,25,0,0,-50:INPUT #1,I:REM  HIT 2ND SPOT
85  PRINT #1,"@STEP 221",0,-109,0,0,0,-50:INPUT #1,I:REM  RETRACT TOOL
90  PRINT #1,"@STEP 221",-89,57,-59,0,0,-50:INPUT #1,I:REM  MOVE OVER 3RD SPOT
95  PRINT #1,"@STEP 221",0,115,35,0,0,-50:INPUT #1,I:REM  HIT 3RD SPOT
100 PRINT #1,"@STEP 221",0,-154,25,0,0,-50:INPUT #1,I:RETRACT TOOL
105 PRINT #1,"@STEP 221",-65,270,-125,0,0,-50:INPUT #1,I:REM MOVE OVER 4TH SPOT
110 PRINT #1,"@STEP 221",0,25,0,0,0,-50:INPUT #1,I:REM HIT 4TH SPOT
115 PRINT #1,"@STEP 221",0,-67,0,0,0,-50:INPUT #1,I:REM RETRACT TOOL
120 PRINT #1,"@STEP 221",125,210,-136,0,0,-50:INPUT #1,I:REM MOVE OVER
125 PRINT #1,"@STEP 221",0,-220,194,0,0,-50:INPUT #1,I:REM  5TH SPOT
130 PRINT #1,"@STEP 221",0,30,0,0,0,-50:INPUT #1,I:REM  HIT 5TH SPOT
135 PRINT #1,"@STEP 221",0,-50,0,0,0,-50:INPUT #1,I:REM  RETRACT TOOL
140 PRINT #1,"@STEP 221",-68,-26,22,0,0,-50:INPUT #1,I:REM  MOVE OVER 6TH SPOT
145 PRINT #1,"@STEP 221",0,42,0,0,0,-50:INPUT #1,I:REM  RETRACT TOOL
150 PRINT #1,"@STEP 221",0,-658,0,0,0,-50:INPUT #1,I:REM  MOVE AWAY FROM PIECE
155 PRINT #1,"@STEP 221",0,0,0,-479,-479,-50:INPUT #1,I:REM  MOVE BACK TO
160 PRINT #1,"@STEP 221",614,500,250,0,0,-50:INPUT #1,I:REM  TOOL HOLDER
165 PRINT #1,"@STEP 221",0,92,57,0,0,-50:INPUT #1,I
170 PRINT #1,"@STEP 221",0,0,0,0,0,600:INPUT #1,I:REM  REPLACE TOOL
173 PRINT #1,"@STEP 221",0,-376,0,0,0,-50:INPUT #1,I:REM  MOVE AWAY FROM
185 PRINT #1,"@CLOSE 240":INPUT #1,I
196 PRINT #1,"@READ":REM  READ POSITION RIGHT NOW
197 INPUT #1,I
198 INPUT #1,H,I,J,K,L,M,N:REM PUT INTERNAL REGISTERS VALUES INTO VARIABLES
199 PRINT H,I,J,K,L,M,N
200 PRINT #1,"@STEP 221",-H,-I,-J,-K,-L,-50:INPUT #1,I:REM  MOVE TO SETPOINT
205 NEXT I:REM RUN AGAIN
210 INPUT "DO YOU WANT TO DO MORE";S$
215 IF S$="YES" GOTO 5
220 STOP
```

Figure #3 - Documentation

CAD/CAM techniques in our Manufacturing Engineering Program will be incorporated in a similar manner.

As one can see, updating current curricula with this method of including these new techniques results in providing the students with an improved education and an enlightened faculty.

Robotics and CAD/CAM Education: How Much and What Kind?

Charles Chrestman
Assistant Director Program Operations

Itawamba Junior College
653 Eason Blvd.
Tupelo, MS 38801

The strong emergence of robotic and CAD/CAM equipment as viable contributors to increased productivity in manufacturing has brought on a wave of new challenges. One of them is education. The more important questions seem to deal with how much and what kind of education will be needed by factories of tomorrow. To compound the question further, John Nesbitt, writing in Megatrends, asserts that scientific and technical information now doubles every five and a half years but will soon double every twenty months. When we add this to the dilemmas already facing our training institutions, there is a potential for current problems to grow exponentially.

In the face of today's competition, U.S. industry must automate or liquidate. Education has much the same fate as it moves into a knowledge intensive age. In contrasting education's choices, institutions must provide training programs that produce quality graduates with broad-based skill mixes or industry will implement its own in-house training programs.

These views suggest that we really take a look at robotics and CAD/CAM education and ask the questions: "How much?" and "What kind?". To do this let's examine a continuum of educational programs that will provide the skills required for the new jobs of the future.

The role of secondary education can be perceived as bifold. From early indications, it appears automation will create long term demands for engineers, scientist and professionals whose education is grounded in college preparatory coursework at the high school level. Most experts agree that this coursework needs to be tempered with applied math and science, communications skills, problem solving skills, and human relation techniques that are associated with more broadly defined jobs being brought on by automation.

The other role of secondary education should be that of providing high-order basic education supplemented with occupational education. Courses taken in this option should prepare the student for technical training programs at the local community college or for immediate employment. The projections that ground this in reality are that the demand for technicians, mechanics, repairers, and installers on the whole are on the rise. It is anticipated that the need for technical sales and service personnel will also increase.

It is important to note, however, that due to the increasing percentages of the labor foce being employed in functions related to automation, that educational options for students be available if or when needed.

"CAD/CAM" and "robotics" are not just fasionable words. They are proliferating facts of life in every corporate endeavor. The community college system of education is the sector of American education that has pioneered educational options and innovative techniques to offer the initial robotics and CAD/CAM education in this country. The community colleges will and should continue to be the source of robotics and CAD/CAM education for technicians and service personnel. In addition, they should continue to provide the first two years of education leading to the baccalaureate degree for the engineers and scientists.

Because of the close institution-community relationship with business, industry, public employers, small businesses and high schools, the community colleges are the logical sites for retraining workers who will need the new skills associated with continued advancements in robotic and CAD/CAM systems.

The universities have traditionally been the sources of our engineers and scientists. This should not change. However, there is a concept that needs nurturing at the university level as it relates to robotic and CAD/CAM education. This concept would allow two year graduates of community college technician programs to continue their education under a "2+2" plan should they aspire to do so. This could potentially increase the available pool of engineers and technologists.

As automation improves, the factories of the future will improve the quality of their products and increase their productivity. This will likely provide a basis for advanced research and development by corporate groups. Universities should remain cognizant of this and be prepared to provide the coursework needed by business and industry in these pursuits. In essence, they should be the continuing education agents for engineers and professional personnel.

In order to illustrate the perceived role of the community college in the continuum just presented, let's consider these components of the robotics and CAD/CAM programs at Itawamba Junior College in Tupelo, Mississippi.

The instructional programs for robotics and CAD/CAM education are both two year Associate of Applied Science degree programs. The Robotics Technology program trains robotic and automated systems technicians, whereas, CAD/CAM is a rapidly growing component of the Drafting & Design Technology program. Both programs are designed to give the students a broad based education for greater flexibility as job tasks change with advancements in robotics and CAD/CAM technology. Listed below are the curricula for the two programs.

ROBOTICS TECHNOLOGY

FIRST YEAR

First Semester	*Semester Hours*
Technical Math II	*3*
Technical Physics I	*4*
Basic Electricity for Electronics	*6*
Pneumatics	*3*
Hydraulics	*3*
Total Hours	*19*

Second Semester	*Semester Hours*
Technical Math III	*3*
Robotics I	*5*
Electronic Devices	*6*
English Composition I	*3*
Computer Concepts & Applications	*3*
Total Hours	*20*

SECOND YEAR

First Semester	*Semester Hours*
Robotics II	*5*
Linear Integrated Circuits	*4*
Microprocessor Fundamentals	*6*
Electro-Servo Systems	*3*
Total Hours	*18*

Second Semester	*Semester Hours*
Robotic Systems Design	*4*
Interface & Control Systems	*6*
Air Logic	*3*
Vision & Sensing Systems	*3*
Approved Elective	*3*
Total Hours	*19*

** Electives must be approved by program advisor.*

DRAFTING AND DESIGN TECHNOLOGY

FIRST YEAR

First Semester	Semester Hours
Fundamentals of Drafting	3
Technical Math I	3
English Composition I	3
Elementary Surveying	3
Computational Methods	3
Industrial Psychology	3
Total Hours	18

Second Semester	Semester Hours
Machine Drafting I	3
Architectural Drafting I	3
CAD Concepts	3
Technical Math II	3
Route Surveying & Photogrammetry	3
Oral Communications	3
Total Hours	18

SECOND YEAR

First Semester	Semester Hours
Machine Drafting II	3
Architectural Drafting II	3
Electrical, Piping & Sheet Metal Drafting	3
Descriptive Geometry	3
Technical Math III	3
Technical Communications I	3
Total Hours	18

Second Semester	Semester Hours
Structural Drafting	3
Architectural Design	3
Technical Elective	3
CAD Applications or Approved Elective	3
Strength of Materials	3
Total Hours	15

To illustrate the "broad based" feature further, the robotic student receives training in the following areas during the two-year program:

- *Fluid Power*
  * *Hydralics*
  * *Pneumatics*
  * *Electro-Servo Systems*
  * *Advanced Air Logic*
- *Electronics*
  * *Basic Electricity & Electronics*
  * *Electronic Devices*
  * *Digital Principles*
  * *Linear Integrated Circuits*
  * *Microprocessor Fundamentals*
- *Industrial Electricity*
  * *Commercial & Industrial Wiring*
  * *Microcomputer Concepts & Applications*
- *Programming*
  * *CNC Programming*
  * *Programmable Controller Programming*
  * *VAL Programming*
  * *FORTH Programming*
  * *RCL Programming*
  * *Other Robot Programming*
- *Robotics*
  * *Robotics I*
  * *Robotics II*
  * *Vision & Sensing Systems*
  * *Robotic Systems Design*

By examining these areas, the question of "how much" can be put into perspective for this segment of education.

The Associate of Applied Science degree programs certainly address one "kind" of education, but there are others. One of the fastest growing services of the community college in recent years has been the continuing education programs where emphasis has been placed on upgrading and retraining of present employees. Growth rates in these programs have been phenomenal as more companies have gradually begun to automate. If 10% or less of our domestic industrial plants now operating have not implemented computer integrated manufacturing, then the potential for an increasingly larger pool of dislocated workers may further accelerate these types of training programs.

For those thousands of companies who will gut their complexes and automate to survive, and for the new companies entering the market, start-up training will be essential. These short term training options evolve quickly and possess much flexibility. Oftentimes pre-employment training and on-the-job training are a part of these programs.

It would be accurate to suggest then that the Associate Degree programs, continuing education programs and start-up training programs illustrate in part the role of education at the community college level.

If in some way we can better plan for the robotic and CAD/CAM education needed by factories of the future, then the problems related to a skilled work force in some ways diminish. It is obvious that the United States cannot survive as an economic power without improving on programmable automation to reduce manufacturing costs and improve product quality. Neither can other countries desiring to establish strong manufacturing based economies.